HARMONIC ANALYSIS IN EUCLIDEAN SPACES

Part 1

PROCEEDINGS OF SYMPOSIA
IN PURE MATHEMATICS
Volume XXXV, Part 1

HARMONIC ANALYSIS IN EUCLIDEAN SPACES

AMERICAN MATHEMATICAL SOCIETY
PROVIDENCE, RHODE ISLAND
1979

PROCEEDINGS OF THE SYMPOSIUM IN PURE MATHEMATICS
OF THE AMERICAN MATHEMATICAL SOCIETY

HELD AT WILLIAMS COLLEGE
WILLIAMSTOWN, MASSACHUSETTS
JULY 10–28, 1978

EDITED BY

GUIDO WEISS
STEPHEN WAINGER

Prepared by the American Mathematical Society
with partial support from National Science Foundation grant MCS 77-23480

Library of Congress Cataloging in Publication Data

Symposium in Pure Mathematics, Williams College, 1978.
Harmonic analysis in Euclidean spaces.
(Proceedings of symposia in pure mathematics; v. 35)
Includes bibliographies.
1. Harmonic analysis—Congresses. 2. Spaces, Generalized—Congresses. I. Wainger, Stephen, 1936– II. Weiss, Guido L., 1928– III. American Mathematical Society. IV. Title. V. Series.
QA403.S9 1978 515'.2433 79-12726
ISBN 0-8218-1436-2 (v.1)

AMS (MOS) subject classifications (1970). Primary 22Exx, 26A51, 28A70, 30A78, 30A86, 31Bxx, 31C05, 31C99, 32A30, 32C20, 35–XX, 42–XX, 43–XX, 44–XX, 47A35, 47B35

CONTENTS OF VOLUME

PART 1

PART 2

These Proceedings are dedicated to

NESTOR M. RIVIÈRE

1940–1978

NESTOR M. RIVIÈRE
JUNE 10, 1940
JANUARY 3, 1978

It is striking to contemplate the influence of Nestor Rivière upon so many areas in analysis, and even more striking to think of his influence upon so many people. His graciousness was reflected in his mathematical work. He loved to work with people and to share his ideas. Many of us attending this conference have benefited in our own work from these ideas and from the breadth of his mathematical knowledge. His collaborations with others were always marked by a brilliance, a willingness to listen, and an optimism that created an unending flow of ideas.

Born and raised in Buenos Aires, Argentina, Nestor entered the University at the age of 16. He received his Licenciatura in mathematics in 1960, married Marisa Renda in 1961 and taught in Buenos Aires and Bariloche until December 1962. At that time, with the help of A.P. Calderón, he came to the University of Chicago to pursue his mathematical studies. Nestor received his Ph.D. degree in 1966 and in the Fall of that year became a member of the faculty at the University of Minnesota. In April 1974 Marisa and Nestor's daughter, Melisa, was born.

Nestor was naturally influenced by his education at Chicago. Real and harmonic analysis always remained his primary mathematical interest. At Minnesota the environment was perfect for the development of this interest and for the application of his knowledge to problems in other areas, especially to certain areas of partial differential equations. From 1966 Nestor's work in real and harmonic analysis went along hand-in-hand with his work in P.D.E. . Below we review some of Nestor's work in four major areas: Singular Integrals, Multiplier Theory, Interpolation Theory, and Partial Differential Equations.

Nestor's love for mathematics and his desire to share ideas made him an exceptional teacher. During his years at Minnesota he supervised the theses of a number of students, among them were Eleonor Harboure de Aquilera, Nestor Aquilera, Norberto Fava, Robert Hanks, Wally Madych, and Felipe Zo.

Singular Integrals

At the time Nestor was a student in Chicago the Calderón-Zygmund theory of singular integral operators of elliptic type had already arrived to a well understood stage. The techniques of the 1952 paper, "On Certain Singular Integrals", were being used by B.F. Jones to study the L^p-continuity of singular integrals arising from parabolic equations. Rivière realized that the entire theory could be placed under one general setting dependent on a fixed notion of dilation, namely

$$\lambda^{\alpha} x = (\lambda^{\alpha_1} x_1, \ldots, \lambda^{\alpha_n} x_n) \ , \ \lambda > 0$$

where $\alpha_1, \ldots, \alpha_n$ are given positive numbers.

Associated with the above (nonisotropic) dilation is the metric, $r(x)$, defined for $x \neq 0$ as the unique positive number satisfying

$$\sum_{\alpha=1}^{n} \frac{x_j^2}{r^{2\alpha_j}} = 1 \ .$$

$r(x)$ has the homogeneity property, $r(\lambda^{\alpha} x) = \lambda r(x)$ and there is a polar decomposition of R^n relative to r, i.e. each $x \neq 0$ can be written as

$$x = r^{\alpha} \sigma \ , \ |\sigma| = 1$$

and

$$dx = r^{(\Sigma\alpha_i)-1} J(\sigma) \, dr d\sigma$$

with $0 < J(\sigma) \in C^{\infty}(\Sigma)$, $\Sigma = \{\sigma : |\sigma| = 1\}$.

In this setting one can mimic the techniques of the 1952 paper of Calderón-Zygmund and prove the L^p continuity, $1 < p < \infty$, of convolution singular integrals of the form

$$(*) \qquad \lim_{\epsilon \to 0} \int_{r(x-y) > \epsilon} k(x-y) f(y) dy$$

where

i) $k(x) \in C^1(R^n \setminus \{0\})$,

ii) $k(\lambda^{\alpha} x) = \lambda^{-\Sigma\alpha_i} k(x)$, $\lambda > 0$, $x \neq 0$,

iii) $\int_{\Sigma} k(\sigma) J(\sigma) d\sigma = 0$ where $d\sigma$ is area measure on Σ.

The proofs of the above results appeared in article [1] and the extensions of the results to certain nonconvolution type operators were given in [3].

Nestor went on to considerably generalize the setting in which one could consider convolution singular integral operators. In [13] he attaches the notion of a singular kernel with a one parameter family, $\{U_r : r > 0\}$, of open bounded neighborhoods of the origin satisfying the conditions:

i) $U_r \subset U_s$, $r < s$, $\bigcap_{r>0} \overline{U}_r = \{0\}$,

ii) The algebraic difference $U_r - U_r \subset U_{\phi(r)}$, with $\phi:(0,\infty)\to(0,\infty)$ nondecreasing, continuous, $\phi(r)\uparrow\infty$ as $r\uparrow\infty$.

iii) The Lebesque measure of $U_{\phi(r)}$, denoted by $m(U_{\phi(r)})$, is $\leq Am(U_r)$, A independent of r.

Associated to such a family Rivière defined the notion of a singular kernel as a function, $k(x)\in L^1_{loc}(R^n\setminus\{0\})$ with the properties,

i) $\int_{U_s\setminus U_r} k(x)dx$ is bounded independently of s and r and

$\lim_{r\to 0+} \int_{U_s\setminus U_r} k(x)dx$ exists for each fixed $s>0$,

ii) $\int_{U_{\phi(r)}\setminus U_r} |k(x)|\,dx \leq A$, independent of $r>0$

iii) There exists $A>0$ such that $\int_{R^n\setminus U_{\phi(r)}} |k(x-y)-k(x)|\,dx \leq A$

for all $y\in U_r$ and for all r.

He then naturally defined the doubly truncated singular integral operator

$$K_{r,s}(f)(x) = \int_{U_s\setminus U_r} k(y)f(x-y)dy$$

and proved the following theorem, which was new even for the elliptic case, i.e. $r(x)=|x|$ and $\phi(r)=2r$.

<u>Theorem</u>. The operator $K_*f(x) = \sup_{r,s}|K_{r,s}(f)(x)|$ is bounded from $L^p\to L^p$, $1<p<\infty$, and from $L^1\to$ weak L^1.

In particular for $f\in L^p$, $1\leq p<\infty$ $\lim_{\substack{s\to\infty\\ r\to 0+}} K_{r,s}(f)(x)$ exist pointwise for almost every $x\in R^n$

In 1973 Nestor, together with Steve Wainger and Alex Nagel, returned to a problem in singular integrals which was first formulated in the study of the singular integral operators involving "mixed homogeneous" kernels defined earlier. The problem was to find a "method of rotation" for these operators corresponding to that developed by Calderón and Zygmund. In the latter case the L^p-continuity of a singular integral of elliptic type arising from an odd kernel was reduced to the continuity of the one dimensional Hilbert transform. The problem was to find the appropriate one-dimensional operator for the mixed homogeneous operators coming from an odd kernel. A candidate was formulated as early as 1966, namely for $x\in R^n$

$$T_\gamma f(x) = \lim_{\epsilon \to 0} \int_{|t|>\epsilon} f(x_1 - \operatorname{sgn} t\,|t|^{\alpha_1}, \dots, x_n - \operatorname{sgn} t\,|t|^{\alpha_n}) \frac{dt}{t} ,$$

$\alpha_i > 0$, $i = 1,\dots,n$.

This operator was called by Nagel, Rivière, and Wainger, the Hilbert transform of f along the curve, $\gamma(t) = (\operatorname{sgn} t\,|t|^{\alpha_1},\dots,\operatorname{sgn} t\,|t|^{\alpha_n})$. In [24] they prove the continuity of T_γ on $L^p(R^n)$, $1 < p < \infty$, and as a consequence, the continuity on $L^p(R^n)$ of the mixed homogeneous operators in the form (*) where the smoothness of $k(x)$ is replaced by the condition

$$\int_\Sigma |k(\sigma)| \log^+ |k(\sigma)| d\sigma < \infty .$$

Multiplier Theory

Rivière's interest in the theory of Fourier multipliers began as a graduate student in Chicago. In [1] there appears the extension of the Hörmander multiplier theorem to the case of multipliers, $m(x)$, behaving like smooth functions with mixed homogeneity zero. More precisely if $r(x)$ denotes the metric, discussed in the previous part, corresponding to the given dilation

$$\lambda^\alpha x = (\lambda^{\alpha_1} x_1, \dots, \lambda^{\alpha_n} x_n) , \alpha_i > 0 ,$$

then $m(x)$ is a Fourier multiplier on all L^p , $1 < p < \infty$ provided m is bounded and

$$R^{2(\alpha\cdot\beta) - |\alpha|} \int_{R/2 \le r(x) \le 2R} |D^\beta m(x)|^2 dx \le C , \text{ independent of } R ,$$

for all β , $|\beta| < N$ with $N > \frac{|\alpha|}{2}$ $(|\alpha| = \sum_i \alpha_i)$.

When Nestor joined the faculty at Minnesota in 1966 he immediately began working with Walter Littman and Charles McCarthy on refinements of the Marcinkiewicz multiplier theorem in R^n ([6]). At this same time he started studying a problem posed to him earlier by A.P. Calderón who asked if a bounded rational function on R^d , $d > 1$, was a Fourier multiplier on L^p for some interval of p's around 2 . Already Littman, McCarthy, and Rivière had given in [7] an example of a bounded rational function on R^2 which was not a multiplier on $L^p(R^2)$, $1 < p < \frac{4}{3}$. The example was $\frac{1}{x^2 - y + i}$. In [13] Nestor extended the Marcinkiewicz multiplier result of [6] to operator valued multipliers and proved that any bounded rational function on R^d is a multiplier on the space of functions

$$L^p(R^{d-1}, L^2(R)) = \{f(t,x), t \in R, x \in R^{d-1} \text{ such that } \left(\int_{R^{d-1}} \left(\int |f(t,x)|^2 dt\right)^{p/2} dx\right)^{1/p} < \infty\} \quad (1<p<\infty) .$$

Finally in [13] Nestor proved a version of the Hörmander multiplier theorem that not only considerably generalized the setting of the theorem but added an original twist which even for the elliptic case gave a very interesting result. It is in this setting that we would like to state the result.

Theorem. Let β_j, $j=1,\dots,d$, be positive integers such that $\sum_{j=1}^{d} \beta_j^{-1} < 2$. Assume $m \in L^\infty$, and

$$\sup_{\substack{n=0,\pm1,\pm2,\\ j=1,\dots,d}} 2^{(2\beta_j - n)d} \int_{2^{n-1}<|x|<2^n} |D_{x_j}^{\beta_j} m(x)|^2 dx < \infty .$$

Then m is a Fourier multiplier on $L^p(R^d)$, $1 < p < \infty$.

The novelty of the above reslut is the "trade-off" of smoothness of the individual variables. One may assume a weak smoothness in one or several of the variables by requiring sufficient smoothness in the remaining ones.

Interpolation

Nestor began his studies at the University of Chicago in the area of interpolation. His unpublished thesis extended the Riesz-Thorin or Complex method of interpolation from Banach spaces to topological vector spaces, B, with a metric topology defined through an s-norm, $0 < s \leq 1$, i.e. a function $\| \ \|_s : B \to [0,\infty)$ such that

i) $\|x\|_s = 0 \iff x = 0$

ii) $\|x+y\|_s \leq \|x\|_s + \|y\|_s$

iii) $\|\lambda x\|_s = |\lambda|^s \|x\|_s$.

The metric is of course defined as $d(x,y) = \|x-y\|_s$. These spaces are called s-Banach spaces and prime examples are the Lebesque and Hardy spaces, $L^s(X,d\mu)$ and $H^s(R^n)$, $0 < s < 1$. In the thesis Nestor identifies, via the complex method, the intermediate spaces of various s-Banach spaces of functions and in particular shows that

$$[L^{p_1}(X,d\mu), L^{p_2}(X,d\mu)]_\alpha = L^p(X,d\mu)$$

where $\frac{1}{p} = \frac{\alpha}{p_1} + \frac{(1-\alpha)}{p_2}$, $0 < \alpha < 1$, $0 < p_1$, $p_2 \leq \infty$.

In [14] Nestor extends the techniques of the Marcinkiewicz interpolation theorem and as a consequence proves that any sublinear operator mapping boundedly

$$L^\infty(R^n) \to BMO(R^n)$$

and

$$L^1(R^n) \to L(1,\infty)$$

must also map boundedly $L^p \to L^p$ for $1 < p < \infty$. Here BMO denotes the space of functions with bounded mean oscillation, as defined by F. John and L. Nirenberg, and $L(1,\infty)$ is the Lorentz space of functions commonly called "weak L^1". This work of Nestor's, published in 1971, was his one mathematical paper written in Spanish.

Together with Yoram Sagher in [17], Nestor calculated the intermediate spaces, $(H^1, C_\omega)_{\theta,p}$, for the Lions-Peetre or real method of interpolation. Here H^1 denotes the classical Hardy space of functions defined on R^n and C_ω denotes the class of continuous functions on R^n vanishing at ∞. They proved

$$(H^1, C_\omega)_{\theta,p} = L^p \quad \text{for} \quad \frac{1}{p} = (1-\theta)\ , \quad 0 < \theta < 1 .$$

As a consequence, if M = space of finite measures then

$$(\mathrm{BMO}, M)_{\theta,q} = L(p',q)\ , \ \frac{1}{p} = 1-\theta\ , \ \frac{1}{p} + \frac{1}{p'} = 1 .$$

($L(p',q)$ denotes the usual Lorentz space.)

The above results were extended in [19] where the equalities

$$(H^{p_0}, L^\infty)_{\theta,p} = H^p \quad \text{and} \quad (H^{p_0}, H^{p_1})_{\theta,p} = H^p$$

were proved respectively for $\frac{1}{p} = \frac{1-\theta}{p_0}$ and $\frac{1}{p} = \frac{1-\theta}{p_0} + \frac{\theta}{p_1}$, $0 < p_0, p_1 < \infty$, $0 < \theta < 1$.

Nestor's interest in interpolation remained throughout his career. In 1976 his student, Robert Hanks, identified in his thesis the intermediate space

$$(H^p, \mathrm{BMO})_{\theta,p} \quad \text{as} \quad H^p \quad \text{for} \quad p = \frac{p_0}{1-\theta}\ , \ 0 < p_0 < \infty .$$

As a consequence Nestor's result on sublinear operators described above was extended to the case

$$T: L^\infty \to \mathrm{BMO}$$
$$T: H^1 \to L(1,\infty) .$$

Partial Differential Equations

As a graduate student Nestor was very interested in the use of singular integral operators as a general tool to study existence, uniqueness, and regularity for a large class of equations modeled mostly from elliptic operators. The use of the symbolic calculus, developed by A.P. Calderón and A. Zygmund in the elliptic case was adapted to parabolic operators in [4]. Any such operator, say $L = \sum_{|\alpha|=2b} A_\alpha(x,t) D_x^\alpha - D_t$, was decomposed on smooth functions with support in R_+^{n+1} as

$$L = S((-1)^{b+1}\Delta^b - D_t)$$

with S a parabolic singular integral operator. Assuming boundedness and uniform continuity on the coefficients the operator S was shown to be invertible on $L^p(S_T)$ with $S_T = R^n \times (0,T)$ and $1 < p < \infty$. From this followed easily the existence and uniqueness for the problem

$$Lu = f \text{ in } S_T , \ u(x,0) = 0$$

with $f \in L^p(S_T)$ and u in the class of functions having spatial derivatives of orders $\leq 2b$ and one time derivative in $L^p(S_T)$. Also in papers [28] and [29] one can again find the development of a symbolic calculus designed to give general algebraic conditions for the solvability of initial boundary value problems associated with the Navier-Stokes equations.

The final three years of Nestor's life were dedicated to problems in partial differential equations and some of his best work was done at this time. Together with Luis Caffarelli very precise regularity results in two dimensions were obtained for the free boundaries arising from the solution of the minimal energy problem above a given obstacle and from the solution of the minimal surface problem staying above an obstacle.

In the above situations we are given a bounded, connected domain $D \subset R^n$ and a function φ, the obstacle, defined on $\overline{D}$, satisfying:

a) $\varphi < 0$ on ∂D

b) $\Delta\varphi$ and $\nabla(\Delta\varphi)$ do not vanish simultaneously.

We let $v(x)$, $x \in D$, be the solution of a specific variational inequality satisfying $v \geq \varphi$ in D. For example in the case of minimizing energy

$$\int_D |\nabla v|^2 dx = \inf \{\int_D |\nabla u|^2 dx : u\big|_{\partial D} = 0 , \ u \geq \varphi \text{ in } D\} .$$

In [26] it was shown that the set of coincidence,

$$\Lambda = \{x \in D : v(x) = \varphi)x)\} ,$$

has the following structure in 2 dimensions:

<u>Theorem</u>. If $\Delta\varphi \in C^{k,\alpha}$, $0 < \alpha < 1$, $k \geq 1$, then each component of the interior of Λ is composed of a finite number of Jordan arcs each having a nondegenerate $C^{k+1,\alpha}$ parametrization. Moreover if $\Delta\varphi$ is real analytic the Jordan arcs are real analytic.

It was later shown by Caffarelli (even for the general n-dimensional case) that if $x_0 \in \Lambda$ is a point of positive density of Λ then there exists a ball, $B(x_0)$, about x_0 such that $\partial\Lambda \cap B(x_0)$ is a C^1 curve. In [31] Rivière and Caffarelli studied the case when x_0 is a point of zero density and showed the existence of a neighborhood, $B(x_0)$, of x_0 in which $\Lambda \cap B(x_0)$ is contained between two tangent C^1 curves. In fact, with a proper choice of coordinates $\Lambda \cap B(x_0)$ is contained between the curves

$$y = \pm C_1 |x| \exp\{-C_2(\log|x|)^{1/2}\} .$$

The final work submitted for publication by Rivière and his co-authors was [34]. Here A.P. Calderón's recent results concerning the Cauchy integral over a C^1-curve were used to solve the Dirichlet and Neumann problems for Laplace's equation in a C^1-domain, D, contained in R^n. The data were assumed to belong to $L^p(\partial D)$, $1 < p < \infty$, and the solutions were written respectively in the form of the classical double and single layer potentials. In the Dirichlet case the nontangential maximal function associated with the solution was shown to belong to $L^p(\partial D)$ and, as a consequence, the solution converged nontangentially to the data at almost every point of the boundary. Similarly in the Neumann problem the nontangential maximal function associated to the gradient of the solution was shown to belong to $L^p(\partial D)$ and again the data was assumed in a pointwise nontangential sense at almost every point of the boundary.

On November 23, 1977, during an informal gathering of harmonic analysts from the Midwest at the University of Chicago, Nestor spoke of some open problems which he considered exceptionally interesting. These problems are listed in this proceedings.

The last three years of Nestor's life were years of great personal growth. For each new crisis in his illness he found in himself new resources of courage. His sensitivity to other people increased, and his mathematical work continued unabated to the end. The grace he had shown under the most relentless pressure one has to face was his last, and greatest achievement.

Alberto Calderón
University of Chicago
Chicago, Illinois 60637

Eugene Fabes
University of Minnesota
Minneapolis, Minnesota 55455

Yoram Sagher
University of Illinois at Chicago Circle
Chicago, Illinois 60680

Bibliography of N.M. Rivière

1. Singular integrals with mixed homogeneity (with E.B. Fabes), Studia Math. 27(1966), 19-38.

2. Commutators of singular integrals (with E.B. Fabes), Studia Math. 26 (1966), 225-232.

3. Symbolic calculus of kernels with mixed homogeneity (with E.B. Fabes), Proc. Sump. Pure Math. 10, Singular Integrals, (1967), 106-127.

4. Systems of parabolic equations with uniformly continuous coefficients (with E.B. Fabes), J. D'Anal. Math. 17 (1966), 305-334.

5. The converse of Wiener-Levy-Marcinkiewicz theorem (with Y. Sagher), Studia Math. 27 (1966), 133-138.

6. L^p-multiplier theorems (with W. Littman and C. McCarthy), Studia Math. 30 (1968), 193-217.

7. The nonexistence of L^p-estimates for certain translation-invariant operators (with W. Littman and C. McCarthy), Studia Math. 30 (1968), 219-229.

8. Vector valued multipliers and applications, Bull. Amer. Math. Soc. 74 (1968), 946-948.

9. Some recursive formulas on free distributive lattices, J. Comb. Theory 5 (1968), 229-234.

10. On singular integrals, Bull. Amer. Math. Soc. 75 (1969), 843-847.

11. Multipliers of trigonometric series and pointwise convergence (with Y. Sagher), Trans. Amer. Math. Soc. 140 (1969), 301-308.

12. L^p-estimates $(1 < p \leq \infty)$ near the boundary for solutions of the Dirichlet problem (with E.B. Fabes), SIAM J. Adv. Diff. Eqns. (special issue)(1969), Proofs appeared in Ann. de Pisa 24 (1970), 491-553.

13. Singular integrals and multiplier operators, Arkiv Math. 9 (1971), 243-278.

14. Interpolacion a la Marcinkiewicz, Revista de la Union Matematica Argentina, Volume dedicated to Professor G. Dominquez, 25 (1971), 363-377.

15. The initial value problem for the Navier-Stokes equations with data in L^p (with E.B. Fabes and B.F. Jones), Arch. Rat. Mech. Anal. 45 (1972), 222-240

16. Estimates for translation invariant operators on spaces with mixed norms (with C.S. Herz), Studia Math. 44 (1972), 511-515.

17. Interpolation between H^1 and L^∞, the real method (with Y. Sagher), J. Funct. Anal. 14 (1973), 401-409.

18. On two theorems of Paley (with Y. Sagher), Proc. Amer. Math. Soc. 42 (1974), 238-242.

19. Interpolation between H^p spaces, the real method (with C. Fefferman and Y. Sagher), Trans. Amer. Math. Soc. 19 (1974), 75-82.

20. On Hilbert transformations along curves (with A. Nagel and S. Wainger), Bull. Amer. Math. Soc. 80 (1974), 106-108.

21. Maximal smoothing operators (with C.P. Calderon and E.B. Fabes), Indiana Math. J. 23 (1974), 889-898.

22. Commutators of singular integral operators with C^1-kernels (with E.B. Fabes and W. Littman), Proc. Amer. Math. Soc. 48 (1975), 397-402.

23. Multipliers of the Hölder classes (with W. Madych), J. Funct. Anal. 21 (1976), 369-379.

24. On Hilbert transforms along curves II (with A. Nagel and W. Wainger), Amer. J. Math. 98 (1976), 395-403.

25. On the rectifiability of domains with finite perimeter (with L.A. Caffarelli), Ann. Scuola Norm. Sup. Pisa, Series IV, 3 (1976), 177-186.

26. Smoothness and analiticity of free boundaries (with L.A. Caffarelli), Ann. Scuola Norm. Sup. Pisa, Series IV, 3 (1976), 289-310.

27. A maximal function associated to the curve (t,t^2) (with A. Nagel and S. Wainger), Proc. Nat. Acad. Sci. 73 (1976), 1416-1417.

28. Singular integrals and hydrodynamical potentials (with E.B. Fabes and J. Lewis), Amer. J. Math. 99 (1977), 601-625.

29. Boundary value problems for the Navier-Stokes equation (with E.B. Fabes and J. Lewis), Amer. J. Math. 99 (1977), 626-668.

30. The smoothness of the elastic-plastic free boundary of a twisted bar (with L.A. Caffarelli), Proc. Amer. Math. Soc., Vol 63, No. 1 (1977), 56-58.

31. Asymptotic behavior of free boundaries at their singular points (with L.A. Caffarelli), Ann. of Math., 106 (1977), 309-317.

32. The Cauchy integral in Lipschitz domains and applications (with A.P. Calderon, C.P. Calderon, E.B. Fabes, and M. Jodeit) Bull. Amer. Math. Soc., Vol. 84, No. 2, (1978), 287-290.

33. On the Lipschitz character of the stress tensor, when twisting an elastic-plastic bar (with L.A. Caffarelli), Arch. Rat. Mech. Anal., to appear.

34. Potential techniques for boundary value problems on C^1-domains (with E.B. Fabes, M. Jodeit), Acta Mathematica, to appear.

Some Open Questions

1. Let $\{\mathcal{U}_t : t>0\}$ be a family of open bounded convex sets containing 0 such that $\mathcal{U}_s \subset \mathcal{U}_t$ for $s<t$, $\cap \overline{\mathcal{U}}_t = \{0\}$. If μ and ν are finite regular Borel measures is it true that

$$\lim_{t \to 0+} \frac{\mu(x+\mathcal{U}_t)}{\nu(x+\mathcal{U}_t)}$$ exists almost everywhere with respect to ν ?

2. Consider the fundamental solution, $m(x,y) = \dfrac{1}{x^2 - y + i}$ of the Schrödinger operator $-\dfrac{\partial^2}{\partial x^2} + i\dfrac{\partial}{\partial y} + i$. As a Fourier multiplier on $L^p(R^2)$ $m(x,y)$ is unbounded on L^p for $p>4$. What can be said for the range $4/3 \le p \le 4$?

3. Suppose $P(x)$ and $Q(x)$ are polynomials on R^n such that $\dfrac{P(x)}{Q(x)}$ is bounded. Does it follow that $\dfrac{P(x)}{Q(x)}$ is Fourier multiplier on $L^p(R^n)$ for some intervals of p's around 2.

4. Assume $k_1(x)$ and $k_2(x)$, $x \in R^n$, are smooth functions on $R^n \setminus \{0\}$ such that $k_1(x)$ is an elliptic singular kernel and $k_2(x)$ is a parabolic kernel, i.e. $k_1(\lambda x) = \lambda^{-n} k_1(x)$, $\lambda > 0$, $x \ne 0$ and its mean value over the unit sphere is zero; $k_2(\lambda x_1, \dots, \lambda x_{n-1}, \lambda^2 x_n) = \lambda^{-n-1} k_2(x)$ and its appropriate mean value on the unit sphere is zero. Set $K_i f = k_i * f$. Does the composition $K_1 K_2$ map $L^1 \to L^{1,\infty}$? (see [23]).

5. Suppose T is a translation invariant operator mapping $L^p \to L^{p,\infty}$ for a given p, $1<p\le 2$. Does this imply $T : L^p \to L^{p,p'}$, $\frac{1}{p} + \frac{1}{p'} = 1$? (The unknown cases are $1<p<2$.)

6. Let $Kf = k * f$ where $k(x)$, $x \in R^n$, is homogeneous of degree $-n$, mean value zero over the unit sphere, Σ, in R^n, and in $L \log^+ L(\Sigma)$. Does $K : L^\infty \to BMO$?

7. For a given bounded <u>C^1-domain</u> $D \subset R^n$, <u>consider</u> the boundary value problems

$$\Delta^2 u(x) = 0 , \; x \in D , \text{ with } u|_{\partial D} , \; \frac{\partial u}{\partial n}\Big|_{\partial D} \text{ given}$$

$$\Delta^2 u(x) = 0 , \; x \in D , \text{ with } \frac{\partial u}{\partial n}\Big|_{\partial D} , \; \frac{\partial^2 u}{\partial n^2}\Big|_{\partial D} \text{ given}$$

$$\Delta^2 u(x) = 0 , \; x \in D , \text{ with } \frac{\partial^2 u}{\partial n^2}\Big|_{\partial D} , \; \frac{\partial^3 u}{\partial n^3}\Big|_{\partial D} \text{ given.}$$ Here $\dfrac{\partial^j u}{\partial n^j}\Big|_{\partial D}$ denotes the jth normal derivative of u on ∂D. Prescribe classes of boundary data which give existence and uniqueness.

Since the meeting in Williamstown Carlos Kenig and Peter Tomas have answered problem 2 and, as a consequence, also problem 3. They have proved that $m(x,y)$ is <u>only</u> a multiplier on $L^2(R^2)$.

CONTENTS OF PART 1

Eli Stein and some of his students attending the Conference.

First row (left to right): David Goldberg, Eli Stein, Steve Wainger, Juan Peral;
Second row: Steve Krantz, Mitchell Taibleson, Daryl Geller, David Jerison;
Top row: Bill Beckner, Norman Weiss, Bob Fefferman, Charlie Fefferman.

A. Zygmund and some of his students attending the Conference.

First row (left to right): Antoni Zygmund, Mischa Cotlar, Eli Stein, Marshall Ash;
Second row: Guido Weiss, Eugene Fabes, Marvin Kohn;
Third row: Ben Muckenhoupt and Yoram Sagher;
Top row: Bill Connett, Dick Wheeden, Dan Waterman.

Conference participants who have been students or faculty at Washington University in St. Louis

First row: Yoram Sagher, Alberto de la Torre, Jon Cohen;
Second row: Ronald Coifman and John Chao;
Third row: Michael Cwickel and Dick Hunt;
Fourth row: Yves Meyer, Antoni Zygmund (standing), Roberto Macias;
Fifth row: Guido Weiss, Pepe Garcia-Cuerva, Richard Rochberg, Ray Kunze;
Sixth row: Michel de Guzman and Ken Gross;
Seventh row: Rich Rubin, Al Baernstein, Mike Hemler, Fulvio Ricci, Mitch Taibleson;
Top center: Richard Bagby.

PREFACE

A considerable development of harmonic analysis in $\mathbb{R}^n$, and related fields, occurred in the decade of the 1970's. New ideas emerged, old techniques were applied in novel ways, and the types of interactions of harmonic analysis with other parts of mathematics increased considerably. A comprehensive description of this activity would be quite long. It would have to include the following items:

A new function space, the functions of bounded mean oscillations (BMO), became as important as the Lebesgue and Hardy spaces. The latter, which once represented much of what was labelled as the "complex methods" in harmonic analysis, were given several real-variable characterizations and were studied by "real methods". This permitted one to introduce "Hardy spaces" in very general settings (spaces of homogeneous type). On the other hand, complex variable methods played a role in the study of real variable problems concerning Lipschitz domains. The Fourier transform became a powerful tool for obtaining estimates for operators which are neither of convolution type nor linear, such as Littlewood-Paley-Stein functions, various maximal operators, pseudo-differential operators, commutators and Fourier integral operators. Weighted norm inequalities became increasingly more significant. The increased interaction between harmonic analysis and other fields of mathematics involved the theory of martingales, nilpotent groups, questions of hypoellipticity for linear differential operators and for the study of function theory in strictly pseudo-convex domains in $\mathbb{C}^n$.

The 1978 American Mathematical Society Summer Institute in Williamstown was devoted to this development and these topics. These two volumes make up the proceedings of this conference; they contain three types of presentations: (i) papers suggesting problems for future research; (ii) expository papers that explain some of the new ideas and techniques introduced in harmonic analysis during the last decade; (iii) technical papers giving the latest reports on progress in various different topics.

We have organized this material in six chapters distributed over two volumes. The first volume contains, for the most part, papers dealing with analysis in $\mathbb{R}^n$ and is divided into Chapter I, Real Harmonic Analysis, Chapter II, Hardy Spaces and BMO and Chapter III, Complex Harmonic Functions, Potential Theory and Functions of One Complex Variable. The second volume contains mostly papers on analysis in other settings. These papers give us a good picture of the strong interaction that has occurred between the various fields we mentioned above. This volume, again, is divided into three chapters: Chapter IV, Several Complex Variables, Chapter V, Pseudo Differential Operators and Partial Differential Equations and Chapter VI, Harmonic Analysis in Other Settings: Probability, Martingales, Local Fields, Lie Groups and Functional Analysis. The organization could well have been different since, in many cases, an individual article is connected with several topics discussed in other chapters so that it could easily have been placed elsewhere. We hope, however, that our selection makes these volumes reasonably accessible. We have also placed articles of a more expository nature at the beginning of the various sections or subsections.

We wish to thank the various people that have made this Summer Institute a success: Donald Sarason, Elias M. Stein, Mitchell H. Taibleson and Richard Wheeden, who have organized the institute's seminars; Dottie Smith, who did everything; Anna Pauline Bailey, who was particularly helpful during the organization period preceding the three weeks of the institute; our colleague, Ronald R. Coifman, who helped us throughout the organization and during the running of the institute; the very helpful staff supplied by Williams College: Peter Andrews, Marie Seitz, Eileen Sprague, and Jim Peck; the American Mathematical Society, who sponsored this meeting and the National Science Foundation who provided funds for the institute.

Stephen Wainger
Department of Mathematics
University of Wisconsin
Madison, Wisconsin

Guido Weiss
Department of Mathematics
Washington University in St. Louis
St. Louis, Missouri

CHAPTER 1

Real harmonic analysis

Proceedings of Symposia in Pure Mathematics
Volume XXXV, Part 1, 1979

SOME PROBLEMS IN HARMONIC ANALYSIS

E. M. Stein

Anyone who is foolish enough to try to list a series of "significant" problems in a rapidly developing field of mathematics should be prepared to face some serious criticism. I will try to reduce this risk by saying, first of all, that obviously some of the problems given below might well have been posed by anyone preparing such a list, and that others have been specially suggested to me by my friends; but I admit that a certain number have not been formulated before. To those who would accuse me of having neglected important parts of harmonic analysis I can only offer a weak defense: I am reluctant to speak about those areas where I have not had some recent personal interest. Finally I should say that it is not the particular problems *per se* to which one should attach importance, but to their spirit taken as a whole, and to the directions of possible research which they convey.

A few words about my point of view. The broad and mighty river which is harmonic analysis is fed by many streams, and in particular three are of great importance:
The Fourier transform; *real variable methods*, including differentiation theorems, other almost everywhere convergence theorems, martingales and Brownian motion; and *complex methods*, in which one would have to include harmonic functions, multiply harmonic functions, etc.

In the last 8 years or so our preception of the interplay of these three sources has undergone considerable change. There has been our loss of naiveté about the Fourier transform in $\mathbb{R}^n$; the recent insights connecting differentiation theory with the Fourier transform; and a further realization of how closely intertwined are the "real" and "complex" methods. It is in the perspective of this deeper understanding that I would like to pose the problems below.

1. Problems related to the disc multiplier

We begin by recording two well-known problems. We let S_R^δ denote the generalized partial sum operator given by

$$(S_R^\delta f)^\wedge(\xi) = (1 - |\xi|^2/R^2)_+^\delta \hat{f}(\xi)$$

and we write $S^\delta = S_1^\delta$, $S = S_1^0$.

Problem 1

Is S^δ, $\delta > 0$, bounded on $L^p(\mathbb{R}^n)$, when $n \geq 3$, and $\frac{2n}{n+1} \leq p \leq \frac{2n}{2-1}$?

This problem is suggested by the counter-example of C. Fefferman [6] for S (when $n \geq 2$, $p \neq 2$), and the positive result of Carleson-Sjolin, when $n = 2$, see [3].

A problem which as it turns out is related to this one is the restriction problem for the Fourier transform.

Problem 2

Does the a-priori inequality

$$\left(\int_{|\xi|=1} |\hat{f}(\xi)|^q \, d\sigma(\xi) \right)^{1/q} \leq A \|f\|_p , \quad f \in \mathcal{S}$$

<u>hold when</u> $1 < p < \frac{2n}{n+1}$, $q = (\frac{n-1}{n+1})\ p'$, <u>and</u> $n \geq 3$?

The analogue of this is known for $n = 2$, as well as the case $q \geq 2$, when $n \geq 3$, as a result of work of C. Fefferman, Tomas, Zygmund and the author. For a further discussion see the paper of Tomas [24] in these proceedings.

We pass now to a homogeneous variant of problem 1 . We let Γ denote the circular cone in $\mathbb{R}^3 = \{(\xi_1, \xi_1, \xi_3)\}$, given by $\Gamma = \{(\xi_1, \xi_2, \xi_3), \xi_3 > 0, \xi_3^2 > \xi_1^2 + \xi_2^2\}$. We write S_Γ^δ for the operator corresponding to the multiplier

$$\left(\frac{\xi_3^2 - \xi_1^2 - \xi_2}{\xi_3^2 + \xi_1^2 + \xi_2^2}\right)^\delta \chi_\Gamma .$$

Problem 3

(a) <u>Is</u> S_Γ^δ <u>bounded on</u> $L^p(\mathbb{R}^3)$, <u>when</u> $4/3 \leq p \leq 4$, $\delta > 0$?

By de Leeuw's theorem this would give a stronger version of the Carleson-Sjolin theorem. In fact it would also imply a positive response to

(b) <u>Is it true that</u>

$$\| (\sum_j |S_{R_j}^\delta (f_j)|^2)^{1/2} \|_p \leq A \| (\sum_j |f_j|^2)^{1/2} \|_p$$

<u>when</u> $4/3 \leq p \leq 4$, $\delta > 0$?

Of course similar problems can be posed in higher dimensions, but these await the solution of problem 1 .

The above question involved a vector-valued version (i.e. ℓ^2) of the partial sum operator. One can also ask for the ℓ^∞ version, and this would involve the maximal operator corresponding to summability. We write

$$S_*^\delta(f) = \sup_{R>0} |S_R^\delta(f)| .$$

Problem 4

(a) Prove that $\|S_*^\delta(f)\|_p \leq A \|f\|_p$ when $n = 2$, $\frac{4}{3} \leq p \leq 4$, and $\delta > 0$.

(b) Is it true that $\|S_*^0(f)\|_2 \leq A \|f\|_2$, when $n = 2$?

As far as part (a) all that is known is an old result which holds when $\delta > |\frac{1}{p} - \frac{1}{2}|$. A positive response to (b) would give circular almost everywhere convergence for L^2, generalizing the result of Carleson (in the form put by Hunt), $n = 1$. This last problem may be connected with problem 12 below.

Since we know that the operator S is not bounded on $L^p(\mathbb{R}^n)$, when $p \neq 2$, $n > 2$, it is tempting to seek as a substitute boundedness in terms of weighted L^2 spaces. The next two problems express this desire.

Problem 5

(a) Find a sufficient (and/or necessary) condition on φ, $\varphi > 0$, so that

$$\int |S(f)(x)|^2 \varphi(x)\,dx \leq A \int |f(x)|^2 \varphi(x)\,dx . \tag{1.1}$$

The special case when $\varphi(x)$ is of the form $|x|^{-\alpha}$ was setteled by Hirschman [10] some time ago.

Our experience with respect to weight functions (the class A_2 etc.) leads us to the following two variants of this problem.

(b) <u>Does there exist a "natural" maximal function</u> M , <u>so that if</u> $M\varphi \leq c\varphi$, <u>then</u> (1.1) holds?

This problem is admittedly vague, but it may be clarified by the following two remarks. First, when $n = 1$, (1.1) holds when M is the Hardy-Littlewood maximal function. For $n = 2$ a reasonable guess is the following variant of a Besicovitch-Nykodim (alias Kakeya) maximal function. Let Ω_δ be the homogeneous function defined by $\Omega_\delta(re^{i\theta}) = \delta\,|1 - e^{i\theta}|^{-1+\delta}$, $\delta > 0$, and write

$$M_\delta(f)(x) = \sup_{\theta,\, 0<r} \frac{1}{r^2} \left| \int_{|y| \leq r} \Omega_\delta(\rho_\theta(y))\, f(x-y)\, dy \right|$$

where ρ_θ denotes rotation by angle θ .

Notice that the limiting case M_0 corresponds to averages over rays in all directions (which is not bounded on any L^p !) , and M_1 is the Hardy-Littlewood-Wiener maximal function. Using work of Cordoba [4] one can prove that

(1.2) $\|M_\delta(f)\|_p \leq A\, \|f\|_p$, when $n = 2$, if $\delta > (2/p) - 1$, and $1 \geq \delta > 0$.

(c) <u>Is it true that</u>

$$\int |S^\varepsilon(f)|^2\, \varphi(x)\, dx \leq A \int |f|^2\, M_\delta(\varphi)\, dx$$

<u>with</u> $0 \leq \delta < 2\varepsilon$ <u>or when</u> $\varepsilon = \delta = 0$?

Observe that if this is so, then (1.2) would imply a positive solution of problem 3(b).

Problem 6

Find the analogue of (1.2) when $n \geq 3$.

Here the critical L^p space occurs when $p = n$. It is likely that a positive answer here would also solve problem 1 .

2. Problems related to differentiation theory.

We shift our focus now to maximal functions arising in differentiation theory; our problems are suggested by recent results whose proofs use the Fourier transform in a decisive way. This leads us directly to the next problem, which while somewhat philosophic in nature and vague reflects the special role of L^2 in arguments requiring the Fourier transform. I owe its formulation to S. Wainger.

Problem 7

Are there "natural" maximal functions which have L^p inequalitites valid for some $p > 2$, but not for $p = 2$?

What one might mean by such maximal functions are those of the form

$\sup_{A \in \mathcal{A}} \frac{1}{m(A)} \left| \int_A f(x-y)\, dy \right|$, where $\mathcal{A}$ is a given collection of sets of some particular geometric nature. Two examples where this problem arises are, first, the circular means in $\mathbb{R}^2$, i.e.

$$\sup_{t>0} \left| \int_{|y|=1} f(x-ty)\, d\sigma(y) \right| .$$

It is known that this maximal function is unbounded on L^2 , and the problem is to determine what happens when $p > 2$. For $n \geq 3$ it is known that the corresponding maximal function is bounded when $p > n/_{n-1}$, and unbounded when $p \leq {}^{n}/_{n-1}$ (See Stein [18]). The second example arises in the maximal function alluded to in problem 6 , where we might expect to have inequalities valid for $p > n$, but failing when $p < n$.

The background to the next two problems is as follows. Let $t \longrightarrow \gamma(t)$, $0 \leq t \leq 1$ be a smooth curvein $\mathbb{R}^n$ which satisfies the appropriate condition of "curvature" : for some $\varepsilon > 0$, $\{\gamma(t) , 0 \leq t \leq \varepsilon\}$ lies in the span of $\{\gamma^{(j)}(0)\}_{j=1}^{\infty}$. Then for f which are locally in L^p , $p > 1$, we have that

$$\lim_{h\to 0} \frac{1}{h} \int_0^h f(x - \gamma(t))\, dt = f(x)$$ almost everywhere. Moreover if

$$(Mf)(x) = \sup_{0<h\leq 1} \frac{1}{h} \left| \int_0^h f(x - \gamma(t))\, dt \right| \tag{2.1}$$

then

$$\|M(f)\|_p \leq A_p \|f\|_p \quad , \quad 1 < p \leq \infty . \tag{2.2}$$

(For further details see Stein and Wainger [21]). The proof of (2.2) uses the Fourier transform in a crucial way, but gives no information for $p = 1$. This motivates the following question.

Problem 8

(a) Find a geometric proof (without the use of the Fourier transform) of (2.2).

(b) Determine wether there is an L^1 theory for $M(f)$.

There is a host of other possibilities that arise when we consider maximal functions analogous to (2.1) as they might appear in different forms. Three of these closely related variants are the subject of the following problem. (For some recent partial results in this direction se Nagel, Stein, and Wainger [16]).

Problem 9

(a) Let $t \to \gamma_x(t)$ depend smoothly on x. Study the maximal function

$$Mf(x) = \sup_{0 < h < a} \frac{1}{h} \left| \int_0^h f(x - \gamma_x(t))\, dt \right| .$$

Next, fix n positive numbers $a_1, a_2, \dots a_n$. Let $\mathfrak{F} = (e_1, e_2, \dots e_n)$ be an arbitrary frame in $\mathbb{R}^n$, and let $B_h(\mathfrak{F}) = \{ x = \sum x_j e_j$, with $|x_j| \leq h^{a_j} \}$. Let $x \longrightarrow \mathfrak{F}(x)$ be a smooth mapping from $\mathbb{R}^n$ to frames in $\mathbb{R}^n$. Let $B_h(x)$ be the "ball" $B_h(\mathfrak{F}(x))$, centered at the origin.

(b) If $Mf(x) = \sup_{0 \leq h \leq 1} \frac{1}{m(B_n(x))} \left| \int_{B_h(x)} f(x - y)\, dy \right|$,

then is there on L^p theory for M ?

It can be shown that when $\max\{a_j\} \leq 2 \min\{a_j\}$ no conditions on $x \longrightarrow \mathfrak{F}(x)$ (except smoothness) are needed, and a covering lemma of Vitali holds. In general a Vitali-type covering theorem does not hold; but ispossible that a "curvature" conditions on the mapping $x \longrightarrow \mathfrak{F}(x)$ play an essential role. (For a discussion of these points see Nagel and Stein [15] ; in fact the condition $\max\{a_j\} \leq 2 \min\{a_j\}$ always holds in the "Step 2" case of the theory of pseudo-differential operators considered there.)

(c) Let $x \longrightarrow \underset{\sim}{v}(x)$ be a smooth non-zero vector field in $\mathbb{R}^n$. Let

$$M(f)(x) = \sup_{0 < h < a} \frac{1}{h} \left| \int_0^h f(x - \underset{\sim}{v}(x)t)\, dt \right| .$$

Is there an L^p theory for M? Is a curvature condition on $\underset{\sim}{v}$ necessary?

3. Problems related to harmonic and analytic functions - (symmetric spaces).

We now turn to some problems suggested by recent developments in the theory of H^p spaces. To put matters in their correct setting we shall briefly describe some of these results in $\mathbb{R}^n$ in its "standard" setting.

For this purpose we consider the upper half-space, $\mathbb{R}_+^{n+1}$ and a harmonic function $u(x, y)$, $x \in \mathbb{R}^n$, $y > 0$, defined on it. One associates to it the non-tangential maximal function

$$N(u)(x) = \sup_{(x',y)\in\Gamma(x)} |u(x', y)| ,$$

where $\Gamma(x)$ is the one $\{(x', y), |x - x'| < \alpha y\}$ with fixed aperture α. One also considers the area integral

$$A(u)(x) = \left(\int_{\Gamma(x)} |\nabla u(x', y)|^2 \, y^{1-n} \, dy \, dx' \right)^{1/2} .$$

Then for any $p > 0$ the following two conditions are equivalent

(3.1) $$N(u) \in L^p(\mathbb{R}^n)$$

(3.2) $$A(u) \in L^p(\mathbb{R}^n)$$

When (3.1) (and thus (3.2)) hold we say $u \in H^p$. There is also a purely real-variable formulation, (not involving harmonic functions) for the tempered distribution f which arises as $\lim_{y\to 0} u(x, y)$. Suppose $\varphi \in \mathcal{S}$, with $\int \varphi \, dx \neq 0$, and let $\varphi_\varepsilon(x) = \varepsilon^{-n} \varphi(x/\varepsilon)$, $\varepsilon > 0$. Then $u \in H^p$ is equivalent with

(3.3) $$\sup_{\varepsilon>0} |(f * \varphi_\varepsilon)(x)| \in L^p(\mathbb{R}^n).$$

At various stages of this theory (and in particular when $n = 1$ as in the work of Burkholder, Gundy and Silverstein [2]), the theory of Brownian motion and the resulting martingales here played a decisive role, and we shall return to this point below.

We may ask what happens when we consider a different variant of the above, namely when $\mathbb{R}_+^{n+1}$ is replaced by (say) $\mathbb{R}_+^2 \times \mathbb{R}_+^2$ and harmonic functions are replaced by biharmonic functions, i.e. $u(z_1, z_2)$ with $z_j \in \mathbb{R}_+^2$, $z_j = x_j + iy_j$, with $u(z_1, z_2)$ harmonic seperately in z_1 and z_2. In this case it is natural to let $\Gamma(x) = \Gamma_1(x_1) \times \Gamma(x_2)$ and define $N(u)$, now with respect to this product cone. The area integral A has then the modified definition $A(u)(x) =$

$$\left(\int_{\Gamma(x)} |\nabla_1 \nabla_2 u|^2 \, dx_1' \, dx_2' \, dy_1 \, dy_2 \right)^{1/2}$$

when $$|\nabla_1 \nabla_2 u|^2 = \left|\frac{\partial^2 u}{\partial x_1 \partial x_2}\right|^2 + \left|\frac{\partial^2 u}{\partial x_1 \partial y_2}\right|^2 + \left|\frac{\partial^2 u}{\partial y_1 x_2}\right|^2 + \left|\frac{\partial^2 u}{\partial y_1 \partial y_2}\right|^2.$$

Basing themselves in part on earlier work of the Malliavins [14], Gundy and the author [9] proved the equivalence of (3.1) and (3.2) in this set-up. ((3.3) also has an appropriate variant).

Closely connected with these conditions are (double time) Brownian motion analogues of N and A, for which equivalent conditions hold. We shall not describe the result have but we pass to another problem in the bidisc (where the formulation is easier than in $\mathbb{R}_+^2 \times \mathbb{R}_+^2$) — but in this case for <u>single-time</u> Brownian motion.

We let $z_t = z_t(\omega) = (z_t^1(\omega) = (z_t^1(\omega), z_t^2(\omega))$ $0 \leq t < \infty$ be the

Brownian motion in the bi-disc $\{z\ ,\ |z_1| < 1\ ,\ |z_2| < 1\}$ whose corresponding "infinitesmal generator" is $\Delta_B = (1-|z_1|^2)^2 \frac{\partial^2}{\partial z_1 \partial \bar{z}_1} + (1-|z_2|^2)^2 \frac{\partial^2}{\partial z_2 \partial \bar{z}_2}$

If $u(z_1\ ,\ z_2)$ is bi-harmonic in the bi-disc we let

$u^*(\omega) = \sup_{0 \leq t} |u(z_t)|$, and $S(u)$ denote the "square function" for the corresponding martingale. Then as is known (see Burkholder and Gundy [1]) $u^* \in L^p$ if and only if $S(u) \in L^p$, $0 < p < \infty$. Thus if (either) condition holds we may say that $u \in H^p$.

Problem 10

Interpret the resulting H^p theory. In particular,

(a) Relate the (Brownian) condition $u^* \in L^p$ with "restricted" non-tangential convergence.

(b) Interpert the (probabilistic) duality of H^1 and BMO.

(c) What are the distributions which arise as boundary values of such H^p function? Are they the same as those that arise in the standard $(\mathbb{R}^3_+)$ theory?

The theory of H^p spaces probably has some significance in any symmetric space (of which the disc and bi-disc are but the simplest examples; the other example for which an H^p theory is known is the complex ball in $\mathbb{C}^n)^*$. So one may raise the following general problem

Problem 11

Develop a "real" theory of H^p spaces on any symmetric space of non-compact type.

---

*Footnote. See Koranyi [12] , Putz [17], Stein [20] , Geller [8] , and Garnett and Latter [7] .

It is to be noted that when the symmetric space has complex structure (i.e. is a bounded domain of Cartan), then the holomorphic H^p functions have already been studied for some time (see e.g. Stein and Weiss [22 , pp. 124-127] and the references given there). However before one can realistically expect to make much progress in this general direction, there would seem to be a host of matters which would have to be clarified. In this spirit we would suggest three test problems.

We recall first a basic example of a symmetric space — the Siegel generalized upper-half space — and discuss several problems for Poisson integrals on these spaces. Here $\mathbb{R}^n$ is realized as M_m , the space of $m \times m$ real symmetric matricies, with $n = \frac{m(m+1)}{2}$. The symmetric space we consider is the tube domain $T_\Gamma = \{z = x + iy , \text{ with } x \in M_n , y \in \Gamma \subset M_n\}$, where Γ is the cone of positive definite symmetric matricies. The Poisson kernel is

$$P_y(x) = c \left(\frac{\det (y)}{|\det (x + iy)|^2} \right)^{\frac{n+1}{2}}$$

For any (reasonable) f defined on $\mathbb{R}^n$ we set $u(x , y) = (f * P_y)(x)$. These are roughly three types of convergence $u(x , y) \to f(x)$, as $y \to 0$, that can be envisaged, and we indicate their corresponding maximal functions

Restricted: $\sup_{y \in \Gamma_0} |u(x , y)| = M_R(f)(x)$ where Γ_0 is a proper sub-cone of Γ .

Unrestricted: $\sup_{y \in \Delta} |u(x , y)| = M_U(f)(x)$ where Δ is the collection of diagonal metrices in Γ .

Hyper-unrestricted: $\sup_{y \in \Gamma} |u(x , y)| = M_H(f)(x)$.

(For a detailed general setting of the restricted and unrestricted modes of convergence in any symmetric space see Koranyi [13]).

For $f \geq 0$ it can be shown that

$$M_R(f) \leq c_1 M_U(f) \leq c_2 M_H(f) .$$

Now it is known (See Stein and N. Weiss [23]) that

(a) $f \longrightarrow M_R(f)$ is of weak-type (1 , 1) , but

(b) $f \longrightarrow M_H(f)$ is unbounded on any L^p space, $p < \infty$.

In retrospect one might say that this last result (in particular when $n = 3$) was a fore-runner of the unboundedness of S discussed at the beginning, since both have as their source the unboundedness of the Besicovitch-Nykodim maximal function. There remains the question of unrestricted convergence.

Problem 12.

Show that $M_U(f)(x)$ is finite almost everywhere for any $f \in L(\log^+ L)^{m-1}$, (with f having bounded support).

Note that a recent result of Cordoba [5] is a strong indication that this problem has an affirmative solution. He is able to deal with the case of $\mathbb{R}^3$, i.e. $m = 2$.

A further problem that arises is whether anything might be salvaged for the "hyper-unrestricted" mode of convergence. The success of the Fourier transform in proving maximal theorems, and the fact that roughly speaking for $f \in L^p(\mathbb{R}^n)$ the condition that it arises as boundary values of a holomorphic function on T_Γ is equivalent with its Fourier transform vanishing outside the dual cone to Γ — all these things tempt one to put the following question :

Problem 13

Suppose u is holomorphic e.g. $u \in H^2(T_\Gamma)$, then is $\sup_{y \in \Gamma} |u(x,y)|$

finite a.e.?

This problem may well have some realtion with problem 4(b) above.

If we pass from symmetric spaces of tube-type to the general class, then the behavior of the Poisson integral near the boundary involves new difficulties. The "restricted" convergence theorem in this generality (for L^p, $1 \leq p$) was proved recently (see Stein [19]). We are therefore lead to stating the following problem

Problem 14

(a) Prove unrestricted convergence for Poisson integrals on any symmetric space for $f \in L^p$, all $p > 1$.

Assuming a positive response to this question one might raise the general form of problem 12, but one must take care because the several different notions of Poisson integrals that one might attach to any symmetric space (each to a given compactification of that symmetric space) could be expected to lead to different critical $L(\log^+ L)^k$ classes.

(b) For each compactification of a symmetric space determine the critical $L(\log^+ L)^k$ class for which unrestricted convergence holds.

4. Problems related to smooth domains in $\mathbb{C}^n$.

We now turn to the possibility of refining the known theory of holomorphic functions in H^p in domains of $\mathbb{C}^n$ with smooth boundary, as described in Stein [20].

Let $\mathcal{D}$ be a smooth bounded domain in $\mathbb{C}^n$, but do not assume that it is necessarily (strictly) pseudo-convex. Then for each w on the boundary of $\mathcal{D}$, one can define an admissible approach region $\mathcal{A}_\alpha(w)$ whose vertex is w (with "aperture" α), which is parabolic in the "complex tangential" directions, but conical in the "normal" directions. With this approach region we can define the maximal function

$$F^*(w) = \sup_{z \in \mathcal{A}_\alpha(w)} |F(z)|$$

and the "area integral"

$$A(F)(w) = \left\{ \int_{\mathcal{A}_\alpha(w)} (\delta^2(z)\,|F_1'|^2 + \delta(z)\,|F_T'|^2)\,\delta_{(z)}^{-n-1}\,dV(z) \right\}^{1/2} .$$

Here δ denotes the distance to the boundary of $\mathcal{D}$, F_1' is the derivative of F in the normal direction, and $|F_T|^2$ is the sum of the squares of the absolute values of the derivatives in the complementary complex tangential directions.

Now th following facts are known (with F holomorphic in $\mathcal{D}$)

(4.1) $$\|F^*(w)\|_{L^p} \leq A_p \|F\|_{H^p} , \quad 0 < p < \infty$$

(4.2) $$\|A(F)\|_p \approx \|F\|_{H^p} , \quad 0 < p < \infty$$

if we assume that $F(z_0) = 0$, for some fixed $z_0 \in \mathcal{D}$.

(4.3) If $\mathcal{D}$ is strictly pseudo-convex, then the sets whose $F^*(w) < \infty$ and $A(F)(w) < \infty$ are equivalent almost everywhere.

(The results (4.1) and (4.3) are in [20] ; the result (4.2) is unpublished).

These results are the "right" ones (in the sense of best possible) if the domain $\mathcal{D}$ is strictly pseudo-convex. When $\mathcal{D}$ the region is not strictly pseudo-convex (i.e. has flat points on the boundary) then it is indicated that the approach region should be broader near these flat points, and should roughly have a degree of tangency reflecting this flatness. The search for the precise definition of these approach regions might involve some interesting geometric and/or algebraic questions.

Problem 15

Formulate precisely the larger approach regions.

One might guess that a reasonable solution to this problem might be found for those domains $\mathcal{D}$ for which the $\bar{\partial}$ problem on $(0, 1)$ forms satisfies local "sub-elliptic" estimates. Kohn has characterized these domains when $n = 2$, and also found interesting sufficient conditions for general n (see Kohn [11]).

Once one has found a reasonable definition of these larger approach regions (which we will call here $\widetilde{\mathcal{A}}_\alpha(\omega)$), one can define the corresponding maximal function, $\widetilde{F}^*(\omega) = \sup_{z \in \widetilde{\mathcal{A}}_\alpha(\omega)} |F(z)|$. The area integral A would have to be modified accordingly. (The precise definition of this modified $\widetilde{A}$ is also an interesting question). The question that one could then raise would be

Problem 16

Show that

(a) $\|\widetilde{F}^*\|_p \leq A_p \|F\|_{H^p}$, $0 < p < \infty$

(b) $\|\widetilde{A}(F)\|_p \approx \|F\|_{H^p}$, if $F(z_0) = 0$.

(c) the sets where $\widetilde{F}^*(\omega) < \infty$ and $\widetilde{A}(F)(\omega) < \infty$ are equivalent almost everywhere.

References

1. D. L. Burkholder and R.F. Gundy, "Extrapolation and interpolation of quasi-linear operators on martingales", Acta Math. 124(1970), 249-304.

2. ______________, and M.L. Silverstein, "A maximal function characerization of H^p", Trans. Amer. Math. Soc. 157, (1971), 137-153.

3. L. Carleson and P. Sjolin, "Oscillatory integrals and a multiplier problem for the disc", Studia Math. 44(1972), 288-299.

4. A. Cordoba, "The multiplier problem for the polygon", Annals of Math. 105, (1977), 581-588.

5. __________, article in the proceedings of this symposium.

6. C. Fefferman, "The multiplier problem for the ball", Annals of Math. 94 (1971), 330-336.

7. J.B. Garnett and R.H. Latter, "The atomic decomposition for Hardy spaces in several complex variables", to appear.

8. D. Geller "Some results for H^p theory for the Heisenberg group", to appear.

9. R.F. Gundy and E.M. Stein, "H^p theory for the polydisc", to appear in Proc. Nat. Acad. of Sci., U.S.A.

10. I.I. Hirschman, "Multiplier transformations, II", Duke Math. J. 28(1961), 45-56.

11. J.J. Kohn, "Sub-ellipticity for the $\bar{\partial}$-Neumann problem on pseudo-convex domains; Sufficient conditions, to appear in Acta Math.

12. A. Koranyi, "Harmonic functions on hermitian hyperbolic space", Trans. Amer. Math. Soc. 135(1969), 507-516.

13. _________, "Harmonic functions on symmetric spaces", in Symmetric Spaces, eds. W.M. Boothby and G. Weiss, Marcel Dekker, New York, 1972.

14. M.P. Malliavin and P. Malliavin, "Integrales de Lusin-Calderon pont les fonctions biharmoniques", Bull. Sciences Math. 101(1977), 357-384.

15. A. Nagel and E.M. Stein, "Some new classes of pseudo-differential operators", proceedings of this symposium.

16. A. Nagel, E.M. Stein, and S. Wainger, "Hiebert transforms and maximal functions related to variable curves", proceedings of this symposium.

17. R. Putz, "A generalized area theorem for harmonic functions on hermitian hyperbolic space", Trans. Amer. Math Soc. 168 (1972) 243-258.

18. E.M. Stein, "Maximal functions: Spherical means", Proc. Nat. Acad. Sci. 73 (1976) 2174-2175.

19. _________, "Maximal functions: Poisson integrals on symmetric spaces", Proc. Nat. Acad. Sci. 73(1976), 2547-2549.

20. _________, Boundary behavior of holomorphic functions of several complex variables, Mathematical Notes, #11, Princeton University Press, 1972.

21. E.M. Stein and S. Wainger "Problems in harmonic analysis related to curvature", Bull. of Amer. Math. Soc. 84, (1978), 1239-1295.

22. E.M. Stein and G. Weiss, Introduction to Fourier Analysis on Euclidean spaces, Princeton Univ. Press, 1971.

23. E.M. Stein and N.J. Weiss, "On the convergence of Poisson integrals", Trans. Amer. Math. Soc. 140 (1969), 34-54.

24. P. Tomas, article in the proceedings of this symposium.

Additional references relevant to section 3

25. R. R. Coifman and G. Weiss, "Extensions of Hardy spaces and their use in analysis", Bull. Amer. Math. Soc. 83(1977), 569-645.

26. R. R. Coifman, R. Rochberg, and G. Weiss, "Factorization theorems for Hardy spaces in several variables", Ann. of Math. 103(1976), 611-635.

27. C. Fefferman and E.M. Stein, "H^p spaces of several variables", Acta Math 129 (1972), 137-193.

Proceedings of Symposia in Pure Mathematics
Volume XXXV, Part 1, 1979

ON OPERATORS OF HARMONIC ANALYSIS WHICH ARE NOT CONVOLUTIONS

R. R. Coifman[1]

The purpose of this talk is to describe recent progress in Harmonic analysis involving operators which do not commute with translations or which are not even linear.

More specifically, we would like to first describe some of the known results involving operators of the following type:

$$T(f)(x) = \text{p.v.} \int_{-\infty}^{\infty} k(x,t) f(x-t) \frac{dt}{t} .$$

We are mainly interested in the question of L^2 boundedness of T. (Other estimates and properties usually follow by the well known methods of Calderón and Zygmund).

The following examples are of interest in Harmonic analysis:

a) $k(x,t) = e^{in(x)t}$, where $n(x)$ is integer valued. This is equivalent to the linearized version of the maximal partial sum operator for Fourier series. L^2 boundedness is Carleson's result. We will not discuss this case, but point it out merely to indicate how difficult some of these problems can be.

b) $k(x,t) = \dfrac{w^{\frac{1}{2}}(x)}{w^{\frac{1}{2}}(x-t)}$.

Here the necessary and sufficient condition for L^2 boundedness was

---

AMS(MOS) subject classifications (1970). Primary 42A40; Secondary 47-02, 47B47

[1]Research supported by the National Science Foundation under grant MCS75-02411 A03.

obtained by Helson and Szëgo in [6]. Their condition basically is that $w(x) = \exp(b_1+\tilde{b}_2)$ where b_1 is bounded and $\|b_2\|_\infty < \frac{\pi}{2}$ ($\tilde{b}_2$ is the Hilbert transform of b_2). More recently Hunt, Muckenhoupt and Wheeden found a more useful condition on w (a condition which admits also an extension to L^p estimates). This is the so called A_2 class of Muckenhoupt (see [8]); i.e.,

$$w \in A_2 \quad \text{iff} \quad \sup_I \left(\frac{1}{|I|}\int_I w\,dx\right)\left(\frac{1}{|I|}\int_I \frac{1}{w}dx\right) < \infty$$

where I denotes an interval. This condition is equivalent to the condition of Helson and Szegö, but has the advantage that it is easy to check.

To indicate the degree of generality and irregularity of these weights we point out that $w = (f^*)^\delta$ $\delta < 1$ is such a weight, where f is any locally integrable function and $*$ denotes the Hardy-Littlewood maximal function.

c) $k(x,t) = a(x) - a(x-t)$ where a is in B.M.O. These kernels were studied in [5] in connection with factorization theorems for holomorphic functions. This class of kernels is related to b).

d) $k_n(x,t) = \left(\frac{A(x) - A(x-t)}{t}\right)^n$ with $A' \in L^\infty$.

The case $n = 1$ was treated by Calderón [1] in 1965 by an ingenious method involving the characterization of H^p spaces in terms of the Lusin area function. A proof using Fourier transforms was obtained by Yves Meyer and the author in [3] yielding the result for $n \geq 1$.

e) More generally one can take

$$k(x,t) = F\left(\frac{A(x)-A(x-t)}{t}\right), \text{ where } F \text{ is analytic on}$$

$|z| < 1$ and $\|A'\|_\infty < \eta_0$ for η_0 sufficiently small.

The most important case here is $F(Z) = \frac{1}{1+iZ}$.

The L^2 boundedness in this case was established by A. P. Calderón in [2] by the use of a very ingenious complex variable argument as well as the various known results involving the weights of example b). We will return to this case shortly.

f) $k_\pm(x,t) = \frac{A(x+t)-2A(x)+A(x-t)}{t}(1 \pm \operatorname{sgn} t)$

see [4].

A variety of other examples can be obtained by replacing the Hilbert kernel $\frac{1}{t}$ by others such as $\frac{1}{|t|^{1+i\gamma}}$, $\gamma \in \mathbb{R}$, or, more generally, by a kernel of a pseudo differential operator of order 0. These can be treated by the Fourier transform methods which we shall sketch briefly later.

It is to be pointed out that in the above list the functions $k(x,t)$ are a.e. $O(1)$ as $t \to 0$ but are quite irregular. If certain assumptions are made concerning the regularity in (x,t) of k then we are dealing with pseudo-differential operators. These are discussed by R. Beals, Y. Meyer, A. Nagel, and M. Taylor in these proceedings.

Although we restricted ourselves to $\mathbb{R}^1$ most estimates extend to the obvious generalizations of these operators to $\mathbb{R}^n$. See [3], [4].

We now would like to discuss briefly various questions involving examples d) and e). As you well know, example e) is related to the following famous problem in the theory of functions of a complex variable:

Let Γ be a rectifiable simple closed curve and f an integrable function defined on the curve, we form the Cauchy integral

$$F(Z) = \frac{1}{2\pi i} \int_{\Gamma} \frac{f(\xi)}{\xi - Z} d\xi \quad .$$

The question arises whether $F(Z)$ has non tangential boundary values as $Z \to \xi_0 \in \Gamma$. It was observed by A. Zygmund that an affirmative answer for C^1 curves implies the corresponding result for rectifiable curves.

Now it is quite easy by localizing the problem to reduce the question to the study of the following integral

$$\text{p.v.} \frac{1}{2\pi i} \int \frac{f(t)(1+iA'(t))}{s-t+i(A(s)-A(t))} dt = C(f)$$

(which is the Cauchy integral for the curve $t + iA(t)$).

As mentioned in e) (this operator belongs to the class described there) A. P. Calderón proved an L^2 estimate assuming that $\|A'\|_\infty \geq \eta_0$ for some small number η_0. This, of course, settles, by the use of real variable techniques, the existence a.e. in the C^1 case, and hence in the general case.

It is an open problem whether the operator $C(f)$ satisfies L^2 estimates when $\|A'\|_\infty \geq \eta_0$ or, more generally, when $A' \in L^p$ $1 \leq p \leq \infty$. We have, however, some indication that such an estimate should hold, provided a certain weight factor is introduced. To be specific there is a theorem of

Nikisin-Stein (which John Gilbert describes in these proceedings) stating that any operator $C(f)$ defined on L^p, $1 \leq p \leq 2$, and mapping into measurable functions can be factored as a weak type (p,p) operator followed by multiplication by a fixed measurable function.

In the case of the Cauchy integral on a rectifiable curve this implies of course that

$$\frac{1}{\Phi_\Gamma(Z)} \text{ p.v. } \frac{1}{2\pi i} \int_\Gamma \frac{f(\xi)}{\xi - Z}\, d\xi \qquad Z \in \Gamma$$

should be of weak type $(2,2)$ (or (p,p)) for some appropriate choice of function $\Phi_\Gamma(Z)$ on Γ.

We digress for a moment in order to indicate how the function $\Phi_\Gamma(Z)$ might be obtained. We consider a fixed function $a \in L^{2+\varepsilon}(\mathbb{R})$ and the operator

$$\int_{-\infty}^{\infty} \frac{f(t)a(t)}{x-t}\, dt = D(f)$$

for $f \in L^2$.

Since fa is in L^p for some $p > 1$, we find that $D(f) \in L^p$. By the Nikisin-Stein theorem there should be a function $\Phi_a(x)$ such that $\frac{1}{\Phi_a(x)} D(f)$ is a weak type $(2,2)$. We claim that $\Phi_a(x) = (a^{2+\delta})^{*\frac{1}{2+\delta}}$ $0 < \delta < \varepsilon$ does the job. In fact, we write

$$\frac{1}{\Phi_a(x)} \int \frac{f(t)a(t)}{x-t}\, dt = \int \frac{f(t)\left(\frac{a(t)}{\Phi_a(t)}\right)}{(x-t)} \cdot \frac{\Phi_a(t)}{\Phi_a(x)}\, dt ,$$

since $\frac{a(t)}{\Phi_a(t)} \leq 1$ and $\Phi_a^2(t)$ is a weight in A_2 we are dealing here with example b) and obtain an L^2 estimate.

This leads one to expect $\Phi_\Gamma(Z)$ to be related to some maximal function, and that the Calderón estimate for $C(f)$ is valid without the restriction $\|A'\|_\infty < \eta_0$.

As mentioned above Calderón's method involves a complex variable argument, which exploits the special properties of the Cauchy kernel. His result is easily seen to be equivalent to the following estimates for operators in example d) i.e.:

$$(*)\qquad \left\| \int_{-\infty}^{\infty} \left(\frac{A(x)-A(x-t)}{t}\right)^n f(x-t)\frac{dt}{t} \right\|_2 \leq C^n \|A'\|_\infty^n \|f\|_2$$

$(C \cong \frac{1}{\eta_0})$.

These estimate can be recombined to give other kernels as in e) or extended to R^n by the so called method of rotations. They can then be applied to study boundary value problems on Lipschitz domains (see the articles by Jodeit and Fabes in these Proceedings).

It would be of great interest to be able to extend estimates (*) for kernels other than $\frac{1}{t}$. This has been done in [3][4], directly in R^n ; unfortunately the constant obtained there is of the order of $n!$ (rather than C^n) . Surprisingly, these operators, though not linear in A' , can be studied effectively using Fourier transforms. Let us indicate briefly how this can be achieved in a simple case; for more details see [4].

We consider an operator on $L^2(R^1)$ defined by

$$M(f) = (m\hat{f})^\vee(x) \qquad \text{where } |m^{(k)}(\xi)| \leq C_k|\xi|^{-k}$$

and an operator of multiplication by a fixed function $A(x)$ which we denote as $A(f)$. The result of Yves Meyer and the author is the following:

$$\|[M,A]f'\|_2 \leq C \|A'\|_\infty \|f\|_2 \ ;$$

i.e. the commutator of M and A is smoothing of order 1.

Briefly the idea of the proof is as follows. We write

$$4\pi^2[M,A](f')(x) = \iint e^{ix(\xi+\alpha)}(m(\xi+\alpha)-m(\xi))\frac{\xi}{\alpha}\hat{f}(\xi)\hat{a}(\alpha)d\alpha d\xi \ ,$$

$$a = A' \in L^\infty \ .$$

This could be handled easily if somehow one could represent the "symbol" $(m(\xi+\alpha)-m(\xi))\frac{\xi}{\alpha}$ as a product or sum of products of a bounded function in ξ with a multiplier for L^∞ . Unfortunately, this cannot be achieved globally in (ξ,α) so that we are led to consider a partition of unity in R^2 with the property that separation of variables can be carried out on its components. Such a procedure leads us to consider operators of the following form as being "elementary"

$$\sum_{-\infty}^{\infty} \eta_i f*\varphi_i \; a*\psi_i ,$$

where η_i is a bounded sequence of numbers,

$$\varphi_i = 2^i\varphi(2^i x) , \psi_i = 2^i\psi(x2^i) \quad \text{with} \quad \varphi \; \psi \in$$

and either $\int \varphi dx = 0$ or $\int \psi dx = 0$; these, in turn, are studied using the Littlewood Paley function

$$g(a,f) = (\sum |f*\varphi_i|^2 |a*\psi_i|^2)^{\frac{1}{2}} ,$$

which can be analysed using the fact that $|a*\psi_i|^2$ is a Carleson measure on an appropriate set. (see [4]). In summary, the idea here as well as for the other examples in d) and f) is to proceed along the same line as in the theory of pseudo-differential operators by microlocalization. The Fourier transform being as effective in the multilinear case as for convolutions.

We now would like to describe briefly another situation where the Fourier transform is an important tool. E. M. Stein [9] recently has shown that the following maximal operator

$$M(f)(x) = \sup_t \left| \int_{|y'|=1} f(x+ty')d\sigma(y') \right|$$

is bounded on $L^2(R^n)$ if $n > 2$ (he has various other results on L^p) .

If we try to extend this result, say to a compact Riemannian manifold, we encounter the difficulty that, viewed in local coordinates, the corresponding averages do not commute with translations, and the Fourier transforms seem useless. In local coordinates the averages over Riemannian spheres on a manifold of dimension n can be expressed as

$$A_r(f)(x) = \int_{\Sigma_{n-1}} f(x+\lambda_x^r(\xi))d\sigma_x(\xi) ,$$

where $\lambda_x^r(\xi)$ is a paramatrization of the spheres given by the inverse of the Gauss map (here we assume that the curvature does not vanish). If we write

$$f(x) = \int_{R^n} e^{ix\eta} \hat{f}(\eta)d\eta$$

we get

$$A_r(f)(x) = \int_{R^n} \left\{ \int_{\Sigma_{n-1}} e^{i(x+\lambda_x^r(\xi))\cdot\eta} d\sigma_x(\xi) \right\} \hat{f}(\eta)d\eta \quad .$$

A simple application of the method of stationary phase shows that the term in curly brackets can be written as a finite sum of terms of the form:

$$e^{ix\cdot\eta + \lambda_x^r(\eta)\cdot\eta} \frac{w_x^r(\eta)}{|\eta|^{\frac{n-1}{2}}} + \epsilon(x,\eta)e^{ix\cdot\eta}$$

with $w_x^r(\eta) \equiv 0$ for $|\eta| \leq \frac{1}{r}$.

The term involving $\epsilon(x,\eta)$ can be shown to be dominated by the Hardy-Littlewood maximal operator. To estimate the first term let us consider the simpler case $r = 2^{-1}$ $i = 0,1,2\cdots$. We estimate

$$\sup_i \left| \int_{\Sigma_{n-1}} e^{i[x+\lambda_x^{2^{-i}}(\eta)]\cdot\eta} \frac{w_x^{2^{-i}}(\eta)}{|\eta|^{\frac{n-1}{2}}} \hat{f}(\eta)d\eta \right| = \sup_i |B_i(f)|$$

by $\left(\sum_{i=1}^{n} |B_i(f)|^2 \right)^{\frac{1}{2}} = g(f)$.

Now the operators B_i are Fourier integral operators whose theory was developed in [7]. L^2 estimates, however, can be obtained directly by microlocalization and a reduction to Plancherel's theorem using a change of variables.

It is straightforward using these estimates, to obtain $\|g(f)\|_2 \leq C\|f\|_2$ which, of course, implies

$$\sup_i |A_{2^{-i}}(f)| \quad \text{is bounded on } L^2 . \qquad (\text{also } L^p \ 1 \leq p \leq \infty)$$

As mentioned above this method is a natural extension of Stein's method and was essentially developed in discussions between Nagel, Wainger, Stein, Y. Meyer, El-Kohen and the author.

REFERENCES

1. A. P. Calderón, "Commutators of singular integral operators", Proc. Nat. Acad. Sci. 53, No. 5.

2. A. P. Calderón, "Cauchy integral on Lipschitz curves and related operators", Proc. Nat. Acad. Sci. 74 (1977).

3. R. Coifman and Y. Meyer, "Commutateurs d'intégrales singulières", Ann. Ist. Fourier (1978).

4. R. Coifman and Y. Meyer, "Opérateurs pseudodifferentiels ...", Astérisque 57 (1978).

5. R. Coifman, R. Rochberg, and G. Weiss, "Factorization theorems for Hardy spaces in several variables", Ann. of Math. 103 (1976).

6. H. Helson, G. Szegö, "A problem in prediction theory", Ann. Math. Pura Appl. 51 (1960).

7. Hormander, "Fourier integral operators", Acta Math. 127 (1971).

8. B. Muckenhoupt, "Weighted norm inequalities for the Hardy maximal function", Trans. A.M.S. 165 (1972).

9. E. M. Stein, "Maximal spherical averages", Proc. Nat. Acad. Sci. 73 (1976).

DEPARTMENT OF MATHEMATICS, WASHINGTON UNIVERSITY, ST. LOUIS, MISSOURI 63130

Proceedings of Symposia in Pure Mathematics
Volume XXXV, Part 1, 1979

Maximal functions, covering lemmas and Fourier multipliers

by

Antonio Córdoba

I would like to begin this talk presenting, as an example of the general theory, a proof of a conjecture of Professor A. Zygmund: given a positive function Φ on $\mathbb{R}^2$, monotonic in each variable separately, let us consider the collection B_Φ defined by the two parameter family of parallelepipeds in $\mathbb{R}^3$, whose sides are parallel to the rectangular coordinate axes and whose dimensions are given by $s \times t \times \Phi(s, t)$, s, t real numbers.

Theorem 1.

(a) B_Φ differentiates integrals of functions which are locally in $L(\mathrm{Log}^+ L)(\mathbb{R}^3)$, that is:

$$\lim_{B_\Phi \ni R \Rightarrow x} \frac{1}{\mu\{R\}} \int_R f(y)d\mu(y) = f(x) \;, \quad \text{a.e.}x$$

so long as f is locally in $L(\mathrm{Log}^+ L)(\mathbb{R}^3)$, where μ denotes Lebesgue measure in $\mathbb{R}^3$.

(b) The associated maximal function

$$M_\Phi f(x) = \sup_{x\varepsilon R\varepsilon B_\Phi} \frac{1}{\mu\{R\}} \int_R |f(y)| \, d\mu(y)$$

satisfies the inequality

$$\mu\{x: M_\Phi f(x) > \alpha > 0\} \leq C \int_{\mathbb{R}^3} \frac{|f(x)|}{\alpha} \{1 + \log^+ \frac{|f(x)|}{\alpha}\} \, d\mu(x)$$

for some universal constant $C < \infty$.

The proof of this theorem will be based on the following geometric lemma.

Covering Lemma. Let B be a family of dyadic parallelepipeds in $\mathbb{R}^3$ satisfying the following monotonicity property: if $R_1, R_2 \in B$ and the horizontal dimensions of R_1 are both strictly smaller than the corresponding dimensions of R_2, then the vertical dimension of R_1 must be not bigger than the vertical dimension of R_2 .

Under these circumstances the family B satisfies the exponential type covering property, that is:

Given $\{R_\alpha\} \subset B$ one can select a subfamily $\{R_j\} \subset \{R_\alpha\}$ such that,

(1) $\mu\{\cup R_\alpha\} \leq C\mu\{\cup R_j\}$

(2) $\int_{\cup R_j} \exp(\sum \chi_{R_j}(x))\, dx \leq C\mu\{\cup R_j\}$

for some universal constant $C < \infty$.

Before presenting some of the details of the proof, let us consider one application of Theorem 1 (more information about this application was provided by E. Stein in his talk two days ago).

Consider $\mathbb{R}^3 = \{X = \begin{pmatrix} x_1, & x_3 \\ x_3, & x_2 \end{pmatrix}$, real, symmetric, 2×2 - matrices$\}$

and the cone $\Gamma = \{X \in \mathbb{R}^3$, positive definite$\}$.

Then T_Γ = tube over Γ = Siegel's upper half-space =

= $\{X + iY$, Y positive definite$\}$.

For each integrable function f in $\mathbb{R}^3$ we have the "Poisson Integral",

$$U(X + iY) = P_Y * f(X) \ , \quad Y \in \Gamma$$

where $$P_Y(X) = c^{te} . \frac{[\det Y]^{3/2}}{|\det(X + iY)|^3}$$

And we may ask the following question: For which functions f is it true that $U(X + iY) \longrightarrow f(X)$, a.e. X, when $Y \to 0$?

It is a well-known fact, see E. Stein and N. Weiss [*], that if $Y = c \times I = \begin{pmatrix} c, & o \\ o, & c \end{pmatrix} \longrightarrow 0$, then $U(X + iY) \longrightarrow f(X)$ a.e. X for integrable functions f. On the other hand if $Y \longrightarrow 0$ without any restriction then the a.e. convergence fails for every class $L^p(\mathbb{R}^3)$.

Here we can settle the case $Y = \begin{pmatrix} y_1, & 0 \\ 0, & y_2 \end{pmatrix} \longrightarrow 0$ because an easy computation shows that

$$Mf(X) = \operatorname*{Sup}_{Y=\begin{pmatrix} y_1, 0 \\ 0, y_2 \end{pmatrix}} |U(X + iY)|$$

is majorized, in a suitable sense, by $M_\phi f$ with $\phi(s, t) = (s, t)^{1/2}$. Therefore we have convergence for $L(\log^+ L)(\mathbb{R}^3)$ and, since $Mf \geqq c\, M_\phi f$ is also true with some universal constant $c > 0$, $L(\log^+ L)(\mathbb{R}^3)$ is the best class for which almost everywhere convergence holds.

<u>Proof of the covering lemma.</u>

We can assume that the given family satisfies the condition $\mu\{\cup R_\alpha\} < \infty$ because otherwise there is nothing to be proved. Therefore we can also assume that $\{R_\alpha\}$ is finite and no one of its members is contained in the union of the others.

Let us choose R_1 to be an element of $\{R_\alpha\}$ with biggest vertical side. Assuming that we have chosen $R_1, \ldots, R_{j-1}$, let R_j be an element of $\{R_\alpha\}$ such that its vertical side is the biggest possible among the α's that satisfy

$$\frac{1}{\mu\{R_\alpha\}} \int_{R_\alpha} \exp\left(\sum_{k=1}^{j-1} \chi_{R_k}(x)\right) dx \leqq 1 + e^{-1}.$$

[*] On the convergence of Poisson integrals, Trans. Amer. Math. Soc. 1969.

The subfamily $\{R_j\}_{j=1,\dots,M}$ obtained in this way satisfies

$$\int_{\bigcup_{j=1}^{M} R_j} \exp\Big(\sum_{j=1}^{M} \chi_{R_j}(x)\Big)\,dx = \int_{\bigcup_{j=1}^{M-1} R_j - R_M} \exp\Big(\sum_{j=1}^{M-1} \chi_{R_j}(x)\Big)\,dx$$

$$+\ e \int_{R_M} \exp\Big(\sum_{j=1}^{M-1} \chi_{R_j}(x)\Big)\,dx$$

$$\leq \int_{\bigcup_{j=1}^{M-1} R_j} \exp\Big(\sum_{j=1}^{M-1} \chi_{R_j}(x)\Big)\,dx + (1+e)\,\mu\{R_M\}$$

$$\leq (1+e)\sum \mu(R_j) \leq \frac{(1+e)\ (e-1)}{e-1\ -\ e^{-1}}\ \mu(\cup R_j)\ .$$

Next given $R \in \{R_\alpha\} - \{R_j\}$ we have

$$\frac{1}{\mu\{R\}} \int_R \exp\Big(\sum{}' \chi_{R_j}(x)\Big)\,dx \geq 1 + e^{-1}$$

Where $\sum'$ is extended over all parallelepipeds R_j with bigger vertical dimension than R .

Let us rearrange the rectangles appearing in $\sum'$ in the following way

$$\sum{}' \chi_{R_j} = \chi_{R_1} + \dots + \chi_{R_p} + \chi_{R_{p+1}} + \dots + \chi_{R_q}$$

where the R_j's

$j = 1\dots p$ have s-dimension $\geq$ s-dimension of R

$j = p+1,\dots q$, have t-dimension $\geq$ t-dimension of R .

Then

$$1 + e^{-1} \le \frac{1}{\mu\{R\}} \int_R \exp(\textstyle\sum' \chi_{R_j}) \, dx =$$

$$= \sum_{r,s=0}^{\infty} \frac{1}{r!} \frac{1}{s!} \sum_{k_1 \dots k_r = 1}^{p} \sum_{e_1 \dots e_s = p+1}^{q} \frac{1}{\mu\{R\}} \int_R \chi_{R_{k_1}} \cdots \chi_{R_{e_s}} \, dx =$$

$$= \sum_{r,s=0}^{\infty} \frac{1}{r!} \frac{1}{s!} \sum_{k_1 \dots k_r = 1}^{p} \sum_{e_1 \dots e_s = p+1}^{q} \frac{\mu\{R_{k_1} \cap \dots \cap R_{k_r} \cap R_{e_1} \cap \dots \cap R_{e_s} \cap R\}}{\mu\{R\}}$$

If $\mu\{R_{k_1} \cap \dots \cap R_{k_r} \cap R_{e_1} \cap \dots \cap R_{e_s} \cap R\} \neq 0$ then the intersection must be of the form shown in the figure where "the block corresponding to ε" = = intersection of the R_j's whose t-dimension is bigger than t-dimension of R and analogously for the block δ.

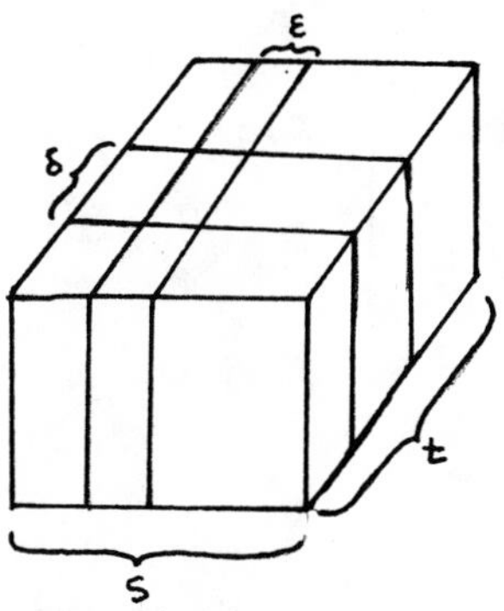

therefore

$$\frac{\mu\{R_{k_1} \cap \dots \cap R_{e_s} \cap R\}}{\mu\{R\}} = \frac{\varepsilon \times \delta}{s \times t}$$

Given $P = (x_o, y_o, z_o) \in R$ consider

$$I_P^1 = \{(x, y_o, z_o) \in R\} ,$$

$$I_P^2 = \{(x_o, y, z_o) \in R\}$$

It happens, again by the monotonicity, that

$$\frac{|R_{k_1} \cap \cdots \cap R_{k_r} \cap I_P{}^1|}{|I_P{}^1|} = \frac{\varepsilon}{s}$$

$$\frac{|R_{e_1} \cap \cdots \cap R_{e_s} \cap I_P{}^2|}{|I_P{}^2|} = \frac{\delta}{t}.$$

Therefore

$$1 + e^{-1} \le \frac{1}{\mu\{R\}} \int_R \exp(\textstyle\sum' \chi_{R_j})$$

$$\le \sum_{r,s=0}^{\infty} \frac{1}{r!} \frac{1}{s!} \sum_{k_1 \ldots k_r = 1}^{p} \sum_{e_1 \ldots e_s = p+1}^{q} \frac{|R_{k_1} \cap \cdots \cap R_{k_r} \cap I_P{}^1|}{|I_P{}^1|}$$

$$\cdot \frac{|R_{e_1} \cap \cdots \cap R_{e_s} \cap I_P{}^2|}{|I_P{}^2|}$$

$$\le \left[\frac{1}{|I_P{}^1|} \int_{I_P{}^1} \exp(\Sigma \chi_{R_j})\right] \cdot \left[\frac{1}{|I_P{}^2|} \int_{I_P{}^2} \exp(\Sigma \chi_{R_j})\right];$$

here $|\ \ |$ denotes 1-dimensional Lebesgue measure.

Therefore if M_x, M_y denote the extensions to $\mathbb{R}^3$ of the one-dimensional Hardy-Littlewood maximal function in the x and y directions respectively, we have:

$$R \subset \{M_x(\exp(\Sigma \chi_{R_j})) \cdot M_y(\exp(\Sigma \chi_{R_j})) \ge 1 + e^{-1}\}$$

$$\subset \{M_x(\exp(\Sigma \chi_{R_j}) \ge \sqrt{1+e^{-1}}\} \cup \{M_y(\exp(\Sigma \chi_{R_j})) \ge \sqrt{1+e^{-1}}\}.$$

Thus,

$$\mu(\cup R_\alpha) \le \frac{2}{\sqrt{1+e^{-1}}} \int_{\cup R_j} \exp(\Sigma \chi_{R_j}) \le C\, \mu(\cup R_j).$$

I. Maximal functions and covering lemmas.

Let us begin with a very general setting. Given a measure space $(X, d\mu)$ and a collection of pairs of measurable sets with finite measure $A = \{(R_\alpha, E_\alpha)\}$ we can define

$$Mf(x) = \sup_{x \in E_\alpha} \frac{1}{\mu\{R_\alpha\}} \int_{R_\alpha} |f(y)|\, d\mu(y) \quad \text{if} \quad x \in \cup E_\alpha$$

and $Mf(x) = 0$ otherwise.

Examples. $X = \mathbb{R}^n$, μ = Lebesgue measure.

(1) Hardy, G.H. and Littlewood, J. E. [1930], A maximal theorem with function-theoretic applications, Acta Math.

$R_\alpha = E_\alpha$ = interval of the real line. The main result is contained in the inequality $\mu\{Mf(x) > \alpha > 0\} \leq c\alpha^{-1} \|f\|_1$, and provides a quantitative version of the Lebesgue differentiation theorem. This result can be extended to $\mathbb{R}^n$ if we consider balls or cubes as substitutes for intervals. The proof is based on the Vitali-Besicovitch-Wiener covering lemma.

(2) Jessen, B., Marcinkiewicz, J. and Zygmund, A. [1935], Note on the differentiability of multiple integrals, Fund. Math.

$R_\alpha = E_\alpha$ = parallelepipeds with sides parallel to the coordinate axis in $\mathbb{R}^n$. Here the crucial estimate is the following:

$$\mu\,\{Mf(c) > \alpha\} \leq C \int_{\mathbb{R}^n} \frac{|f(x)|}{\alpha} \{1 + \log^+ \frac{|f(x)|}{\alpha}\}^{n-1}\, dx\ .$$

In [9] we obtained the covering lemma satisfied by these parallelepipeds producing, therefore, a geometric proof of the result of Jessen, Marcinkiewicz and Zygmund.

(3) M_Φ, the maximal function that we have considered before and which appears naturally after the work of Jessen, Marcinkiewicz and Zygmund.

(4) Rectangles with arbitrary direction. Here the differentiation theorem

is false even for characteristic functions of measurable sets, and the max. function may be infinite a.e. for functions on every L^p-class, $1 \leq p < \infty$.

There is an interesting remark by J. Littlewood in The Scientist Speculates, 1962. He says that, in order to prove some theorem in mathematics, the right move is to find the worst enemy of what one wants to prove and then, introduce the hypothesis necessary to force him to change sides. Littlewood gives a nice example of this principle: Suppose that we have an analytic function $F(z)$ on the half-strip $-1/2\pi \leq x \leq 1/2\pi$, $y \geq 0$ which satisfies $|F(z)| \leq 1$ in the boundary. Is it true that $|F(z)| \leq 1$ inside the strip? The function $F_0(z) = \exp(e^{-iz}-1)$ shows that the answer is no. On the other hand, if we add the condition $|F(z)| \leq K \exp(e^{(1-\varepsilon)\cdot y})$ then the answer is yes and the enemy plays an important role in the proof of this fact. (It is curious to note that Littlewood used this philosophy to claim, in 1962, that Lusin's conjecture was false).

For the Hardy-Littlewood maximal function as well as for the strong maximal function the enemy is the point mass, but, when we consider rectangles with arbitrary directions, the enemy has a more sophisticated personality and it is described by the Nikodym set and by the Besicovitch solution set of the Kakeya needle problem. Our present understanding of the role played by these two sets is not good enough and, following C. Fefferman, we can state that they introduce an element of contradiction with respect to one of the more important paradigms of Fourier Analysis, namely, Lebesgue Integral.

However, by studying very carefully the properties of the Besicovitch's set some positive results have been obtained: maximal functions with respect to lacunary sets of directions (which have been considered at length by Robert Fefferman in the previous talk), and the case of a uniformly distributed set of N directions which I would like to look at more carefully at the end of this talk.

(5) The examples so far considered correspond to the so called Busemann-Feller maximal functions (very imprecisely that means that $E_\alpha = R_\alpha$ for every α). But there are many other interesting cases, for example: (a) R_α = annulus of thickness $\delta > 0$ in $\mathbb{R}^n$ and E_α = ball of radius δ centered at the center of the annulus, $n > 1$. (b) R_α = arc of a thick parabola and E_α = square of the same thickness placed at its vertex, etcetera. The problem is to get estimates for these maximal functions independently of the thickness. Fourier transform methods and complex interpolation have been rather successful in handling these problems and S. Wainger will describe to us, very soon, some of the beautiful ideas involved in their proofs. It is an interesting problem, however, to find directly the geometric arguments needed to establish those results [12], [13], [15]; probably that approach would settle some of the limiting cases, like $n = 2$ in (a), where the Fourier transform methods do not seem to work.

(6) Related with the previous examples are the maximal functions associated to vector fields: given in $\mathbb{R}^n$ a continuous field of directions $v(x)$ and a positive valued functions $p(x)$, consider the operator,

$$Mf(x) = \sup_{0<t<p(x)} \frac{1}{2t} \int_{-t}^{t} |f(x + sv(x))| \, ds$$

As is usual in Differentiation theory, the behavior of these maximal functions produces quantitative versions of the differentiation theorems. If $v(x)$ is constant then we have differentiability for locally integrable functions. However, the general problem is more difficult and the covering techniques have been applied to very particular cases only. Fourier transform methods seem to be better adapted to this problem. In [8], page 131, is presented an example of a continuous field of directions for which the Lebesgue differentiation theorem is false, even for characteristic functions of measurable sets, by building an adapted Nikodym set. However, A. Casas has shown recently that Nikodym sets do not appear for C^1-vector fields. I would like to present several positive observations of Charles Fefferman, Robert Fefferman and the author, probably of other persons too, for some particular vector fields. Let us take $n = 2$ for the sake of simplicity:

(a) Let $v(x) = \frac{x}{\|x\|}$ and $p(x) = 1/2\ \|x\|$ then M is weak-type $(1, 1)$.

(b) $v(x) = \frac{x}{\|x\|}$ and $p(x) = \infty$, then M is bounded in $L^p(\mathbb{R}^2)$, $p > 2$ and unbounded if $p < 2$.

Proof. Using polar coordinates, $x = re^{i\theta}$, we have

$$Mf(x) = \sup_{t>0} \frac{1}{2t} \int_{-t}^{t} |f((r+s)e^{i\theta})|ds$$

$$= \sup_{t>0} \frac{1}{2t} \int_{-t}^{+t} |f_\theta(r+s)|\ ds, \text{ where } f_\theta(u) = f(ue^{i\theta}).$$

Therefore

$$Mf(x) = f_\theta^*(r), \quad x = re^{i\theta}, \quad * = \text{H-L max. function}$$

Then

$$\int_{\mathbb{R}^2} |Mf(x)|^p\ dx = \int_0^{2\pi} \int_0^\infty |Mf(re^{i\theta})|^p\ rdrd\theta$$

$$= \int_0^{2\pi} \left[\int_0^\infty |f_\theta^*(r)|^p\ rdr \right] d\theta \leq C_p \int_0^{2\pi} \int_0^\infty |f_\theta(r)|^p\ rdrd\theta$$

$$= C_p \|f\|_p^p, \text{ because } w(r) = |r| \text{ satisfies condition } A_2.$$

(c) $v(x, y) = \frac{(y, -x)}{\|(x, y)\|}$, $p(x,y) = 1/2\|(x,y)\|$, then M is bounded on $L^p(\mathbb{R}^2)$, $1 < p \leq \infty$.

Proof. The change of variable $w = \log(z)$ maps the straight line $R_o e^{i\theta_o}(1 + ir)$, $-\infty < r < \infty$, into the curve $\log R_o + i\theta_o + \log(1+ir)$, and this curve is the result of translating the fixed curve $w = \log(1+ir)$ to the point $\log R_o + i\theta_o$. Therefore, in the w-plane, the maximal function is given by

$$\tilde{M}g(w) = \sup_{t>0} \frac{1}{2t} \int_{-t}^{t} |g(w) + \gamma(s))| \, ds$$

where $r(t) = \log(1+it)$. We are then in condition to invoke the theorem of E. Stein and S. Wainger [15], to finish the proof.

(d) $v(x, y) = \frac{(1, x)}{(1+x^2)^{1/2}}$, $p(x,y) = \infty$, then M is bounded on $L^p(\mathbb{R}^2)$, $1 < p \leq \infty$.

Proof. The change of variables $\phi(x, y) = (u, v)$, $u = x$, $v = x^2 - 2y$, give us

$$f((x,y) + s(1,x)) = f\circ\Phi^{-1}(x+s, y+sx)) =$$
$$= f\circ\Phi^{-1}(\phi(x, y) + (s, s^2)),$$ and again the problem is reduced to the maximal function associated to a fixed parabola.

* This idea can be used to produce families of curves for which the differentiation theorem is true. For example:

$$\gamma_{(x,\ y)}(s) = (x+s,\ y+3x^2s+3xs^2)$$

$$\gamma_{(x,\ y)}(s) = (x+s,\ y+4x^3s+6x^2s^2+4xs^2)$$

or for vector fields of the form $v(x+iy) = c(x+iy)^n$, where c is a complex number.

Is it true that for a smooth vector field v, there is a positive function P s.t. the associated max. function is bounded in L^p, $1 < p \leq \infty$? What happens near L^1 ?

We need to introduce some notation and definitions. For example, we will use the letters Φ , Ψ to represent Orlicz's spaces defining functions, like t^p , $1 \leq p \leq \infty$, $\exp(t^{1/n})$, $t(\log^+ t)^k$ etcetera.

Definition. $A = \{(R_\alpha, E_\alpha)\}$ has the covering property V_Φ if for each subcollection $B = \{(R_\beta, E_\beta)\} \subset A$ one can select a subsequence $(R_1, E_1), (R_2, E_2), \ldots$ such that:

i) $\mu\{\cup E_\beta\} \leq C\ \mu\{\cup E_j\}$

ii) The overlapping function $\mu(x) = \sum_j \frac{\mu\{E_j\}}{\mu\{R_j\}} \chi_{R_j}(x)$

satisfy $\int_{\cup R_j} \Phi(\nu(x))\, d\mu(x) \leq C\ \mu\{\cup R_j\}$

for some universal constant $C < \infty$.

The particular case $\Phi(t) = t^p$, $1 \leq p \leq \infty$ will be denoted by V_p . We have the following result:

Theorem 2.

(a) If A has the covering property V_q then M is of weak type (p, p), $1/p + 1/q = 1$.

(b) If M is weak type (p, p) and $\tilde{M}$, the maximal function associated to $\tilde{A} = \{(E_\alpha, E_\alpha)\}$, satisfies $\mu\{\tilde{M}\chi_E(x) \geq 1/2\} \leq C\ \mu\{E\}$ for measurable sets E and some universal constant C , then A has the covering property V_q, $1/p + 1/q = 1$, $q < \infty$.

II. Maximal functions and Fourier multipliers.

I will consider now another important paradigm of our school of mathematics, namely the Calderón-Zygmund theory: "the Hardy-Littlewood maximal function controls the Hilbert transform." I will use one of the more recent formulations of this principle, which is contained in the following inequality (see [10]):

$$(H)\quad \int_{\mathbb{R}^n} |Tf(x)|^p\, w(x)dx \leq C_{p,s} \int_{\mathbb{R}^n} |f(x)|^p\, A_s w(x)dx \ ^{\dagger}$$

where $A_s w(x) = [(w^s)^*(x)]^{1/s}$, $1 < p,\ s < \infty$, $*$ denotes the H-L max. function and T is a Calderón-Zygmund singular integral.

Now, some of the best multiplier theorems that I know contain the following strategy: to decompose the kernel, the multiplier or both into pieces, in such a way that each piece is majorized by a positive operator, that the sup. of these positive operators is bounded and that we are able to put together the different estimates by some orthogonality argument, like a g-function estimate for example. In order to perform that program one has to take into account the smoothness properties of the multiplier, its geometric properties or both, and the uncertainty principle which force us to compromise between the localization that we can make at the multiplier side or at the kernel side of the problem. It is not a surprise that these ideas about the L^p-boundedness of constant coefficient operators are useful to analyse the L^2-boundedness of the variable coefficient case: Cotlar's lemma, the theorem of Calderón and Vaillancourt about exotic symbols or the Wave-packets theory of Fourier Integral operators,

---

† In order to prove (H) it is shown in [10] that (H') $\int |f(x)|^p g(x)dx \leq$ $\leq C_{p,s} \int |f^{\#}(x)|^p A_s\omega(x)dx$ holds. An extension of this inequality to (H") $\int f\cdot g \leq C \int f^{\#}\cdot g^*$, with $*$ = non-tangential max. function has been obtained by R. Fefferman and the author, independently, for the dyadic case, and by E. Stein in general.

are, naturally, in that line of thought. So, it is not the first time, even in this meeting, that the following has been said, but I will do it again: "each multiplier has an adequate maximal function which controls it", or taking the expression from [3], the Calderón-Zygmund theory may be phrased as "el culto a la función maximal idónea" by Fourier analysts.

Robert Fefferman has described before how inequality (H) allows us to push the Calderón-Zygmund theory in order to handle kernels with more complicated sets of singularities, like multipliers associated to polygons with a lacunary set of directions. I would like to describe another case where the program has been carried out with total success [16]: Let S_j, j=1, 2, ... , be the angle in $\mathbb{R}^2$ defined by $z = re^{i\theta} \in S_j$ iff

$$\frac{\pi}{2^{j+1}} \leq \theta < \frac{\pi}{2^j} \quad \text{and let} \quad S_o = \mathbb{R}^2 - \bigcup_{j=1}^{\infty} S_j .$$

Define $\widehat{T_j f}(\xi) = \chi_{S_j}(\xi)\, \hat{f}(\xi)$,

and $g(f)(x) = (\Sigma\, |T_j f(x)|^2)^{1/2}$, then we have,

<u>Theorem 3</u>. For every p , $1 < p < \infty$, there are finite positive constants A_p, B_p such that $B_p\|f\|_p \leq \|g(f)\|_p \leq A_p\|f\|_p$.

The Bochner-Riesz spherical summation multiplier $m_\alpha(x) = (1 - \|x\|^2)_+^\alpha$ provides us with an example of the limitations of the paradigm and, at the same time, with the challenge of extending its validity to cover this and similar cases. I would like to consider the problem in $\mathbb{R}^2$, even if, probably, the more interesting universes to be discovered will happen to be related with the remaining cases.

Here is what we know,

i) T_o is only bounded on $L^2(\mathbb{R}^2)$, C. Fefferman.

ii) T_α , $0 < \alpha \leq 1/2$ is bounded on $L^p(\mathbb{R}^2)$, $\frac{4}{3+2\alpha} < p < \frac{4}{1-2\alpha}$, L. Carleson and P. Sjölin.

iii) T_α , $\alpha > 1/2$ is bounded on every $L^p(\mathbb{R}^2)$.

Let me sketch an alternative proof of the Carleson-Sjölin result which, at the same time, explains the differences between T_o and T_α , $\alpha > 0$, for details see [19]. Consider a family of smooth functions $\{\Phi_n\}$ such that $\operatorname{Supp} \Phi_n \subset [1 - 2^{-(n-1)} , 1 - 2^{-(n+1)}]$, $|D^k\Phi_n| \leq$ $\leq C_k\, 2^{+nk}$, $k = 0, 1, 2,\ldots$ and $\Sigma\Phi_n(x) \equiv 1$ on $[1/2, 1]$. Then,

$$m_\alpha(x) = \overline{m}_\alpha(x) + \sum_{n=1}^{\infty} \Phi_n(|x|)\, m_\alpha(x) ,$$

and observe that

$$\Phi_n(|x|) m_\alpha(x) \simeq 2^{-n\alpha}\, \Phi_n(|x|) .$$

Therefore the problem is reduced to get good estimates for the norm of the following multipliers: with $\delta > 0$ let Φ be a smooth function supported on the interval $[1-\delta, 1+\delta]$ and such that $|D^k\Phi| \leq \delta^{-k}$, $k = 0, 1, 2,\ldots$ and define $\widehat{S_\delta f}(\xi) = \Phi(|\xi|)\, \hat{f}(\xi)$.

<u>Theorem 4</u>. $\|S_\delta\|_{(L^p,L^p)} \simeq |\log\delta|^{a(p)}$, $4/3 \leq p \leq 4$. (Since $a(p) = 0$ iff $p = 2$, this theorem explains why T_α , $\alpha > 0$, is bounded and T_o is not).

<u>The strategy</u>. Let $\{\omega_j\}_{j=1,\ldots,\delta^{-1/2}}$ be a smooth partition of unity in the

circle such that:

i) $\omega_1(\theta) = \omega(\delta^{-1/2}\theta)$ where ω is a smooth function supported on the unit interval.

ii) $\omega_j(\theta) = \omega_1(\theta - j(2\pi\delta^{1/2}))$

Then $\Phi = \sum_j \omega_j \cdot \Phi$ and $\widehat{\omega_j \Phi}$ lies, basically, in a rectangle of dimensions $\delta^{-1/2} \times \delta^{-1}$ and direction $(\cos(2\pi\delta^{1/2}j), \sin(2\pi\delta^{1/2}j))$, furthermore $|\widehat{\omega_j \Phi}| \leq \delta^{3/2}$ there. Therefore,

$$|\widehat{\omega_j \Phi} * f(x)| \leq M_\delta f(x) ,$$

where M_δ is the maximal function associated to the family of rectangles of dimension $\delta^{-1/2} \times \delta^{-1}$ and arbitrary direction.

Theorem 5. There exists a constant C, independent of $\delta > 0$ such that $\| M_\delta f \|_2 \leq C |\log 2\delta|^{1/2} \|f\|_2$.

Conjecture 2. $\int |S_\delta f(x)|^2 w(x)dx \leq C \int |f(x)|^2 N_\delta w(x)dx$,

where N_δ is some operator related to M_δ, like in [10].

If true, this conjecture will provide us with a vector-valued Carleson-Sjölin theorem and, therefore, we could prove a multiplier theorem for the cone

$$m_\alpha(x, y, z) = \left(\frac{z^2 - x^2 - y^2}{z^2 + x^2 + y^2}\right)_+^\alpha , \quad \alpha > 0 .$$

Of course one can make the same conjecture in higher dimensions, but there the corresponding maximal theorem result is still unknown.

Proof of Theorem 5. We will prove a slightly more general result and one of the interests of this proof [6] is that it shows, in a very precise manner, the relationship between the dual of the linearized max. function and the covering properties of those rectangles.

With $N >> 1$ and $d > 0$ define

$$R_N^d = \{\text{rectangles of dimensions } d\times dN \text{ , arbitrary direction}\}$$

and

$$Mf(x) = \operatorname*{Sup}_{x\varepsilon R\varepsilon R_N^d} \frac{1}{|R|} \int_R |f(y)|\, dy$$

then we have the following results: there exists $C < \infty$, independently of N and d, such that

$$\|Mf\|_2 \leq C(\log 3N)^{1/2} \|f\|_2 \ .$$

To show it we decompose the interval of directions $[0, 2\pi]$ into eight equal subintervals and, without any loss of generality, we can assume that our rectangles have directions in the interval $[0, \pi/4]$. We divide the plane, by vertical and horizontal lines, into a grid of squares of side dN. The maximal function M acts "independently" on the squares of the grid and, therefore, we can simplify the problem by considering only functions f supported on one of these squares. So let Q be a square with sides parallel to the coordinate axis and length dN and suppose that $f \varepsilon L^2(Q)$.

Then $Mf(x) = 0$ if $x \notin Q^*$ (where Q^* is the double square). We decompose Q^* into $9N^2$ small squares $\{Q_{in}\}$ of side d, by vertical and horizontal lines (assume that $\{Q_{i,n}\}$, $i = 1, \ldots, 3N$, $n=1, \ldots 3N$ are ordered from left to right and from top to bottom). The point is that for every square $Q_{i,n}$ we can find a rectangle R_{in} with direction in the interval $[0, \pi/4]$ and dimensions $d\times dN$, such that:

i) $Q_{in} \cap R_{in} \neq \phi$

ii) $Mf(x) \leq 2 \frac{1}{|R_{in}|} \int_{R_{in}} |f(y)|dy$, for $x \varepsilon Q_{in}$

Therefore if we define the linear operator, f is fixed,

$$T_f(g)(x) = \sum_{i,n} \frac{1}{|R_{in}|} \int_{R_{in}} g(y)dy \cdot \chi_{Q_{in}}(x)$$

it happens that $Mf(x) \leq 2T_f(|f|)(x)$. Therefore, in order to prove our result, it is enough to show that $\|T_f(g)\|_2 \leq C(\log 3N)^{1/2} \|g\|_2$, with C independent of f, d and N.

Thus we have linearized the problem and we can consider the adjoint of T_f, T_f^*. Given $h \in L^2(Q^*)$ we have that

$$T_f^*(h)(x) = \sum_{i,n} \frac{1}{|R_{in}|} \int_{Q_{in}} h(y)\, dy \cdot \chi_{R_{in}}(x) .$$

If $h = h_1 + h_2 + \dots + h_{3N}$, where $h_i = h/E_i$ is the restriction of h to the vertical strip E_i of width d, it is then enough to show that $\|T_f^*(h_i)\|_2 \leq C\, N^{-1/2} (\log 3N)^{1/2} \|h_i\|_2$, $i=1,\dots,3N$.

So, suppose that the function h lies on the strip E_i. We decompose E_i into $3N$ squares, $\{Q_{in}\}_{n=1,\dots,3N}$ ordered from top to bottom, of side d and also we decompose the function $h = h_1+\dots+h_{3N}$, $h_n = h/Q_{in}$.

$$|T_f^*(h)(x)| \leq \sum_n \frac{1}{|R_{in}|} \int_{Q_{in}} |h_n(y)|\, dy \cdot \chi_{R_{in}}(x) \leq \frac{1}{dN} \sum_n \|h_n\|_2\, \chi_{R_{in}}(x) \quad \text{and}$$

$$\|T_f^*(h)\|_2^2 \leq \frac{1}{d^2N^2} \sum_{n,m} \|h_n\|_2\, \|h_m\|_2\, |R_{in} \cap R_{im}| \leq \frac{1}{N} \sum_{n,m} \frac{\|h_n\|_2\, \|h_m\|_2}{1 + |n-m|} \leq \frac{1}{N} (\log 3N) \|h\|_2^2$$

q.e.d.

It is curious to observe that the exponent 1/2 can not be improved in the statement of the theorem and also that, by looking to T_f^* instead of T_f, one does not need to use the Kakeya set to show the sharpenest of this result. Theorem 5 may be extended in several directions; for example, it provides the best L^2 result in higher dimensions for the analogous max.

function, but also the hypothesis can be relaxed: Let us consider N uniformly distributed directions in the plane i.e. given by the angles $\theta_1, \theta_2, \ldots, \theta_N$ s.t. $A|j-k| \leq |\theta_j - \theta_k| \leq B|j-k|$, $0 < A, B$, $j, k = 1, \ldots N$, and let us denote by M the maximal function associated to the family of rectangles with these directions but arbitrary dimensions. Then there exists a constant $C(B/A)$, but independent of N, such that:
$\mu\{Mf(x) > \alpha\} \leq C(B/A)\,(\log 3N)\alpha^{-2}\,\|f\|_2^2$.

There are basically two proofs of this result, one uses covering lemmas and may be founded in [21], [20] and [19], the other has been obtained, independently, by S. Wainger and the author and uses Fourier transform methods. But I would like to point out that both approaches have their remote origins in relation with the same problem: the spherical summation multipliers.

To finish, let us state and sketch the proof of a more sophisticated version of Theorem 5. Let us consider a lacunary sequence of directions θ_j, and, on each angle, $[\theta_j, \theta_{j-1}]$, a uniformly distributed set of N directions $\theta_j^1, \ldots, \theta_j^N$, with uniform constants A and B. Let us denote by M the maximal function associated to the family of rectangles in $\mathbb{R}^2$ whose direction belong to that set $\{\theta_j^k\}_{\substack{j=1,2,\ldots\\ k=1,\ldots,N}}$.

<u>Theorem 6</u>. There exists a constant $C(B/A)$, independent of N, such that:

$$\mu\{Mf(x) > \alpha > 0\} \leq C(B/A)(\log 3N)\,\alpha^{-2}\,\|f\|_2^2 .$$

<u>Proof</u>. We may assume that the rectangles belong to some dyadic decomposition in the corresponding direction. We have $\{Mf(x) > \alpha\} = \bigcup_\lambda R_\lambda$ where $\frac{1}{\mu\{R_\lambda\}} \int_{R_\lambda} |f| > \alpha$. We select in $\{R_\lambda\}$ a

subsequence $\{R_j\}_{j=1},\ldots$ with the following strategy: at each step j we choose a rectangle with biggest diameter among the R_λ that satisfy $\sum_{k<j} |R_k \cap R_\lambda| \leq 1/2\, |R_\lambda|$.

Obviously $\|\Sigma \chi_{R_j}\|_2 \leq C\, \mu\{R_j\}^{1/2}$. On the other hand if $R \in \{R_\lambda\} - \{R_j\}$ then two possibilities may happen:

i) $\Sigma' |R_j \cap R_\lambda| \geq 1/4\, |R_\lambda|$, where Σ' is extended over all the rectangles of the sequence $\{R_j\}$ with diameter bigger than or equal to R_λ and direction belonging to a different lacunary angle than R_λ.

ii) $\Sigma'' |R_j \cap R_\lambda| \geq 1/4\, |R_\lambda|$, where Σ'' means now that diameter $(R_j) \geq$ diameter (R_λ) but their directions belong to the same set of N uniformly distributed directions within the same lacunary angle.

Therefore

$$\bigcup R_\lambda \subset \{M_s(\Sigma \chi_{R_j}) \geq \tfrac{1}{16}\} \cup \bigcup_k \{M^k(\Sigma^{(k)} \chi_{R_j}) \geq 1/4\}$$

where M_s denotes the strong max. function in $\mathbb{R}^2$, M^k is the max. function associated to the N-uniformly distributed direction within the lacunary angle $[\theta_k, \theta_{k+1}]$ and $\Sigma^{(k)}$ is extended over all the rectangles of the sequence with direction in the interval $[\theta_k, \theta_{k+1}]$. Therefore,

$$\mu(\cup R_\lambda) \leq C(B/A) \log N \cdot \mu(\cup R_j)$$

and finally

$$\mu(\cup R_j) \leq \sum_j \mu(R_j) \leq \frac{1}{\alpha} \sum_j \int_{R_j} |f| \leq$$

$$\leq \frac{1}{\alpha} \|f\|_2 \, \|\Sigma \chi_{R_j}\|_2 \leq \alpha^{-1} \|f\|_2 \, \mu\{\cup R_j\}^{1/2}$$

q.e.d.

Remark. There is quite a number of mathematicians who are related with the results considered in this paper. Taking into account only the more recent developments we must mention the names of C. Fefferman, R. Fefferman, A. Nagel, N. Rivière, E. Stein, J. Stromberg, S. Wainger, and a long etcétera. I have tried to present the situation of this theory without interfering too much with the talks of R. Fefferman and S. Wainger, but I have been very sloppy associating names with results. I apologize for that and also because my list of references is rather incomplete, however, rephrasing Gonzalo de Berceo, I hope that my presentation "aunque no ha sido en roman paladino, si se merece, al menos, un vaso de buen vino."

Williamstown, July of 1978.

References

[1] C. Fefferman [1970], Inequalities for strongly singular convolution operators, Acta Math., 124.

[2] ------------ [1972], The Multiplier problem for the ball, Ann. of Math., 94.

[3] G. Torrente Ballester [1972], La saga/fuga de J. B. Ediciones Destino. Ancora y Delfin, 338.

[4] L. Carleson and P. Sjölin [1972], Oscillatory integrals and a multiplier problem for the disc, Studia Math., 44.

[5] C. Fefferman [1973], A note on spherical summation multipliers, Israel J. Math., 15.

[6] A. Córdoba [1974], Thesis, University of Chicago.

[7] ---------- [1974], Los operadores de Bochner-Riesz Universidad Complutense.

[8] M. de Guzman [1975], Differentiation of integrals, Springer Lecture Notes, 481.

[9] A. Córdoba and R. Fefferman [1975], A geometric proof of the strong maximal theorem, Ann. Math., 102.

[10] A. Córdoba and C. Fefferman [1976], A weighted norm inequality for singular integrals, Studia Math, T. LVII.

[11] A. Córdoba [1976], On the Vitali covering properties of a differentiation basis, Studia Math., T. LVII.

[12] A. Nagel, N. Rivière and S. Wainger [1976], A maximal function associated to the curve (t, t^2), Proc. Natl. Acad. Sci., USA, 73.

[13] E. Stein [1976], Maximal functions: spherical means, Proc. Natl. Acad. Sci., USA, 73.

[14] -------- [1976], Maximal functions: homogeneous curves, Proc. Natl. Acad. Sci., USA, 73.

[15] E. Stein and S. Wainger [1976], Maximal functions associated to smooth curves, Proc. Natl. Acad. Sci., USA, 73.

[16] A. Córdoba and R. Fefferman [1977], On the equivalence between the boundedness of certain classes of maximal and multiplier operators in Fourier Analysis, Proc. Natl. Acad. Scil, USA, 74.

[17] J. Strömberg [1977], Maximal functions for rectangle with given directions, Mittag-Leffler Institute.

[18] A. Córdoba and R. Fefferman [1977], On differentiation of integrals, Proc. Natl. Acad. Sci., USA, 74.

[19] A. Córdoba [1977], The Kakeya maximal function and the spherical summation multipliers, Amer. J. Math.

[20] ---------- [1977], The multiplier problem for the polygon, Ann. of Math., 105.

[21] J. Strömberg [1978], Maximal functions associated to uniformly distributed families of directions, Ann. of Math., 108.

[22] A. Nagel, E. Stein and S. Wainger, [1978], Differentiation in lacunary directions, Proc. Natl. Acad. Sci., USA.

[23] A. Córdoba [1978], $s \times t \times \phi(s, t)$, Mittag-Leffler Institute, Report No. 9.

[24] A. Córdoba and C. Fefferman, Wave packets and F.I.O., to appear.

Universidad Complutense

and

Princeton University

Proceedings of Symposia in Pure Mathematics
Volume XXXV, Part 1, 1979

COVERING LEMMAS, MAXIMAL FUNCTIONS AND MULTIPLIER OPERATORS IN FOURIER ANALYSIS

R. Fefferman[1]
University of Chicago

In recent years, the subject of differentiation of integrals has undergone quite a drastic change. Nowadays, using the new methods developed in differentiation theory, theorems on multiplier and singular integral operators may be obtained as well as results on differentiating integrals.

In this article, I would like to discuss both a method for obtaining new maximal theorems (or what is the same thing, theorems on differentiating integrals) and also the interplay that these maximal theorems have with other problems in Fourier analysis.

Basically, we now know of two systematic methods for proving maximal theorems. The first, due to Nagel, Rivière, Stein, and Wainger has, as its main tool, the Fourier transform. The second, and method of Cordoba and the author, uses covering lemmas. Since Wainger, in his talk, gives a thorough account of the first method, we shall limit ourselves to a discussion of the second.

Perhaps the best way to understand the technique of covering lemmas is to examine two simple examples and to try to see clearly the relation between them. We wish to particularly stress here that it is the common thread which runs through both examples that is of main interest rather than the specific details of either example treated in isolation.

Our first example is the theorem of Jessen, Marcinkiewicz, and Zygmund [8] concerning the so-called strong maximal theorem in R^n. Suppose we consider $\mathcal{R}$, the class of all rectangles in R^n whose sides are in the directions of the coordinate axes. Form the maximal operator corresponding to the sets in $\mathcal{R}$:

$$M_S(f)(x) = \sup_{x \in R \in \mathcal{R}} \frac{1}{|R|} \int_R |f(y)|\,dy.$$

Then the result we are considering states that for each $\alpha > 0$ and f defined on R^n,

$$|\{M_S f > \alpha\}| \le \frac{C_n}{\alpha} \|f\|_{L(\log^+ L)^{n-1}}.$$

Of course, from this result, it follows immediately that differentiation of the integral holds for functions in the class $L(\log^+ L)^{n-1}$ when the differentiation is with respect to the rectangles in $\mathcal{R}$:

$$\lim_{\substack{\operatorname{diam}(R)\to 0\\ x\in R\in\mathcal{R}}} \frac{1}{|R|}\int_R f(y)\,dy = f(x), \text{ a.e. on } R^n.$$

AMS (MOS) subject classifications (1977). Primary 42-XX; Secondary 42A18, 42A40

[1]Research supported by the National Science Foundation under grant NSF MCS78-02128.

In order to prove this result by a covering lemma, it turns out that what is needed is the following:

Given an arbitrary collection of rectangles $\mathcal{R}$, $\{R_\beta\}$, there is a subcollection $\{\tilde{R}_i\}_{i=1,2,\ldots}$ of $\{R_\beta\}$ such that

$$1. \quad |\cup \tilde{R}_i| \geq c|\cup R_\beta|$$

and

$$2. \quad \|\exp(\Sigma \chi_{\tilde{R}_i})^{\frac{1}{n-1}} - 1\|_{L^1} \leq C|\cup \tilde{R}_i|,$$

where c and C are constants depending only on the dimension n. Notice that condition (1) insures that our subcollection captures at least a fixed fraction of the measure of all the original rectangles, while condition (2) says that the subcollection is suitably "sparse." Also notice that the norm appearing on the left side of (2) is really just the norm dual to the $L(\log^+ L)^{n-1}$ norm. This is no coincidence, of course.

Our first attempt at proving such a covering lemma proceeds as follows: First we give a rule for selecting the rectangles $\tilde{R}_i$. (Clearly we may assume that $\{R_\beta\}$ is countable and we refer to the original family of rectangles as $\{R_j\}_{j=1,2,\ldots}$).

Take $\tilde{R}_1 = R_1$. To find $\tilde{R}_2$, examine $R_2, R_3, \ldots$ etc. until we come to the first R_j, $j > 1$ with the property that $|R_j \cap \tilde{R}_1| \leq \frac{1}{2}|R_j|$. Call this first R_j $\tilde{R}_2$. To find $\tilde{R}_3$ continue along the list of R_j until we first find an R_j such that

$$|R_j \cap [\bigcup_{i=1}^{2} \tilde{R}_i]| \leq \frac{1}{2}|R_j|$$

and let that first such R_j be $\tilde{R}_3$. Continuing this process ad infinitum we arrive at the whole subcollection $\{\tilde{R}_i\}$. And we now try to show that the $\{\tilde{R}_i\}$ satisfy (1) and (2) above.

First observe that if R_j has been discarded in our selection process, we must have

$$|R_j \cap [\bigcup_i \tilde{R}_i]| > \frac{1}{2}|R_j|$$

so that

$$M_S(\chi_{\cup \tilde{R}_i}) > \frac{1}{2} \text{ on } \bigcup_j R_j$$

and so

$$|\cup R_j| \leq C|\cup \tilde{R}_i|$$

because the operator M_S is bounded on the L^p spaces. This proves (2). To show (1) we use duality, namely, consider the following estimate:

$$\int_{R^n} \sum_i \chi_{\tilde{R}_i} \cdot g\, dx =$$

$$= \sum_i |\tilde{R}_i| \left(\frac{1}{|\tilde{R}_i|} \int_{\tilde{R}_i} g\, dx \right) \leq 2 \sum_i |\tilde{E}_i| \left(\frac{1}{|\tilde{R}_i|} \int_{\tilde{R}_i} g\, dx \right)$$

where $\tilde{E}_i = \tilde{R}_i - [\bigcup_{j<i} \tilde{R}_j]$. Since the $\tilde{E}_i$ are disjoint and

$$\frac{1}{|\tilde{R}_i|} \int_{\tilde{R}_i} g\, dx \leq M_S(g)(x), \text{ for all } x \in \tilde{E}_i$$

we see that

$$\Sigma |\tilde{E}_i| \left(\frac{1}{|\tilde{R}_i|} \int_{\tilde{R}_i} g\, dx \right) \leq \int_{\cup \tilde{R}_i} M_S(g)(x)\, dx.$$

This last integral will be finite when $M_S(g) \in L^1_{\text{locally}}$ or when $g \in L(\log^+ L)^n$ (by Jessen-Marcinkiewicz-Zygmund). So our estimate becomes

$$|\int \Sigma \chi_{\tilde{R}_i} \cdot g| \leq C\|g\|_{L(\log^+ L)^n}$$

which means that

$$\|\exp(\Sigma \chi_{\tilde{R}_i})^{\frac{1}{n}}\|_{L^1} < \infty.$$

At this point we must raise two objections. First we obtained control only of $\int \exp(\Sigma \chi_{\tilde{R}_i})^{\frac{1}{n}} dx$ and not $\int \exp(\Sigma \chi_{\tilde{R}_i})^{\frac{1}{n-1}} dx$, which we wanted. And much more serious is the objection that in trying to prove the strong maximal theorem by means of covering lemmas, we have used this very theorem in our arguments. When later, we deal with maximal operators whose behavior is unknown it will be of vital importance to eliminate the second objection. Fortunately we may easily deal with these problems as follows: Think of ordering the rectangles $\{R_j\}$ so that their side length in the x_n coordinate direction decreases. Repeat the selection process as before, discarding a rectangle R_j whenever the selected rectangles chosen so far occupy at least $\frac{1}{2}$ its volume. Now suppose that $x \in R_j$, a discarded rectangle, and we pass a hyperplane through x perpendicular to the x_n coordinate direction, calling T_i the intersection of any rectangle R_i with the hyperplane. Then it is easy to see that not only do we have

$$|R_j \cap [\bigcup_i \tilde{R}_i]| > \frac{1}{2}|R_j|$$

but also

$$|T_j \cap [\bigcup_i \tilde{T}_i]| > \frac{1}{2}|T_j|.$$

This means that already $M_{n-1}(\chi_{\cup \tilde{R}_i}) > \frac{1}{2}$ on $\cup R_j$, where M_{n-1} is the n-1 dimensional strong maximal function. (Here $M_{n-1}(f)(x) = \sup_{x \in T \in \mathcal{T}} \frac{1}{|T|} \int_T |f(y)|dy$, and $\mathcal{T}$ is the collection of all n-1 dimensional rectangles with sides parallel to the $x_1, x_2, \ldots, x_{n-1}$ axes in a hyperplane throughout x, and perpendicular to the x_n direction). Also, let us observe that if we slice the selected rectangles $\tilde{R}_i$, with a hyperplane perpendicular to the x_n direction, the $\tilde{T}_i$ also have the property that $|\tilde{T}_j - \bigcup_{i<j} \tilde{T}_i| > \frac{1}{2}|\tilde{T}_j|$. So applying the same reasoning as was used to estimate the norm $\|\exp(\Sigma\chi_{\tilde{R}_i})^{\frac{1}{n}}\|_1$, we see that

$$\int(\sum_i \chi_{\tilde{T}_i})g\,dx_1,\ldots,dx_{n-1} \leq \int M_{n-1}(g)\,dx_1,\ldots,dx_{n-1}.$$

Integrating this inequality in x_n we have

$$\int(\Sigma\chi_{\tilde{R}_i})g\,dx \leq \int_{\cup\tilde{R}_i} M_{n-1}(g)\,dx$$

and an application on duality, and the fact that M_{n-1} maps $L(\log+L)^{n-1}$ into L^1 gives the desired estimate that $\|\exp(\Sigma\chi_{\tilde{R}_i})^{\frac{1}{n-1}}\|_1 < \infty$.

By controlling the strong maximal operator in n dimensions by the "better" operator M_{n-1}, it is clear that we have proven the strong maximal theorem by induction on n, without any circular reasoning.

Next, we shall give a somewhat similar argument which proves a new maximal theorem. Suppose, in R^2, we let $\mathcal{D}$ denote the collection all rectangles with arbitrary side lengths, oriented in a direction making an angle of 2^{-k} (for some $k \geq 1$) with the positive x axis. Let

$$M_{\mathcal{D}}(f)(z) \cdot \sup_{z\in R\in\mathcal{D}} \frac{1}{|R|}\int_R |f(y)|\,dy.$$

We intend to show the inequality

$$m\{M_D f > \alpha\} \leq \frac{C}{\alpha^2}\|f\|_{L^2(R^2)}, \text{ for all } \alpha > 0.$$

Some remarks on the recent history of the operator M_D are in order here. In [13] J. Strömberg obtained the result one could differentiate the integrals of functions in $L^2(\log+L)^{4+\varepsilon}$, $\varepsilon > 0$ with respect to rectangles in $\mathcal{D}$. In [5], Cordoba and the author extended this result to L^2. Both works involve the method we have discussed above at length. Then, using the Fourier transform, Nagel, Stein, and Wainger [11] extended these results to get differentiation for L^p, $p > 1$. Again, since we are interested in describing a general method, rather than an isolated result, we content ourselves with showing that M_D is of weak type (2,2).

To do this, assume given a family $\{R_j\}$ of rectangles in D. We shall show that it is possible to select from $\{R_j\}$ a subfamily $\{R_j\}$ so that

1. $|\cup\tilde{R}_i| \geq c|\cup R_j|$ for some constant c.

2. $\|\Sigma\chi_{\tilde{R}_i}\|_2^2 \leq C|\cup R_j|$ for some C.

First assume $\{R_j\}$ has been arranged so that the longest side of R_j decreases. The rule for selecting the R_i is only slightly different from the case of the strong maximal operator:

Choose $\tilde{R}_1 = R_1$. Go along in the sequence considering $R_2, R_3, \ldots$ until finally hitting an R_j such that $|R_j \cap \tilde{R}_1| \leq \frac{1}{2}|R_j|$. When this happens call that R_j, $\tilde{R}_2$. Now given $\tilde{R}_1, \tilde{R}_2, \ldots, \tilde{R}_{i-1}$, continue along the sequence of R_j's until hitting a rectangle R_ℓ such that

$$\sum_{k \le i-1} |R_\ell \cap \tilde{R}_k| \le \frac{1}{2} |R_\ell|,$$

and call this rectangle $\tilde{R}_i$. Notice that since

$$\| \Sigma \chi_{\tilde{R}_i} \|_2^2 = 2 \int \sum_{j<i} \chi_{\tilde{R}_i} \chi_{\tilde{R}_j} + \sum_i |\tilde{R}_i|$$

$$= 2 \sum_i \sum_{j<i} |\tilde{R}_j \cap \tilde{R}_i| + \sum_i |\tilde{R}_i| \le C | \cup \tilde{R}_i |,$$

condition (2) above is already trivially satisfied, just because of the way we chose the R_i. To show that $|\cup R_i| \ge c |\cup R_j|$, assume that R_j has been discarded. Then it must be the case that

$$\sum_{\substack{\ell(\tilde{R}_i) \ge \\ \ell(R_j)}} |\tilde{R}_i \cap R_j| > \frac{1}{2} |R_j|$$

(Here $\ell(R)$ denotes the length of the longest side of R.) Now we must resort to geometry. Because the orientations of the rectangles in question are dyadically convergent, we can make a simple observation which will reduce our problem to a question about the strong maximal operator, M_S.

Suppose that S is the smallest rectangle *with sides parallel to the axes* which contains R_j (see Fig.).

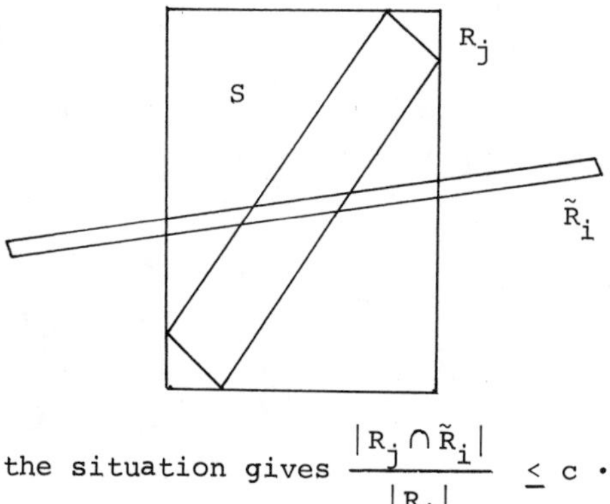

Then the geometry of the situation gives $\dfrac{|R_j \cap \tilde{R}_i|}{|R_j|} \le c \cdot \dfrac{|\tilde{R}_j \cap S|}{|S|}$, for some absolute constant c. Since

$$\frac{1}{2} < \frac{1}{|R_j|} \sum_{\ell(\tilde{R}_i) \ge \ell(R_j)} |\tilde{R}_i \cap R_j| \le \frac{1}{|S|} \sum_{\text{all } i} |\tilde{R}_i \cap S|$$

so that $M_S(\Sigma \chi_{\tilde{R}_i}) > \frac{1}{2}$ on R_j. But j is arbitrary so that

$$\cup R_j \subseteq \{ M_S(\Sigma \chi_{\tilde{R}_i}) > \tfrac{1}{2} \}$$

and because M_S is known to be bounded in L^2,

$$|\cup R_j| \leq |\{M_S(\Sigma \chi_{R_j}) > \frac{1}{2}\}| \leq c\|\sum_i \chi_{\tilde{R}_i}\|_2^2 \leq C|\cup \tilde{R}_j|,$$

concluding the proof of the covering lemma for M_D. (We point out that we have assumed that in our figure the rectangle $\tilde{R}_i$ is in a *different* orientation than R_j. If this is not the case, things are even easier, and we may conclude that on R_j $M(\Sigma\chi_{\tilde{R}_i}) > \frac{1}{2}$ where M is the Hardy-Littlewood maximal operator in R^2.

Now it is clear what our program for proving maximal theorems is. We control a maximal operator about which we known nothing by a better understood maximal operator through the geometry of covering lemmas. For an extension of the ideas used to handle the operator M_D, we refer the reader to Cordoba's talk.

At this point we shall show that the maximal operators which have been dealt with above are intimately related to certain natural multiplier operators of Fourier analysis. Through the fundamental work of Charles Fefferman [6] it is known that the characteristic function of the unit disk Δ in R^2 is not an L^p multiplier whenever $p \neq 2$.

In other words C. Fefferman's theorem is that if we define an operator T_Δ by

$$\widehat{T_\Delta(f)}(\xi) = \chi_\Delta(\xi)\cdot\hat{f}(\xi),$$

then T_Δ is not bounded on $L^p(R^2)$ for any $p \neq 2$. If $S \subseteq R^2$ we may define in the same manner an operator T_S, and ask for which p is T_S a bounded operator on L^p? If the set S is open and sufficiently tame then Fefferman's proof shows that the boundary of S cannot contain any "curved part" if T_S is to be bounded on L^p for some $p \neq 2$. So the natural sets to look at are polygonal sets.

Let $\theta_1 > \theta_2 > \theta_3 \cdots$ be a sequence of angles tending to 0. Let P_θ denote the polygonal region in the figure below.

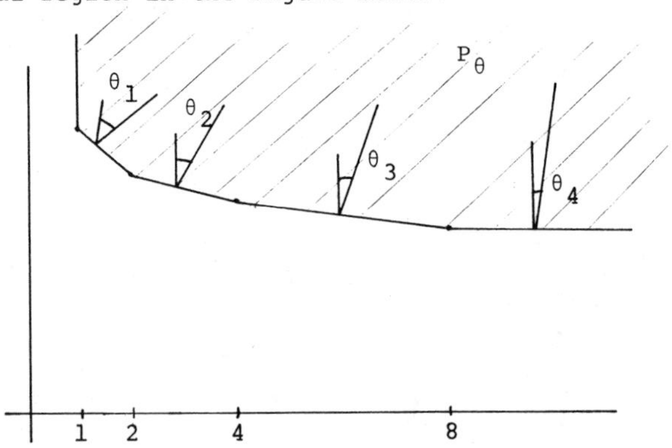

Suppose M_θ is the maximal operator corresponding to averages over rectangles in the plane whose side lengths are arbitrary and which are oriented in one of the directions θ_i for some i. Then we intend here to sketch a proof that, for $p \geq 2$ if the operator T_θ is bounded on L^p, then M_θ is weak type on $L^{(p/2)'}$:

$$|\{M_\theta f > \alpha\}| \leq [\frac{C}{\alpha} \|f\|_{(\frac{p}{2})'}]^{(\frac{p}{2})'},$$

and conversely, if M_θ is bounded on $L^{(p_0/2)'}$, for some $p_0 > p$, then T_θ is bounded on L^p. (Here if $1 < p < \infty$, p' denotes the exponent dual to p.)

First, assume then that M_θ is known to be bounded on $L^{(p_0/2)'}$ where $p_0 > p$. Then we write $T_\theta f = \sum_k H_k S_k f$ where

$$\widehat{S_k f}(\xi) = \chi_{2^k < \xi_1 \leq 2^{k+1}}(\xi) \cdot \hat{f}(\xi)$$

and

$$\widehat{H_k f}(\xi) = \chi_{\pi_k}(\xi) \cdot \hat{f}(\xi),$$

where π_k is a half plane containing the region P_θ and tangent to P_θ on the $k^{\underline{th}}$ side.

By Littlewood-Paley theory,

$$\|T_\theta f\|_{L^p} \sim \|(\Sigma |H_k S_k f|^2)^{\frac{1}{2}}\|_{L^p}.$$

To estimate this L^p norm, consider

$$\int \Sigma \, |H_k S_k f|^2 \cdot \varphi \, dx = \Sigma \int |H_k S_k f|^2 \varphi \, dx.$$

Since H_k is basically just a Hibert transform in the direction orthogonal to the $k^{\underline{th}}$ side of P_θ, we may use the following duality inequality (see [2]):

$$\int_{R^1} |Hf|^2 g \, dx \leq C_\varepsilon \int_{R^1} |f|^2 \, M(g^{1+\varepsilon})^{\frac{1}{1+\varepsilon}} \, dx, \quad \varepsilon > 0.$$

(Here H is the Hilbert transform on the line, and M is the classical Hardy-Littlewood maximal function.)

Applying this inequality, we have

$$(*) \qquad \int_{R^2} \Sigma |H_k S_k f|^2 \cdot \varphi \, dx \leq c_\varepsilon \int_{R^2} \Sigma |S_k f|^2 [M_\theta(\varphi^{1+\varepsilon})]^{\frac{1}{1+\varepsilon}} dx.$$

If $\varphi \in L^{(p/2)'}$, since M_θ is known to be bounded on $L^{(\frac{p}{2})' \cdot \frac{1}{1+\varepsilon}}$ if $\varepsilon > 0$ is small enough, and by Littlewood-Paley theory $\Sigma|S_k f|^2 \in L^{p/2}$ the right hand side of (*) must be finite, concluding the proof of the boundedness of T_θ on L^p.

On the other hand suppose we know that the operator T_θ is bounded on $L^p(R^2)$. We show that given the weakest information about M_θ it follows that M_θ is of weak type $(p/2)',(p/2)'$. More precisely suppose that it is given that $|\{M_\theta(\chi_E) > \frac{1}{2}\}| \leq C|E|$ for any set E. (An example of this situation arose prior to the work of Nagel, Stein, and Wainger on the maximal operator M_θ where $\theta_i = 2^{-i}$. M_θ was known to be of weak type (2,2) but nothing was known about the operator M_θ on L^p, $p < 2$, so in this case the Tauberian condition $|\{M_\theta\chi_E > \frac{1}{2}\}| \leq C|E|$ was known to be true.)

In order to show that M_θ is weak type $(p/2)',(p/2)'$, it will suffice to prove the following covering lemma:

Given rectangles $\{R_j\}$ it is possible to select out a subcollection $\{\tilde{R}_i\}$ so that

1. $|\cup\tilde{R}_i| > c|\cup R_j|$ and

2. $\|\Sigma\chi_{\tilde{R}_i}\|_{p/2}^{p/2} \leq C|\cup\tilde{R}_i|$.

Select the $\tilde{R}_i$ in order to have

$$|\tilde{R}_i - {}_{j<i}\tilde{R}_j| > \frac{1}{2}|\tilde{R}_i|,$$

as in the proof of the strong maximal theorem. Then, because $\cup R_j \subseteq \{M_\theta(\chi_{\cup\tilde{R}_i}) > \frac{1}{2}\}$ we have (1) satisfied. To show (2), we use a technique in [6]. Since T_θ is a bounded operator, the following Littlewood-Paley inequality holds:

$$(**) \qquad \|(\Sigma|H_kf_k|^2)^{\frac{1}{2}}\|_p \leq \|(\Sigma|f_k|^2)^{\frac{1}{2}}\|_p,$$

where H_k is a Hilbert transform in the direction in which $\tilde{R}_k$ is oriented. Let $\tilde{E}_k = \tilde{R}_k - \bigcup_{j<k}\tilde{R}_j$, and $f_k = \chi_{\tilde{E}_k}$. If we apply (**) several times we finally wind up with the desired inequality $\|\Sigma\chi_{\tilde{R}_k}\|_{p/2}^{p/2} \leq |\cup R_k|$. To see this consider the following picture:

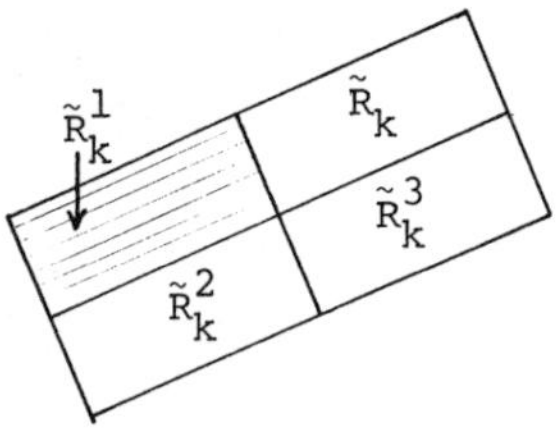

On at least half of the lines parallel to the longest side of $\tilde{R}_k^1$, whose union we shall call Q_k^1, we easily see that $H_kf_k \geq \frac{1}{2}$. Now let $f_k^1 = H_kf_k \cdot \chi_{Q_k^1}$ for each k. Apply (**) once more, only this time in the directions orthogonal to the ones in which the $\tilde{R}_k$ are oriented . We see that if $H_{k^\perp}$ is a Hilbert transform in the direction orthogonal to that of $\tilde{R}_k$ we have

$H_k f_k^1 \geq \frac{1}{2}$ on all of $\tilde{R}_k^2$, so that using (**)

$$\|(\Sigma\chi_{\tilde{R}_k^2})^{\frac{1}{2}}\|_p \leq c\|(\Sigma|f_k^1|^2)^{\frac{1}{2}}\|_p \leq c\|(\Sigma|H_k f_k|^2)^{\frac{1}{2}}\|_p \leq c\|(\Sigma|f_k|^2)^{\frac{1}{2}}\|_p \leq c|\cup\tilde{R}_k|^{\frac{1}{p}}.$$

Repeating the procedure twice again we have

$$\|(\Sigma\chi_{\tilde{R}_k})^{\frac{1}{2}}\|_p \leq C\|(\Sigma\chi_{\tilde{R}_k^3})^{\frac{1}{2}}\|_p \leq C\|(\Sigma\chi_{\tilde{R}_k^2})^{\frac{1}{2}}\|_p$$

we obtain the desired estimate of $\Sigma\chi_{\tilde{R}_k}$.

To conclude this article we wish to give another application of the methods of differentiation theory to Fourier analysis. The theorem we have in mind is from the theory of singular integrals:

Suppose $K(x) = \dfrac{\Omega(x)}{|x|^n}$ is a Calderón-Zygmund kernel in R^n, $n \gneq 1$. (This means that Ω is homogeneous of degree 0 on R^n and satisfies some mild smoothness condition on the surface of the unit sphere S^{n-1}, and $\int_{S^{n-1}} \Omega d\sigma = 0$.) Suppose we multiply $K(x)$ by an arbitrary bounded radial function, and call the resulting kernel $H(x)$. Then the operator $f \to f*H$ is bounded on $L^p(R^n)$ for all p such that $1 < p < \infty$.

This theorem is an application of the theory of Hilbert transforms and maximal functions on curves of Nagel, Rivière and Wainger [9], [10]. In fact in order to prove this result, we write the kernel $H(x)$ in polar coordinates:

$$Hdx = \int_0^\infty [\Omega_r(x')d\sigma_r(x')]\frac{dr}{r},$$

where the functions Ω_r are uniformly smooth on S^{n-1} and $d\sigma_r$ is the unit mass on rS^{n-1} which is rotationally invariant. If we set $d\mu_r = \Omega_r(x')d\sigma_r(x')$, after some simple estimates we have

$$|\widehat{d\mu_r}(\xi)| \leq J(r|\xi|) \text{ where } J(r) \leq \begin{cases} cr & r \text{ small} \\ \omega(\frac{1}{r^{1/2}})+\frac{1}{r^{1/2}} & r \text{ large} \end{cases}$$

(ω is the modulus of continuity of Ω on S^{n-1}). The singular integral in question therefore has exactly the same form as a singular integral along a curve, i.e., the kernel is written as

$$\int_0^\infty d\mu_r \cdot \frac{dr}{r}$$

where $\widehat{d\mu_r}$ has good decay at $r = 0$ and ∞. Then $\hat{H} = \int \widehat{d\mu_r}\,\frac{dr}{r}$ is clearly bounded, so the operator is bounded on L^2, and the L^p results follow from this observation by a complex interpolation (see [7]).

We would like to conclude by saying that the results on differentiation and the applications given in this article represent just a small fraction of the new ideas in this area, and we want to take this opportunity to refer the reader to the articles in this volume of Cordoba and Wainger for more on this subject.

BIBLIOGRAPHY

[1] A. P. Calderon and A. Zygmund, On the existence of certain singular integrals, Acta Math. 88 (1952), 85.

[2] A. Cordoba and C. Fefferman, A weighted norm inequality for singular integrals, Studia Math. 57 No. 1 (1976), 97.

[3] A. Cordoba and R. Fefferman, A geometric proof of the strong maximal theorem, Annals of Math. 102 (1975), 95.

[4] A. Cordoba and R. Fefferman, On the equivalence between the boundedness of certain classes of maximal and multiplier operators in Fourier analysis, Proc. N.A.S. 74 No. 2 (1977), 423.

[5] A. Cordoba and R. Fefferman, On differentiation of integrals, Proc. N.A.S. 74 no. 6 (1977).

[6] C. Fefferman, The multiplier problem for the ball, Annals of Math. 94 No. 2 (1971), 330.

[7] R. Fefferman, A note on singular integrals, to appear.

[8] B. Jesson, J. Marcinkiewicz, and Z. Zygmund, Note on the differentiability of multiple integrals, Fund. Math. 25 (1935), 217.

[9] A. Nagel, N. Rivière and S. Wainger, On Hilbert transforms along curves II, Amer. Journal of Math. 98 (1976), 395.

[10] A. Nagel, N. Rivière and S. Wainger, A maximal function associated to the curve (t,t^2), Proc. N.A.S. 73 (1976), 1416.

[11] A. Nagel, E. M. Stein, S. Wainger, Differentiation in lacunary directions, Proc. N.A.S. 75 No. 3 (1978), 1060.

[12] E. M. Stein, S. Wainger, Problems in harmonic analysis related to curvature, manuscript.

[13] J. Strömberg, Weak estimates for maximal functions with rectangles in certain directions, Inst. Mittag-Leffler, report #10, also to appear in Arkiv für Math.

Proceedings of Symposia in Pure Mathematics
Volume XXXV, Part 1, 1979

BESICOVITCH THEORY OF LINEARLY MEASURABLE SETS AND FOURIER ANALYSIS

Miguel de Guzmán

ABSTRACT. Besicovitch was one of the founders of the modern local theory of differentiation. He was also one of the founders of a not so well known theory, that of the geometric structure of linearly measurable sets in $\mathbb{R}^2$. This was no accident. The local differentiation theory in $\mathbb{R}^2$ is essentially a study of the geometric structure of two dimensional measurable sets. Some of the techniques in differentiation theory, as it is now, come from the earlier developed theory of linearly measurable sets. Such is, for example, the Besicovitch set and it seems that a better understanding of the techniques used in the theory of linearly measurable sets will lead to interesting developments in some topics in differentiation and in harmonic analysis such as maximal operators and multipliers.

1. BESICOVITCH LINEAR MEASURE IN $\mathbb{R}^2$. The linear measure in $\mathbb{R}^2$ is the 1-dimensional Hausdorff measure. For $E \subset \mathbb{R}^2$ and $\rho > 0$ one first considers

$$\Lambda_\rho(E) = \inf \{\Sigma\delta(A_n) : \cup A_n \supset E, \quad \delta(A_n) \leq \rho\}$$

and $\Lambda^*(E) = \lim \Lambda_\rho(E)$ as $\rho \to 0$. Then Λ^* is an outer measure. Caratheodory's process gives us the associated measure Λ. Besicovitch considered sets $E \subset \mathbb{R}^2$ such that $0 \leq \Lambda(E) < \infty$ and asked for their geometrical structure, i.e. for their properties related to density, tangency, projection, polarity,...

2. DENSITY. REGULAR AND IRREGULAR SETS. If $0 \leq \Lambda(E) < \infty$ and $x \in \mathbb{R}^2$, one can consider the lim sup y lim inf as $r \to 0$ of

$$\frac{\Lambda(E \cap B(x,r))}{2r}$$

where $B(x,r)$ is a ball centered at x and of radius r. They are respectively the upper and lower density (the density, if they coincide) of E at x. We denote

---

AMS (MOS) subject classifications (1970). Primary 28A75, 26A63, 49F20, 42A18; Secondary 26A15.

them by $\bar{D}(E,x)$, $\underline{D}(E,x)$ ($D(E,x)$ if they are equal). One can prove that $D(E,x)$ is 0 at Λ-almost every $x \notin E$. But in contrast to what happens in the ordinary two-dimensional differentiation theory, it can happen that $1 > \bar{D}(E,x) > \underline{D}(E,x)$ on subsets of E of positive Λ-measure. So the situation is much more complicated here.

If for $x \in E$ we have $D(E,x) = 1$ we say that x is a regular point of E. If a set F is such that Λ-almost all of its points are regular, then F is said to be a regular set. Parallelly, if $x \in E$ is such that $\underline{D}(E,x) < 1$ (so that $D(E,x)$ does not exist or it is not 1) then x is said to be an irregular point of E. If a set G is such that Λ-almost all of its points are irregular then G is said to be an irregular set.

For E with $\Lambda(E) < \infty$. We take R = {regular points of E}, I = {irregular points of E} and then it turns out that R is a regular set and I is an irregular set. So the study of the structure of E is reduced to the study of regular sets and of irregular sets. Regular and irregular sets have sharply contrasting geometric properties.

3. TANGENCY. Regular sets behave very much like ordinary, well-bred curves. The tangent at $x \in E$ to a set with $\Lambda(E) < \infty$ is defined as that unique line t, if it exists, such that for each $\varepsilon > 0$, the density at x of the set $E - R_{t,\varepsilon}$ is zero, where $R_{t,\varepsilon}$ is the shaded portion in the figure below.

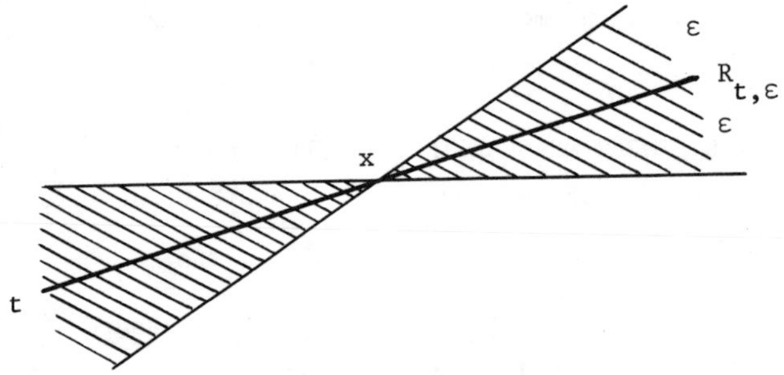

It is then a theorem that E is regular if and only if at Λ-almost all $x \in E$ the tangent exists.

Another theorem affirms, on the other hand, that E is irregular if an only if the tangent exists at Λ-almost no $x \in E$.

4. PROJECTION. Like a well-behaved curve regular sets have orthogonal projection of positive linear measure on every straight line with the possible exception of those lines in a single direction. An irregular set, on the

contrary, is such that for almost all direction ("almost all" in the sense of Lebesgue measure on the unit circle) the projection in that direction over any straight line is of linear measure zero.

As a matter of fact, it can be proved that each regular set is contained, except perhaps for a Λ-null subset, in a countable union of rectifiable arcs.

5. POLAR LINES. It is also a useful fact to know that the union of the polar lines of the points of a regular set E ($0 < \Lambda(E) < \infty$) with respect to any fixed circle is a set E* of positive two-dimensional measure, whereas the corresponding set E* for an irregular set E has always measure zero. So one can say that although every set E with $0 < \Lambda(E) < \infty$ is a null set with respecto to two-dimensional measure, the irregular sets are more null than the regular ones.

These are some of the features of the theory of linearly measurable sets. We shall now try to look at their connections with certain useful tools for the solution of some problems in recent Fourier analysis.

6. THE BESICOVITCH SET AND THE PERRON TREE. For the solution of a certain problem about the Riemann integral in $\mathbb{R}^2$ Besicovitch constructed in 1917 a compact null set F in $\mathbb{R}^2$ containing unit segments in all directions. Later on this set became better known as the tool for a straightforward solution of the so-called "needle" or "Kakeya" problem.

The easiest construction of a such set is the one obtained by means of the Perron tree. Given any triangle ABC and $\varepsilon > 0$ it is possible to partition the basis BC into segments $I_1, I_2, \ldots, I_n$ and to move the triangles determined by A and $I_1, \ldots, I_n$ parallelly to BC so that the union (Perron tree) of the translated triangles has area less than ε times the area of ABC.

A nice and important feature in the classical construction of the Perron tree is that the small triangles end up with the vertices in reversed order, as the figure indicates

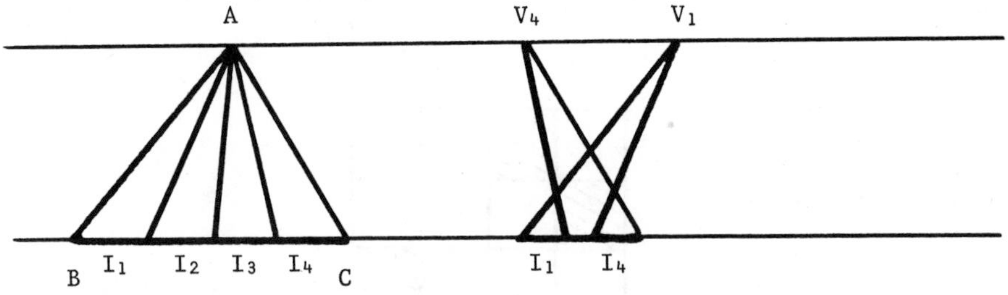

7. C. FEFFERMAN'S LEMMA FOR THE MULTIPLIER PROBLEM FOR THE BALL. The solution of the multiplier problem for the ball is rather simple by means of a lemma due to Y. Meyer once one has the following result: For each $\varepsilon > 0$ there exists a measurable set $E \subset \mathbb{R}^2$ and a finite sequence of pairwise disjoint triangles $\{T_j\}$ so that if $\tilde{T}_j$ is the triangle symmetric to T_j with respect to one of its vertices one has

$$\text{(i)} \quad |E \cap \tilde{T}_j| > \frac{1}{10} |T_j|$$

$$\text{(ii)} \quad |E| < \varepsilon \, \Sigma |T_j|$$

This result is obvious from the above figure by taking as E the Perron tree and as T_j the triangles symmetric to the small ones of the Perron tree with respect to the vertices V_j.

8. GENERALIZATION. So the Perron tree is a good tool for the solution of the multiplier problem. Once one has a Perron tree one can formulate a multiplier theorem. For example it is rather easy to construct a Perron tree in such a way that the sides of the small triangles form angles $\Theta_k = 1/k$ with a fixed direction. So one can immediately state: The characteristic function of a polygon with denumerable infinite number of consecutive sides ℓ_k forming an angle $1/k$ with a fixed direction is not a multiplier in $L^p(\mathbb{R}^2)$ for any $p \neq 2$.

The question now is how to construct Perron trees. The Perron tree was initially the tool for constructing the Besicovitch set. But it turns out that from a Besicovitch set one can construct a Perron tree and Besicovitch sets are easily obtained by means of the theory of linearly measurable sets. For this one first constructs an irregular set P of points so that joining the points of P to a fixed point 0 one obtains lines in all directions. The set B of points in the polar lines of the points of P with respect to a circle centered at 0 has null measure and contains lines in all directions.

Sets P as above are easily obtained. Divide the closed unit square as indicated in the figure below and select the shaded small squares. On each of them

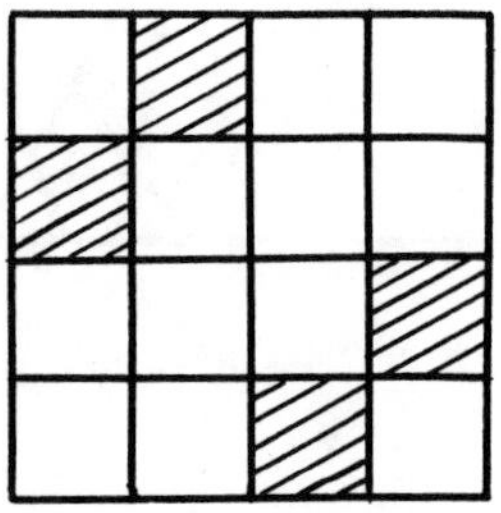

perform the same operation, obtaining a set L. Consider the set P that in polar coordinates is

$$P = \{(\rho,\Theta): \rho = x + 1,\ \Theta = y,\ (x,y) \in L\}$$

The set P is irregular and joining its points to 0 one obtains all directions in $[0, \frac{\pi}{2}]$.

9. THE NIKODYM SET. In 1927 Nikodym constructed a measurable set N contained in the unit cube Q of $\mathbb{R}^2$ such that $|N| = 1$ and for each $x \in N$ there is a straight line $\ell(x)$ such that $\ell(x) \cap N = \{x\}$. The construction has been greatly simplified and generalized by Davies and other authors.

One of the easy constructions of the Nikodym set goes again through the Perron tree. When the small triangles of the Perron tree are extended below their bases $I_1, I_2, \ldots, I_n$, these extended triangles cover pretty much area on the strip below BC of length equal to the height of ABC. This fact leads easily to the following result: Given two rectangles R_1 and R_2 as in the figure below and $\varepsilon > 0$ one can substitute the strip S by strips $S_1, \ldots, S_N$ so that

$$R_2 \subset \bigcup_1^N S_j$$

and $|(\bigcup_1^N S_j) \cap (R_1 - R_2)| \leq \varepsilon.$

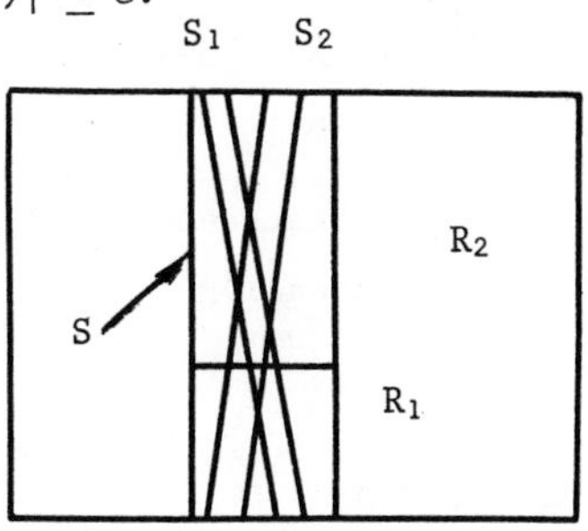

With this lemma the construction of the Nikodym set is straightforward.

10. A PROBLEM IN DIFFERENTIATION. For each $x \in \mathbb{R}^2$ let $d(x)$ be a given direction. Consider all rectangles $B(x)$ with a side in that direction containing x and for $f \in L^1(\mathbb{R}^2)$ we ask ourselves whether the limit of the means

$$\frac{1}{|B(x)|} \int_{B(x)} f(y)\,dy$$

exists and is finite as $\delta(B(x)) \to 0$.

If there is a Nikodym set N such that the corresponding line $\ell(x)$ to the point $x \in N$ has the direction $d(x)$, it is easy to see that the answer is negative, even if $f \in L^\infty(\mathbb{R}^2)$. Therefore one has to ask rather for additional conditions on $d(x)$ and f in order to obtain a positive answer. This question is connected with some of the problems considered by Stein and Wainger in their papers in this volume.

In this way one is led to ask which fields of directions $d(x)$ allow the construction of a Nikodym set and which ones do not. If other words, how good can be the field of directions $\ell(x)$ in the Nikodym set?

One can show that there exists a continuous field of directions $d(x)$ and a set N of positive measure so that for $x \in N$ the line $\ell(x)$ passing through x with direction $d(x)$ satisfies $\ell(x) \cap N = \{x\}$.

However, according to a result of A. Casas (1978) there cannot exist a set N of full measure in $\mathbb{R}^2$ (i.e. $|N'| = 0$) and a Lipschitz field of directions $d(x)$ such that for each $x \in N$, $\ell(x) \cap N = \{x\}$. The proof is rather simple with the results of the theory of linearly measurable sets. Assume there exists such a set N. Intersect N with a line p so that $p \cap N$ has full measure in the line. For each $x \in p \cap N$ take $\ell(x)$ and its pole with respect to a circle centered at a fixed point of p. The set of poles plus a set of null linear measure form a rectifiable curve. So the points in the union of all lines $\ell(x)$ form a set of positive linear measure, which contradicts $|N'| = 0$.

These considerations seem to indicate that a better knowledge of the Nikodym sets may lead us to a better understanding of some interesting problems in differentiation theory.

BIBLIOGRAPHY

1. Besicovitch, A.S., Sur deux questions d'intégrabilité des fonctions, J. Soc. Phys.-Math. (Perm') 2(1919-1920), 105-123.

2. Besicovitch, A.S., On Kakeya's problem and a similar one, Math. Z. 27 (1928), 312-320.

3. Casas, A., Aplicaciones de la teoría de medida lineal, Tesis (Univ. Complutense de Madrid, 1978).

4. Cunningham, F., Three Kakeya problems, Amer. Math. Monthly 81 (1974), 582-592.

5. Davies, R.O., Accessibility of plane sets and differentiation of functions of two variables (Ph. D. Dissertation, Cambridge Univ., 1953).

6. Fefferman, C., The multiplier problem for the ball, Ann. of Math. 94 (1971), 330-336.

7. de Guzmán, M., Differentiation of Integrals in $\mathbb{R}^n$, Springer Lecture Notes 481 (1975).

8. Nikodym, O., Sur la mesure des ensembles plans dont tous les points sont rectilinéairement accessibles, Fund. Math. 10 (1927), 116-168.

9. Perron, O., Ueber einen Satz von Besicovitch, Math. Z. 28 (1928), 383-386.

10. Rademacher, H., A new construction of the Perron tree, in Studies in Mathematical Analysis (edited by Gilbarg, Solomon and other) (Stanford, 1962).

Miguel de Guzmán
Facultad de Matemáticas
Universidad Complutense
Madrid.

Proceedings of Symposia in Pure Mathematics
Volume XXXV, Part 1, 1979

WEIGHTED NORM INEQUALITIES FOR CLASSICAL OPERATORS

Benjamin Muckenhoupt[1]

ABSTRACT. This is a discussion of various weighted norm inequalities. The problem is to characterize the weight functions for which these inequalities are true. The emphasis is on recent one dimensional results and unsolved problems.

1. INTRODUCTION. The principle problem to be discussed here is the determination of all pairs of non-negative functions $U(x)$, $V(x)$ such that

$$\int |Sf(x)|^p\, U(x)dx \le C \int |Tf(x)|^p\, V(x)dx \quad , \tag{1.1}$$

where $1 \le p < \infty$, S and T are two given operators and C is a constant independent of f. Typical of the operators used for S and T are the identity, the indefinite integral $\int_0^x f(t)dt$, the Hardy-Littlewood maximal function

$$f^*(x) = \sup_{y \ne x} \frac{1}{y-x} \int_x^y |f(t)|\, dt \quad , \tag{1.2}$$

the Hilbert transform

$$\tilde{f}(x) = \lim_{\varepsilon \to 0^+} \frac{1}{\pi} \int_{|y|>\varepsilon} \frac{f(x-y)}{y}\, dy \tag{1.3}$$

and various Littlewood-Paley operators.

Obvious variations on the basic problem include considering the inequality

$$\left[\int |Sf(x)|^q\, U(x)dx\right]^{1/q} \le C \left[\int |Tf(x)|^p\, V(x)dx\right]^{1/p} \quad , \tag{1.4}$$

AMS(MOS) subject classifications (1970). Primary 26A84, 26A86, 42A40, 44A15; Secondary 30A78, 46E30, 47A30, 47G05.

[1]Research supported in part by NSF grant MCS 78-04800.

where $p \neq q$. This is the natural approach when S is a fractional integral operator and T is the identity and can also be done for an analogue of Hardy's inequality. Fractional integrals are considered in [27] and [39] but will not be discussed further here. Weak type inequalities of the form

$$(1.5) \qquad \int_{U(x)|Sf(x)|^p > a} V(x)dx \leq \frac{C}{a} \int |Tf(x)|^p \, W(x)dx,$$

where U, V and W are non-negative functions $a > 0$ and C is independent of f and a, are also of interest. They are substitutes for (1.1) when (1.1) is false and they are often the basis for proving inequalities of the form (1.1) by use of an interpolation theorem. Another variation of (1.1) is to replace $U(x)dx$ and $V(x)dx$ by measures. This has not been particularly popular in the literature because in some cases it can be shown that only absolutely continuous measures can be used, and in other cases the results are just complicated but trivial consequences of (1.1).

There are many reasons for interest in inequalities of the form (1.1). With S the Hilbert transform, T the identity and $U(x)=V(x)=1$, (1.1) was used originally, [32], to prove mean convergence for Fourier series. The results with $U(x)$ and $V(x)$ not identically 1 give weighted mean convergence results for Fourier series. They also imply various mean convergence theorems for other orthogonal expansions, examples for $U(x)$ and $V(x)$ of particular forms appear in [22] and [23]. Results of the form (1.1) with S the Hardy-Littlewood maximal function and T the identity are useful for proving mean summability results as in [25] and are often needed as part of the proof of mean convergence theorems. Results for the Littlewood-Paley operators are useful for proving multiplier theorems as in [20], and as a basis for a weighted H^p theory as in [31], [36], and [41]. Applications of weighted norm inequalities have appeared in a surprising number of places such as prediction theory [15], estimation of Cauchy integrals [5], and in estimating the commutator of a singular integral and multiplication by a function [7].

This paper is a survey of various results of the types described above. Space requirements dictate that only a small part of the field be covered. In particular, recent results for mixed homogeneity operators such as those in [1], [21] and [36] are not covered. The discussion is confined primarily to one dimensional problems with an emphasis on recent results and unsolved problems. The discussion demonstrates two facts that may surprise many readers familiar with the subject. First, even in the simplest cases and with the most familiar operators there are many unsolved problems. Second, the A_p condition appears to be much less universal than it did at one time. The first problems considered all had the A_p condition characterizing their weights. This has not been true, however, of many other natural problems considered since.

In §2 some theorems related to Hardy's inequality are discussed; §3 contains inequalities for the Hardy-Littlewood maximal function. The

Hilbert transform is the subject of §4. Finally, in §5 some results are described for other operators.

2. HARDY'S INEQUALITY. The form to be discussed here is

$$\left[\int_0^\infty \left|\int_0^\infty f(t)dt\right|^q U(x)dx\right]^{1/q} \leq C \left[\int_0^\infty |f(x)|^p V(x)dx\right]^{1/p}, \tag{2.1}$$

where C is independent of f. The original equation of this type by Hardy [12] in 1928 has $p=q$, $U(x)=x^a$ and $V(x)=x^{a+p}$. Hardy showed that (2.1) was valid in this case for $1 \leq p < \infty$ provided $a < -1$. Talenti [37] and Tomaselli [38] in 1969 showed that (2.1) holds in the case $p=q$, $1 \leq p < \infty$ if and only if

$$\sup_{r>0} \left[\int_r^\infty U(x)dx\right]^{1/q} \left[\int_0^r V(x)^{-1/(p-1)} dx\right]^{(p-1)/p} < \infty . \tag{2.2}$$

Recent work of Bradley [4] shows that (2.2) is a necessary and sufficient condition for (2.1) provided $1 \leq p \leq q \leq \infty$.

Recently in joint work with K. Andersen [3], the corresponding weak type result has also been considered. The inequality in this case is

$$\left[\int_{x^{-b}\left|\int_0^x f(t)dt\right| > \alpha, x>0} x^{bq} U(x)dx\right]^{1/q} \leq \frac{C}{\alpha} \left[\int |f(x)|^p V(x)dx\right]^{1/p}, \tag{2.3}$$

where C is independent of f and $\alpha > 0$. If $b \leq 0$ and $p \leq q$, (2.2) characterizes the solutions of (2.3). The case $b \leq 0$ is, however, not a particularly interesting one; note that if $f(x)$ is non-negative, then the set where $x^{-b} \int_0^x f(t)dt > \alpha$ is just an interval of the form (c,∞).

The case $b=1$ is in many ways the most natural one for (2.3) since, as shown by the original Hardy's inequality, the operator $x^{-1} \int_0^x f(t)dt$ is a bounded operator between L^p spaces with the same weight. If $b=1$, $p=q$ and $V(x)=U(x)$, then the weights for which (2.3) holds are characterized by (2.2). Therefore, in this single weight function problem, the strong type problem (2.1) and the weak type problem (2.3) have the same solution. More generally, however, if we only require that $q \geq p$ and $b > 0$, then (2.3) holds if and only if for every $a > 0$

$$\sup_{r>0} \left[\int_r^\infty \left[\frac{r}{x}\right]^a U(x)dx\right]^{1/q} \left[\int_0^r V(x)^{-1/(p-1)} dx\right]^{(p-1)/p} < \infty . \tag{2.4}$$

If (2.2) holds for a pair U,V, it is immediate that (2.4) is true. In general, however, it is easy to find pairs U, V for which (2.4) is true but (2.2) is not. This phenomenon that the one weight function problem has the same solution for weak type and strong type while the two weight function problem does not is a common one. Other examples are described later in this paper.

Surprisingly, the higher dimensional analogues of Hardy's inequality seem to be considerably harder. In two dimensions the obvious analogue of (2.1) is

$$(2.5)\qquad \int_0^\infty\int_0^\infty \left|\int_0^x\int_0^y f(s,t)\,dt\,ds\right|^p U(x,y)\,dy\,dx \le C\int_0^\infty\int_0^\infty |f(x,y)|^p V(x,y)\,dy\,dx.$$

This is a useful inequality since it appears in various two dimensional proofs in much the same way that (2.1) appears in one dimensional proofs. If (2.5) is true, then the analogue of (2.2),

$$(2.6)\qquad \sup_{r,s>0}\left[\int_r^\infty\int_s^\infty U(x,y)\,dy\,dx\right]\left[\int_0^r\int_0^s \left[V(x,y)\right]^{-1/(p-1)}dy\,dx\right]^{p-1} < \infty ,$$

is true. The converse, however, is false; there are examples of pairs U,V that satisfy (2.6) but not (2.5). What does characterize the weights satisfying (2.5) is an interesting open question.

3. THE HARDY-LITTLEWOOD MAXIMAL FUNCTION. With f^* as defined in (1.2), the basic problem is, given p, to find all pairs U,V of non-negative functions such that

$$(3.1)\qquad \int_{-\infty}^\infty |f^*(x)|^p U(x)\,dx \le C\int_{-\infty}^\infty |f(x)|^p V(x)\,dx$$

The first result by Hardy and Littlewood in 1930 [13] showed that if $U(x)=V(x)=1$, then (3.1) holds for $1 < p < \infty$.

As is often the case, it is easier to characterize the pairs U, V for the corresponding weak type problem. This was done in 1972 in [25]. The result is that if $1 \le p < \infty$, then

$$(3.2)\qquad \int_{f^*(x)>a} U(x)\,dx \le \frac{C}{a^p}\int_{-\infty}^\infty |f(x)|^p\, V(x)\,dx$$

holds for $a > 0$ if and only if there is a constant B such that for every interval I

$$(3.3)\qquad \left[\frac{1}{|I|}\int_I U(x)\,dx\right]\left[\frac{1}{|I|}\int_I V(x)^{-1/(p-1)}dx\right]^{p-1} \le B.$$

If p=1, (3.2) holds if and only if there is a B such that for every interval I

$$(3.4) \qquad \left[\frac{1}{|I|}\int_I U(x)dx\right]\left[\operatorname*{ess\,sup}_I \frac{1}{V(x)}\right] \le B$$

or equivalently

$$(3.5) \qquad U^*(x) \le BV(x)$$

almost everywhere. The condition (3.3) for $p > 1$ is known as the A_p condition and (3.5) is known as the A_1 condition. A single non-negative function $U(x)$ is said to satisfy the A_p condition if the appropriate condition holds with $V(x)=U(x)$.

The A_p condition is useful because the solution to a number of weight function problems is the class of A_p functions. Some of these will be described below. One interesting problem concerning the class A_p itself is whether every function in A_p can be written in the form $U(x)V(x)^{1-p}$, where $U(x)$ and $V(x)$ are in A_1. Holder's inequality shows easily that a function of this form is in A_p. The converse was proved by Peter Jones during the conference. This converse will undoubtedly be useful in proving weighted norm inequalities.

If $V(x)=U(x)$ and $1 < p < \infty$, then (3.1) was shown in [25] to hold if and only if $U(x)$ is an A_p function. Without the restriction $U(x)=V(x)$, (3.1) still implies (3.3) but the converse is false. It was shown in [29] that (3.1) implies the stronger condition

$$(3.6) \qquad \left[\int_{-\infty}^{\infty}\frac{U(x)\,dx}{|I|^p+|x-x_I|^p}\right]\left[\int_I V(x)^{-1/(p-1)}dx\right]^{p-1} \le B ,$$

where I is an arbitrary interval and x_I denotes the center of I. It was also shown that (3.6) and an additional condition imply (3.1) and conjectured that (3.6) implies (3.1) without the additional condition. The characterization of the pairs U,V for which (3.1) holds is, however, still an open question.

A related question of interest is that of finding a V for which (3.1) holds given a non-negative function $U(x)$. It is trivial to show that $U(x) \le CV(x)$ almost everywhere is a necessary condition. Fefferman and Stein in [10] showed that $V(x)=U^*(x)$ will work, but examples show that functions significantly smaller than U^* may also work. Similarly, given a V, there is the problem of finding the U. Simpler but also unsolved problems are the characterization of the U's for which there are non-trivial V's and the characterization of the V's for which there are non-trivial U's. Little is known about any of these problems.

Oddly enough, although A_p characterizes the weights for both strong and weak type in the single weight function problem and for weak type for the general problem, this is no longer true for restricted weak type. If χ_E denotes the characteristic function of the set E, the inequality is

$$(3.7)\qquad \int_{\chi_E^*(x)>a} U(x)dx \le \frac{C}{a^p}\int_E V(x)dx$$

for $a > 0$. It was shown by R. Kerman in [18] that (3.7) holds if and only if for every interval I and every subset E of I

$$(3.8)\qquad \int_I V(x)dx \le C\left[\frac{|I|}{|E|}\right]^p \int_E U(x)dx ,$$

where $|E|$ denotes the measure of E and C is independent of I and E. Restricted weak type results are of interest because they can be used to obtain strong type results by interpolation. This result may provide a new way of approaching sufficiency proofs.

Another problem closely related to the weak type inequality (3.2) for which A_p does not characterize the weights is the problem discussed in [30] of determining all functions U such that for $a > 0$.

$$(3.9)\qquad \int_{U(x)[f^*(x)]^p>a} dx \le \frac{C}{a}\int |f(x)|^p U(x)dx.$$

This inequality is also a weak type version of (3.1). It is of interest primarily as a first step in understanding the mixed type norm inequalities described below. Since (3.1) with $V(x)=U(x)$ implies (3.9), it follows that if $1 < p < \infty$, then (3.9) is true for any U in A_p. There are, however, functions U not in A_p for which (3.8) is also true; a simple example is $U(x)=1/x$. The inequality (3.9) does imply that for every interval I

$$(3.10)\qquad \left[\frac{1}{|I|}\,||U\chi_I||_{1,\infty}\right]\left[\frac{1}{|I|}\int_I U^{-1/(p-1)}dx\right]^{p-1} \le C,$$

where C is independent of I and the norm $||\ \ ||_{1,\infty}$ is defined by $||g||_{1,\infty}=\sup_{y>0} y|\{x: |f(x)|>y\}|$. It is conjectured in [30] that (3.10) implies (3.9); the characterization of the weights for (3.9) is an open question.

Finally, there is the general mixed type norm inequality

$$(3.11)\qquad \int_{U(x)|f^*(x)|^p>a} V(x)dx \le \frac{C}{a}\int_{-\infty}^{\infty} |f(x)|^p W(x)dx$$

for $a > 0$. Proving such inequalities may be essential to prove (3.1) under the most general hypothesis. This is because the interpolation with change of measure theorem in [35] that uses weak type inequalities requires a mixed type inequality like (3.11). Since the strong type version of (3.11) is

$$(3.12)\qquad \int_{-\infty}^{\infty} [f^*(x)]^p\, U(x)V(x)dx \le C\int_{-\infty}^{\infty} |f(x)|^p W(x)dx,$$

it would seem natural to conjecture that if $U(x)$ is smooth, then a condition relating the product UV and the function W would characterize the weights for which (3.11) is true. This is not true, however, as the example above for (3.9) shows. With $W(x)=1/x$, $U(x)=1/x$ and $V(x)=1$, (3.11) is true, but with $W(x)=1/x$, $U(x)=1$ and $V(x)=1/x$, it is false. Another example in [3] shows that if U is in A_1 and $d\neq 1$, then

$$(3.13) \qquad \int_{|x|^d f^*(x)>a} |x|^{-d} U(x)dx \leq \frac{C}{a}\int_{-\infty}^{\infty} |f(x)|U(x)dx ,$$

but the result fails if $d=1$. Nothing more is known about (3.11) except for the cases mentioned above when either $U(x)=1$ or $V(x)=1$.

4. THE HILBERT TRANSFORM. With $\tilde{f}$ as defined in (1.3), the basic problem is, given p, to find all paris U, V of non-negative functions such that

$$(4.1) \qquad \int_{-\infty}^{\infty} |\tilde{f}(x)|^p U(x)dx \leq C \int_{-\infty}^{\infty} |f(x)|^p V(x)dx.$$

The first result by Reisz in 1927 [32] showed that if $U(x)=V(x)=1$, then (4.1) holds for $1 < p < \infty$. There are many results concerning (4.1) with $V(x)=U(x)$; it was shown in [17] in 1973 that if $1 < p < \infty$ and $V(x)=U(x)$, then (4.1) holds if and only if $U(x)$ satisfies the A_p condition defined in §3. Furthermore, the weak type inequality for $a > 0$,

$$(4.2) \qquad \int_{|\tilde{f}(x)>a} U(x)dx \leq \frac{C}{a^p} \int_{-\infty}^{\infty} |f(x)|^p V(x)dx$$

with $V(x)=U(x)$ and $1 \leq p < \infty$, is true if and only if $U(x)$ satisfies the A_p condition. To this extent the Hilbert transform results resemble the Hardy-Littlewood maximal function results. Beyond this, however, the similarity ends. The A_p condition though implied by (4.2) for $1 \leq p < \infty$ in the general case does not imply (4.2). In [29] it is shown that (4.2) implies that for every interval I

$$(4.3) \qquad \left[\int_I U(x)dx\right] \left[\int_{-\infty}^{\infty} \frac{V(x)^{-1/(p-1)}dx}{(I)^{p'}+|x-x_I|^{p'}}\right]^{p-1} \leq B ,$$

where x_I denotes the center of I and $p'=p/(p-1)$. Condition (4.3) is stronger than the A_p condition. With an additional condition (4.3) implies (4.2), but it remains an open question whether (4.3) implies (4.2) without the additional condition.

Similarly, the A_p condition does not imply (4.1) in the general case. The inequality (4.1) implies both (3.6) and (4.3). Here again (3.6) and (4.3) and an additional assumption imply (4.1) but whether this implication is true without the additional assumption is unknown. Recent work of Cotlar and Sadosky [8] gives a necessary and sufficient condition for (4.1) to hold for $p=2$. To check whether this condition is satisfied for a pair U, V requires finding an auxiliary function, and, as a result, it may be difficult to determine whether a given pair satisfies the condition.

To summarize, the characterization of the weight functions by easily verified inequalities remains an open question for (3.1), (4.1) and (4.2). Examples show, however, that these problems all have different solutions and that the A_p condition is not sufficient for any of them.

The related question of finding the U's that satisfy (4.1) given a V or finding the V's given the U is also largely unknown though there are some results by Koosis [19] for the first of these for the case $p=2$. Koosis, however, restricts the class of functions to be used for f and assumes that V is integrable. This assumption on V is a rather strong restriction since it excludes, for example, all the solutions of the general problem for which $V(x)=U(x)$.

The inequality

$$(4.4)\qquad \int_{-\infty}^{\infty} |\tilde{f}(x)|^p U(x)dx \le C \int_{-\infty}^{\infty} [f^*(x)]^p U(x)dx$$

is the basis for the proof given by Coifman and Fefferman in 1974 [6] of the fact that the A_p condition characterizes the weight functions for (4.1) with $V(x)=U(x)$. For this they used the A_∞ condition; a non-negative function $U(x)$ is said to be in A_∞ if there exist positive constants C and ε such that for every interval I and every subset E of I

$$(4.5)\qquad \int_E U(x)dx \le C \left[\frac{|E|}{|I|}\right]^\varepsilon \int_I U(x)dx.$$

Their result is that if $U(x)$ satisfies the A_∞ condition, then (4.4) is true. Since (4.4) is a basic inequality, it is natural to ask if A_∞ contains all the weight functions for which (4.4) is true. In [26] this question is considered; it is shown that (4.4) does not imply that U is in A_∞. In fact $U=\chi_{(0,\infty)}$ satisfies (4.4) but is not in A_∞. The inequality (4.4) does imply a condition called the C_p condition that is similar to the A_∞ condition. A non-negative function U is said to satisfy the C_p condition if there exist positive constants C and ε such that for every interval I and every subset E of I

$$(4.6)\qquad \int_E U(x)dx \le C \left[\frac{|E|}{|I|}\right]^\varepsilon \int_{-\infty}^{\infty} \frac{|I|^p U(x)dx}{|I|^p+|x-x_I|^p},$$

where x_I denotes the center of I. Whether this condition implies (4.4), however, remains an open question.

Another class of questions concerning the Hilbert transform is the determination of the weight functions for (4.1) if the set of functions f used in the inequality is restricted. For example, K. Andersen in [2] showed that if $V(x)=U(x)$ and f is odd, then (4.1) holds if and only if $|x|^{-1/2}U(|x|^{1/2}\operatorname{sgn} x)$ is in A_p. If $U(x)=|x|^a$, this produces an old result of Hardy and Littlewood [14] that (4.1) holds if and only if $-1 < a < 2p-1$. Andersen also obtained results for even f and weak type inequalities for these cases, see [2] and [3].

A result closely related to Andersen's for odd functions concerns (4.1) when $V(x)=U(x)$ and it is assumed that $\int_{-\infty}^{\infty} f(x)dx=0$. Inequality (4.1) certainly holds in this case if U satisfies the A_p condition. Furthermore, since in this case

$$\tilde{f}(x)=\lim_{\varepsilon\to 0^+}\frac{1}{\pi}\int_{|x-y|>\varepsilon} f(y)\left[\frac{1}{x-y}-\frac{1}{x}\right]dy=\lim_{\varepsilon\to 0^+}\frac{1}{\pi x}\int_{|x-y|>\varepsilon}\frac{yf(y)}{x-y}\,dy, \tag{4.7}$$

it is easy to show that (4.1) holds if $U(x)$ equals $|x|^p$ times an A_p function. These two facts show that if $V(x)=U(x)=|x|^a$ and $\int f=0$, then (4.1) holds for $-1 < a < p-1$ and $p-1 < a < 2p-1$; furthermore, it is not hard to show that the result fails for $a=p-1$ and that these two ranges for a give all the weights of the form $|x|^a$. This shows in particular that Andersen's condition for weights for odd f is not the right condition for this problem. Since the sum of two weight functions is also a weight function, we know that any U that can be written as the sum of an A_p function and $|x|^p$ times an A_p function is a weight function for this problem. Characterizing the weight functions for this problem remains an open question. In addition to being an interesting question by itself, it seems that the answer will be important in characterizing the weights for various classes of multiplier transformations.

Other interesting open questions are the characterization of the weight functions for the Hilbert transform version of (3.9); some results are in [30]. Mixed type inequalities like (3.11) are also largely unknown except for the analogue of (3.13).

5. OTHER OPERATORS. For these operators the theory is even less developed than for those described above. For each operator various results and unsolved problems are described below.

The g_λ^* operator for a function $f(x)$ is defined for $1 < \lambda \le 2$ by

$$g_\lambda^*(x) = \left[\int_{-\infty}^{\infty}\int_0^{\infty}\left[\frac{y}{|x-z|+y}\right]^{\lambda}|\nabla f(z,y)|^2\,dy\,dz\right]^{1/2}, \tag{5.1}$$

where the Poisson integral $f(z,y)$ is defined by

(5.2) $$f(z,y) = \frac{1}{\pi}\int_{-\infty}^{\infty} \frac{yf(z-x)}{x^2+y^2}\,dx$$

It was shown in [28] that if $2/\lambda < p \le 2$ and $U(x)$ is in $A_{\lambda p/2}$, then

(5.3) $$\int_{-\infty}^{\infty} \left[g_\lambda^*(x)\right]^p U(x)dx \le C\int_{-\infty}^{\infty} |f(x)|^p U(x)dx$$

It is also easy to show that if (5.3) holds, then $U(x)$ is in A_p and $B_{\lambda p/2}$; a function $U(x)$ is said to satisfy the condition B_p if for every interval I

(5.4) $$\int_{-\infty}^{\infty} \frac{|I|^{p-1}U(x)dx}{|I|^p+|x-x_I|^p} \le \frac{C}{|I|}\int_I U(x)dx ,$$

where x_I denotes the center of I.

It is shown in [17] that if a function satisfies the A_p condition, then it satisfies the B_p condition. The converse is false; the example in [9] is actually an example of a function which is in some B_p class but in no A_p class.

The problem then is to close the gap between the sufficient condition $A_{\lambda p/2}$ for (5.3) and the necessary condition of A_p plus $B_{\lambda p/2}$. If $p=2$, it is easy to show that A_2 and B_λ imply (5.3); this solves the problem for this case. Recent work of Stromberg and Torchinsky [36] shows that there is a real gap between the results. They showed that there are functions satisfying A_p and $B_{\lambda p/2}$ that do not satisfy $A_{\lambda p/2}$. They also came close to closing the gap by showing that if U is in A_p and B_q with $q < \min(\lambda p/2,\lambda)$ then (5.3) holds. The complete solution, however, remains an open question.

The γ operator is defined as

(5.5) $$\gamma(f) = \left[\sum_n \left|\int_{2^n}^{2^{n+1}} \hat{f}(t)e^{2\pi ixt}dt\right|^2\right]^{1/2}$$

where $\hat{f}$ denotes the Fourier transform of f defined by

(5.6) $$\hat{f}(x) = \int_{-\infty}^{\infty} e^{-2\pi ixt}f(t)dt .$$

It was shown by D. Kurtz in [20] that if $1 < p < \infty$ and $U(x)$ satisfies the A_p condition, then

(5.7) $$\int_{-\infty}^{\infty} |\gamma(f)|^p U(x)dx \le C\int_{-\infty}^{\infty} |f(x)|^p U(x)dx .$$

This result is important since, when it is compined with the g_2^* result described above, it shows that if T is a multiplier operator satisfying the Marcinkiewicz condition, $1 < p < \infty$ and U is in A_p, then

$$(5.8) \qquad \int_{-\infty}^{\infty} |Tf(x)|^p U(x)dx \le C \int_{-\infty}^{\infty} |f(x)|^p U(x)dx .$$

This generalizes a result of Hirshman [16] where the case $U(x) = |x|^a$ is considered. Whether the A_p condition is necessary for (5.7) to hold is unknown.

The area function S_λ is defined for $\lambda > 0$ by

$$(5.9) \qquad S_\lambda(f) = \left[\iint_{|x-z|\le\lambda y, y>0} |\nabla f(z,y)|^2 dzdy\right]^{1/2}$$

where $f(z,y)$ is the Poisson integral defined in (5.2). The norm inequality

$$(5.10) \qquad \int_{-\infty}^{\infty} |S_\lambda(f)|^p U(x)dx \le C \int_{-\infty}^{\infty} |f(x)|^p U(x)dx$$

was considered by Gundy and Wheeden in [11] and in the periodic case by Segovia and Wheeden in [33]. If $1 < p < \infty$, $\lambda > 0$ and $U(x)$ is in A_p, (5.10) is true, and with the additional assumption $\int_{-\infty}^{\infty} f(x)dx=0$ the inequality

$$(5.11) \qquad \int_{-\infty}^{\infty} |f(x)|^p U(x)dx \le C \int_{-\infty}^{\infty} |S_\lambda(f)|^p U(x)dx$$

is also valid. Here again, whether A_p is a necessary condition is an open question.

The non-tangential maximal function N_b is defined by

$$(5.12) \qquad N_b(f) = \sup_{|z|\le by} |f(x-z,y)| ,$$

where $f(x-z,y)$ is as defined in (5.2). The norm inequality

$$(5.13) \qquad \int_{-\infty}^{\infty} |S_\lambda(f)|^p U(x)dx \le C \int_{-\infty}^{\infty} [N_b(f)]^p U(x)dx$$

was considered by Gundy and Wheeden in [11]. If $\lambda > 0$, $b > 0$, $1 < p < \infty$ and $U(x)$ satisfies the A_∞ condition, then (5.13) is true. It seems likely, however, that other weights can be used in (5.13). The characterization of the weights for which (5.13) is true is another open problem.

Another interesting inequality is

$$(5.14)\qquad \int_{-\infty}^{\infty} [N_b(f)]^p\, U(x)dx \le C \int_{-\infty}^{\infty} [N_0(f)]^p\, U(x)dx$$

In [40] Wheeden showed that if $b > 0$ and $1 < p < \infty$, then (5.14) is true for any U satisfying the doubling condition

$$(5.15)\qquad \int_{2I} U(x)dx \le C \int_I U(x)dx$$

where I is an arbitrary interval and $2I$ has the same center as I but is twice as long. As shown in [9], the doubling condition is weaker than A_∞. Here again it is not known whether (5.14) implies the sufficient condition (5.15).

Finally, a natural inequality on which almost no progress has been made is

$$(5.16)\qquad \int_{-\infty}^{\infty} |\hat{f}(x)|^p\, U(x)dx \le C \int_{-\infty}^{\infty} |f(x)|^p\, V(x)dx$$

where $\hat{f}(x)$ is the Fourier transform as defined in (5.6). If U and V are powers of $|x|$ the results are known as Pitt's inequality, see [34]. Results of this type would be valuable for proving weighted norm inequalities in the same way that Plancherel's identity is useful for proving unweighted inequalities. They would also be useful for proving weighted multiplier theorems. Inequality (5.16) may well be the most rewarding of all the unsolved problems described here.

REFERENCES

[1] N. Aguilera and C. Segovia, "Weighted norm inequalities relating the g_λ^* and the area functions." Studia Math. 61 (1977), 293-303.

[2] K. Andersen, "Weighted norm inequalities for Hilbert transforms and conjugate functions of even and odd functions," Proc. Amer. Math. Soc. 56 (1976), 99-107. MR 53 #3581.

[3] K. Andersen and B. Muckenhoupt, "Weighted weak type Hardy inequalities with applications to Hilbert transforms and maximal functions," to appear.

[4] S. Bradley, "Hardy inequalities with mixed norms," Can. Math. Bull., to appear.

[5] A. Calderón, "Cauchy integrals on Lipschitz curves and related operators," Proc. Nat. Acad. Sci. U.S.A. 74 (1977), 1324-1327.

[6] R. Coifman and C. Fefferman, "Weighted norm inequalities for maximal functions and singular integrals," Studia Math. 51 (1974), 241-250. MR 50 #10670.

[7] R. Coifman, R. Rochberg and G. Weiss, "Factorization theorems for Hardy spaces in several variables," Ann. of Math. (2) 103 (1976), 611-635. MR 54 #843.

[8] M. Cotlar and C. Sadosky, "Characterization of two measures satisfying Riesz's inequality for the Hilbert transform in L^2," to appear in these proceedings.

[9] C. Fefferman and B. Muckenhoupt, "Two non-equivalent conditions for weight functions," Proc. Amer. Math. Soc. 45 (1974), 99-104. MR 50 #13399.

[10] C. Fefferman and E. Stein, "Some maximal inequalities," Amer. J. Math. 93 (1971), 107-115. MR 44 #2026.

[11] R. Gundy and R. Wheeden, "Weighted norm inequalities for the nontangential maximal function, Lusin area integral, and Walsh-Paley series," Studia Math. 49 (1974), 107-124. MR 50 #5340.

[12] G. Hardy, "Notes on some points in the integral calculus (LXIV)," Messenger of Math. 57 (1928), 12-16.

[13] G. Hardy and J. Littlewood, "A maximal theorem with function-theoretic applications," Acta Math. 54 (1930), 81-116.

[14] ________________, "Some theorems on Fourier series and Fourier power series," Duke J. 2 (1936), 354-381.

[15] H. Helson and G. Szegö, "A problem in prediction theory," Ann. Math. Pura Appl. 51 (1960), 107-138. MR 22 #12343.

[16] H. Hirshman, "The decomposition of Walsh and Fourier series," Mem. Amer. Math. Soc. No. 15 (1955), 65pp. MR 17, p. 257.

[17] R. Hunt, B. Muckenhoupt, and R. Wheeden, "Weighted norm inequalities for the conjugate function and Hilbert transform," Trans. Amer. Math. Sco. 176 (1973), 227-251. MR 45 #2461.

[18] R. Kerman, "Restricted weak type inequalities with weights," to appear.

[19] P. Koosis, "Moyennes quadratiques de transformées de Hilbert et fonctions de type exponentiel," C. R. Acad. Sci. Paris Ser. A-B 276 (1973), A1201-A1204. MR 48 #9241.

[20] D. Kurtz, "Littlewood-Paley and multiplier theorems on weighted L^p spaces," Doctoral Dissertation, Rutgers University, New Brunswick, N. J., 1978, 56pp.

[21] R. Macias and C. Segovia, "Weighted norm inequalities for parabolic fractional integrals," Studia Math. 61 (1977), 277-291.

[22] B. Muckenhoupt, "Mean convergence of Hermite and Laguerre series II, Trans. Amer. Math. Soc. 147 (1970), 433-460. MR 41 #711.

[23] ______________, "Mean convergence of Jacobi series," Proc. Amer. Math. Soc. 23 (1969), 306-310. MR 40 #628.

[24] ______________, "Hardy's inequality with weights," Studia Math. 44 (1972), 31-38. MR 47 #418.

[25] ______________, "Weighted norm inequalities for the Hardy maximal function," Trans. Amer. Math. Soc. 165 (1972), 207-226. MR 45 #2461.

[26] ______________, "Norm inequalities relating the Hilbert transform to the Hardy-Littlewood maximal function," to appear.

[27] B. Muckenhoupt and R. Wheeden, "Weighted norm inequalities for fractional integrals," Trans. Amer. Math. Soc. 192 (1974), 261-274. MR 49 #5275.

[28] ______________, "Norm inequalities for the Littlewood-Paley function g_λ^*," Trans. Amer. Math. Soc. 191 (1974), 95-111. MR 52 #8810.

[29] ______________, "Two weight function norm inequalities for the Hardy-Littlewood maximal function and the Hilbert transform," Studia Math. 55 (1976), 279-294. MR 54 #5720.

[30] ______________, "Some weighted weak-type inequalities for the Hardy-Littlewood maximal function and the Hilbert transform," Indiana Univ. Math. J. 26 (1977), 801-816.

[31] ______________, "On the dual of weighted H^1 of the half space," Studia Math., to appear.

[32] M. Riesz, "Sur les fonction conjugees," Mat. Zeit. 27 (1927), 218-244.

[33] C. Segovia and R. Wheeden, "Weighted norm inequalities for the Lusin area integral," Trans. Amer. Math. Soc. 176 (1973), 103-123. MR 47 #483.

[34] E. Stein, "Interpolation of linear operators," Trans. Amer. Math. Soc. 83 (1956), 482-492. MR 18 p. 575.

[35] E. Stein and G. Weiss, "Interpolation of operators with change of measures," Trans. Amer. Math. Soc. 87 (1958), 159-172. MR 19, p. 1184.

[36] J. Stromberg and A. Torchinsky, "Weighted H_w^p spaces, $0 < p < \infty$," to appear.

[37] G. Talenti, "Osservazioni sopra una Classe di Disuguaglianze," Rend. Sem. Mat. e Fis. Milano 39 (1969), 171-185. MR 43 #6380.

[38] G. Tomaselli, "A class of inequalities," Boll. Un. Mat. Ital. 21 (1969), 622-631. MR 41 #411.

[39] G. Welland, "Weighted norm inequalities for fractional integrals," Proc. Amer. Math. Soc. 51 (1975), 143-148. MR 52 #8785.

[40] R. Wheeden, "On the radial and non-tangential maximal functions for the disc," Proc. Amer. Math. Soc. 42 (1974), 418-422. MR 48 #11519.

[41] ________, "On the dual of weighted $H^1(|z|<1)$, Studia Math.

RUTGERS UNIVERSITY
NEW BRUNSWICK, NEW JERSEY

Proceedings of Symposia in Pure Mathematics
Volume XXXV, Part 1, 1979

APPLICATIONS OF FOURIER TRANSFORMS TO AVERAGES OVER LOWER DIMENSIONAL SETS

Stephen Wainger[1]

Let $f(x)$ be a locally integrable function on R^n, and let $B_r = \{y \mid |y| \le r\}$ be the ball in R^n of radius r with center at the origin. Then according to a classical theorem of Lebesgue

(1) $$\lim_{r \downarrow 0} \frac{1}{|B_r|} \int_{B_r} f(x-y)\,dy = f(x) \quad \text{a.e.}$$

(Here $|E|$ denotes the Euclidean measure of E, for any set E.)

E. M. Stein [S 1961] has shown that almost everywhere convergence results such as (1) are equivalent to certain estimates for maximal functions and convergence on a dense set. In (1) and throughout this paper that dense set can be taken to be the continuous functions with compact support in R^n. Hence our problems will be estimates for the appropriate maximal functions.

The maximal function appropriate for (1) is the Hardy-Littlewood Maximal function

(2) $$\mathfrak{M}f(x) = \sup_{1 \ge r > 0} \frac{1}{|B_r|} \int_{B_r} |f(x-y)|\,dy .$$

The basic estimates for $\mathfrak{M}f$ are the weak type 1-1 and type p-p inequalities

(3a) $$|\{x \mid \mathfrak{M}f(x) > \lambda\}| \le \frac{C_n}{\lambda} \|f\|_1 ,$$

and

(3b) $$\|\mathfrak{M}f\|_p \le A_{p,n} \|f\|_p , \quad 1 < p \le \infty .$$

The estimates (3) are important not only for (1), but (as is well known) also for the study of Calderón-Zygmund singular integrals.

I would also like to recall one of the fundamental geometric properties of balls that is used in the proof of (3) as it is given for example in Stein [S. 1970] chapter I.

(4) If B_{r_1} and B_{r_2} are 2 balls of radii r_1 and r_2 respectively with $B_{r_1} \cap B_{r_2} \neq 0$ and $|B_{r_1}| > \frac{1}{2} |B_{r_2}|$, then $B_{r_1}^* \supset B_{r_2}$,

[1] Supported in part by N.S.F. grant 144-L581.

where $B_{r_1}^*$ is the ball with the same center as B_{r_1} and radius $C_n r_1$ where C_n is a constant depending only on the dimension n.

Here we wish to discuss averages over lower dimensional sets. We will discuss in particular 3-types of maximal functions: (In order not to get bogged down in technicalities we shall always assume f is a C^∞ function with compact support and seek a priori estimates analogous to (3).)

$$\mathfrak{m}_1 f(x) = \sup_{1 \geq h > 0} \frac{1}{h} \int_0^h |f(x - \gamma(t)| \, dt \tag{5}$$

where $\gamma(t)$ is a C^∞ curve in R^n with $\gamma(0) = 0$.

$$\mathfrak{m}_2 f(x) = \sup_{r > 0} \int_{S_r} |f(x-y)| \, d\sigma_r(y) \tag{6}$$

where S_r is the unit sphere centered at the origin having radius r, and $d\sigma_r$ is the rotationally invariant singular measure on S_r having total mass 1.

$$\mathfrak{m}_3 \Sigma \; f(x) = \sup_{\substack{h > 0 \\ v \in \Sigma}} \frac{1}{h} \int_0^h |f(x - t\vec{v})| \, dt \tag{7}$$

where Σ is a set of unit vectors in R^n.

We can see the difficulty of these problems by noting how badly property (4) fails. For example look at 2 "slightly fattened" parabolas of equal measure which intersect.

Clearly no bounded multiple of one contains the other.

We would like to concentrate on the progress made on the study of these maximal functions by Fourier transform methods. The people mainly responsible for these developments are Nagel, Riviere, Stein, and Wainger.

The first application of Fourier transform methods to these problems came in the study of $\mathfrak{m}_1$ in the special case (t, t^2) in R^2 by Nagel, Riviere, and Wainger [NRW 1976] where it was shown that in this case

$$\| \mathfrak{m}_1 f \|_p \leq A_p \| f \|_p, \quad 1 < p \leq \infty . \tag{8}$$

I would like to recall the basic ideas of the proof in the case $p = 2$. It suffices to consider

$$\sup_{h > 0} |M_h f(x, y)|$$

where

$$M_h f(x, y) = \frac{1}{h} \int_{-\infty}^{\infty} e^{-t^2/h^2} f(x - t, y - t^2) \, dt . \tag{9}$$

Then

(10) $$M_h f = \mu_h * f$$

where μ_h is a measure, and for a test function g

$$\mu_h(g) = \frac{1}{h} \int_{-\infty}^{\infty} e^{-(t^2/h^2)} g(t, t^2)\,dt$$

Thus

$$\hat{\mu}_h(\xi, \eta) = \frac{1}{h} \int_{-\infty}^{\infty} e^{-(t^2/h^2)} e^{i\xi t} e^{i\eta t^2}\, dt.$$

This integral can be easily evaluated, and one obtains

$$\hat{\mu}_h(\xi, \eta) = (\text{nice smoothly decaying function}) \cdot \exp i \frac{h^4 \xi^2 \eta}{1 + h^4 \eta^2} .$$

Now it would not be unreasonable to guess that the singularity of the measures μ_h is reflected in the oscillatory behavior of the factor

$$\exp i \frac{h^4 \xi^2 \eta}{1 + h^4 \eta^2} .$$

Furthermore we note that at least for $h^4 \eta^2$ large

$$\exp i \frac{h^4 \xi^2 \eta}{1 + h^4 \eta^2}$$

is roughly $\exp\{i\, \xi^2/\eta\}$ which is independent of h. Thus one might be tempted to write

$$\hat{\mu}_h(\xi, \eta) = \underbrace{\hat{\mu}_h(\xi, \eta) \exp\{-i\, \xi^2/\eta\}}_{=\ m_h(\xi,\, \eta)} \quad \underbrace{\exp i(\xi^2/\eta)\, \hat{f}(\xi, \eta)}_{\hat{g}(\xi,\, \eta)}$$

g is an L^2 function with the same norm as f and one might hope that m_h is a nice family of measures which lies within the scope of the classical theory. Roughly this is the case.

Thus in effect we have used the Fourier transform to change our problem concerning a family of singular measures to a problem of nice measures that can be treated by the classical theory.

Unfortunately, the method requires us to know a great deal about the Fourier transform of the singular measures μ_h — much more than one can calculate for general curves. We would like to have a methods which uses only the decay of $\hat{\mu}_h(\xi, \eta)$ and doesn't require knowledge of the oscillatory factor. To this end Stein [S 1976] and [S 1976b] introduced g-functions into the theory, and all further progress on the maximal functions (5) and (6) make use of appropriate g-functions.

Let us see how the g-function method works for the L^2 estimate we just described above. Of course we wish to obtain that result using only the fact that $\hat{\nu}(\xi, \eta)$ decays at ∞ for a "nice" singular measure ν on (t, t^2) rather than the rather complete information used above. We define

$$N_h f(x,y) = \frac{1}{h}\int_h^{2h} f(x-t, y-t^2)\,dt.$$

Let $\phi(x) \in C_0^\infty$ with $\hat{\phi}(0) = 1$. Set $\phi_h(x,y) = \frac{1}{h^3}\phi(\frac{x}{h}, \frac{y}{h^2})$. We set

$$g(f)(x,y) = \left\{\int_0^\infty |N_h f(x,y) - \phi_h * f(x,y)|^2 \frac{dh}{h}\right\}^{1/2}.$$

It is known that

$$\| \sup_{h>0} |\phi_h * f| \|_2 \leq C\|f\|_2 .$$

Thus, we would show

$$\|\mathfrak{m}_2 f\|_2 \leq C\|f\|_2$$

if we could show

$$(11) \qquad \|g(f)\|_2 \leq C\|f\|_2$$

and

$$(12) \qquad \sup_{h>0} \frac{1}{h}\int_0^h f(x-t, y-t^2)\,dt < C(gf(x,y) + \sup_{h>0} |\phi_h * f|$$

(at least for non-negative f). Let us first show (12). (This part of the argument holds for any curve). For any $\varepsilon > 0$

$$g^2 f(x,y) \geq \frac{1}{\varepsilon}\int_0^\varepsilon |N_h f(x,y) - \phi_h * f(x,y)|^2\,dh .$$

So by Schwarz's inequality

$$\frac{1}{\varepsilon}\int_0^\varepsilon |N_h f(x,y) - \phi_h * f(x,y)|\,dh \leq g(f)(x,y).$$

Thus

$$\frac{1}{\varepsilon}\int_0^\varepsilon N_h f(x,y)\,dh \leq g(f)(x,y) + \sup_{h>0} |\phi_h * f(x,y)|.$$

Finally, note that for non-negative f

$$\frac{1}{\varepsilon}\int_0^\varepsilon N_h f(x,y)\,dh = \frac{1}{\varepsilon}\int_h^\varepsilon \frac{1}{h}\int_0^\varepsilon f(x-t, y-t^2)\,dt\,dh$$

$$\geq \frac{1}{\varepsilon}\int_0^\varepsilon f(x-t, y-t^2)\int_{t/2}^t \frac{dh}{h}\,dt > \frac{C}{\varepsilon}\int_0^\varepsilon f(x-t, y-t^2)\,dt.$$

Since this is true for any ε we have (12). We turn now to the proof of (11).

$$\iint g^2(f)(x,y)\,dxdy = \iint dxdy \int_0^\infty \frac{dh}{h}|N_h f(x,y) - \phi_h * f(x,y)|^2$$

Interchanging the order of integration and applying Parseval's relation we see

$$\|g(f)\|_2^2 = \int_0^\infty \iint |\widehat{N_h f}(\xi,\eta) - \hat{\phi}(h\xi, h^2\eta)\hat{f}(\xi,\eta)|^2\,d\xi\,d\eta\,\frac{dh}{h}.$$

But

$$\widehat{N_h f}(\xi,\eta) = \hat{\nu}(h\xi, h^2\eta)\,\hat{f}(\xi,\eta)$$

where

$$\hat{\nu}(\xi,\eta) = \int_1^2 \exp(it\xi + it^2\eta)\,dt$$

is the Fourier transform of a "nice" singular measure supported on the curve (t, t^2). Thus, interchanging orders of integration again

$$\|g(f)\|_2^2 = \int |\hat{f}(\xi,\eta)|^2 d\xi\, d\eta \int_0^\infty \frac{dh}{h} |\hat{\nu}(h\xi, h^2\eta) - \hat{\phi}(h\xi, h^2\eta)|^2 .$$

Hence by Parseval's relation it suffices to show that the inner integral is uniformly bounded. Note first that we may assume (ξ,η) is on the unit sphere since the measure $\frac{dh}{h}$ is invariant under dilations $h \to \lambda h$. Then since $\hat{\phi}(0) = 1$ and $\hat{\nu}(0) = 1$, it is clear that

$$\int_0^1 \frac{dh}{h} |\hat{\nu}(h\xi, h^2\eta) - \hat{\phi}(h\xi, h^2\eta|^2 dh \le C \int_0^1 \frac{h^2}{h} dh \le C .$$

It is also clear that

$$\int_1^\infty \frac{dh}{h} |\hat{\phi}(h\xi, h^2\eta)| \le C ,$$

since the Fourier Transform of ϕ decays rapidly at ∞. Thus we need only show

$$\int_1^\infty \frac{dh}{h} |\hat{\nu}(h\xi, h^2\eta)|^2 dh < C < \infty .$$

This follows because of the decay of $\hat{\nu}$ at ∞. Actually $|\hat{\nu}(h\xi, h^2\eta)| \le C/h^{1/2}$ for $\xi^2 + \eta^2 = 1$. Intuitively, we might expect decay of $\hat{\nu}(\xi,\eta)$ as $\xi^2 + \eta^2$ get large if $\exp\{i(\xi t + \eta t^2)\}$, $1 \le t \le 2$, oscillates rapidly as $\xi^2 + \eta^2$ becomes large; that is if some derivative of

$$u(t) = \xi t + \eta t^2 , \quad 1 \le t \le 2,$$

is large. But

$$u'(t) = \xi + 2\eta t ,$$

and

$$u''(t) = 2\eta .$$

So if $\eta \ge \xi/5$, $|u''(t)| > C(\xi^2 + \eta^2)^{1/2}$, and if $\eta < \xi/5$, $|u'(t)| > C(\xi^2 + \eta^2)^{1/2}$, for some $C > 0$.

Stein [S 1976b] used this idea to treat curves $\gamma(t) = (t^{a_1}, \dots, t^{a_n})$, $a_i \ge 0$, and the ideas were modified by Stein and Wainger [SW 1976] to treat C^∞ curves with appropriate curvature. We consider curves having the following property:

$\gamma(t)$ lies in the span of the vectors

$$\gamma'(0),\ \gamma''(0),\ \ldots,\ \gamma^{(j)}(0),\ \ldots$$

Theorem 1: If $\gamma(t)$ satisfies (13) then

$$\| \mathfrak{m}_1 f \|_p \le A_p \| f \|_p, \quad 1 < p \le \infty,$$

and for any f locally in L^p, $p > 1$,

$$\lim_{h \to 0} \frac{1}{h} \int_0^h f(x - \gamma(t)| \, dt = f(x).$$

The curvature assumption (13) is needed for the proof in order to have decay for "nice" singular measures supported on $\gamma(t)$. (13) is satisfied for any real analytic curve, and the conclusion of theorem 1 is false for arbitrary C^∞ curves. Full details may be found in Stein and Wainger [SW 1978].

We turn now to the so called spherical maximal function of (6).

Here Stein [S 1976b] proved the following

Theorem 2: Let $p > \frac{n}{n-1}$ and $n \ge 3$. Then

$$\| \mathfrak{m}_2 f_2 \|_p \le A_{p,n} \| f \|_p .$$

Note that the analogue of Theorem (2) is obviously false if we would use boundaries of cubes instead of spheres. Simple examples show that the conclusion of Theorem 2 is false if $n = 1$ or $p \le \frac{n}{n-1}$. Note that we don't know what happens if $n = 2$ and $p > 2$, presumably because Fourier transform and other orthogonality methods break down for $p \ne 2$.

We shall give a variant of Stein's g-function which can be used to prove Theorem 2. However we are not going to use it to prove Theorem 2, but we are going to use it to study the averages

$$M_r f(x) = \int_{S_r} f(x-y) \, d\sigma(y)$$

in 2 dimensions. For x in R^2, set

$$\mathfrak{m}^\ell f(x) = \sup_{j=0,1,\ldots,2^\ell} \left| \int_{S_{1+j/2^\ell}} f(x-y) \, d\sigma_{1+j/2^n}(y) \right| .$$

Theorem 3: In R^2

$$\| \mathfrak{m}^\ell f \|_2 \le C\ell \, \| f \|_2 .$$

One reason that we are giving the proof of Theorem 3 is that it gives the simplest application of the g-function I know of.

Let $j(x)$ be any measurable function taking values $0, 1, 2, \ldots, 2^n$. Set

$$M^{\ell}_{j(x)} f(x) = \int_{S_{1+j(x)/2^n}} f(x-y)\, d\sigma_{1+j(x)/2^n}.$$

To prove Theorem 3, it suffices to show

$$\| M^{\ell}_{j(x)} f(x) \|_2 \le C\ell \, \| f \|_2. \tag{14}$$

We shall produce a function $k(x)$ taking the values $0, 1, 2, \ldots 2^{\ell-1}$ such that

$$\| M^{\ell}_{j(x)} f - M^{\ell-1}_{k(x)} f \|_2 \le C \| f \|_2. \tag{15}$$

We can then add ℓ inequalities of this type to obtain (14). ($M^{1}_{j(x)}$ is clearly bounded.) We define $k(x)$ so that

$$\frac{k(x)}{2^{n-1}} = \begin{cases} \dfrac{j(x)}{2^n} & \text{if } j(x) \text{ is even} \\[2ex] \dfrac{j(x)-1}{2^n} & \text{if } j(x) \text{ is odd.} \end{cases}$$

Then clearly

$$| M^{\ell}_{j(x)} f(x) - M^{\ell-1}_{k(x)} f(x) | \le \left\{ \sum_{j=1}^{2^{\ell}} | M_{1+\frac{j}{2^{\ell}}} f(x) - M_{1+\frac{j-1}{2^{\ell}}} f(x) |^2 \right\}^{1/2}.$$

So we set

$$gf(x) = \left\{ \sum_{j=1}^{2^{\ell}} | M_{1+j/2^{\ell}} f(x) - M_{1+\frac{j-1}{2^{\ell}}} f(x) |^2 \right\}^{1/2},$$

and remark that (15) will follow from the estimate

$$\| g(f) \|_2 \le C \| f \|_2. \tag{16}$$

Thus we could prove Theorem 3 if we could show that the operator norm (in L_2) of

$$M_{1+j/2^n} - M_{1+\frac{(j-1)}{2^n}}$$

is at most $C 2^{-n/2}$. But $M_r f$ is $f * d\sigma_r$. Thus to prove Theorem 3, it would suffice to prove

$$\| \widehat{d\sigma}_{1+j2^{-n}} - \widehat{d\sigma}_{1+(j-1)2^{-n}} \|_{\infty} \le C 2^{-n/2}. \tag{17}$$

(17) follows easily from the fact that

$$\widehat{d\sigma_r}(\xi) = J_0(r|\xi|).$$

Thus Theorem 3 is proved.

E. M. Stein and I recently used similar ideas to prove

$$\sup_{\substack{r \\ r \in \text{cantor set}}} \|M_{1+r} f\|_2 \leq C\|f\|_2 .$$

As we mentioned, one reason we chose to prove Theorem 3 in the above manner is its extreme simplicity. A second reason is that we wish to point out a similarity between $\mathfrak{m}_2$ and $\mathfrak{m}_3$ is 2 dimensions. Thus we set

$$\overline{M}^{\ell}_{j,h} f(x,y) = \frac{1}{h} \int \phi\left(\frac{t}{h}\right) f\left(x-t, y - \frac{j}{2^\ell} t\right) dt,$$

and

$$\mathfrak{m}^{\ell}_3 f(x,y) = \sup_{\substack{j = 0, 1, \ldots 2^\ell \\ h > 0}} |\overline{M}^{\ell}_{j,h} f(x,y)| .$$

Here $\phi \geq 0$ is to be positive near 0, non-negative, and $\hat{\phi}$ should be C^∞ with compact support. Then we have Theorem 4:

$$\| \mathfrak{m}^{\ell}_3 f(x)\|_2 \leq C\ell \|f\|_2 .$$

This result was previously obtained by A. Cordoba [C 1977] (perhaps with some higher power of ℓ) and by J. Stromberg [St 1978] using delicate covering lemmas.

We shall give a proof by Fourier transform methods which is very similar to the proof of Theorem 3. Here for simplicity we shall assume $h = 1$ and so the sup is only on j. As in the proof of Theorem 3 it suffices to show that for any function $j(x,y)$ taking the values $0, 1, \ldots 2^\ell$,

$$\| \overline{M}^{\ell}_{j(x,y)} f(x,y)\|_2 \leq C\ell \|f\|_2 .$$

Thus again it would suffice to find, for any such $j(x,y)$, a function $k(x,y)$ taking the values $0, 1, \ldots 2^{\ell-1}$ such that

$$\| \overline{M}^{\ell}_{j(x,y)} - \overline{M}^{\ell-1}_{h(x,y)} \|_2 \leq C\|f\|_2 .$$

We define $k(x,y)$ just as above

$$k(x,y) = \begin{cases} j(x,y)/2^\ell & \text{if } j(x,y) \text{ is even} \\ (j(x,y)-1)/2^\ell & \text{if } j(x,y) \text{ is odd.} \end{cases}$$

Once again, $\overline{M}^{\ell}_{j(x,y)} - \overline{M}^{\ell-1}_{k(x,y)}$ is dominated by a g function

$$gf(x,y) = \left\{ \sum_{j=1}^{2^n} |\overline{M}^{\ell}_j f(x,y) - \overline{M}^{\ell}_{j-1} f(x,y)|^2 \right\}^{1/2} .$$

It is not true that the norm is $\| \overline{M}^{\ell}_j - \overline{M}^{\ell}_{j-1} \|$ is small. However computing a few Fourier Transforms and arguing as in the proof of (11), showing

$$\|g(f)\|_2 \leq C\|f\|_2$$

comes down to the uniform boundedness of (in ξ and η) of

$$\sum_{j=1}^{2^\ell} |\hat{\phi}(\xi + \frac{j}{2^\ell}\eta) - \hat{\phi}(\xi + \frac{j-1}{2^\ell}\eta)|^2$$

and this is not difficult. Thus Theorem 4 is proved if $h = 1$ and the sup is only on j.

To prove Theorem 4 in general (i.e., taking a sup in h and j) by the Fourier transform method we must introduce cones as was done in Nagel, Stein, and Wainger [NSW 1977]. One must also introduce a more complicated g function to take care of the dependence on h as well as on j.

We thus see that Theorems 3 and 4 can be given very similar proofs. Now it is known that

$$\|\mathfrak{m}_3^\ell f(x,y)\|_p \leq C_p\|f\|_p$$

is false for every $p < \infty$ (even if the sup is taken only over j and $h = 1$). This might make one suspect of the conjecture

$$\|\mathfrak{m}_2^\ell f\|_p \leq A_p\|f\|_p, \quad p > 2 \quad \text{in} \quad R^2.$$

We would like to point out another similarity. Both the spherical averages and the averaging over equally spaced lines have "lacunary" versions for which the maximal functions are bounded in all L^p, $p > 1$. Precisely

(18) $$\|\sup_j\left|\int_{S_{2^{-j}}} f(x-y)\,d\sigma_{2^{-j}}(y)\right|\|_p \leq A_p\|f\|_p \qquad 1 < p \leq \infty,$$

and

(19) $$\|\sup_{\substack{j \\ h>0}} \frac{1}{h}\left|\int_0^h f(x-t, y-2^{-j}t)\,dt\right|\| \leq A_p\|f\|_p$$

(18) was proved independently by Coifman and Weiss [CW 1978] and C. Calderón [Ca 1978]. (19) was proved by Nagel, Stein, and Wainger [NSW 1977], after partial results had been obtained by J. Strömberg [St 1977] and A. Cordoba and R. Fefferman [CF 1977].

One final reason for introducing the g-functions used in the arguments for Theorems 3 and 4 was to indicate that one might consider introducing "square functions" into the theory to greatly unify the arguments.

<u>Note:</u> A. Cordoba has independently found a proof of Theorem 4 similar to the proof given above.

Bibliography

Ca 1978 C. Calderón. To appear.

CW 1978 R. Coifman and G. Weiss. Review of Littlewood-Paley and multiplier theory by Edwards and Gaudry, Bulletin of the AMS vol. 84, 1978 pp 242-250.

C 1977 A. Cordoba. "The multiplier problem for the polygon", Annals of Math. vol. 105, 1977 pp 581-588.

CF 1977 A Cordoba and R. Fefferman. "On differentiation of integrals", Pro. Nat. Acad. Sci., vol. 74, 1977 pp 2211-2213.

NRW 1976 A. Nagel, N. M. Riviere, and S. Wainger. "A maximal function associated to the curve (t, t^2)", Proc. of the Nat. Acad. of Sci., vol. 73 pp 1416-1417.

NSW 1978 A. Nagel, E. M. Stein, and S. Wainger. "Differentiation in lacunary directions", Proc. of the Nat. Acad. of Sci. vol. 75, 1978 pp 1060-1062.

S 1961 E. M. Stein. "On limits of sequences of operators", Ann. of Math. 74, 1961 pp 140-170.

S 1970 E. M. Stein. <u>Singular Integrals and Differentiability Properties of Functions</u>, Princeton University Press, 1970.

S 1976a E. M. Stein. "Maximal functions: Spherical means", Proc. of Nat. Acad. Sci. vol. 73, 1976 pp 2174-2175.

S 1976b E. M. Stein. "Maximal functions; Homogeneous curves", Proc. Nat. Acad. Sci. vol. 73, 1976 pp 2176-2177.

SW 1976 E. M. Stein and S. Wainger. "Maximal functions associated to smooth curves", Proc. Nat. Acad. Sci. vol. 73, 1976 pp 4295-4296.

SW 1978 E. M. Stein and S. Wainger. "Problems in harmonic analysis related to curvature", to appear in the Bulletin of the AMS.

St 1977 J. Strömberg. "Weak estimates for maximal functions with rectangles in certain directions", Arkiv. för Mat. vol. 15, 1977 pp 229-240.

St 1978 J. Strömberg. "Maximal functions associated to rectangeles with uniformly distributed directions", Annals of Math. vol. 107, 1978 pp 399-402.

University of Wisconsin
Madison, WI 53706

Proceedings of Symposia in Pure Mathematics
Volume XXXV, Part 1, 1979

HILBERT TRANSFORMS AND MAXIMAL FUNCTIONS RELATED TO VARIABLE CURVES

by

Alexander Nagel[1], Elias M. Stein[1], and Stephen Wainger[1]

We wish to announce results on the L^2 boundedness of Hilbert Transforms and Maximal functions related to variable curves in R^2 and applications of these results to differentiation of averages along vector fields.

We shall assume $r(x, y, t)$ is a C^∞ curve starting at (x, y), that is $r(x, y, 0) = 0$. We also assume $\frac{d}{dt} r(x, y, t)\Big|_{t=0} = 0$ and $\frac{d^2}{dt^2} r(x, y, t)\Big|_{t=0} \neq 0$.

For f in C^∞, we set

1) $$Hf(x, y) = \Phi(x, y) \int \frac{\Psi(t)}{t} f(x-t, y-r(x, y, t))\, dt$$

and

2) $$Mf(x, y) = \Phi(x, y) \sup_{h>0} \left| \frac{1}{h} \int_0^h \Psi(t)\, f(x-t, y-r(x, y, t)\, dt \right|.$$

Here Φ and Ψ are appropriate cutoff functions.

We prove the following theorem

THEOREM: A) $\|Hf\|_2 \leq c\|f\|_2$, B) $\|Mf\|_2 \leq c\|f\|_2$ <u>for some constant c</u>.

Assertion B has the following corollary.

---

1) Supported in part by N.S.F. grants.

COROLLARY: *Let* $v(z)$ *be a real analytic vector field in* R^2. *Then for every* $f(z)$ *locally in* L^2,

$$\lim_{h\to 0} \frac{1}{h}\int_0^h f(z - tv(z))\,dt = f(z) \qquad \text{a.e.}$$

When r is independent of x and y stronger versions of this theorem are known, see [sw] and the references given there; however we would like to consider again the case $r(x, y, t) = t^2$ to motivate our approach.

We set

$$Kf(x, y) = \int f(x - t, y - t^2)\frac{dt}{t}$$

dropping the cutoff factors. Then knowing that K is bounded in L^2, we might hope that KK^*f is convolution of f with a non-isotropic Calderòn Zygmund kernel. This is not the case. Let us see by looking at the Fourier transform if we can understand why not. It is known that

$$\widehat{Kf}(\xi, \eta) = m(\xi, \eta)\hat{f}(\xi, \eta)$$

where $m(\xi, \eta)$ is "nice" except near $\eta = 0$. Near $\eta = 0$, roughly

3) $$m(\xi, \eta) = \frac{\sqrt{\eta}}{\xi}e^{i\xi^2/\eta} + \text{constant}\,(\operatorname{sgn}\xi).$$

Thus

$$m(\xi, \eta)\overline{m(\xi, \eta)} = \frac{\sqrt{\eta}}{\xi^2}e^{i\xi^2/\eta}e^{-i\xi^2/\eta}$$

$$+ \text{const}\frac{\sqrt{\eta}}{3}e^{i\xi^2/\eta}\operatorname{sgn}\xi + \text{const}\frac{\sqrt{\eta}}{\xi}e^{-i\xi^2/\eta}\operatorname{sgn}\xi.$$

In other words all of the oscillation in $m(\xi, \eta)\overline{m(\xi, \eta)}$ is caused by the constant in 3). Hence one might subtract from Kf Rf, where R is an operator such that R is known to be bounded on L^2, and such that

$$\widehat{Rf}(\xi, \eta) = \text{const}\operatorname{sgn}\xi\,\hat{f}(\xi, \eta).$$

We might then hope that

$$(K - R)(K - R)^*$$

is basically a Calderon-Zygmund operator: It would then follow that K is bounded on L^2. The simplest such R is given by

$$Rf(x, y) = \int f(x-t, y) \frac{dt}{t} .$$

This actually works and suggests trying to prove $(H-T)(H-T)^* f$ is given by convolution of f with a Calderòn-Zygmund kernel where here H is the operator 1), and

$$Tf(x, y) = \Phi(x, y) \int \frac{\Psi(t)}{t} f(x-t, y)\, dt .$$

In fact we show this by rather elaborate calculations and thus obtain a proof of part A.

We might now suspect that a similar idea would work for part B. It turns out that it is more convenient to deal with averages

$$N_h f(x, y) = \Phi(x, y) \frac{1}{h} \int_{-\infty}^{\infty} \Psi(t)\, \Theta\left(\frac{t}{h}\right) f(x-t, y-r(x, y, t)),$$

where $\Theta(A)$ is even, positive, C^1, L^1, and has the property that $\Theta(2s) > C\,\Theta(s)$ for $s > 0$. The analogue of R is

$$S_h f(x, y) = \Phi(x, y) \frac{1}{h} \int_{-\infty}^{\infty} \Psi(t)\, \Theta\left(\frac{t}{h}\right) f(x-t, y)\, dt .$$

If one now replaces h by a function $h(x, y)$, one can show that for $f \geq 0$, one has a pointwise inequality.

$$N_{h(x,y)} \left(N_{h(x,y)}\right)^* f \leq C (N_{h(x,y)} \left(S_{h(x,y)}\right)^* f$$

$$+ S_{h(x,y)} \left(N_{h(x,y)}\right)^* f + Qf$$

where Q is the Hardy-Littlewood Maximal function. This allows one to prove part B. The argument is a modification of ideas used by Kolmogorov and Seliverstov in studying almost everywhere convergence of Fourier series. (See [z] Vol. 2, pp. 161-163.)

REFERENCES

[sw] E. M. Stein and S. Wainger "Problems in harmonic analysis related to curvature" to appear in the Bulletin of the Amer. Math. Soc.

[z] A. Zygmund <u>Trigonometric Series</u> Cambridge University Press 1959.

ALEXANDER NAGEL
UNIVERSITY OF WISCONSIN
MADISON, WISCONSIN 53706

ELIAS M. STEIN
PRINCETON UNIVERSITY
PRINCETON, NEW JERSEY 08540

STEPHEN WAINGER
UNIVERSITY OF WISCONSIN
MADISON, WISCONSIN 53706

Proceedings of Symposia in Pure Mathematics
Volume XXXV, Part 1, 1979

REGULARITY OF SPHERICAL MEANS

Jacques Peyrière

This lecture is a presentation of a work done by P. Sjölin and myself [1]. Here is an abstract of it.

Let $\mathbf{R}^n$ denote the n-dimensional euclidean space and $|x|$ the norm of an element x of $\mathbf{R}^n$. For $f \in L^1_{loc}(\mathbf{R}^n)$, $\beta \in \mathbf{R}$, $x \in \mathbf{R}^n$ and almost every t in $\mathbf{R}$, set

$$\mathfrak{M}_{\beta,x} f(t) = |t|^{\beta} \int_{S^{n-1}} f(x-ty)\, d\sigma(y)$$

and

$$\mathfrak{M}_x f(t) = \begin{cases} 0 & \text{if} \quad t < 0 \\ \mathfrak{M}_{0,x} f(t) & \text{if} \quad t \geq 0, \end{cases}$$

where σ is the surface measure on $S^{n-1} = \{x \in \mathbf{R}^n \; ; \; |x| = 1\}$.

We deal with the regularity of the functions $\mathfrak{M}_{\beta,x} f$ and $\mathfrak{M}_x f$. This regularity is expressed in terms of the Besov spaces $B_p^{\alpha,q}$. Our results are somewhat related to one of E. M. Stein [2].

THEOREM 1. Assume $n \geq 3$, $n/(n-1) < p < \infty$, $\alpha < 1/p$ and $f \in L^p(\mathbf{R}^n)$, then $\mathfrak{M}_x f \in B_p^{\alpha,1}(\mathbf{R})$ for almost every $x \in \mathbf{R}^n$.

THEOREM 2. Assume $n \geq 2$, $\alpha > 0$ and $-1 < \beta < (n-2)/2$. Then there exists C such that

$$\int_{\mathbf{R}^n} \|\mathcal{M}_{\beta,x} f\|^2_{B_2^{\alpha,2}(\mathbf{R})} \, dx \le C \int_{\mathbf{R}^n} |f * \check{f}(x)| \, |x|^{2(\beta-\alpha)-n+1} (1+|x|^{2\alpha}) \, dx$$

for every continuous f with compact support. (As usual, $\check{f}(x) = \bar{f}(-x)$).

COROLLARY. Assume $n \geq 3$, $n/(n-1) < q \leq 2$, $\alpha < n(1 - \frac{1}{q}) - 1$ and $f \in L^q_{loc}(\mathbf{R}^n)$. Then for almost every $x \in R^n$ the function $\mathcal{M}_x f$ coincides almost everywhere on $]0, \infty[$ with a function in $\Lambda_\alpha(]0,\infty[)$.

[1] PEYRIERE, J. and SJÖLIN, P. Regularity of spherical means. Ark. för Mat. 16 (1978), 117-126.

[2] STEIN, E. M. Maximal functions : spherical means. Proc. Nat. Acad. Sc. USA 73 (1976), 2174-2175.

Proceedings of Symposia in Pure Mathematics
Volume XXXV, Part 1, 1979

RESTRICTION THEOREMS FOR THE FOURIER TRANSFORM TO SOME MANIFOLDS IN $\mathbb{R}^n$

Elena Prestini

ABSTRACT. Let σ be a smooth compact manifold of dimension d in $\mathbb{R}^n$ such that n=2d and $d \geq 1$ or d=1 and $n \geq 2$. Let T be the restriction operator defined by $Tf = \hat{f}_{|\sigma}$. Sufficient conditions are stated for T to be a bounded operator from $L^p(\mathbb{R}^n)$ to $L^q(\sigma)$, for certain values of p and q.

1. Introduction. It was first proved by E. Stein and C. Fefferman [1] that the following sharp inequality holds

$$\|\hat{f}\|_{L^q(\sigma)} \leq C_p \|f\|_{L^p(\mathbb{R}^2)}$$

where $S^1 = \{\vec{x} \in \mathbb{R}^2 : \|\vec{x}\| = 1\}$, whenever $\frac{1}{p} + \frac{1}{3q} \geq 1$ and $1 \leq p < \frac{4}{3}$. This implies that it makes sense to restrict $\hat{f}$ to the unit circle S^1 for $f \in L_p$ $1 \leq p < \frac{4}{3}$ even though in principle the Fourier transform is defined only up to sets of measure zero.

In [2] we proved a restriction theorem for compact space curves with curvature and torsion never vanishing. In this paper we observe that the same method can be used to prove analogous results for compact manifolds of dimension d in $\mathbb{R}^n$ whenever n=2d and $d \geq 1$ or d=1 and $n \geq 2$.

The interested reader should see [4]-[9] where other results are proved.

2. The Case of Manifolds of Dimension n in $\mathbb{R}^{2n}$ $(n \geq 1)$. Let σ be a compact manifold of class C^k $(k \geq 2)$ in $\mathbb{R}^{2n}$. We shall show that a sufficient condition for a restriction theorem to hold is that the set of vectors $\left\{\frac{\partial \vec{X}}{\partial x_i}(\vec{x}), \frac{\partial^2 \vec{X}}{\partial x_i \partial x_j}(\vec{x})\right\}$ span $\mathbb{R}^{2n}$ whenever $\vec{X}(\vec{x}) = \left\{X_i(x_1, \ldots, x_n)\right\}$ i=1,...,2n is a local chart around $\vec{x}$. The problem is a local one, hence without loss of generality we may work in a small neighborhood of the origin, assuming $\vec{X}(\vec{0}) = \vec{0}$. We choose the n tangent vectors of the manifold at the origin to be

AMS (MOS) subject classification (1970). Primary 42A68, 42A72; Secondary 60E05.

the first n vectors of the frame of $\mathbb{R}^{2n}$. Then locally σ has equation

$$(1)\qquad \begin{cases} X_i = x_i + \mathcal{O}(\|\vec{x}\|^2) & i=1,\dots,n \\ X_j = \sum_{i=1}^{n} a^j_{ii} x_i^2 + 2 \sum_{\substack{i\neq k \\ i,k=1,\dots,n}} a^j_{ik} x_i x_k + \mathcal{O}(\|\vec{x}\|^3) & j=n+1,\dots,2n \end{cases}$$

($a^j_{ik} = a^j_{ki}$). We consider the truncated equation

$$\begin{cases} \tilde{X}_i = x_i & i=1,\dots,n \\ \tilde{X}_j = \sum_{i=1}^{n} a^j_{ii} x_i^2 + 2 \sum_{\substack{i\neq k \\ i,k=1,\dots,n}} a^j_{ik} x_i x_k & j=n+1,\dots,2n \end{cases}$$

and we write $\tilde{J}(\vec{x}) = \det \left[\frac{\partial \tilde{X}_j(\vec{x})}{\partial x_i} \right]$ $i=1,\dots,n$ $j=n+1,\dots,2n$. $\tilde{J}(\vec{x})$ is a homogeneous polynomial of degree n in n variables. Unless $\tilde{J}(\vec{x})$ is identically zero, the zero set of $\tilde{J}(\vec{x})$ is a cone of dimension n hence it has Lebesgue measure zero in $\mathbb{R}^{2n}$.

PROPOSITION. If the set of vectors $\left\{ \frac{\partial \vec{X}(\vec{0})}{\partial x_i}, \frac{\partial^2 \vec{X}(\vec{0})}{\partial x_i \partial x_k} \right\}$ $i,k=1,\dots,n$ spans $\mathbb{R}^{2n}$ then $\tilde{J}(\vec{x})$ is not identically zero.

PROOF. Suppose

$$\tilde{J}(\vec{x}) = 2^n \cdot \begin{vmatrix} \sum_{k=1}^{n} a^1_{1k} x_k & \sum_{k=1}^{n} a^1_{2k} x_k & \cdots & \sum_{k=1}^{n} a^1_{nk} x_k \\ \vdots & \vdots & & \vdots \\ \sum_{k=1}^{n} a^n_{1k} x_k & \sum_{k=1}^{n} a^n_{2k} x_k & \cdots & \sum_{k=1}^{n} a^n_{nk} x_k \end{vmatrix} \equiv 0$$

Letting $x_1 = \cdots = x_{h-1} = x_{h+1} = \cdots = x_n = 0$ and $x_h = 1$, we have that the (n,h) entry of the above matrix is a linear combination of the other (i,h) entries $i=1,\dots,n-1$. Since $\tilde{J}(\vec{x}) \equiv 0$ the coefficients of the linear combination for different h's are the same. This contradicts the assumption.

Now we are going to prove the following:

THEOREM 1. Let σ have equation (1) in a neighborhood $\mathcal{U}$ of the origin and let us assume that the set of vectors $\left\{\frac{\partial \vec{X}(\vec{0})}{\partial x_i}, \frac{\partial^2 \vec{X}(\vec{0})}{\partial x_i \partial x_k}\right\}$ $i,k=1,\dots,n$ spans $\mathbb{R}^{2n}$. If $|\tilde{J}(\vec{x})|^{-a}$ is integrable in $\mathcal{U}$ for $a < p_0$ then the following inequality holds:

$$\|\hat{f}\|_{L^q(\sigma)} \leq C_p \|f\|_{L^p(\mathbb{R}^{2n})} \tag{2}$$

when $\frac{1}{q} > \left(2 + \frac{1}{p_0}\right)\left(1 - \frac{1}{p}\right)$ and $p < \frac{2(1+p_0)}{2+p_0}$. The inequality need not hold if $\frac{1}{q} < 3(1-\frac{1}{p})$ or $p \geq \frac{4}{3}$.

PROOF. For $f \in C_0^\infty(\mathbb{R}^{2n})$ let $Tf = \hat{f}_{|\sigma}$. Then the adjoint operator is

$$T^* f(\vec{\xi}) = \int_{\mathcal{U}} e^{i\vec{\xi}\vec{X}(\vec{x})} f(\vec{x}) d\vec{x}.$$

We consider

$$(T^* f)^2(\vec{\xi}) = \int_{\mathcal{U}\times\mathcal{U}} e^{i\vec{\xi}[\vec{X}(\vec{t}) + \vec{X}(\vec{u})]} f(\vec{t}) f(\vec{u}) d\vec{t}\, d\vec{u}.$$

We make the following change of variables: $V_i = t_i + u_i$, $i=1,\dots,n$ and

$$V_j = \sum_{i=1}^{n} a^j_{ii} (t_i^2 + u_i^2) + 2 \sum_{\substack{i \neq k \\ i,k=1,\dots,n}} a^j_{ik}(t_i t_k + u_i u_k), \quad j = (n+1),\dots,2n.$$ Then

$$(T^* f)^2(\vec{\xi}) = \int_D e^{i\vec{\xi}\vec{V}} |J(\vec{t},\vec{u})|^{-1} f(\vec{t}) f(\vec{u}) d\vec{V}$$

where D is the image of $\mathcal{U}\times\mathcal{U}$ under the change of variables and J is the Jacobian of the transformation. Now we apply the Hausdorff-Young inequality to obtain

$$\|T^* f\|^2_{L^s(\mathbb{R}^{2n})} = \|(T^* f)^2\|_{L^{s/2}(\mathbb{R}^{2n})} \leq \tag{3}$$

$$\leq \left(\int_D (|f(\vec{t}) f(\vec{u})| \; |J(\vec{t},\vec{u})|^{-1})^{s/s-2} d\vec{V}\right)^{s-2/s}$$

We change variables back and we obtain

$$\|T^* f\|^2_{L^s(\mathbb{R}^{2n})} \leq \left(\int_{\mathcal{U}\times\mathcal{U}} |f(\vec{t}) f(\vec{u})|^{s/s-2} \; |J(\vec{t},\vec{u})|^{(-s/s-2)+1} d\vec{t}\, d\vec{u}\right)^{s-2/s}$$

Instead of checking that the above changes of variables are admissible we refer to [10] Lemma 1.8. We set $F = |f|^{s/s-2}$ and $a = \frac{s}{s-2} + 1 = \frac{2}{s-2}$. Then

$$(4) \qquad \|T^{*}f\|^{2}_{L^{s}(\mathbb{R}^{2n})} \leq \left(\int_{\mathcal{U}\times\mathcal{U}} \frac{F(\vec{t}\,)\,F(\vec{u})}{|J(\vec{t},\vec{u})|^{a}}\, d\vec{t}\, d\vec{u} \right)^{s-2/s}$$

Now we are going to compute the lowest order term of $J(\vec{t},\vec{u})$.

$|J(\vec{t},\vec{u})| \geq c\,2^{n}$.

$$\left|\begin{array}{ccccccc}
1 & 1 & 0 & 0 & \cdots & 0 & 0 \\
0 & 0 & 1 & 1 & \cdots & 0 & 0 \\
\vdots & \vdots & \vdots & \vdots & & \vdots & \vdots \\
0 & 0 & 0 & 0 & \cdots & 1 & 1 \\
\sum_{k=1}^{n} a^{1}_{1k} t_k & \sum_{k=1}^{n} a^{1}_{1k} u_k & \sum_{k=1}^{n} a^{1}_{2k} t_k & \sum_{k=1}^{n} a^{1}_{2k} u_k & \cdots & \sum_{k=1}^{n} a^{1}_{nk} t_k & \sum_{k=1}^{n} a^{1}_{nk} u_k \\
\vdots & \vdots & \vdots & \vdots & & \vdots & \vdots \\
\sum_{k=1}^{n} a^{n}_{1k} t_k & \sum_{k=1}^{n} a^{n}_{1k} u_k & \sum_{k=1}^{n} a^{n}_{2k} t_k & \sum_{k=1}^{n} a^{n}_{1k} u_k & \cdots & \sum_{k=1}^{n} a^{n}_{nk} t_k & \sum_{k=1}^{n} a^{n}_{nk} u_k
\end{array}\right|_{(2n,2n)} = c2^{n}.$$

$$\left|\begin{array}{cccc}
\sum_{k=1}^{n} a^{1}_{1k}(t_k - u_k) & \sum_{k=1}^{n} a^{1}_{2k}(t_k - u_k) & \cdots & \sum_{k=1}^{n} a^{1}_{nk}(t_k - u_k) \\
\vdots & \vdots & & \vdots \\
\sum_{k=1}^{n} a^{n}_{1k}(t_k - u_k) & \sum_{k=1}^{n} a^{n}_{2k}(t_k - u_k) & \cdots & \sum_{k=1}^{n} a^{n}_{nk}(t_k - u_k)
\end{array}\right|_{(n,n)} =$$

$$= c2^{n}\,|\tilde{J}(\vec{t} - \vec{u})|.$$

The equality between the two determinants above is obtained subtracting from each even column the next one. Now we go back to (4) and we set $\vec{v} = \vec{t}$, $\vec{w} = \vec{t} - \vec{u}$. Then

$$\|T^{*}f\|^{2}_{L^{s}(\mathbb{R}^{2n})} \leq \left(\int_{\mathcal{U}\times\mathcal{U}} \frac{F(\vec{v})\,F(\vec{v} - \vec{w})\, d\vec{v}\, d\vec{w}}{|\tilde{J}(\vec{w})|^{a}} \right)^{s-2/s}$$

Since $\int_{\mathcal{U}} \frac{d\vec{w}}{|\tilde{J}(\vec{w})|^{a}} < \infty$ for $a < p_0$ then

$$\|T^*f\|^2_{L^s(\mathbb{R}^{2n})} \le C_s \|F\|^{2(s-2)/s}_{L^2(\sigma)} = C_s \|f\|^2_{L^{2s/s-2}(\sigma)}$$

when $2/s-2 < p_0$. By duality we have the inequality

$$\|\hat{f}\|_{L^{2p/3p-1}(\sigma)} \le C_p \|f\|_{L^p(\mathbb{R}^{2n})}$$

when $1 \le p < \frac{2(1+p_0)}{2+p_0}$. The first part of the theorem is proved interpolating this estimate with $p = \frac{2(1+p_0)}{2+p_0} - \varepsilon$ $(\varepsilon > 0)$ and the trivial one $\|\hat{f}\|_{L^\infty(\sigma)} \le \|f\|_{L^1(\mathbb{R}^{2n})}$.

We now give a counterexample following an idea of T. Knapp. We take $f \in C_0^\infty(\mathbb{R}^{2n})$ such that its Fourier transform is supported on a parallelopiped P centered at the origin of dimensions $(\delta, \ldots, \delta, \delta^2, \ldots, \delta^2)$ along the $x_1, \ldots, x_n, x_{n+1}, \ldots, x_{2n}$ axis respectively and such that $\hat{f}$ is identically 1 on the parallelopiped obtained by P under dilation by $\frac{1}{2}$. Then

$$\|\hat{f}|_\sigma\|_q \approx \delta^{(n(n+1)/2)\cdot 1/q} \quad \text{and} \quad \|f\|_p \approx \delta^{(3n(n+1)/2)(1-1/p)}.$$

Hence (2) does not hold if $\frac{1}{q} < 3(1-\frac{1}{p})$.

Finally if in (3) we take $s=4$ and we apply Plancherel's theorem we see that the desired inequality does not hold if $p \ge \frac{4}{3}$ since $\int_{\mathcal{U}} \frac{d\vec{w}}{|\tilde{J}(\vec{w})|} = \infty$. □

Note that for $n=1$ Theorem 1 gives the sharp result of [1]. In the particular case of a manifold of dimension 2 in $\mathbb{R}^4$ we will compute explicitly $\tilde{J}$ to understand the geometrical implications of the fact that $|\tilde{J}|^{-a}$ can be locally integrable for different values of a. We go back to (1) and we write $a_{ik} = a^1_{ik}$, $b_{ik} = a^2_{ik}$. We suppose that $\left\{\frac{\partial \vec{X}(\vec{0})}{\partial x_i}, \frac{\partial^2 \vec{X}(\vec{0})}{\partial x_1^2}\right\}$ $i=1,2$ are linearly independent (the other cases can be worked out similarly). Then the equation of σ is

$$\begin{cases} X_i = x_i + \mathcal{O}(\|\vec{x}\|^2 & i=1,2 \\ X_3 = a_{11}x_1^2 + {}_\llcorner a_{12}x_1x_2 + \mathcal{O}(\|\vec{x}\|^3) \\ X_4 = 2b_{12}x_1x_2 + b_{22}x_2^2 + \mathcal{O}(\|\vec{x}\|^3) \end{cases}$$

with $a_{11} \neq 0$ and $b_{22} \neq 0$. Hence $\tilde{J}(x_1, x_2)/4 = a_{11}b_{12}x_1^2 + a_{11}b_{22}x_1x_2 + a_{12}b_{22}x_2^2$. Let us denote by $\Delta = a_{11}^2b_{22}^2 - 4a_{11}a_{12}b_{12}b_{22}$. Then $\Delta \neq 0$ implies $\int \frac{d\vec{x}}{|\tilde{J}(\vec{x})|^a} < \infty$ if and only if $a < 1$, $\Delta = 0$ implies the same inequality if and only if $a < \frac{1}{2}$. If $\Delta = 0$ then $a_{11}b_{22} = 4a_{12}b_{12}$ and so the last two equations defining σ become $X_3 = a_{11}x_1(x_1 + \alpha x_2)$, $X_4 = b_{12}x_2(x_1 + \alpha x_2)$ where $\alpha = 2\alpha_{12}/a_{11} = b_{22}/2b_{12}$. Therefore the intersection of σ with the hyperplane $X_1 + \alpha X_2 = 0$ is a curve in $\mathbb{R}^3$ with zero curvature at the origin. Conversely if there exists a hyperplane $X_1 + \alpha X_2 = 0$ $(X_2 + \alpha X_1 = 0)$ such that its intersection with σ is a curve with zero curvature at the origin then if $\alpha = 0$ we have $b_{22} = 0$ $(a_{11} = 0)$, a contradiction; if $\alpha \neq 0$ we have $\alpha a_{11} + 2a_{12} = 0$ $(a_{11} + 2\alpha a_{12} = 0)$ and $2\alpha b_{12} + b_{22} = 0$ $(2b_{12} + \alpha b_{22} = 0)$ i.e., $\Delta = 0$.

3. The Case of Curves in $\mathbb{R}^n$ $(n \geq 2)$. We begin with the following

LEMMA. Let $J(x_1, \dots, x_n) = \prod_{i=2}^{n} x_i \prod_{\substack{i>j \\ i,j=2}}^{n} (x_i - x_j)$ be defined in a neighborhood of the origin $\mathcal{U}^{(n)} = \{\vec{x} \in \mathbb{R}^n : 0 < x_1 < x_2 < \cdots < \delta\}$ $(\delta > 0)$. Then $|J|^{-a}$ is integrable in $\mathcal{U}^{(n)}$ if and only if $a < \frac{2}{n}$ where $n \geq 2$.

PROOF. We write

$$I_2 = \int_{\mathcal{U}^{(n)}} |J(\vec{x})|^{-a} \, d\vec{x} = \delta \int_{\mathcal{U}^{(n-1)}} \frac{dx_3 \dots dx_n}{|\prod_{i=3}^{n} x_i \prod_{\substack{i>j \\ i,j=3}}^{n} (x_i - x_j)|^a} \int_0^{\delta} \frac{dx_2}{|x_2 \prod_{i=3}^{n} (x_i - x_2)|^a}$$

To evaluate $\int_0^{\delta} \frac{dx_2}{|x_2 \prod_{i=3}^{n} (x_i - x_2)|^a}$ we break the interval of integration into three parts $(0, \frac{x_3}{2})$, $(\frac{x_3}{2}, 2x_3)$, $(2x_3, \delta)$. Then I_2 is the sum of three integrals. We will evaluate only the first one, that we denote by I_3. The estimate for the other two terms are similar. We have

$$I_3 = c_3\delta\int_{\mathcal{U}^{(n-2)}} \frac{dx_3 \ldots dx_n}{\left|x_3^{3a-1} \prod_{i=4}^{n} x_i^a \prod_{i=4}^{n} (x_i - x_3)^{2a} \prod_{\substack{i>j\\ i,j=4}}^{n} (x_i - x_j)^a\right|}$$

where c_3 is a suitable constant. As before we break I_3 into the sum of three integrals. Call the first one I_4 and so on. Finally at the last step we obtain

$$I_n = c_n \delta \int_0^{\delta} \frac{dx_n}{x_n^{n(n-1)a/2 - (n-2)}}$$

where c_n is a constant depending only on n. This proves the lemma.

THEOREM 2. Let γ be a curve in $\mathbb{R}^n$ of class C^k $(k \geq n)$ having equation $\vec{X} = \vec{X}(t)$ $0 \leq t \leq 1$. Suppose that for every $t_0 \in [0,1]$ the set of vectors $\left\{\frac{\partial^k \vec{X}(t)}{\partial t^k}\Big|_{t=t_0}\right\}$ $k=1,\ldots,n$ are linearly independent. Then

$$\|\hat{f}\|_{L^q(\gamma)} \leq C_p \|f\|_{L^p(\mathbb{R}^n)} \tag{5}$$

whenever $\frac{1}{q} \geq \frac{n(n+1)}{2}\left(1 - \frac{1}{p}\right)$ and $1 \leq p < \frac{n^2 + 2n}{n^2 + 2n - 2}$. The inequality need not hold if $\frac{1}{q} < \frac{n(n+1)}{2}\left(1 - \frac{1}{p}\right)$ or $p \geq \frac{2n}{2n-1}$.

PROOF. Let P_0 be any point on γ. We may assume P_0 to be the origin and $\vec{X}(0) = \vec{0}$. Then in a neighborhood of the origin γ has equation

$$X_i(t) = a_i t^i + \mathcal{O}(t^{i+1}) \qquad i=1,\ldots,n$$

where $\prod_{i=1}^{n} a_i \neq 0$ and $0 < t < \delta$ $(\delta > 0)$. Let T^* be the adjoint of the restriction operator. We consider

$$(T^*f)^n(\vec{\xi}) = \int_{\mathcal{U}^{(n)}} e^{i\left(\xi_1 \sum_{i=1}^{n} X_1(t_i) + \cdots + \xi_n \sum_{i=1}^{n} X_n(t_i)\right)} f(t_1)\cdots f(t_n)\,dt_1 \cdots dt_n \,.$$

We change variables in the following way $u_j = \sum_{i=1}^{n} X_j(t_i)$ $j=1,\dots,n$ and we compute the Jacobian of the transformation using the Vandermonde determinant.

$$|J(t_j,\dots,t_n)| \geq c\left|\prod_{i=1}^{n} ia_i\right| \begin{vmatrix} 1 & 1 & \dots & 1 \\ t_1 & t_2 & \dots & t_n \\ t_1^2 & t_2^2 & \dots & t_n^2 \\ \vdots & \vdots & & \vdots \\ t_1^{n-1} & t_2^{n-1} & \dots & t_n^{n-1} \end{vmatrix} = b\left|\prod_{\substack{i>j\\ i,j=1}}^{n} (t_i - t_j)\right|$$

where c is a suitable constant and $b = c\left|\prod_{i=1}^{n} ia_i\right|$. We change variables again by setting $x_1 = t_1$, $x_i = t_1 - t_i$ $i=2,\dots,n$. Then

$|J(x_1,\dots,x_n)| \geq b\left|\prod_{i=2}^{n} x_i \prod_{\substack{i>j\\ i,j=2}}^{n} (x_i - x_j)\right|$. Therefore the preceding lemma and computations similar to those in [2] give

$$\|T^*f\|_{L^s(\mathbb{R}^n)} \leq C_s \|f\|_{L^{ns/s-n}(\gamma)} \tag{6}$$

whenever $\frac{n}{s-n} < \frac{2}{n}$ i.e., $s > \frac{n(n+2)}{2}$. Hence

$$\|\hat{f}\|_{L^q(\gamma)} \leq C_p \|f\|_{L^p(\mathbb{R}^n)}$$

if $p < p_0 = \frac{n(n+2)}{n^2+2n-2}$ and $q = \frac{np}{(2n-1)p-n}$. Inequality (5) is proved interpolating this estimate with $p = p_0 - \varepsilon$ $(\varepsilon > 0)$ and the trivial one $\|\hat{f}\|_{L^\infty(\gamma)} \leq \|f\|_{L^1(\mathbb{R}^n)}$.

To show that (5) does not hold if $\frac{1}{q} < \frac{n(n+1)}{2}\left(1 - \frac{1}{p}\right)$ we consider a function $f \in C_0^\infty(\mathbb{R}^n)$ such that $\hat{f}$ is supported on a parallelopiped centered at the origin, of dimensions $\delta, \delta^2, \dots, \delta^n$ along the directions of the Frenet frame of γ at the origin. Moreover choose f such that $\hat{f}$ is identically 1 on the dilation by $\frac{1}{2}$ of the above parallelopiped. Then $\|\hat{f}\|_q \approx \delta^{1/q}$ and $\|f\|_p \approx \delta^{(n(n+1)/2)(1-\frac{1}{p})}$. This shows that $\frac{1}{q} > \frac{n(n+1)}{2}(1-\frac{1}{p})$ must be satisfied for (5) to hold. Finally if in (6) we take $s = 2n$ and we use Plancherel's theorem we see that T^*f does not belong to $L^{2n}(\mathbb{R}^n)$, hence T is not a bounded operator from $L^p(\mathbb{R}^n)$ to $L^q(\gamma)$ for any q if $p \geq \frac{2n}{2n-1}$. □

Note that in [1] Theorem 2 is proved for $n=2$ and in [2] for $n=3$.

REFERENCES

1. C. Fefferman, Inequalities for strongly singular convolution operators, Acta Math. 124 (1970), 9-35.
2. E. Prestini, A restriction theorem for space curves, Proc. Amer. Math. Soc. 70 (1978), 8-10.
3. __________, On the phase problem for space curves, Boll. Un. Mat. It. (to appear).
4. A. Zygmund, On Fourier coefficients and transforms of functions of two variables, Studia Math. 50 (1974), 189-201.
5. L. Hormander, Oscillatory integrals and multipliers on FL^p, Ark. Math. 11 (1973), 1-11.
6. L. Carleson and P. Sjolin, Oscillatory integrals and a multiplier problem for the disc, Studia Math. 44 (1972), 287-299.
7. P. Sjolin, Fourier multipliers and estimates of Fourier transform of measures carried by smooth curves in $\mathbb{R}^2$, Studia Math. 51 (1974), 169-182.
8. P. Tomas, A restriction theorem for the Fourier transform, Bull. Am. Math. Soc. 81 (1975), 476-478.
9. R. S. Strichartz, Restrictions of Fourier transforms to quadratic surfaces and decay of solutions of wave equations, Duke Math. J. 44 (1977), 705-713.
10. D. Finch, Ph.D. Thesis, M.I.T. (1977).

Università di Milano, Italia

Proceedings of Symposia in Pure Mathematics
Volume XXXV, Part 1, 1979

RESTRICTION THEOREMS FOR THE FOURIER TRANSFORM

Peter A. Tomas*

We shall discuss a result of Stein [2] and some of its extensions. If $n > 1$ there is a $p(n)$, $1 < p(n) < 2$ such that for every Schwartz function f,

$$(*) \qquad \int_{S^{n-1}} |\hat{f}(\theta)|^2 d\theta \le C\|f\|^2_{p(n)} .$$

Once we know results like this <u>can</u> exist, it is rather simple to find the optimal $p(n)$. We note first that in the given range of p, $\hat{f}$ is defined merely a.e., and we are restricting it to a set of measure zero; for this reason inequalities such as (*) are called "restriction" theorems. S^{n-1} possesses some curvature to distinguish it from arbitrary sets of measure zero (restriction to a straight line segment clearly implies restriction of one-dimensional Fourier transforms to points; this amounts to defining $\hat{f}(0) = \int f$ for non-integrable f).

The role of curvature may be clarified by using it to find those (p,q) for which $\|\hat{f}|_{S^1}\|_q \le C\|f\|_p$. These were originally determined by C. Fefferman and E.M. Stein (unpublished) using spherical harmonics, but A.W. Knapp (unpublished) discovered the following argument which clarifies the role curvature plays. As in the proof that the indices in the Hausdorff-Young inequality are optimal, Knapp uses homogeneity ideas. Let $\hat{f}_\delta$ denote the characteristic function of a rectangle centered at $(1,0)$ and of dimensions $\delta \times \delta^k$. We write $\hat{f}_\delta(x,y) = \hat{f}_1(x/\delta, y/\delta^k)$ (this is accurate up to translations, which do

*Supported in part by a grant from the National Science Foundation.

not affect $\|f_\delta\|_p = C\delta^{\frac{1+k}{p'}}$. To minimize $\|f_\delta\|_p$ and keep $\hat{f}_\delta|_{S'}$ large, we choose k so that $\hat{f}_\delta|_{S'}$ intersects the circle in an arc of length δ, but so that the rectangle has minimal area. As the circle behaves like $y = \sqrt{1-x^2} \sim 1 - x^2/2$, when x is small, $k = 2$. Then $C\delta^{1/q} \le C_p\delta^{3/p'}$ or $p' \ge 3q$. Knapp's reasoning applies to very general surfaces in R^n; in particular for S^{n-1}, $p' \ge \frac{n+1}{n-1}q$ (if $p = 2$, $p(n) = \frac{2(n+1)}{n+3}$). But this cannot be the entire answer; we must have some requirement $p \le 2$. In fact, p must be much less than 2. By examining the adjoint of the restriction operator, we see restriction implies that for g in $L^{q'}(S^{n-1})$, $\|\widehat{gd\theta}\|_{p'} \le C\|g\|_{q'}$. If $g \equiv 1$, we have $\|\widehat{d\theta}\|_{p'} < \infty$; $\widehat{d\theta}$ may be computed explicitly as

$$\frac{J_{\frac{n-2}{2}}(2\pi|x|)}{|x|^{n-2/2}}$$

which is in $L^{p'}$ iff $1 \le p < 2n/n+1$. Restriction theorems are valid for non-compact surfaces as well; Cordoba and Stein [3] established the result for cones in R^3; Strichartz [5] independently for more general quadratic surfaces. In such circumstances arguments other than the above are necessary. Cruder results can be given if we realize the size of $\widehat{d\theta}$ is a reflection of its curvature: let $\hat{f}_j$ denote $\hat{f}_\delta$ above, but rotated by an angle of $\arcsin(j\delta)$; as j varies from 0 to δ^{-1}, the $\hat{f}_j$ cover S' disjointly, and $\|\Sigma\hat{f}_j|_{S'}\|_q \ge 1$. To compute f_j, note they are "essentially" supported in a rectangle of size $\delta^{-1} \times \delta^{-2}$, and therefore a simple computation of angles shows that the supports of the f_j are "essentially" disjoint. Then $C \le \|\Sigma\widehat{f_j}\|_p^p = \Sigma\int |f_j|^p \le \Sigma\delta^{3p-3} = \delta^{3p-4}$; as δ is small, $1 \le p \le 4/3$ is required. This idea comes from Cordoba [1] where it is made rigorous.

We have seen that the size of $\widehat{d\theta}$ controls the existence of restriction theorems; a result of Stein [3] shows that size alone implies restriction:

$$\int |\hat{f}(\theta)|^2 d\theta = \int \widehat{f * \tilde{f}}\, d\theta = \int f * \tilde{f}\, \widehat{d\theta}$$
$$\le \|\widehat{d\theta}\|_{S'} \|f * \tilde{f}\|_S \le C\|f\|_p^2 \quad \text{if}$$

$$s' > \frac{2n}{n-1} \quad \text{and} \quad 1+\frac{1}{s} = \frac{2}{p'}, \quad \text{or} \quad 1 \le p < \frac{4n}{3n+1} .$$

(We remark that the use of L^2 here is critical: to date no restriction theorems have been established without using the Plancherel theorem.)

$\frac{4n}{3n+1}$ tends to 4/3 as n tends to infinity, while the conjectured optimal result, $\frac{2(n+1)}{n+3}$, tends to 2. To improve the above result, we must avoid using Holder's inequality on $\widehat{d\theta}$. Interchanging integrals,

$$\int f * \tilde{f}\, \widehat{d\theta} = \int f\, \widehat{d\theta} * \tilde{f} \le \|f\|_p \|\widehat{d\theta} * \tilde{f}\|_p ,$$

and this is bounded by $C\|f\|_p^2$ if convolution with $\widehat{d\theta}$ is bounded from L^p to $L^{p'}$. If we bound $|\widehat{d\theta}(x)| \le |x|^{1-n/2}$ and use the theory of fractional integration, we get only the range $1 \le p \le \frac{4n}{3n+1}$; we still have not utilized the cancellation in $\widehat{d\theta}$. We must also analyze convolution operators from L^p to $L^{p'}$. Recall $\|k * f\|_\infty \le \|k\|_\infty \|f\|_1$, and $\|k * f\|_2 = \|\hat{k}\hat{f}\|_2 \le \|\hat{k}\|_\infty \|f\|_2$; both of these results are optimal. $\widehat{d\theta}$ is in L^∞, but $d\theta$ is not; other ideas are needed. There are two standard approaches to the analysis of operators on L^p, $1 < p < 2$; the first is Riesz-Thorin interpolation, the second is complex interpolation (see Stein and Weiss [4]). The first method was employed by the author in [6] to get $1 \le p < \frac{2(n+1)}{n+3}$, but the method inherently is inaccurate at endpoints.

E.M. Stein (unpublished) employed complex interpolation ideas to yield the endpoint $\frac{2(n+1)}{n+3}$. The essence of Stein's proof follows.

We see that not only $\widehat{d\theta}$, but $|x|^{1-n/2}\widehat{d\theta}$ is in L^∞; we can worsen $\widehat{d\theta}$ considerably and still get an operator bounded from L' to L^∞. At L^2, we must improve $\widehat{d\theta}$ considerably to get a bounded Fourier transform. In between lies $\widehat{d\theta}$, which presumably has boundedness from L^p to $L^{p'}$ for p between 1 and 2. Stein chooses a complex analytic family of operators given by convolution with

$$K_z(x) = C_z \frac{J_{\frac{n-2}{2}+z}(2\pi|x|)}{|x|^{n-2/2}+z}$$

To get L' to L^∞, we choose z as negative as possible -- intuitively $z = \frac{1-n}{2}$ is correct -- and indeed, this is the last time K_z is in L^∞. We also want to find z as close to zero as possible for K_z to be bounded; $K_z(q) = (1-|q|^2)_+^{z-1}$. We must choose $z = 1$, and by interpolation we find K_z maps L^p to $L^{p'}$ for $z = \frac{n+3}{2} - \frac{n+1}{p}$. At $z = 0$, $\widehat{d\theta}$ maps L^p to $L^{p'}$ for $p = \frac{2(n+1)}{n+3}$, which is optimal.

Bibliography

1. A. Cordoba, The Kakeya maximal function and the spherical summation multipliers, Thesis, University of Chicago, 1974.

2. C. Fefferman, Inequalities for strongly singular convolution operators, Acta Math., 124(1970), 9-36.

3. E.M. Stein, Lecture notes, Princeton University, 1976.

4. E.M. Stein and G. Weiss, Introduction to Fourier Analysis on Euclidean Spaces, Princeton Univ. Press, Princeton, N.J., 1971.

5. R. Strichartz, Restrictions of Fourier transforms to quadratic surfaces and decay of solutions of wave equations, to appear, Duke J. Math.

6. P.A. Tomas, A restriction theorem for the Fourier transform, Bulletin AMS, 81(1975), 477-478.

DEPARTMENT OF MATHEMATICS, THE UNIVERSITY OF TEXAS AT AUSTIN, AUSTIN, TEXAS, 78712.

Proceedings of Symposia in Pure Mathematics
Volume XXXV, Part 1, 1979

RIESZ POTENTIALS AND FOURIER MULTIPLIERS

Richard J. Bagby

ABSTRACT: Continuity properties of the multiplier operator T_m defined by $(T_m f)^\wedge = m\hat{f}$ can be inferred if m is a convolution product with factors in certain Lorentz spaces. In particular, if $m = I_\alpha g$ is a Riesz potential, then the norm of g in a Lorentz space may be used to bound T_m. A useful characterization of Riesz potentials is presented, as well as a method for inverting the Riesz potential operator which gives an a priori bound for the inverse operator.

For $0 < \alpha < n$, the Riesz potential operator in R^n may be defined either by an integral

$$I_\alpha f(x) = c_{n,\alpha} \int f(y)|x-y|^{\alpha-n} dy$$

or by means of Fourier transforms

$$(I_\alpha f)^\wedge(x) = |x|^{-\alpha} \hat{f}(x)$$

for suitably restricted functions f. The standard Hardy-Littlewood-Sobolev theorem (see Stein [9] for a simple proof) states that I_α maps $L^p(R^n)$ into $L^q(R^n)$ continuously for $1 < p < q < \infty$ and $1/p - 1/q = \alpha/n$. A sharper version may be given in terms of Lorentz spaces: I_α is continuous from $L(p,r)$ into $L(q,r)$ for p and q as above and $1 \le r \le \infty$ as well as from $L(n/\alpha,1)$ into L^∞. See Hunt [7].

Fourier multipliers are operators of the form $(Tf)^\wedge = m\hat{f}$. Hörmander [6] proved that Fourier multipliers can be continuous operators from L^p to L^q only when $p \le q$, and that only when $p \le 2 \le q$ can a condition $|m| \le F$ imply continuity. In cases with $p > 2$ or $q < 2$ sufficient conditions on m for continuity of the multiplier operator have been obtained by a number of authors.

In [1] we proved that certain classes of convolution products give rise to continuous multiplier operators. Since the kernel for I_α is in $L(n/(n-\alpha),\infty)$, we obtain

---

AMS(MOS) subject classifications: Primary 26A33, 42A18, 46E30; Secondary 42A96.

THEOREM 1. For $m = I_\alpha f$, the Fourier multiplier corresponding to m is a continuous operator from $L^p(R^n)$ to $L^q(R^n)$ provided either

(a) $f \in L(n/\alpha, 1)$ and $p = q$ where $1 < p < \infty$ with $|1/p - 1/2| < \alpha/n$

(b) $f \in L(r,\infty)$ and $1/p - 1/q = 1/r - \alpha/n$ where $1 < p < q < \infty$ with both $1/2 - 1/q < \alpha/n$ and $1/q - 1/2 < \alpha/n$.

In order to use this result, we need a characterization of Riesz potentials. This was our goal in [2]. Following Strichartz [10], we defined

$$S_\alpha^{(k)} f(x) = \left\{\int_0^\infty \left[\int_{|y|\le 1} \Delta_{ry}^k f(x)\, |dy\right]^2 r^{-2\alpha-1} dr\right\}^{1/2}$$

where $$\Delta_{ry}^k f(x) = \sum_{j=0}^\infty C(k,j)(-1)^{k-j} f(x + jry)$$

is the usual k-th order difference and $0 < \alpha < k$. We proved the following:

THEOREM 2. Suppose $g \in L(p,r)$ and either $0 < \alpha/n < 1/p < 1$ with $1 \le r \le \infty$ or $0 < \alpha/n = 1/p < 1$ with $r = 1$. Then for $f = I_\alpha g$, $\|S_\alpha^{(k)} f\|_{p,q}$ and $\|g\|_{p,q}$ are comparable.

In order to use Theorem 1, we need a converse to Theorem 2 which we state below:

THEOREM 3. Suppose $S_\alpha^{(k)} f \in L(p,r)$ where either $0 < \alpha/n < 1/p < 1$ with $1 \le r \le \infty$ or $0 < \alpha/n = 1/p < 1$ with $r = 1$. Suppose further that f vanishes at infinity in the sense that

$$\lim_{x\to\infty} \int_{|y|\le 1} |f(x+y)|\, dy = 0 .$$

Then $f = I_\alpha g$ for some $g \in L(p,r)$.

The problem here is that $S_\alpha^{(k)} f$ is not g and does not tell us how to find it. Presumably g could be obtained from a hypersingular integral as in Stein [8], but the integral for $S_\alpha^{(k)} f$ does not seem to control the corresponding hypersingular integral. In order to prove Theorem 3 we develop a new formula for the inverse of I_α which is controlled by $S_\alpha^{(k)} f$. We sketch the proof below; details will appear in [3]. Our formula is based on a formula of Herz [5] for inverting difference operators in Lipschitz spaces.

Fix $k > \alpha$ and choose $\phi \in C_0^\infty(0,\infty)$ with

$$\int (e^{2\pi i y_1} - 1)^k \phi(|y|)\, dy = 1 . \tag{1}$$

Define K on $R^n \times R$ by

$$K(x,r) = -r^{n+1} \int e^{2\pi i x\cdot z} |z|^{n+1} \phi'(r|z|)\, dz \tag{2}$$

and an operator T by

(3) $$Tf(x) = \int_0^\infty r^{-1}dr \int K(x-z,r)dz \int_{|y|\le 1} \Delta_{ry}^k f(z)dy \ .$$

Taking the Fourier transform in R^n,

$$\hat{K}(x,r) = -r^{n+1}|x|^{n+1}\phi'(r|x|) \ .$$

Since ϕ' vanishes near 0 and ∞ we have $K(\cdot,r) \in S(R^n)$. Since also $K(x,r) = r^{-n}K(x/r,1)$ we may easily prove

(4) $$|K(x,r)| \le Cr^{-\alpha}|x|^{\alpha-n}\min[(|x|/r)^\varepsilon,(r/|x|)^\varepsilon] \ .$$

Hence

(5) $$\int_0^\infty |K(x,r)|^2 r^{2\alpha-1}dr \le C|x|^{2(\alpha-n)} \ .$$

From (3), we see

$$|Tf(x)| \le \int dz \int_0^\infty (r^\alpha|K(x-z,r)|)(r^{-\alpha}\int_{|y|\le 1}|\Delta_{ry}^k f(z)|dy)r^{-1}dr \ .$$

Schwartz's inequality and (5) yields

(6) $$|Tf(x)| \le c\int |x-z|^{\alpha-n}S_\alpha^{(k)}f(z)dz = c\ I_\alpha S_\alpha^{(k)}f(x) \ .$$

From (6) and our assumptions on $S_\alpha^{(k)}f$ we see that the iterated integral defining Tf is absolutely convergent a.e., and that Tf vanishes at infinity in the same sense as f.

Formally the Fourier transform of Tf is given by

$$\begin{aligned}(Tf)^\wedge(x) &= \int_0^\infty \hat{K}(x,r)r^{-1}dr \int_{|y|\le 1}(\Delta_{ry}^k f)^\wedge(x)dy \\ &= -\hat{f}(x)\int_0^\infty (r|x|)^{n+1}\phi'(r|x|)dr \int_{|y|\le 1}(e^{2\pi i r y\cdot x}-1)^k dy \\ &= \hat{f}(x) \qquad \text{by (2) and (1)} \ .\end{aligned}$$

On a rigorous level, if $\psi \in C^\infty(R^n)$ with compact support not containing the origin, then

$$\int Tf(x)\hat{\psi}(x)dx = \int f(x)\hat{\psi}(x)dx$$

so that $(Tf)^\wedge$ and $\hat{f}$ can only differ by a distribution supported at the origin, and hence $Tf - f$ is a polynomial. Since both Tf and f vanish at infinity, we have in fact $Tf = f$.

This representation for f suggests our formula for the inverse Riesz potential of f. Set

(7) $$K_\alpha(x,r) = -r^{n+1+\alpha}\int e^{2\pi i x\cdot z}|z|^{n+1+\alpha}\phi'(r|z|)dz$$

so that $K(x,r) = r^{-\alpha}I_\alpha K_\alpha(x,r)$. Our method for inverting the Riesz potential is given below.

THEOREM 4. Let f satisfy the hypotheses of Theorem 3, and let

$$D_m^\alpha f(x) = \int_{1/m}^m r^{-\alpha-1}dr \int K_\alpha(x-z,r)dz \int_{|y|\le 1} \Delta_{ry}^k f(z)dy \ .$$

Then (i) The iterated integral for $D_m^\alpha f(x)$ converges absolutely a.e.

(ii) $\lim_{m\to\infty} I_\alpha D_m^\alpha f(x) = f(x)$ a.e.

(iii) There is a constant C depending only on n, k, α, and p such that $\|D_m^\alpha f\|_{p,r} \le C\|S_\alpha^{(k)} f\|_{p,r}$.

(iv) There is a function $D^\alpha f \in L(p,r)$ such that $\lim_{m\to\infty} \|D_m^\alpha f - D^\alpha f\|_{p,r} = 0$.

Part (i) is fairly simple once we note that K_α satisfies estimates of the same type that K does; we need only repeat earlier arguments. Part (ii) then follows immediately from our representation of f. To prove (iii) and (iv), we make use of estimates for vector-valued singular integral operators as found in Benedek, Calderón, and Panzone [4].

Fix a,b with $0 < a < b < \infty$, and let H be the Hilbert space $L^2((a,b), r^{-1}dr)$. For each $x \in R^n$, we define a continuous linear functional $\Lambda(x)$ on H by

$$\Lambda(x)u = \int_a^b K_\alpha(x,r)u(r)r^{-1}dr$$
$$= (u,\overline{K}_\alpha(x,\cdot))$$

and an operator A on H-valued functions by

$$AF(x) = \int \Lambda(x-z)F(z)dz \ .$$

Since $\hat{K}_\alpha(x,r) = -(r|x|)^{n+1+\alpha}\phi'(r|x|)$, $\|\hat{K}_\alpha(x,\cdot)\|_H = M < \infty$ and so by means of Fourier transforms we have $\|AF\|_2 \le M\|F\|_{L^2(H)}$. We can also show

$$\int_{|x|\ge 2|z|} \|\Lambda(x+z) - \Lambda(x)\|_H dx \le C \ .$$

It follows that A is continuous from $L^p(H)$ into L^p for $1 < p < \infty$, with norm bounded independently of a and b. Continuity extends to $L(p,r)$ immediately.

If we now take $F(x) = \int_{|y|\le 1} \Delta_{ry}^k f(x)dy$, then we have $\|F(x)\|_H \le S_\alpha^{(k)} f(x)$. Parts (iii) and (iv) follow routinely.

REFERENCES

1. R. J. Bagby, On L^p,L^q multipliers of Fourier transforms, Pacific J. Math. 68, No. 1 (1977), 1-12, M. R. 56 #3544.

2. _____, The range of the Riesz potential, submitted to Pacific J. Math.

3. _____, A method for inverting Riesz potentials, in preparation.

4. A. Benedek, A. P. Calderón, and M. Panzone, Convolution operators on Banach space valued functions, Proc. Nat. Acad. Sc. U.S.A. 48 (1962), 356-365, M. R. 24 #A3479.

5. C. S. Herz, Lipschitz spaces and Bernstein's theorem on absolutely convergent Fourier transforms, J. Math. Mech. 18 (1968), 283-323, M. R. 55 #11028.

6. L. Hörmander, Estimates for translation invariant operators in L^p spaces, Acta Math. 104 (1960), 93-140, M. R. 22 #12389.

7. R. A. Hunt, On L(p,q) Spaces, L'Ens. Math. 12 (1966), 249-275, M. R. 36 #6921.

8. E. M. Stein, The characterization of functions arising as potentials, Bull. Amer. Math. Soc. 67 (1961), 102-104, M. R. 23 #A1051.

9. _____, Singular integrals and differentiability properties of functions, Princeton University Press, Princeton, 1970, M. R. 44 #7280.

10. R. S. Strichartz, Multipliers on fractional Sobolev spaces, J. Math. Mech. 16 (1967), 1031-1061, M. R. 35 #5927.

DEPARTMENT OF MATHEMATICAL SCIENCES, NEW MEXICO STATE UNIVERSITY, LAS CRUCES, NEW MEXICO 88003.

Proceedings of Symposia in Pure Mathematics
Volume XXXV, Part 1, 1979

FOURIER MULTIPLIERS VANISHING AT INFINITY

Leonede De Michele and Ian R. Inglis

1. INTRODUCTION. It has been known for a long time that there exist singular convolution operators on $L^p(R^n)$ which fall outside the scope of the Calderon - Zygmund theory . The nuclei corresponding to multiplier transformations of the form

$$m(\xi) = \psi(\xi) \frac{e^{i|\xi|^\alpha}}{(1+|\xi|^2)^\beta} \tag{1}$$

(where $\alpha,\beta>0$ and $\psi(\xi)$ is smooth, zero on a neighbourhood of 0 and equal to 1 for large ξ) are examples of such operators, and have been studied by Hirschman [6] , Wainger [9] , Stein [8] , Fefferman [3] , and Fefferman and Stein [4] . The problem of determining the range of p for which m is bounded on L^p was completely solved in [4] . The well known result is : m is an L^p multiplier if and only if $\left|\frac{1}{p}-\frac{1}{2}\right| \leqslant \frac{2\beta}{n\alpha}$ (when α and n are not both 1) .

Using, as a starting point, an analogue of the Theorem 1 in [4] , where L^∞ - BMO estimates are given , we have developed in [2] a theory of strongly singular integral operators on spaces of homogeneous type, on which Coifman, de-Guzman and Weiss studied analogues of Calderon-Zygmund type singular integrals [1].

We obtain a theorem which allows us to pass from conditions on the nucleus defining the operator to smoothness conditions for the corresponding Fourier multipliers; of course in those situations in which some "good" dual object is available,as for R^n, SU(2) , the Heisenberg group . Moreover by this result can be obtained useful multiplier theorems which give sufficient conditions for a function to be a Fourier multiplier for p in a closed sub-interval of $(1,\infty)$.

In order to avoid heavy notatios and technicalities, here we state only the multiplier theorem for R^n ; this theorem resembles a dual version of Theorem 6 in [4] .

2. RESULTS. Let R^n denote the n-dimensional Euclidean space with the inner product denoted $\langle\cdot,\cdot\rangle$ and the Lebesgue measure dx . For $p \geqslant 1$ L^p is the usual Lebesgue space of p-integrable functions with the norm $\|\cdot\|_p$. The Fourier transform of a function in the Schwartz space S is defined by :

AMS(MOS) subject classifications (1970). Primary 42A18 .

$$\hat{f}(\xi) = \int f(x)\, e^{-i<x,\xi>}\, dx$$

A function Φ is called L^p-Fourier multiplier if the operator T_Φ defined by

$$(T_\Phi f)^{\wedge} = \hat{f}\,\Phi$$

satisfies $\| T_\Phi f \|_p \leqslant C \| f \|_p$ for all f in S. Then we have the following

THEOREM. Suppose $0 < \alpha \leqslant \theta < 1$, $M(\xi) \in C^\infty(R^n \setminus \{0\})$,

$$|M(\xi)| \leqslant \frac{1}{(1+|\xi|)^{\frac{n\alpha}{2}}}$$

and for $|\gamma| \leqslant [\frac{n}{2}] + 1$ (where γ is a multiindex)

$$|D^\gamma M(\xi)| \leqslant C\, |\xi|^{-|\gamma|} \qquad \text{if } |\xi| \leqslant 1$$

$$|D^\gamma M(\xi)| \begin{cases} C\, |\xi|^{-|\gamma| + \frac{\theta}{2}} & n \text{ odd} \\ C\, |\xi|^{-|\gamma| + \theta} & n \text{ even} \end{cases} \qquad |\xi| > 1 .$$

Then M is an L^p multiplier whenever

$$\left| \frac{1}{p} - \frac{1}{2} \right| \leqslant \begin{cases} \dfrac{(n+1)\alpha}{2(n\alpha + \theta)} & n \text{ odd} \\ \dfrac{(n+2)}{2(n\alpha + 2\theta)} & n \text{ even} \end{cases}$$

(when $\theta = \alpha$ the extreme case $p=\infty$ has to be interpreted as : M is a multiplier between L^∞ and BMO , and the dual result for $p=1$) .

REMARKS. When $\alpha = \theta$ the theorem is an extension of the classical Hormander's theorem , ($\alpha = \theta = 0$) .

By a simple evaluation of partial derivatives the " if part " of the result for the multiplier (1) can be obtained .

A similar theorem,in the context of totally disconnected L.C.A. groups, has been obtained by the second author and G.I.Gaudry in [5] .

The following corollary is an amusing variation of the classical Littlewood-Paley theorem . For sake of simplicity we treat here only the one-dimensional case .

Let

$$I_k = \begin{cases} [-2^k, -2^{k+1}) & k < 0 \\ [-1, 1) & k = 0 \\ [2^k, 2^{k+1}) & k > 0 \end{cases}$$

Suppose $0<\theta<1$ and for $k\neq 0$ split I_k into $2^{[k\theta]}$ equal subintervals $I_{k,j}$, $j = 1, 2, \ldots, 2^{[k\theta]}$.

COROLLARY. If $0<\alpha\leq\theta<1$ we have

$$\left\| \left\{ \sum_{k=-\infty}^{\infty} \frac{1}{2^{|k|\alpha}} \sum_{j=1}^{2^{[k\theta]}} |S_{k,j} f|^2 \right\}^{\frac{1}{2}} \right\|_p \leq C_{p,\alpha,\theta} \|f\|_p$$

where $(S_{I_{k,j}} f)\hat{} = \chi_{I_{k,j}} \hat{f}$ and $\left|\frac{1}{p}-\frac{1}{2}\right| \leq \frac{\alpha}{2\theta}$.

It is natural to ask if similar results hold for multipliers which behave at zero like the above multipliers behave at ∞. The second author has recently shown that the exact analogues of the above results hold for such multipliers [7]. Moreover in both the cases results for L^p-L^q multipliers can be obtained.

The proof of the above results can be found in [2] where, in the general setting of the spaces of homogeneous type, are also given applications to SU(2) and to the Heisenberg group.

BIBLIOGRAPHY

1. R.R.Coifman and G.Weiss, Analyse harmonique non-commutative sur certain espaces homogenes, Lecture notes in Math., vol. 242, Springer-Verlag, Berlin and New York, 1971.
2 L.De Michele and I.R. Inglis, L^p-estimates for strongly singular integrals on spaces of homogeneous type, preprint.
3 C. Fefferman, Inequalities for strongly singular integrals, Acta Math. 124 (1970), 9-36.
4 C.Fefferman and E.M.Stein, H^p spaces of several complex variables, Acta Math. 129 (1972), 137-193.
5 G.I.Gaudry and I.R.Inglis, Weak strong convolution operators on certain disconnected spaces, to appear in Studia Math..
6 I.I.Hirschman Jr., Multiplier transformation I, Duke Math. J., 26 (1956), 222-242.
7 I.R.Inglis, Fourier Multipliers with compact support, preprint.
8 E.M. Stein, Singular integrals, harmonic functions and differentiability properties of function of several variables, Proc. Symp. in Pure Math., 10 (1967), 316- 335
9 S. Wainger, Special trigonometric series in k dimensions, Mem. Am. Math. Soc. 59 (1965).

(LEONEDE DE MICHELE, IAN R. INGLIS, ISTITUTO DI MATEMATICA DELL'UNIVERSIA'
VIA C. SALDINI 50, 20133 MILANO, ITALY.)

Proceedings of Symposia in Pure Mathematics
Volume XXXV, Part 1, 1979

Weighted Norm Inequalities for the Littlewood-Paley Function g^*_λ

Alberto Torchinsky[1]

In this note we extend some results of Muckenhoupt and Wheeden [5] concerning the relationship between weighted L^p norms of the Littlewood-Paley function g^*_λ and some maximal functions. Our results, which hold however for arbitrary functions $u(y,t)$ defined in R^{n+1}_+, can be described as follows. Suppose that

$$M(u,x) = \sup_{\rho(x - y) \leq t} |u(y,t)|$$

is in the class $L^p_w(R^n)$, where $\rho(x)$ is the parabolic distance associated to the group $\{t^P\}_{t > 0}$, γ = trace P (see [2] or simply assume P = Id, $\gamma = n$, $\rho(x) = |x|$) and the measure $w = w(x)dx$ satisfies the A_∞ and B_∞ conditions, or $w \varepsilon A_\infty \cap B_\infty$, with respect to the balls $Q(x,t) = \{y/\rho(x - y) < t\}$. This means that w satisfies a reverse Holder inequality of order $\delta > 1$, $w \varepsilon RH_\delta$, and that $w \varepsilon B_\mu$, i.e., there is a $\mu \geq 1$ such that $w(Q(x,h))/w(Q(y,s)) \leq c(h/s)^{\gamma\mu}$ for all $Q(y,s) \subset Q(x,h)$.

For such a w and f defined in R^n we set

$$H_w(f,x) = \sup_{h > 0} w(Q(x,h))^{-1} \int_{Q(x,h)} |f(y)|w(y)dy ,$$

and if χ denotes the characteristic function of the unit interval and $\chi_{R(x,h)}$ that of the points (y,s) in R^{n+1}_+ for which $\chi(\rho(x - y)/h)\chi(s/h) = 1$, we set for $0 < p_0 < r$

$$T_{p_0,r}(u,x) = \sup_{h > 0} [h^{-\gamma r/p_0} \iint \chi_{R(x,h)} (y,s) s^{\gamma r/p_0 - \gamma} |u(y,s)|^r dy \frac{ds}{s}]^{1/r}$$
$$= \sup_{h > 0} [\Phi_{r,p_0} (u,x,h)]^{1/r}$$

AMS(MOS) subject classifications (1970). Primary 44A15, 30A78.

[1] Research partly supported by grants of the National Science Foundation and a Faculty Research Grant from Indiana University.

Our first result is the following theorem.

Theorem 1. If $p \geq \mu p_0$ and $M(u)$ is in L^p_w, then

$$T_{p_0,r}(u,x) \leq c[H_w(M(u)^{\mu p_0},x)]^{1/\mu p_0} .$$

Consequently, $||T_{p_{0,r}}(u)||_{L^p_w} \leq c\, ||M(u)||_{L^p_w}$, for $p > \mu p_0$ and $w(\{x/T_{p_0,r}(u,x) > \alpha\}) \leq c\, ||M(u)||^p_{L^p_w}/\alpha^p$ for $p = \mu p_0$.

Let now

$$S^2(u_t,x) = \int\int \chi(\rho(x - y)/t)|t\frac{\partial}{\partial t}u(y,t)|^2 t^{-\gamma} dy\frac{dt}{t}$$

and for $\lambda > 1$ define

$$g^*_\lambda(u_t,x)^2 = \int\int |t\frac{\partial}{\partial t}u(y,t)|^2(1 + \rho(x - y)/t)^{-2\lambda} t^{-\gamma} dy\frac{dt}{t} .$$

We will complement Theorem 1 with the following results.

Theorem 2. $T_{p_0,2}(u,x) \leq c(M(u,x) + g^*_\lambda(u_t,x))$, provided $\lambda = \gamma/p_0$.

Theorem 3. For $\alpha > 0$ let $0^\alpha = 0^\alpha_1 \cup 0^\alpha_2 = \{x/S(u_t,x) > \alpha\} \cup \{x/T_{2,p_0}(tu_t,x) > \alpha\}$. Then if $\lambda = \gamma/p_0$ and $w \in B_\mu$, $\mu < 2/p_0$ we have that

$$w(\{x/g^*_\lambda(u_t,x) > \alpha\}) \leq c\ (\alpha^{-2}\int_0^\alpha w(0^\beta_1)\beta d\beta + w(0^\alpha)) .$$

When $u(y,t)$ is the parabolic extension of a tempered distribution in R^n by means of convolutions with the kernel $t^{-\gamma}\exp(-\pi\,|t^{-P}x|^2)$, then the mean value inequality in [6] can be used to interpret these results in terms of the usual Littlewood-Paley function (see [2]). In this case related results have been considered in [7] where also the assumptions on w are discussed.

We pass now to the proofs of the results. The proof of Theorem 1 rests upon the following two observations which extend an inequality due to Hardy and Littlewood [2, Theorem 2.6] to the weighted case and which can be proved in an analogous way. See also [4].

Proposition 4. $|u(x,t)| \le ||M(u)||_{L^p_w} / w(B(x,t))^{1/p}$, $0 < p < \infty$.

Proposition 5. Let $d\mu$ and $d\nu$ be positive measures in R^{n+1}_+ and R^n respectively such that $\nu(R^n) = \infty$ and $\nu(Q(x,2h)) \le c\, \nu(Q(x,h))$. Then for $p_1 \ge p$, $[\int\int |u(y,s)|^{p_1}\, d\mu(y,s)]^{1/p_1} \le c[\int M(u,x)^p\, d\nu(x)]^{1/p}$ if and only if $\mu(R(x,t)) \le c\, \nu(Q(x,t))^{p_1/p}$.

The proof of Theorem 1 may be now readily obtained. Let $w \in RH_\delta$ and set $\delta = p_1/\mu p_0 > 1$. Then either $p_1 < r$ or $p_1 \ge r$. In the second case we may as well assume that $p_1 = r$, and the first case can also be reduced to this situation because by Proposition 4

$$\chi_{R(x,t)}(y,s)|u(y,s)|^r \le \chi_{R(x,t)}(y,s)|u(y,s)|^{p_1}||M(u\chi)||^{r-p_1}_{L^{\mu p_0}_w} / w(B(y,s))^{(r-p_1)/\mu p_0} .$$

Since $w \in B_\mu$ we have that

$$\Phi_{p_1,p_0}(u,x,h) \le c\, w(Q(x,h))^{-\delta} \int\int \chi_{R(x,h)}(y,s)|u(y,s)|^{p_1}\, s^{-\gamma} w(Q(y,s))^\delta dy\frac{ds}{s} .$$

Now since $w \in RH_\delta$ it is readily seen that if $d\mu(y,s) = s^{-\gamma}w(Q(y,s))^\delta dy\frac{ds}{s}$ and $d\nu(x) = w(x)dx$, then $\mu(R(x,h)) \le c\, w(Q(x,h))^\delta$. So from Proposition 5 it follows that

$$\begin{aligned}\Phi_{p_1,p_0}(u,x,h)^{1/p_1} &\le c(w(Q(x,h))^{-1} \int M(u\chi_{R(x,h)},y)^{\mu p_0} w(y)dy)^{1/\mu p_0} \\ &\le c(w(Q(x,h))^{-1} \int_{Q(x,2h)} M(u,y)^{\mu p_0} w(y)dy)^{1/\mu p_0} ,\end{aligned}$$

and $T_{p_0,p_1}(u,x) \le c(H_w(M(u)^{\mu p_0},x))^{1/\mu p_0}$ as we wished to show.

To prove Theorem 2 we may assume that $u(y,t)$ is real valued. Let $0 < \varepsilon < h$ and set $\delta = \gamma\, 2/p_0 - \gamma$. Then

$$\int_\varepsilon^h \frac{\partial}{\partial s}(u(y,s)^2)s^\delta ds = s^\delta u(y,s)^2\Big]_\varepsilon^h - \delta\int_\varepsilon^h u(y,s)^2 s^{\delta-1}\frac{ds}{s}$$
$$= \int_\varepsilon^h 2u(y,s)\frac{\partial}{\partial s}u(y,s)s^\delta ds .$$

Therefore for any $\eta > 0$ we have

$$\delta\int_\varepsilon^h u(y,s)^2 s^\delta\frac{ds}{s} \le h^\delta u(y,h)^2 - \int_\varepsilon^h 2u(y,s)\frac{\partial}{\partial s}u(y,s)s^\delta ds$$
$$\le h^\delta u(y,h)^2 + \eta\int_\varepsilon^h u(y,s)^2 s^\delta\frac{ds}{s} + \eta^{-1}\int_\varepsilon^h (s\frac{\partial}{\partial s}u(y,s))^2 s^\delta\frac{ds}{s} .$$

Whence choosing $\eta < \delta$, setting $\varepsilon = 0$, multiplying through by $h^{-\gamma 2/p_0}\chi(\rho(x-y)/h)$ and integrating it follows that

$$\Phi_{2,p_0}(u,x,h) \le c(h^{-\gamma}\int_{Q(x,h)} u^2(y,h)dy$$
$$+ \int\int \chi_{R(x,h)}(h,s)(s/h)^{\gamma 2/p_0}s^{-\gamma}(s\frac{\partial}{\partial s}u(y,s))^2 dy\frac{ds}{s} .$$

Clearly the first integral above $\le c\, M(u,x)^2$ and since also $\chi_{R(x,h)}(y,s)(s/h)^{\gamma 2/p_0} \le c(1 + \rho(x-y)/s)^{-\gamma 2/p_0}$ the second integral above $\le c\, g_\lambda^*(u_t,x)^2$. The proof of Theorem 2 is thus complete.

To prove Theorem 3 it will suffice to show that

$$I(\alpha) = \int_{R^n/2O^\alpha} g_\lambda^*(u_t,x)^2 w(x)dx \le c(\alpha^2 w(O^\alpha) + \int_0^\alpha w(O_1^\beta)\beta d\beta) .$$

Let $\{I_j\} = \{Q(x_j,h_j)\}$ be a Whitney decomposition of the open set $2O^\alpha$, let $M_j = \{(y,s)/\rho(x_j - y) + s \le h_j\}$ and set $\chi_j(y,s)$ = characteristic function of M_j . Put $U = R_+^{n+1} \setminus \bigcup_j M_j$ and denote with χ_U its characteristic function.

Then we have

$$I(\alpha) = \iint (s\tfrac{\partial}{\partial s}u(y,s))^2 s^{-\gamma} \int \chi_{R^n \setminus 2O^\alpha}(x)(1 + \rho(x-y)/s)^{-2\lambda} w(x)dx\, dy\tfrac{ds}{s}$$

$$\le c \iint \chi_U(y,s)(s\tfrac{\partial}{\partial s}u(y,s))^2 s^{-\gamma} \int (1 + \rho(x-y)/s)^{-2\lambda} w(x)dx\, dy\tfrac{ds}{s}$$

$$+ c \sum_j \iint \chi_j(y,s)(s\tfrac{\partial}{\partial s}u(y,s))^2 s^{-\gamma} \int \chi_{R^n \setminus Q(x_j,2h_j)}(x)(\rho(x-x_j)/s)^{-2\lambda} w(x)dxdy\tfrac{ds}{s}$$

$$\equiv c\, (J(\alpha) + \textstyle\sum_j J_j(\alpha)) .$$

Now since $w \in B_\mu$, $\mu < 2\lambda/\gamma$, it readily follows that

$$\int (1 + \rho(x-y)/s)^{-2\lambda} w(x)dx \le c\, w(Q(y,s)) .$$

Consequently

$$J(\alpha) \le c \iint \chi_U(y,s)(s\tfrac{\partial}{\partial s}u(y,s)^2 s^{-\gamma} w(Q(y,s))dy\tfrac{ds}{s}$$

$$\le c \int_{R^n \setminus O_1^\alpha} S^2(u_t,x)w(x)dx$$

$$\le c \int_0^\alpha w(O_1^\beta)\beta d\beta .$$

Consider now each $J_j(\alpha)$ separately. Observe that

$$\int \chi_{R^n \setminus Q(x_j,2h_j)}(x)\ \rho(x-x_j)^{-2\lambda}\ w(x)\ dx$$

$$\le c\ h_j^{-2\lambda} \int (1 + \rho(x-x_j)/h_j)^{-2\lambda} w(x)dx$$

$$\le c\ h_j^{-2\lambda}\ w(Q(x_j,h_j)) .$$

Furthermore from our definition of 0^α and a well-known property of the Whitney balls I_j we may assume that if $\chi_j(y,s) \neq 0$ there is $\bar{x}_j \in R^n \backslash 0_2^\alpha$ such that $\chi_{Q(\bar{x}_j, ch_j)}(y,s) = 1$. It is possible that $\bar{x}_j = x_j$. Thus combining this remark with the above estimate we obtain that

$$J_j(\alpha) \leq c\ w(Q(x_j,h_j))h_j^{-2\lambda} \int\int \chi_{Q(\bar{x}_j, ch_j)}(y,s)\ s^{2\lambda-\gamma}(s\frac{\partial}{\partial s}u(y,s))^2 dy\ \frac{ds}{s}$$

$$\leq c\ w(Q(x_j,h_j))\ \Phi_{2,p_0}(tu_t,\bar{x}_j,ch_j) \leq c\ w(Q(x_j,h_j))\alpha^2 .$$

Consequently $\sum J_j(\alpha) \leq c\alpha^2 \sum_j w(Q(x_j,h_j)) \leq c\alpha^2 w(0^\alpha)$ and the proof of Theorem 3 is complete.

References

[1] A.P. Calderón, "Inequalities for the maximal function relative to a metric", Stud. Math. 57 (1976), 297-306.

[2] A.P. Calderón and A. Torchinsky, "Parabolic maximal functions associated with a distribution", Advances in Math. 16 (1975), 1-64.

[3] C. Fefferman and E.M. Stein, "H^p spaces of several variables", Acta Math. 129 (1972), 137-193.

[4] R.A. Macías and C. Segovia, "Weighted norm inequalities for parabolic fractional integrals", Stud. Math. 61 (1977), 279-291.

[5] B. Muckenhoupt and R.L. Wheeden, "Norm inequalities for the Littlewood-Paley function g_λ^*", Trans. Amer. Math. Soc. 191 (1974), 95-111.

[6] J.A. Ortiz and A. Torchinsky, "On a mean value inequality", Indiana Univ. Math. J. 26 (1977), 555-566.

[7] J.-O. Strömberg and A. Torchinsky, "Weighted Hardy spaces", preprint.

School of Mathematics
Institute for Advanced Studies
Princeton, N. Jersey 08540

and

Department of Mathematics
Indiana University
Bloomington, Indiana 47405

Proceedings of Symposia in Pure Mathematics
Volume XXXV, Part 1, 1979

WEIGHTED NORM INEQUALITIES FOR MULTIPLIERS

WO-SANG YOUNG[1]

This is an outline of recent work with B. Muckenhoupt and R. L. Wheeden on multipliers for weighted L^p spaces.

Let w be a non-negative measurable function on $\mathbb{R}$. The weighted space $L^p_w(\mathbb{R})$, $1 < p < \infty$, consists of all measurable functions f such that $\int_{\mathbb{R}} |f|^p w dx < \infty$. A weight w satisfies Muckenhoupt's A_p condition [5] if

$$\left(\frac{1}{|I|} \int_I w dx\right) \left(\frac{1}{|I|} \int_I w^{-\frac{1}{p-1}} dx\right)^{p-1} \leq C$$

for all intervals I. If $w(x) = |x|^{pa}$, then it satisfies the A_p condition iff $-\frac{1}{p} < a < 1-\frac{1}{p}$. Suppose w is an A_p weight. A bounded function m is a multiplier for $L^p_w(\mathbb{R})$ if

$$(1) \qquad \int_{\mathbb{R}} |(m\hat{f})^{\vee}|^p w dx \leq C \int_{\mathbb{R}} |f|^p w dx, \qquad \forall f \varepsilon S,$$

where S is the space of Schwartz functions. By the density of S in L^p_w, the mapping $f \mapsto (m\hat{f})^{\vee}$ has a unique bounded extension to L^p_w. For weights $|x|^{pa}$, $a \geq 0$, outside the A_p range, it can be shown that (1) holds for all $f \varepsilon S$ only when m = constant a.e. The reason why this happens can be traced to the fact that $\hat{f}$ and a certain number of its derivatives are not zero at the origin. Hence, for these weights, we have to choose a smaller dense subset. For $a \geq 1-\frac{1}{p}$, we say that m is a multiplier for $L^p_{|x|^{pa}}(\mathbb{R})$ if

$$\int_{\mathbb{R}} |(m\hat{f})^{\vee}|^p |x|^{pa} dx \leq C \int_{\mathbb{R}} |f|^p |x|^{pa} dx, \qquad \forall f \varepsilon S_{00},$$

where S_{00} is the space of functions whose Fourier transforms are in C^∞, have compact support, and vanish near the origin. S_{00} is also dense in $L^p_{|x|^{pa}}$, $a \geq 0$, $1 < p < \infty$.

For $p = 2$, $a \geq \frac{1}{2}$, we identified all the $L^2_{|x|^{2a}}$ multipliers. To do this we need the following definitions. Let m be a measurable function on $\mathbb{R}$. Let $\alpha > 0$ and $k = [\alpha]$, the largest integer less than or equal

AMS (MOS) subject classification (1970). Primary 42A18.

[1]Research was partially supported by NSF grant MCS 77-03980.

to α. We say that m belongs to $H(2,\alpha)$ if it is in L^∞, its kth weak derivative in $\mathbb{R}\setminus\{0\}$, $m^{(k)}$, exists and satisfies

$$\int_{r\le|x|\le 2r}|m^{(\alpha)}(x)|^2\,dx \le C\,r^{1-2\alpha},\ r>0,\ \text{if } \alpha \text{ is an integer}, \tag{2}$$

and

$$\int_{r\le|x|\le 2r}\int_{|h|\le r/2}\frac{|m^{(k)}(x+h)-m^{(k)}(x)|^2}{|h|^{1+2(\alpha-k)}}\,dh\,dx \le C\,r^{1-2\alpha},\ r>0, \tag{3}$$

if α is not an integer. It can be shown that $H(2,\alpha)\subset H(2,\beta)$, $\beta<\alpha$; and that these spaces are the same as those defined in terms of Bessel potentials. (See [2].) We have the following

THEOREM 1. Let $2a\ge 1$, and m be a function locally integrable in $\mathbb{R}\setminus\{0\}$. Then

$$\int_{\mathbb{R}}|(m\hat{f})^\vee|^2|x|^{2a}dx \le C\int_{\mathbb{R}}|f|^2|x|^{2a}dx,\quad \forall f\in S_{00}, \tag{4}$$

iff

(i) $m\in H(2,a)$ if $2a \ne 1,3,5,\ldots$,

(ii) $m =$ constant a.e. if $2a = 1,3,5,\ldots$.

The case $a = 1$ was first observed by Coifman and Weiss [1].

It is natural to ask whether the same is true for the compact case of the circle T. m is said to be a multiplier for $L^2_{|x|^{2a}}(T)$, $a\ge\frac{1}{2}$, if

$$\int_{-\pi}^{\pi}|(m\hat{f})^\vee|^2|x|^{2a}dx \le C\int_{-\pi}^{\pi}|f|^2|x|^{2a}dx$$

for all polynomials f on T with $\hat{f}(j)=0$, $|j|=0,1,\ldots,[a+\frac{1}{2}]-1$. These functions again form a dense subset of $L^2_{|x|^{2a}}(T)$. For $2a\ne 1,3,5,\ldots$, the $L^2_{|x|^{2a}}(T)$ multipliers are the discrete analogues of $H(2,a)$. If $2a$ is an odd integer, however, there are nontrivial multipliers. It can be shown that, for $2a=1$, $\{m_k\}$ is a $L^2_{|x|}(T)$ multiplier iff it is bounded and satisfies

$$\sum_{|k|=2^n}^{\infty}\sum_{\ell=-\infty}^{\infty}\frac{|m_{k+\ell}-m_k|^2}{|\ell|^2}\le\frac{c}{n},\ n=1,2,\ldots .$$

For L^p results, Hörmander showed that every $H(2,1)$ function is an L^p_w multiplier for $w\equiv 1$ [3]. It can be shown that the same is true if w is an A_p weight. This is no longer the case for functions in $H(2,\alpha)$, $\frac{1}{2}<\alpha<1$.

THEOREM 2. Suppose $m\in H(2,\alpha)$, $\frac{1}{2}<\alpha<1$, and $1<p<\infty$. Then

$$\int_{\mathbb{R}} |(m\hat{f})^{\vee}|^p |x|^{pa} dx \leq C \int_{\mathbb{R}} |f|^p |x|^{pa} dx, \quad \forall f \in S,$$

if $\max(-\alpha, -\frac{1}{p}) < a < \min(\alpha, 1-\frac{1}{p})$.

These ranges of a are best possible except for the points $a = \pm\alpha$.

If we allow ourselves to go beyond the A_p boundary, i.e. restrict ourselves to functions in S_{00}, we obtain, for $m \in H(2,\alpha)$, $1 < p < \infty$, the full range of a for which m is a $L^p_{|x|^{pa}}$ multiplier.

THEOREM 3. Suppose $m \in H(2,\alpha)$, $\alpha > \frac{1}{2}$, and $1 < p < \infty$. Then m is a multiplier for $L^p_{|x|^{pa}}$ if $0 \leq a < \min(\alpha+\frac{1}{2}-\frac{1}{p}, \alpha)$, $a \neq j-\frac{1}{p}$, $j = 1,2,\ldots,[\alpha]$.

These ranges of a are best possible except for the end-points $a = \min(\alpha+\frac{1}{2}-\frac{1}{p}, \alpha)$. The points $a = j-\frac{1}{p}$, $j = 1,2,\ldots,$ correspond to the points $2a = 1,3,5,\ldots$ when $p=2$. It can be shown that $m(x) = \operatorname{sgn} x$ is not an $L^p_{|x|^{pa}}$ multiplier for these values of a, even though it belongs to $H(2,\alpha)$, $\forall\alpha$.

We also obtained results for multipliers whose derivatives satisfy other L^r integrability conditions. There are also results where the weights are products of power weights and A_p weights.

Proofs of these theorems and other results will be published elsewhere. See [6], [7], and [8]. In the following we shall outline the proofs of Theorems 1 and 3 for the case $\alpha = 1$ and a outside the A_p range. To avoid all technicalities we shall assume that m is differentiable in $\mathbb{R}\setminus\{0\}$.

PROOF OF THEOREM 1 FOR THE CASE $\alpha = 1$. Suppose $m \in H(2,1)$. We shall show that

$$\int_{\mathbb{R}} |(m\hat{f})'|^2 dx \leq C \int_{\mathbb{R}} |\hat{f}'|^2 dx, \quad \forall f \in S_{00}.$$

Since $\hat{f} = 0$ in a neighborhood of the origin,

$$\begin{aligned}(m\hat{f})'(x) &= m(x)\hat{f}'(x) + m'(x)\hat{f}(x)\\ &= m(x)\hat{f}'(x) + m'(x)(\textstyle\int_0^x \hat{f}'(t)dt).\end{aligned}$$

The boundedness of m yields $||m\hat{f}'||_2 \leq C\,||\hat{f}'||_2$. The L^2 norm of the second term is estimated by the weighted Hardy's inequality [4], which states that for any $U,V \geq 0$,

$$\int_0^\infty U(x) \left|\int_0^x g(t)dt\right|^2 dx \leq C \int_0^\infty V(x)|g(x)|^2 dx$$

for all measurable g iff

$$\left(\int_r^\infty U dx\right) \left(\int_0^r V^{-1} dx\right) \leq C, \quad r > 0.$$

Since we have $\int_r^\infty |m'(x)|^2 dx \leq Cr^{-1}$, we get

$$\int_{\mathbb{R}} |m'(x)|^2 \left|\int_0^x \hat{f}'(t)dt\right|^2 dx \leq C \int_{\mathbb{R}} |\hat{f}'(x)|^2 dx.$$

To prove the converse, assume that (4) holds. Let $\phi \in C^\infty$, $\phi \geq 0$, $\phi(x) = 1$ for $|x| \leq \frac{1}{4}$, $\phi(x) = 0$ for $|x| \geq \frac{1}{2}$, and $\int_{\mathbb{R}} \phi dx = 1$. Let $\phi_r(x) = \frac{1}{r}\phi(\frac{x}{r})$, $r > 0$. We first show that m is bounded. Let $x_0 \neq 0$ be a point such that $m(x_0) \neq 0$. Let $0 < R < |x_0|$ be chosen such that $|m*\phi_R(x_0)| > \frac{1}{2}|m(x_0)|$ and $|m|*\phi_R(x_0) < 2|m(x_0)|$. Define f_{x_0} by $\hat{f}_{x_0}(x) = \phi_R(x-x_0)$. Then

$$\int_{\mathbb{R}} |f_{x_0}|^2 |x|^2 dx = R^{-3} \int_{\mathbb{R}} |\phi^\vee|^2 |x|^2 dx = CR^{-3} .$$

On the other hand,

$$|(m\hat{f}_{x_0})^\vee(x)| \geq \left|\int_{\mathbb{R}} m(t)\, \hat{f}_{x_0}(t) e^{2\pi i x x_0} dt\right|$$

$$- \left|\int_{\mathbb{R}} m(t) \hat{f}_{x_0}(t) \,(e^{2\pi i x t} - e^{2\pi i x x_0}) dt\right| .$$

The first term is larger than $\frac{1}{2}|m(x_0)|$. The second term is less than $\frac{1}{4}|m(x_0)|$ if $|x| \leq \frac{1}{100R}$. Hence, for $|x| \leq \frac{1}{100R}$, $|(m\hat{f}_{x_0})^\vee(x)| \geq \frac{1}{4}|m(x_0)|$. Therefore

$$|m(x_0)|^2 \leq CR^3 \int_{|x| \leq \frac{1}{100R}} |(m\hat{f}_{x_0})^\vee(x)|^2 |x|^2 dx$$

$$\leq CR^3 \int_{\mathbb{R}} |f_{x_0}|^2 |x|^2 dx \leq C.$$

To show that m satisfies (2), define f_r, $r > 0$, by $\hat{f}_r(x) = \phi(\frac{x}{2r} - \frac{3}{4}) + \phi(\frac{x}{2r} + \frac{3}{4})$. Then

$$\int_{\mathbb{R}} |f_r|^2 |x|^2 dx = Cr^{-1}.$$

Since $\hat{f}_r(x) = 1$ for $r \leq |x| \leq 2r$, we have

$$\int_{r \leq |x| \leq 2r} |m'|^2 dx = \int_{r \leq |x| \leq 2r} |(m\hat{f}_r)'|^2 dx$$

$$\leq C \int_{\mathbb{R}} |(m\hat{f}_r)^\vee|^2 |x|^2 dx \leq C \int_{\mathbb{R}} |f_r|^2 |x|^2 dx$$

$$= Cr^{-1}$$

This completes the proof of Theorem 1 for the case $\alpha = 1$.

PROOF OF THEOREM 3 FOR THE CASE $\alpha = 1$, $a > 1 - \frac{1}{p}$. Let $\phi \in C^\infty$ with $\operatorname{supp} \phi \subset \{\frac{1}{2} < |x| < 2\}$ such that $\sum_{j=-\infty}^{\infty} \phi(2^{-j}x) = 1$, $x \neq 0$. Let $\phi_j(x) = \phi(2^{-j}x)$, and $m_j = m\phi_j$. Define $k_j = \check{m}_j$, and $K_N = \sum_{j=-N}^{N} k_j$. Since $\hat{f}$ vanishes near the origin, we have

$$(m\hat{f})^\vee(x) = \int_{\mathbb{R}} f(y)[K_N(x-y) - K_N(x)]dy$$

for some sufficiently large N. We shall need some estimates on K_N. We note that our estimates are independent of N.

LEMMA. Let $m \in H(2,1)$, $r > 0$ and $|y| < r/2$. Then

$$(5) \qquad \left(\int_{r\le|x|<2r} |K_N(x)|^p dx\right)^{1/p} \le C\, r^{1/p-1}, \quad 1 < p < \infty, \qquad \text{and}$$

$$(6) \qquad \left(\int_{r\le|x|<2r} |K_N(x-y) - K_N(x)|^p dx\right)^{1/p} \le \begin{cases} C\, r^{1/p-3/2}|y|^{1/2} & \text{if } 1 < p \le 2 \\ C\, r^{-1}|y|^{1/p} & \text{if } 2 < p < \infty \end{cases}$$

PROOF. To prove (5), we first observe that for $j \in \mathbb{Z}$,

$$(7) \qquad \left(\int_{r\le|x|<2r} |k_j(x)|^p dx\right)^{1/p} \le r^{-1} \left(\int_{r\le|x|<2r} |k_j(x)|^p |x|^p dx\right)^{1/p} \le \begin{cases} C\, r^{1/p-3/2}\, 2^{-j/2}, & 1 < p \le 2 \\ C\, r^{-1}\, 2^{-j/p}, & 2 < p < \infty \end{cases}.$$

This follows directly from Hölder's inequality and Plancherel's theorem for the case $1 < p \le 2$, and from the Hausdorff-Young inequality and Hölder's inequality for the case $2 < p < \infty$. Now the left side of (5) is majorized by

$$\left(\sum_{j=-\infty}^{-[\log_2 r]} + \sum_{j=-[\log_2 r]+1}^{\infty}\right) \left(\int_{r\le|x|<2r} |k_j(x)|^p dx\right)^{1/p}.$$

The first sum is bounded by $Cr^{1/p-1}$ since $|k_j(x)| \le \int_{\mathbb{R}} |m_j| dx \le C\, 2^j$. The second sum is also bounded by the same by (7).

To prove (6), again we estimate $\left(\int_{r\le|x|<2r} |k_j(x-y) - k_j(x)|^p dx\right)^{1/p}$. This is done as in the proof of (7), using the fact that

$$|[(-2\pi it)(k_j(t-y)-k_j(t))]^\wedge(x)| = \left|\frac{d}{dx}[m_j(x)(e^{-2\pi iyx}-1)]\right| \le |m_j'(x)|2\pi|y||x| + |m_j(x)|2\pi|y|.$$

We obtain

$$(8)\qquad \left(\int_{r\le|x|<2r}|k_j(x-y)-k_j(x)|^p dx\right)^{1/p} \le \begin{cases} C\, r^{1/p-3/2}\, 2^{j/2}|y| & \text{if } 1<p\le 2 \\ C\, r^{-1}\, 2^{j(1-1/p)}|y| & \text{if } 2<p<\infty . \end{cases}$$

To finish the proof, we majorize the left side of (6) by

$$\left(\sum_{j=-\infty}^{-[\log_2|y|]} + \sum_{j=-[\log_2|y|]+1}^{\infty}\right)\left(\int_{r\le|x|<2r}|k_j(x-y)-k_j(x)|^p dx\right)^{1/p}.$$

The first sum is bounded by the right side of (6) by (8). For the second sum, we note that

$$\left(\int_{r\le|x|<2r}|k_j(x-y)-k_j(x)|^p dx\right)^{1/p} \le 2\left(\int_{r/2\le|x|<4r}|k_j(x)|^p dx\right)^{1/p},$$

since $|y| < r/2$. Thus, by (7), the second sum is also bounded by the right side of (6). This proves the lemma.

PROOF OF THEOREM. We write

$$\int_{\mathbb{R}}|(m\hat{f})^{\vee}|^p|x|^{pa}dx = \int_{\mathbb{R}}\left|\int_{\mathbb{R}} f(y)\,[K_N(x-y)-K_N(x)]dy\right|^p|x|^{pa}dx$$

$$= \sum_{n=-\infty}^{\infty}\int_{2^{-n-1}\le|x|<2^{-n}}\left|\left(\int_{|y|<2^{-n-2}} + \int_{2^{-n-2}\le|y|<2^{-n+1}} + \int_{|y|\ge 2^{-n+1}}\right) f(y)\,[K_N(x-y)-K_N(x)]dy\right|^p|x|^{pa}dx$$

$$\le \mathrm{I} + \mathrm{II} + \mathrm{III}.$$

We apply Minkowski's integral inequality to I and III to obtain

$$(9)\qquad \mathrm{I} \le \sum_{n=-\infty}^{\infty} 2^{-npa}\left\{\int_{|y|<2^{-n-2}}|f(y)|\left(\int_{2^{-n-1}\le|x|<2^{-n}}|K_N(x-y) - K_N(x)|^p dx\right)^{1/p}dy\right\}^p,$$

and

$$(10)\qquad \mathrm{III} \le \sum_{n=-\infty}^{\infty} 2^{-npa}\left\{\int_{|y|\ge 2^{-n+1}}|f(y)|\left[\left(\int_{|y|/2\le|x|<2|y|}|K_N(x)|^p dx\right)^{1/p} + \left(\int_{2^{-n-1}\le|x|<2^{-n}}|K_N(x)|^p dx\right)^{1/p}\right]dy\right\}^p.$$

By the estimates obtained in the lemma and Hardy's inequality, the right side of (9) and (10) are bounded by $C \int_{\mathbb{R}} |f|^p |x|^{pa} dx$. In the use of Hardy's inequality, we need the restrictions $a < \min(\frac{3}{2} - \frac{1}{p}, 1)$ for I, and $a > 1 - \frac{1}{p}$ for III.

The second term is majorized by the sum of

$$II_1 = C \sum_{n=-\infty}^{\infty} 2^{-npa} \int_{\mathbb{R}} \left| \int_{2^{-n-2} \le |y| < 2^{-n+1}} f(y) K_N(x-y) dy \right|^p dx$$

and

$$II_2 = C \sum_{n=-\infty}^{\infty} 2^{-npa} \left| \int_{2^{-n-2} \le |y| < 2^{-n+1}} f(y) dy \right|^p \int_{2^{-n-1} \le |x| < 2^{-n}} |K_N(x)|^p dx.$$

II_1 is estimated by means of Hörmander's multiplier theorem [3]. We obtain

$$II_1 \le C \sum_{n=-\infty}^{\infty} 2^{-npa} \int_{2^{-n-2} \le |y| < 2^{-n+1}} |f(y)|^p dy \le C \int_{\mathbb{R}} |f|^p |x|^{pa} dx.$$

II_2 is also majorized by $C \int_{\mathbb{R}} |f|^p |x|^{pa} dx$, by Hölder's inequality and (5). This completes the proof of Theorem 3 for the case $\alpha = 1$, $a > \frac{1}{p} - 1$.

REFERENCES

[1] R. R. Coifman and G. Weiss, "Extensions of Hardy spaces and their use in analysis," Bull. Amer. Math. Soc. 83 (1977) 569-645.

[2] W. C. Connett and A. L. Schwartz, "The theory of ultraspherical multipliers," Mem. Amer. Math. Soc. 9 (1977) no. 183. (MR 55 #8666)

[3] L. Hörmander, "Estimates for translation invariant operators in L^p spaces," Acta Math. 104 (1960) 93-139. (MR 22 #12389)

[4] B. Muckenhoupt, "Hardy's inequality with weights," Studia Math. 44 (1972) 31-38. (MR 47 #418)

[5] ______________, "Weighted norm inequalities for the Hardy maximal function," Trans. Amer. Math. Soc. 165 (1972) 207-226. (MR 45 #2461).

[6] B. Muckenhoupt, R. L. Wheeden and W.-S. Young, "Weighted L^2 multipliers," preprint.

[7] ______________, "Weighted L^p multipliers," preprint.

[8] B. Muckenhoupt and W.-S. Young, "Weighted L^2 multipliers for odd power weights," in preparation.

DEPARTMENT OF MATHEMATICS, RUTGERS UNIVERSITY
NEW BRUNSWICK, N. J. 08901

Proceedings of Symposia in Pure Mathematics
Volume XXXV, Part 1, 1979

NON-EQUIVALENCE BETWEEN TWO KINDS OF CONDITIONS ON WEIGHT FUNCTIONS

Jan-Olov Strömberg*

A weight function w on $\mathbb{R}^n$ is said to satisfy the A_p condition $(1 < p < \infty)$ if there is a constant C such that for all cubes Q in $\mathbb{R}^n$ the following yields

$$(A_p) \qquad (|Q|^{-1} \int_Q w(x)\, dx)\, (|Q|^{-1} \int_Q (w(x))^{-1/(p-1)}\, dx)^{p-1} \leq C .$$

This condition was introduced by B. Muckenhoupt [2]. Another condition which has been used on weight functions is the B_q condition $(1 < q < \infty)$:

$$(B_q) \qquad |Q|^{q-1} \int_{\mathbb{R}^n} w(x) \,/\, [|Q|^q + |x|^{nq}]\, dx \leq C \;\; |Q|^{-1} \int_Q w(x)\, dx$$

for all cubes Q in $\mathbb{R}^n$ with the constant C independent of C. For each weight function w we can associate the numbers p_w and q_w as

$$p_w = \inf \{p ,\ 1 < p < \infty ,\ \text{ such that } w \text{ satisfies the } A_p \text{ condition}\}.$$

$$q_w = \inf \{q ,\ 1 < q < \infty ,\ \text{ such that } w \text{ satisfies the } B_q \text{ condition}\}.$$

(We set $p_w = \infty$ and $q = \infty$ if corresponding set of p resp. q is empty). If the set of p is non-empty w is said to satisfy the A_∞ condition and this set of p is the open interval (p_w , ∞). If the set of q is non-empty w is said to satisfy the doubling condition (since this is equivalent to $\int_{2Q} w(x)\, dx < C \int_Q w(x)\, dx$ for every cube Q) and this set of q

---

*This work was done while the author was supported by NSF-Grant #MCS 76-82196.

is the open interval (q_w , ∞) or the closed interval $[q_w , \infty)$.

It is known that the A_p condition are stronger than the B_p condition, i.e. $q_w \leq p_w$. To see this we let $p_w < q < p$ and conclude from (A_p) that $|\{y \in 2^jQ ; w(y) < c\, 2^{-njp} \int_{2^jQ} w(x)\}| < |Q| / 2$ for each $j > 0$ if $c > 0$ is small enough. Hence $\int_Q w(x) \geq c\, 2^{-jnp} \int_{2^jQ} w(x)\, dx$ from which B_q easily follows.

On the other hand Muckenhoupt and C. Fefferman [1] have constructed a weight function which satisfies the doubling condition by not the A_∞ condition.

The purpose of this paper is to give more example of weight functions showing that the B_q conditions are strictly weaker than the A_p condition. The result can be summarized in the following Theorem.

Theorem. Given any a , b with $1 \leq a \leq b \leq \infty$. Then there exists a weight function w with $p_w = b$, $q_w = a$.

This result may be of interest in some situations. For instance some results about weighted Hardy spaces can be proved using the A_∞ and a B_q condition and from the Theorem above we conclude that this is a strictly weaker condition than the A_q condition.

Constructions of weight functions

We assume that $n = 1$. At the end of this paper we will see how we easily can get the weight functions on $\mathbb{R}^n$, $n > 1$ from the weight functions on $\mathbb{R}^1$. Let $\phi(x , \gamma)$, $0 < \gamma < 1$, and ψ periodic functions on $\mathbb{R}^1$ with period 4 defined by

$$\phi(x , \gamma) = \begin{cases} 1 - \gamma & 0 \leq x < 1 \\ 1 + \gamma & 1 \leq x \leq 3 \\ 1 - \gamma & 3 < x < 4 \end{cases}$$

and

$$\psi(x) = \begin{cases} 0 & 0 \le x < 1 \\ 1 & 1 \le x \le 3 \\ 0 & 3 < x < 4 \end{cases}$$

Let $m(x)$ be the non-negative integer such that

$$4^{-m(x)-1} < |x| < 4^{-m(x)} \qquad \text{if} \qquad |x| < 1$$

and

$$m(x) = 0 \qquad \text{for} \qquad |x| \ge 1 .$$

By means of these functions we are now able to write explicit weight functions w for all cases $1 \le q_w \le p_w \le \infty$.

(i) $a = b = 1$ $\qquad w(x) \equiv 1$

(ii) $a = b = \infty$ $\qquad w(x) = e^{|x|}$

(iii) $1 \le a \le b < \infty$

$$w(x) = \prod_{m=1}^{\infty} \tau_m(x) = \prod_{m=1}^{\infty} (1 - \alpha \prod_{j=1}^{m} \psi\,(4^{\beta_j} x)\,)$$

where $\alpha = 1 - 2^{1-a}$ and $\{\beta_j\}_{j=1}^{\infty}$ is a strictly increasing sequence of integers. More precisely let $\beta_j = [j(b - a) / 2(a - 1)]$ if $a > 1$ and $\beta_j = j^2$ if $a = 1$.

(iv) $1 < a < b = \infty$ let $(\gamma + 1) / (\gamma - 1) = 4^{a-1}$

$$w(x) = \prod_{m=1}^{\infty} \tau_m(x) \quad \text{where} \quad \tau_m(x) = \begin{cases} \phi(4^m x \, , \gamma) & \text{if } 2m(x) \ge m \\ 1 & \text{if } 2m(x) < m \end{cases}$$

(v) $a = 1 \; , \; b = \infty$

$$w(x) = \prod_{m=1}^{\infty} \tau_m(x) \quad \text{where} \quad \tau_m(x) = \begin{cases} \phi(4^m x \, , 2^{-\ell}) & \text{if } 2m(x) \ge m \\ 1 & \text{if } 2m(x) < m \end{cases}$$

and $N_{\ell-1} < m < N_\ell$. The sequence $\{N_\ell\}_{\ell=1}^{\infty}$ will be chosen later such that $N_\ell - N_{\ell-1}$ is large enough for each ℓ .

<u>Verification of the A_p and B_q conditions</u>

We will replace the A_p and B_q conditions with other conditions where only dyadic intervals are used. Let $\mathcal{J}_k$ be the family of intervals $(j-1)\,2^{-k} \leq x \leq j\,2^{-k}$, j integer. We will write the C, D_q and E_q conditions as

(C) $\displaystyle\int_{I_1} w(x)\,dx \leq c \int_{I_2} w(x)\,dx$ for any intervals I_1, $I_2 \in \mathcal{J}_k$ with a common endpoints, any k.

(D_q) $\displaystyle\int_{I_1} w(x)\,dx \leq C\,2^{jq} \int_{I_2} w(x)\,dx$ for any $I_2 \subset I_1$

$I_1 \in \mathcal{J}_k$, $I_2 \in \mathcal{J}_{k+j}$, for any $j > 0$, k

(E_q) $\displaystyle[|I| \int_I w(x)\,dx]\;[|I|^{-1} \int_I (w(x))^{-1/(p-1)}\,dx]^{p-1} < C$

for any $I \in \mathcal{J}_k$, any k.

The constants C above is not depending on the intervals.

We leave to the reader to check the following easy implications. (Here $w \in C \cap E_p$ means w satisfies both the C and the E_p condition etc.)

$$w \in C \cap E_p \iff w \in A_p$$

$$w \in C \cap D_q \Rightarrow w \in B_{q'} \quad \text{for } q' > q$$

$$w \in B_{q'} \Rightarrow w \in C \cap D_q \quad \text{for } q' \leq q .$$

<u>Case</u> (i) It is obvious that $w \equiv 1$ satisfies any A_p and B_q conditions thus $p_w = q_w = 1$.

<u>Case</u> (ii) It is obvious that $w(x) = e^{|x|}$ does not satisfy any B_q condition. Thus $q_w = \infty$ since $p_w \geq q_w$ we get $p_w = \infty$.

<u>Case</u> (iii) τ_m is constant on $I \in \mathcal{J}_k$, $k \leq \beta_m$. Moreover at most one of the functions τ_m can be discontinuous at a given point. We observe also that $\sup_x \tau_m(x) = 1$, $\inf \tau_m(x) = 1 - \alpha$ and

$$|I| / 2 \le \int_I \Big(\prod_{m'>m} \tau_m(x)\Big)\, dx \le |I| \quad \text{for } I \in \mathcal{J}_k \quad k \le \beta_m .$$

From this we conclude that (C) is satisfied with the constant $C = 2/(1-\alpha)$. To show that $w \in D_q$ for $q > a$ but not for $q < a$ (when $a > 1$) we observe that

$$\sup_x \prod_{m=m_1+1}^{m_2} \tau_m = 1 \quad \text{and} \quad \inf_x \prod_{m=m_1}^{m_2} \tau(x) = (1-\alpha)^{m_2-m_1} .$$

Thus if $2\beta_{m_1-1} < k \le 2\beta_{m_1}$, $2\beta_{m_2-1} < k+j \le 2\beta_{m_2}$ we get

$$\int_{I_1} w(x)\, dx \le 2(1-\alpha)^{m_1-m_2} \cdot 2^j \int_{I_2} w(x)\, dx \quad \text{for } I_2 \subset I_1 ,$$

$I_1 \in \mathcal{J}_k$, $I_2 \in \mathcal{J}_{k+j}$. On the other hand this is the best possible estimate. It follows that the D_q condition is satisfied for small intervals when $q > a$, but not when $q < a$ $(a > 1)$. To get the D_q condition for $q > a$ for all dyadic intervals we have just to blow up the constant C.

We also conclude that the E_p conditions are satisfied exactly for those p for which

$$(*) \qquad \int_I \prod_{m'>m} [\tau_{m'}(x)]^{-1/(p-1)}\, dx < C , \; I \in \mathcal{J}_{m''} \quad \beta_{m-1} \le m \le \beta_m .$$

But we have for such an interval

$$|\{x \in I ,\ \prod_{m'>m} \tau_{m'}(x) < (1-\alpha)^k\}| \le C\, 2^{-k} |I| \qquad k > 0$$

and this estimate is the best possible for some intervals I. Since $(1-\alpha)^{-1/(a-1)} = 2$, $(*)$ will hold for $p > a$ but not for $p < a$. Thus we have shown the properties in case (iii).

Cases (iv) and (v). To show that the (C) conditions is satisfied could be done with the same argument as in case (iii). The only difference here is at most two functions τ_m can be discontinuous in a given point instead of at most one in case (iii). That the D_q is satisfied in case (iv) could also be done with the same argument as in case (iii). We leave these verifications to the reader.

We will now show that $p_w = \infty$ (i.e. $w \notin A_\infty$) in case (iv) and (v) and that the D_q condition is satisfied for all $q > 1$ in case (v).

The argument to show that $w \notin A_\infty$ is the same as used by Fefferman and Muckenhoupt. (Their example is case (iv) with $\gamma = 1/2$).

We want to find an interval I_ℓ and a set E_ℓ such that $|E_\ell| \leq 2^{-\ell} |I_\ell|$ but $\int_{E_\ell} w(x)\,dx > \frac{1}{2} \int_{I_\ell} w(x)\,dx$. Consider the expression

$$1 \equiv [s + (1 - s)]^N = \sum_{j=0}^{N} \binom{N}{j} s^j (1 - s)^{N-j}$$

where $0 < s \leq \frac{1}{2}$ and N are very large. It is a well known property of this sum that

$$\lim_{N\to\infty} \sum_{Ns(1-\varepsilon)<j<Ns(1+\varepsilon)} \binom{N}{j} s^j (1 - s)^{N-j} = 1$$

for any small $\varepsilon > 0$. Hence if $0 < s < \frac{1}{2}$ we can find N such that

$$\sum_{0\leq j\leq N[\frac{1}{4} + \frac{s}{2}]} \binom{N}{j} 2^{-N} \leq 2^{-\ell} \quad \text{but} \quad \sum_{0\leq j\leq N[\frac{1}{4} + \frac{s}{2}]}^{N} \binom{N}{j} s^j (1 - s)^{N-j} \geq \frac{1}{2}.$$

Now set $s = (1 - \gamma) / 2$ in case (iv) and $s = (1 - 2^{-\ell}) / 2$ in case (v). Let I_ℓ be the interval $\frac{1}{2} \cdot 4^{-N} < x < 4^{-N}$ in case (iv) ($\frac{1}{2} 4^{-N-N_{\ell-1}} < x < 4^{-N-N_{\ell-1}}$ in case (v)). We observe that $m(x) \geq N' = N$ (resp. $N' = N + N_{\ell-1}$) in this interval. Let $N'' = N' + N$. We have

$$|I''|^{-1} \int_{I''} \prod_{m>N''} \tau_m(x)\,dx = 1 \text{ for all } I'' \in \mathcal{J}_{N''}$$

and if E is a union of intervals I'' in I_ℓ, $I'' \in \mathcal{J}_{N''}$ then we have

$$|I_\ell|^{-1} \int_E w(x)\,dx = (|I_\ell|^{-1} \int_{I_\ell} w(x)\,dx)\,(|I_\ell|^{-1} \int_E \prod_{j=N'+1}^{N''} \tau_j(x)\,dx).$$

Set $E_\ell = \{x \in I_\ell ;\ \prod_{j=N'+1}^{N'} \tau_j(x) < 2^N s^{N_s} (1 - s)^{N-N_s}\}$ where

$N_s = N(1 + 2s) / 4$. Then

$$\int_{E_\ell} w(x)\,dx \Big/ \int_{I_\ell} w(x)\,dx = |I_\ell|^{-1} \int_{E_\ell} \prod_{j=N'+1}^{N} \tau_j(x)\,dx$$

$$= \sum_{0\leq j<N_s} \binom{N}{j} s^j (1-s)^{N-j} > 1/2 .$$

On the other hand $|E_\ell| / |I_\ell| = \sum_{0\leq j<N_s} \binom{N}{j} 2^{-N} \leq 2^{-\ell}$.

Since ℓ is an arbitrary integer this shows that w does not satisfy the A_∞ condition. In case (v) we now can choose $N_\ell > N'' = N_{\ell-1} + 2N$ and repeat the argument above for each integer ℓ . This shows that w does not satisfy the A_∞ condition in case (v) .

It remains to show that w satisfies the D_q condition for every $q > 1$. We observe that $\sup_x \tau_m(x) / \inf_x \tau_m(x) < (1 + 2^{-\ell}) / (1 - 2^{-\ell})$ if $m > N_\ell$. Given $q > 1$, the argument in case (iii) shows that the D_q condition is satisfied for intervals which are small enough. To get the D_q condition for all intervals we just have to blow up the constant C in (D_q) . This completes the proof in one dimension.

Higher dimensions

The conditions C , D_q , E_p could be formulated in n-dimensional versions, where the dyadic intervals are replaced by n-dimensional dyadic cubes and the factor 2^{jq} in (D_q) is replaced by 2^{jnq} . It is easy to verify that the same implications holds as in the one-dimensional case. If w_1 is a weight function in $\mathbb{R}^1$ constructed as above it is easy to verify that

$$w_n(x) = \prod_{j=1}^{n} w_1(x_j) \qquad x = (x_1, \ldots, x_n)$$

is a weight function in $\mathbb{R}^n$ with corresponding properties.

References

[1] C. Fefferman and B. Muckenhoupt, Two non-equivalent conditions for weight functions. Proc. Amer. Math. Soc. 45 (1974), 99-104.

[2] B. Muckenhoupt, Weighted Norm Inequalities for the Hardy Maximal functions. Trans. Amer. Math. Soc. 165 (1972), 207-226.

Department of Mathematics
Princeton University
Princeton, New Jersey 08540

Proceedings of Symposia in Pure Mathematics
Volume XXXV, Part 1, 1979

SOME REMARKS ABOUT BECKNER'S INEQUALITY

Daniel W. Stroock*

In his remarkable paper [1], Beckner proved that the best constant in the Hausdorf-Young inequality for R^1 is $A_p = (p^{1/p}/(p')^{1/p'})^{1/2}$ for $1 < p \le 2$ (here and throughout $\frac{1}{p} + \frac{1}{p'} = 1$). His proof begins by reducing the problem to showing that

$$\|e^{z_p H} f\|_{L^{p'}(\gamma)} \le \|f\|_{L^p(\gamma)} \tag{1}$$

for all polynomials f, where

$$e^{-z_p} = \omega_p = i(p-1)^{1/2}, \tag{2}$$

$$H = \frac{\partial^2}{\partial x^2} - x\frac{\partial}{\partial x}, \tag{3}$$

and

$$\gamma(dx) = \frac{1}{(2\pi)^{1/2}} e^{-x^2/2} dx . \tag{4}$$

This reduction is clever but elementary, and so I will take (1) as the primary object of my concern. Indeed, what will concern me here is how to prove (1) in a way which hopefully will elucidate the scope of Beckner's method.

There are two steps in Beckner's proof of (1). The first of these is the observation (non-trivial in itself) that if in (1) one replaces γ by $\mu_0 = 1/2(\delta_{-1} + \delta_1)$ and H by the operator K_{μ_0} where

$$K_{\mu_0} f(x) = \int (f(y) - f(x))\mu_0(dy) \tag{5}$$

then the inequality holds; that is:

* Partially supported by N.S.F. Grant M.C.S. #77-14881 and by Guggenheim Foundation.

(6) $$\|e^{z_p K_\mu} f\|_{L^{p'}(\mu_0)} \le \|f\|_{L^p(\mu_0)} .$$

Unfortunately I can add very little to the understanding of (6) . Beckner's second step is to show that (1) follows from (6) . Although this second step may seem to analysts as the more dramatic of the two, it is not so surprising to a probabilist as I now hope to explain.

Let μ be any probability measure on R^1 such that

(7) $$\int |x|^N \mu(dx) < \infty , \quad N \ge 1 ,$$

(8) $$\int x\mu(dx) = 0 ,$$

and

(9) $$\int x^2 \mu(dx) = 1 .$$

Set $M = (R^1)^{Z^+}$ $(Z^+ = \{1,2,\ldots,n,\ldots\})$ and let $\mathscr{D}$ denote the class of bounded continuous φ from M into $\mathbb{C}$ such that φ depends on only a finite number of coordinates (i.e. there is an $N \ge 1$ such that $\varphi(\eta) = \varphi(\eta')$ whenever $\eta,\eta' \in M$ satisfy $\eta_k = \eta'_k$ for all $k \le N$). On $\mathscr{D}$ define

(10) $$\tilde{K}_\mu \varphi(\eta) = \sum_{k\in Z^+} \int (\varphi(\eta_1,\ldots,\eta_{k-1},x,\eta_{k+1},\ldots) - \varphi(\eta))\mu(dx).$$

Then there is a unique continuous non-negative semi-group $\{e^{t\tilde{K}_\mu} : t \ge 0\}$ on $C_b(M)$ such that

$$e^{t\tilde{K}_\mu}\varphi(\eta) - \varphi(\eta) = \int_0^t e^{s\tilde{K}_\mu}\tilde{K}_\mu \varphi(\eta)ds$$

for all $t \ge 0$ and $\varphi \in \mathscr{D}$. (This can be easily seen by defining K_μ on $C_b(R^1)$ as in (5) with μ_0 replaced by μ , checking that $\{e^{tK_\mu} : t \ge 0\}$ on $C_b(R^1)$ is continuous and non-negative, and setting $e^{t\tilde{K}_\mu} = (e^{tK_\mu})^{Z^+}$.) Furthermore, if $\tilde{\mu}$ on M is the measure μ^{Z^+} , then it is easy to check that

(11) $$\int \varphi e^{t\tilde{K}_\mu}\psi d\tilde{\mu} = \int \psi e^{t\tilde{K}_\mu}\varphi d\tilde{\mu} , \quad t \ge 0 \text{ and } \varphi,\psi \in C_b(M)$$

(this can again be seen by starting with the 1-dimensional case). Finally, using the central limit theorem, one can show that if f and g on R^1 into $\mathbb{C}$ are continuous and slowly increasing, then

(12) $$\lim_{n\to\infty}\int_M \varphi_n e^{t\tilde{K}_\mu}\psi_n d\tilde{\mu} = \int_{R^1} fe^{tH}g d\gamma \quad , \qquad t \geq 0 \quad ,$$

where

$$\varphi_n(\eta) = f(\frac{1}{n^{1/2}}\sum_1^n \eta_k)$$

and

$$\psi_n(\eta) = g(\frac{1}{n^{1/2}}\sum_1^n \eta_k) \quad .$$

To understand the origin of (12), one interprets both sides of (12) probabilistically. In the first place, it is easy to see that

$$\int fe^{tH}g\, d\gamma = \int_{R^2} f(x)g(y)\gamma_t^{(2)}(dx\times dy)$$

where

$$\gamma_t^{(2)}(dx\times dy) = \frac{1}{2\pi(1-e^{-2t})^{1/2}}\exp\left[-\frac{x^2-2e^{-t}xy+y^2}{2(1-e^{-2t})}\right]dx\, dy \quad .$$

The measure $\gamma_t^{(2)}$ on R^2 is characterized by the fact that for all $\alpha,\beta\in R^1$ the function $\alpha x+\beta y$ is distributed under $\gamma_t^{(2)}$ as a Gaussian random variable with mean 0 and variance $\alpha^2+2\alpha\beta e^{-t}+\beta^2$. We now turn to the left hand side of (12). To this end, define $\mu_t^{(2)}$ on R^2 by:

$$\mu_t^{(2)}(dx\times dy) = (1-e^{-t})\mu(dx)\mu(dy) + e^{-t}\delta_x(dy)\mu(dx)$$

and set $\tilde{\mu}_t^{(2)} = (\mu_t^{(2)})^{Z^+}$ on $(R^2)^{Z^+}$. Then one can easily check that for $\varphi,\psi\in L^2(\tilde{\mu})$:

$$\int_{M^2}\varphi(\eta)\psi(\eta')\tilde{\mu}_t^{(2)}(d\eta\times d\eta') = \int_M \varphi e^{t\tilde{K}_\mu}\psi d\tilde{\mu} \quad .$$

By construction, the pairs (η_k,η_k'), $k\in Z^+$, are mutually independent and identically distributed under $\tilde{\mu}_t^{(2)}$. Moreover, for any $\alpha,\beta\in R^1$, one has

$$\int (\alpha\eta_1 + \beta\eta_1')\tilde{\mu}_t^{(2)}(d\eta \times d\eta') = \alpha\int x\mu(dx) + \beta\int y\mu(dy) = 0$$

and

$$\int(\alpha\eta_1 + \beta\eta_1')^2\tilde{\mu}_t^{(2)}(d\eta \times d\eta') = \alpha^2\int x^2\mu(dx) + 2\alpha\beta\int x\,y\,\mu_t^{(2)}(dx \times dy)$$
$$+ \beta^2\int y^2\mu(dy)$$
$$= \alpha^2 + 2\alpha\beta e^{-t} + \beta^2 \quad .$$

Thus, by the central limit theorem, the distribution of $\frac{1}{n^{1/2}}\sum_1^n (\alpha\eta_k + \beta\eta_k')$ tends weakly to the Gaussian distribution with mean 0 and variance $\alpha^2 + 2\alpha\beta e^{-t} + \beta^2$. In particular, if $f, g \in C_1(R^1)$, then:

$$\int\varphi_n e^{t\tilde{K}_\mu}\psi_n d\tilde{\mu} = \int_{M^2} f\left(\frac{\eta_1 + \ldots + \eta_n}{n^{1/2}}\right) g\left(\frac{\eta_1' + \ldots + \eta_n'}{n^{1/2}}\right)\tilde{\mu}_t^{(2)}(d\eta \times d\eta')$$
$$\longrightarrow \int_{R^2} f(x)g(y)\gamma_t^{(2)}(dx \times dy) = \int fe^{tH}g\,d\gamma \quad .$$

This proves (12) for $f, g \in C_b(R^1)$. To complete the derivation of (12) for continuous slowly increasing f and g , one need only check that (7) implies that

$$\sup_{n\geq 1}\int\left|\frac{\eta_1 + \ldots + \eta_n}{n^{1/2}}\right|^N\tilde{\mu}(d\eta) < \infty$$

for all $N \geq 1$. Such estimates are standard in the probability theory literature (when $\mu = \mu_0$, one can use Khincline's inequality for sums of Rademacher functions.) The line of resaoning just used is suggested by L. Gross [2] in connection with Nelson's inequality.

In order to carry out Beckner's transition from (6) to (1) , (12) is not quite enough. Indeed, what one still needs is to extend (12) to read:

(13) $$\lim_{n\to\infty}\int\varphi_n e^{z\tilde{K}_\mu}\psi_n d\tilde{\mu} = \int fe^{zH}g d\gamma \quad , \quad z \in \mathbb{C} \quad ,$$

for all polynomials f and g on R^1 into $\mathbb{C}$. This extension is not obvious on probabilistic grounds since one does not have a probabilistic interpretation $e^{z\tilde{K}_\mu}$ or e^{zH} unless $z \geq 0$. Nevertheless (13) is quite easy to derive from (12) with a little spectral analysis. The idea is the following. From (11) plus the non-negativity of $e^{t\tilde{K}_\mu}$, it is easy to see that $e^{t\tilde{K}_\mu}$ determines a continuous semi-group of self-adjoint contraction operators on $L^2(M;\tilde{\mu})$. Hence $\tilde{K}_\mu$ admits a unique extension as a non-positive definite self-adjoint operator $\tilde{A}_\mu$ on $L^2(M;\tilde{\mu})$. Furthermore, one can even describe the spectral resolution $\{\tilde{E}_\lambda^{(\mu)} : \lambda \geq 0\}$ of $-\tilde{A}_\mu$. In fact, because of the product structure of the set-up, this boils down to studying the spectral properties of $-K_\mu$ on $L^2(R^1;\mu)$. But $-K_\mu f = 0$ if f is constant and $-K_\mu f = f$ if $\int f d\mu = 0$. Thus

$$\tilde{A}_\mu = -\int_0^\infty \lambda d\tilde{E}_\lambda^{(\mu)} \tag{14}$$

where $\tilde{E}_\lambda^{(\mu)}$ is the orthogonal projection operator onto the subspace of $L^2(M;\tilde{\mu})$ spanned by functions φ which depend on at most $[\lambda]$ coordinates (i.e. there exist $1 \leq k_1 < \ldots < k_{[\lambda]}$ such that $\varphi(\eta) = \varphi(\eta')$ whenever $\eta_{k_j} = \eta'_{k_j}$ for $1 \leq j \leq [\lambda]$). But this means that if f and g are mth degree polynomials, then $((\tilde{E}_\lambda^{(\mu)} - \tilde{E}_m^{(\mu)})\varphi_n, \psi_n) = 0$ for $\lambda > m$ and so:

$$\begin{aligned} \left|\int \varphi_n e^{z\tilde{K}_\mu} \psi_n d\tilde{\mu}\right| &= \left|\int_0^m e^{-z\lambda} d(\tilde{E}_\lambda^{\mu}\varphi_n, \psi_n)\right| \\ &\leq e^{m|z|} \|\varphi_n\|_{L^2(\tilde{\mu})} \|\psi_n\|_{L^2(\tilde{\mu})} . \end{aligned}$$

Since, as has been already pointed out, $\sup_{n\geq 1} \|\varphi_n\|_{L^2(\tilde{\mu})} \|\psi_n\|_{L^2(\tilde{\mu})} < \infty$, one now sees that the entire functions $z \to \int \varphi_n e^{z\tilde{K}_\mu} \psi_n d\tilde{\mu}$ are uniformly bounded in n as z varies over compacts. Since for $z = t \geq 0$ (12) holds, it is now clear that (13) follows for all $z \in C$ and all polynomials f and g.

The passage from (6) to (1) is easily made once one has (13). In fact one has the following more general statement.

<u>Theorem</u> (15): Let μ be a probability measure on R^1 satisfying (7), (8), (9). Given $\omega \in \mathbb{C}$, define T_ω on $C_b(R^1)$ by

$$T_\omega^{(\mu)} f = \int f d\mu + \omega(f - \int f d\mu) \tag{16}$$

and let $z = -\log \omega$. Suppose that for some $1 < p \leq q < \infty$ one has the inequality:

$$\|T_{\omega}^{(\mu)} f\|_{L^q(\mu)} \leq \|f\|_{L^p(\mu)}, \quad f \in C_b(R^1). \tag{17}$$

Then one has the inequality:

$$\|e^{zH} f\|_{L^q(\gamma)} \leq \|f\|_{L^p(\gamma)}, \quad f \in C_b(R^1). \tag{18}$$

Proof: Notice first that $T_{\omega}^{(\mu)} f = e^{zK_{\mu}} f$. Now use the product structure and Lemma 2 of [1] to conclude from (17) that $\|e^{z\tilde{K}_{\mu}}\varphi\|_{L^q(\tilde{\mu})} \leq \|\varphi\|_{L^p(\tilde{\mu})}$, $\varphi \in C_b(M) \cap L^2(M;\tilde{\mu})$. In particular, if f and g are polynomials on R^1, then

$$\left|\int \varphi_n e^{z\tilde{K}_{\mu}} \psi_n d\tilde{\mu}\right| \leq \|\varphi_n\|_{L^{q'}(\tilde{\mu})} \|\psi_n\|_{L^p(\tilde{\mu})}.$$

Hence, by (13)

$$\left|\int f e^{zH} g d\gamma\right| \leq \|f\|_{L^{q'}(\gamma)} \|g\|_{L^p(\gamma)},$$

since $\|\varphi_n\|_{L^{q'}(\mu)} \to \|f\|_{L^{q'}(\gamma)}$ and $\|\psi_n\|_{L^p(\tilde{\mu})} \to \|g\|_{L^p(\gamma)}$. Clearly (18) is an immediate consequence of this.

Q. E. D.

In some ways the contra-positive statement of Theorem (15) is the more striking way to phrase this result: if (18) fails then (17) fails for all μ's. This version of Theorem (15) makes it natural to ask whether (18) implies (17) (E.M. Stein raised this question for the case $\mu = \mu_0$). As the following example demonstrates, the answer to this question is in general no. Indeed, let $\mu = \gamma$ and consider Beckner's set-up: $1 < p < 2$, $q = p'$, and $\omega = i(p-1)^{1/2}$. Beckner has shown that (18) holds for this choice of parameters. On the other hand, I will now show that (17) fails in this situation when $\mu = \gamma$. To this end, define

$$f_a(x) = e^{ax - a^2/2}, \quad x \in R^1 \text{ and } a \in C. \tag{19}$$

Then, for any $1 \le r < \infty$:

$$\|f_a\|^r_{L^r(\gamma)} = \frac{e^{-\frac{r}{2}\mathrm{Re}\, a^2}}{(2\pi)^{1/2}} \int e^{r\,\mathrm{Re}\, a} e^{-x^2/2} dx$$

$$= \exp[\tfrac{r}{2}(r(\mathrm{Re}\ a)^2 - \mathrm{Re}(a^2))] \quad ,$$

and so

$$\|f_a\|_{L^r(\gamma)} = \exp[\tfrac{1}{2}((r-1)\alpha^2 + \beta^2)] \tag{20}$$

where $a = \alpha + i\beta$. Also,

$$T^{(\gamma)}_{\omega} f_a = 1 + \omega(f_a - 1) \quad .$$

Hence if (17) held when $\mu = \gamma$, then one would have:

$$\exp[\tfrac{1}{2}((p-1)\alpha^2 + \beta^2)] \ge (p-1)^{1/2}\exp[\tfrac{1}{2}((p'-1)\alpha^2 + \beta^2] - p^{1/2}$$

or equivalently

$$(p-1)^{1/2}\exp[\tfrac{1}{2}(p'-p)\alpha^2] \le 1 + p^{1/2}\exp[-\tfrac{1}{2}((p-1)\alpha^2 + \beta^2)]$$

for all $\alpha,\beta \in R'$. But by letting β and then α tend to ∞ , one sees that this is impossible.

The preceding example does not rule out the possibility that (17) and (18) are equivalent when $\mu = \mu_0$. However, it does show that if this equivalence does hold, then its validity rests on special properties of μ_0 and does not follow from the find of general considerations that I used to prove Theorem (15) . As yet I have been unable to settle this question very satisfactorily. The best that I can do is contained in the next theorem.

<u>Theorem</u> (21): Let $1 < p \le 2 \le q < \infty$, $\omega \in \mathbb{C}$, and $z = -\log \omega$. Then the following are equivalent:

i) $\|e^{zH}f\|_{L^q(\gamma)} \le \|f\|_{L^p(\gamma)}$, $f \in C_b(R^1)$,

ii) $(q-1)\eta^2 + \xi^2 \le 1$ and $(q-2)\xi^2\eta^2 \le [1 - \xi^2 - (q-1)\eta^2][(p-1) -$

$(q-1)\xi^2 - \eta^2]$ where $\xi = \text{Re}\,\omega$ and $\eta = \text{Im}\,\omega$,

iii) $\|T_\omega^{(\mu_0)} f\|_{L^q(\mu_0)} \le \|f\|_{L^p(\mu_0)}$, $f \in C_b(R^1)$.

Proof: Assume i) . Let f_a , $a \in \mathbb{C}$, be the function in (19) . Then $f_a = \sum_0^\infty \frac{a^n}{n!} H_n$ where $H_n = (-1)^n e^{x^2} \frac{d^n}{dx^n} e^{-x^2}$ is the $n^{\underline{th}}$ (unnormalized) Hermite polynomial. Since $\frac{d^2 H_n}{dx^2} - x\frac{dH_n}{dx} = -nH_n$, one sees that $e^{zH} f_a = f_{a\omega}$. Thus, by (20) , i) implies that

$$(22) \qquad (q-1)(\alpha\xi - \beta\eta)^2 + (\alpha\eta + \beta\xi)^2 \le (p-1)\alpha^2 + \beta^2 \ , \ \alpha,\beta \in R^1 \ .$$

Clearly (22) is equivalent to

$$(q-1)(\xi - \eta x)^2 + (\eta + \xi x)^2 \le (p-1) + x^2 \ , \ x \in R^1 \ ,$$

and so ii) now follows from the elementary theory of discriminants.

Next assume ii) . Reversing the reasoning above, one sees that ii) implies (22) . But from Beckner's proof of Lemma (1) in [1] , it is easy to deduce that (22) implies iii) .

Finally, i) follows from iii) by Theorem (15) with $\mu = \mu_0$.

Q. E. D.

In conclusion, I want to express my doubt that the introduction of μ's different from μ_0 does any more than clarify the mechanism by which one can pass from the inequality (17) to inequality (18) . In particular, I believe that (17) may always hold for μ_0 if it holds for any μ . This will certainly be the case if there is a $\Gamma \in \mathcal{B}_{R^1}$ such that $\mu(\Gamma) = 1/2$, since one can then show from (17) that $\|T_\omega^{(\mu)} f\|_{L^q(\mu)} \le \|f\|_{L^p(\mu)}$ for $f = a + b(2\chi_\Gamma - 1)$ and therefore that $\|T_\omega^{(\mu_0)} f\|_{L^q(\mu_0)} \le \|f\|_{L^p(\mu_0)}$ for all $f \in C_b(R^1)$.

References

[1] Beckner, William, "Inequalities in Fourier analysis," Ann. Math. 102, pp. 159-182 (1975).

[2] Gross, Leonard, "Logarithmic Sobolev inequalities," Am. J. Math. vol. XCVII #4, pp. 1061-1083 (1975).

Proceedings of Symposia in Pure Mathematics
Volume XXXV, Part 1, 1979

HYPERCONTRACTIVE ESTIMATES FOR SEMIGROUPS

Fred B. Weissler[1]

ABSTRACT. Necessary and sufficient conditions for the Hermite semigroup e^{-tH} , Re $t \geq 0$, to be bounded from $L^p(R,\mu)$ into $L^q(R,\mu)$, μ being Gauss measure, are given. In many cases these operators are contractions. This information enables one to compute the norm of $e^{t\Delta}$, Re $t \geq 0$, as a map from $L^p(R,dx)$ into $L^q(R,dx)$ in a large number of cases.

1. INTRODUCTION. In this paper we discuss sharp norm estimates for two particular semigroups of operators which are important in harmonic analysis, the Gauss-Weierstrass semigroup and the Hermite semigroup. We begin by recalling the definitions of these semigroups. The Gauss-Weierstrass semigroup on R is given by convolution with a Gauss kernel, $e^{t\Delta}\phi = k_t * \phi$ where $k_t(x) = (4\pi t)^{-1/2}\exp(-x^2/4t)$. The Hermite semigroup on R is given by integration against the Mehler kernel,

$$(e^{-tH}\phi)(x) = [\pi(1-\omega^2)]^{-1/2}\int_R \exp[-(\omega x-y)^2/(1-\omega^2)]\phi(y)\,dy \quad ,$$

where $\omega = e^{-t}$. Alternately, the Hermite semigroup can be expressed as multiplier operators:

$$e^{-tH}\left(\sum_{n=0}^{\infty} a_n h_n\right) = \sum_{n=0}^{\infty} a_n \omega^n h_n$$

where the h_n are the Hermite polynomials forming an orthonormal basis of $L^2(\mu)$, μ being the Gauss measure $d\mu = \pi^{-1/2}\exp(-x^2)dx$. As before, $\omega = e^{-t}$.

If $t > 0$ and $1 \leq p \leq \infty$, it is straightforward to verify that $e^{t\Delta}$ is bounded from $L^p(R,dx)$ into itself with norm 1, and that e^{-tH} is bounded from $L^p(R,\mu)$ into itself with norm 1. In this paper we discuss the answers to the following questions:

AMS(MOS) Subject Classifications (1970). 47D05, 42A18

[1]Research supported by a Dr. Chaim Weizmann Postdoctoral Fellowship for Scientific Research.

I) Let $\operatorname{Re} t \geq 0$ and $1 \leq p,q \leq \infty$. When is $e^{t\Delta} : L^p(R,dx) \to L^q(R,dx)$ bounded and what is its norm?

II) Let $\operatorname{Re} t \geq 0$ and $1 \leq p,q \leq \infty$. When is $e^{-tH} : L^p(R,\mu) \to L^q(R,\mu)$ bounded and what is its norm?

These questions have been answered in some special cases by Nelson, Gross, Beckner, and others; and we will indicate how their results fit in.

2. BOUNDEDNESS. The question as to when the semigroups are bounded turns out to be relatively elementary and can be answered completely in both cases. For the Gauss-Weierstrass semigroup, if $1 \leq p \leq q \leq \infty$ and $\operatorname{Re} t > 0$, the classical Young's convolution inequality ([6], p. 178) implies that $e^{t\Delta} : L^p(R,dx) \to L^q(R,dx)$ is bounded. If $1 \leq p \leq 2$ and $\operatorname{Re} t = 0$ $(t \neq 0)$, the classical Hausdorff-Young inequality ([6], p. 178) implies that $e^{t\Delta} : L^p(R,dx) \to L^{p'}(R,dx)$ is bounded. The classical inequalities give estimates for the norms of these maps, but in general these estimates are not sharp. It is easy to see using computations with Gaussians that $e^{t\Delta}$ can not be bounded except in the above cases.

For the Hermite semigroup, the question of boundedness can be reduced to boundedness for the Gauss-Weierstrass semigroup. We sketch the argument here. The details can be found in Section 2 of [7]. Let $\operatorname{Re} t \geq 0$ $(e^{-t} \neq \pm 1)$, and let non-zero real γ be such that $\operatorname{Re} (\gamma \sinh t) \geq 0$. Then there exist explicit maps T_1 and T_2 , made up of dilations and multiplications by Gaussians, such that

$$e^{-tH} = T_1 \exp[(\gamma \sinh t)\Delta]\, T_2 \quad . \tag{1}$$

Thus, for example, if $1 \leq p \leq q \leq \infty$, $\operatorname{Re} (\gamma \sinh t) > 0$, and $T_2 : L^p(R,\mu) \to L^p(R,dx)$ and $T_1 : L^q(R,dx) \to L^q(R,\mu)$ are bounded; it follows that $e^{-tH} : L^p(R,\mu) \to L^q(R,\mu)$ is bounded. Since T_1 and T_2 are known explicitly in terms of p, q, γ, and t , it can be determined when they are bounded. Carrying out these and other computations, one gets that if $\operatorname{Re} t \geq 0$ $(e^{-t} \neq \pm 1)$, $1 \leq p \leq q < \infty$ and

$$|p-2 - \omega^2(q-1)| \leq p - |\omega|^2 q \quad , \tag{2}$$

then $e^{-tH} : L^p(R,\mu) \to L^q(R,\mu)$ is bounded. (As above, ω will always denote e^{-t} .) If $1 \leq q < p < \infty$ and either $|\omega| = 1$ or $\operatorname{Re} \omega = 0$, then (2) also implies $e^{-tH} : L^p(R,\mu) \to L^q(R,\mu)$ is bounded. For the rest of the cases with $q < p$, the condition needed is (2) with strict inequality. (If $p = \infty$ or $q = \infty$ is considered, a modified form of (2) is used.) If $e^{-t} = \pm 1$, then e^{-tH} is the identity or reflection and so is bounded $L^p(R,\mu) \to L^q(R,\mu)$ if $p = q$. Computations with Gaussians show that e^{-tH} can not be bounded except in the above cases.

3. COMPUTATION OF NORMS. The question of the norms of these operators is much more subtle. We look first at the Hermite semigroup. It is a result of Nelson [4], whose proof was considerably simplified by Gross [3], that if $\omega^2 \leq (p-1)/(q-1)$, then $e^{-tH} : L^p(R,\mu) \to L^q(R,\mu)$ is a contraction. In other words, for real t , $e^{-tH} : L^p(R,\mu) \to L^q(R,\mu)$ has norm 1 whenever it is bounded. Beckner [1] has shown that if $1 < p \leq 2$, $\mathrm{Re}\,\omega = 0$, and $|\omega|^2 = p-1$, then $e^{-tH} : L^p(R,\mu) \to L^{p'}(R,\mu)$ has norm 1. (This is equivalent to the sharp Hausdorff-Young inequality.)

One might hope that e^{-tH} has norm 1 whenever it is bounded. If $q < p$, this is clearly false, because the condition for boundedness is an <u>open</u> condition. In other words, if $e^{-tH} : L^p(R,\mu) \to L^q(R,\mu)$ has norm 1 whenever strict inequality holds in (2), then it has norm 1 whenever equality holds in (2); and this is false if $q < p$. On the other hand, it is reasonable to make the following conjecture:

CONJECTURE: Let $\mathrm{Re}\, t \geq 0$ ($e^{-t} \neq \pm 1$) and $1 < p \leq q < \infty$. Then $e^{-tH} : L^p(R,\mu) \to L^q(R,\mu)$ has norm 1 whenever it is bounded, i.e. whenever (2) holds.

(We remark that if $p = 1$, then t must be real and q must equal 1 in order for $e^{-tH} : L^p(R,\mu) \to L^q(R,\mu)$ to be bounded. Similarly, if $q = \infty$, then we must have t real and $p = \infty$ for boundedness. In both cases, the norm is 1.)

This conjecture has been proved in the cases:

a) $1 < p \leq 2 \leq q < \infty$.
b) $2 \leq p \leq q < \infty$ and $q \geq 3$.
c) $1 < p \leq q \leq 2$ and $p \leq 3/2$.

(See Sections 3 and 4 of [7].) The proof uses the method of "two-point inequalities" developed by Gross and Beckner. Beckner [1] showed that if $1 < p \leq q < \infty$, in order to prove that $e^{-tH} : L^p(R,\mu) \to L^q(R,\mu)$ has norm 1, it suffices to show that

$$\left[\frac{|a+\omega b|^q + |a-\omega b|^q}{2}\right]^{1/q} \leq \left[\frac{|a+b|^p + |a-b|^p}{2}\right]^{1/p} \tag{3}$$

for all complex a and b . Thus, by showing that (2) implies (3), one could prove the conjecture. This is what has been done in the cases listed above. We mention that if $1 < p \leq 2$ and $q = p'$, the conjecture has also been proved using interpolation techniques. (See Coifman, <u>et al.</u> [2].)

Now let us return to the Gauss-Weierstrass semigroup. The formula (1) can be used to compute the norms for $e^{t\Delta}$. Let $1 < p \leq q < \infty$ and suppose $T_1 : L^q(R,dx) \to L^q(R,\mu)$ and $T_2 : L^p(R,\mu) \to L^p(R,dx)$ are not only bounded, but also invertible and simply scale the norms. Suppose further that $e^{-tH} : L^p(R,\mu) \to L^q(R,\mu)$ has norm 1. Then the norm of

$$\exp[(\gamma \sinh t)\Delta] : L^p(R,dx) \to L^q(R,dx)$$

can literally be read off from (1). This gives enough information to compute explicitly the norms of $e^{t\Delta} : L^p(R,dx) \to L^q(R,dx)$ whenever $\operatorname{Re} t > 0$ and p and q satisfy one of a), b), or c) above. (If $\operatorname{Re} t = 0$ and $t \neq 0$, the corresponding norms follow easily from the sharp Hausdorff-Young inequality.)

The computation of these norms is somewhat complicated, and if $p \neq q$ it involves the solution of a cubic equation. We refer the reader to [7] for the details and the explicit results. There is one case, however, that is of some interest and can be stated easily.

THEOREM. Let $3 \leq p < \infty$. Then $e^{t\Delta} : L^p(R,dx) \to L^p(R,dx)$ is a contraction if and only if $|\arg t| \leq \arccos[(p-2)/p]$.

We note that for the special case of the Gauss-Weierstrass semigroup this theorem improves the result given by complex interpolation in [5], p. 67.

Finally, we remark that if the conjecture above is true for all p and q , then the computation of the norm of $e^{t\Delta}$ is valid for all p and q . In particular, we may allow $2 \leq p < \infty$ in the above theorem.

REFERENCES

1. W. Beckner, Inequalities in Fourier analysis, Ann. of Math. (2) 102(1975), 159-182.

2. R. Coifman, M. Cwikel, R. Rochberg, Y. Sagher, and G. Weiss, Complex interpolation of families of spaces, these proceedings.

3. L. Gross, Logarithmic Sobolev inequalities, Amer. J. Math. 97 (1975), 1061-1083.

4. E. Nelson, The free Markoff field, J. Functional Analysis 12 (1973), 211-227.

5. E. M. Stein, Topics in Harmonic Analysis Related to the Littlewood-Paley Theory, Princeton University Press, Princeton, 1970.

6. E. M. Stein and G. Weiss, Introduction to Fourier Analysis on Euclidean Spaces, Princeton University Press, Princeton, 1971.

7. F. B. Weissler, Two-point inequalities, the Hermite semigroup, and the Gauss-Weierstrass semigroup, J. Functional Analysis, to appear.

DEPARTMENT OF MATHEMATICS, THE UNIVERSITY OF TEXAS, AUSTIN, TEXAS, 78712.

Proceedings of Symposia in Pure Mathematics
Volume XXXV, Part 1, 1979

SINGULAR INTEGRALS NEAR L^1

William C. Connett

Let $x \in R^n$, $x' = \frac{x}{|x|} \in \Sigma$, the n - 1 sphere in R^n. Define the function $K(x) = \Omega(x)/|x|^n$, and the truncation $K_\varepsilon(x) = K(x)$ if $\varepsilon < |x| < 1/\varepsilon$, $= 0$ otherwise. The basic concern of the theory of singular integrals is the behavior of the operator $T_\varepsilon f(x) = K_\varepsilon * f(x)$ acting on functions in $L^p(R^n)$. In what follows it always will be assumed that $\Omega(\lambda x') = \Omega(x')$ for all scalar $\lambda > 0$, and that $\int_\Sigma \Omega(x')dx' = 0$.

In 1952, Calderón and Zygmund showed that if Ω satisfied the Dini condition, i.e.,

$$\omega(\Omega,\delta) = \sup_{\substack{u'\in\Sigma \\ |\zeta|<\delta}} |\Omega(u') - \Omega(\zeta u')| \qquad (\zeta \text{ a rotation})$$

and $\int_0^1 \omega(\Omega,\delta)\frac{d\delta}{\delta} < \infty$.

Then two types of results were obtained:

A.) Continuity of the operator -

$\|T_\varepsilon f\|_p \leq C_p \|f\|_p$ for $1 < p < \infty$

(This implies $\lim_{\varepsilon\to 0} T_\varepsilon f = Tf$ in L^p norm), and weak type 1-1.

B.) Pointwise convergence -

$\lim_{\varepsilon\to 0} T_\varepsilon f(x) = Tf(x)$ a.e. x, $1 \leq p < \infty$.

In 1956, Calderón and Zygmund introduced the "method of rotations" and simplified the assumptions above. If $\Omega \in L^1(\Sigma)$ and odd, or $\Omega \in L^q(\Sigma)$, $q > 1$, and even, then A.) and B.) follow for $1 < p < \infty$. The weak type 1-1 and the pointwise convergence for $f \in L^1$ are the only price paid for this dramatic weakening of assumptions on Ω. Of special interest here is that in the case Ω even, the assumption L^q can be weakened to $L \log^+ L$.

AMS(MOS) subject classification (1970). Primary 44A25.

In 1960, Hörmander published an important paper on singular integrals, where, among many other things, he developed the theory of singular integrals using as his basic smoothness assumption on the kernel, not the Dini condition, but a certain inequality on p. 121 in the 1952 paper mentioned above. In particular:

$$\int_{|y|>2|x|} |K(y - x) - K(y)|dy < C. \tag{1}$$

From this and the boundedness of $\hat{K}$ the conclusions of the 1952 paper follow.

In 1965, M. Weiss and Zygmund showed that no metric condition on Ω weaker than $L \log^+ L$ would suffice to obtain pointwise convergence for $1 < p < \infty$. They showed that for any $\phi(t)$ such that $\phi(t) = o(t\log(1 + t))$ there is an $\Omega \in L^\phi$ for which $T_\varepsilon f$ diverges for $f \in L^p$, $1 \leq p < \infty$.

In 1965, Calderón, M. Weiss, and Zygmund showed that the smoothness assumption could be weakened from Dini to integral Dini:

$$\omega_1(\Omega,\delta) = \sup_{|\zeta|<\delta} \int_\Sigma |\Omega(u') - \Omega(\zeta u')|du'$$

and

$$\int_0^1 \omega_1(\Omega,\delta)\frac{d\delta}{\delta} < \infty.$$

If Ω is in $L^1(\Sigma)$ and satisfies the above condition, then the conclusions of the 1952 paper follow. Another important result in this paper is that integral Dini implies both (1) and $\Omega \in L \log^+ L(\Sigma)$.

Now to turn to more recent results. The above theory is essentially complete. The only questions concern what happens when $\Omega \in L^q$ and $f \in L^p$ and p or q are near 1.

A result when q is near 1

The condition $\Omega \in L \log^+ L$ can be improved. Consider the formula for $\hat{K}$:

$$\hat{K}(x') = \int_\Sigma \Omega(y')\{\log \frac{1}{|x'\cdot y'|} - \frac{\pi i}{2} \operatorname{sign} x' \circ y'\}dy'.$$

If T is to be bounded on L^2, then $\hat{K}$ must be in L^∞. If $\Omega \in L \log^+ L$, then the fact that the logarithm is exponentially integrable and an Orlicz space version of Hölder's inequality yields the boundedness of $\hat{K}$.

But the recent results of Fefferman and Stein (1972) that the dual of $H^1(R^n)$ is BMO, and the observation that the log function is the typical example of an unbounded function in BMO, suggests that $\hat{K}$ is really just a bounded linear functional for $\Omega \in H^1(\Sigma)$. This is true, and follows easily from the atomic representation of $H^1(\Sigma)$. But much more is true.

Theorem If $\Omega \in H^1(\Sigma)$, then A.) and B.) follow for $1 < p < \infty$.

One proof of this follows by adopting the "method of rotations" to this setting. The idea is to extend a function in $H_1(\Sigma)$ off the sphere to be in $H^1(R^n)$. The atomic decomposition is used to do this. Then the Riesz transforms can be applied as in the proof for the $L \log^+ L$ case. A different proof has been given by F. Ricci and G. Weiss and will be reported on later this week, so I will skip the details.

Results when p = 1

There seems to be no easy way to obtain pointwise convergence for $f \in L^1(R^n)$, unless some smoothness is assumed about Ω. We might hope to improve the result with a condition on $\omega_1(\Omega,\delta)$ weaker than integral Dini, but this is not possible.

Theorem No condition on $\omega_1(\Omega,\delta)$ weaker than integral Dini implies convergence. The method of proof is by showing that for any reasonable $\phi(\delta)$ such that $\int \phi(\delta)\frac{d\delta}{\delta} = \infty$, there is an Ω with $\omega_1(\Omega,\delta) < \phi(\delta)$, yet $T_\varepsilon f$ diverges for a.e. x.

A final question remains. The condition (1) used by Hörmander is not a condition on the modulus of continuity. Since integral Dini implies (1), is (1) weaker? This is false. In 1977, Calderón and Zygmund showed that if $\Omega \in L^1(\Sigma)$, then (1) implies the integral Dini condition. One consequence of this worth mentioning is that (1) implies that $\hat{K}$ is bounded.

Department of Mathematics
University of Missouri-St. Louis
63121

Proceedings of Symposia in Pure Mathematics
Volume XXXV, Part 1, 1979

ON CARLESON'S CONVERGENCE THEOREM FOR L^2 FUNCTIONS

Bogdan M. Baishanski

ABSTRACT. The strongest estimate on the maximal partial sum of the Fourier series of L^2 functions is shown to be equivalent to its discrete analogue, provided this analogue is taken in a restricted form.

1. Let f be any square integrable function, and let $s_n(f)$ denote the nth partial sum of the Fourier series of f. Each of several known estimates on $\sup_n |s_n(f)|$ implies immediately that $s_n(f)$ converge almost everywhere. The strongest such estimate, due to R. Hunt, is the following one:

$$\|\sup_n |s_n(f)|\,\|_2 \leq C\|f\|_2 , \tag{1}$$

where C is an absolute constant.

We shall present here a statement equivalent to (1), that reveals the combinatorial character of Carleson's result, which remains hidden in formulation (1).

We use the following notation: every absolute constant is denoted by the same symbol C; ϵ stands for $\exp \frac{2\pi i}{N}$; and $\mathfrak{A}(N,n)$ denotes the collection of all $N \times N$ matrices

$$A = \frac{1}{\sqrt{N}} \begin{bmatrix} 1 & 1 & 1 & \cdots\cdots\cdots\cdots & 1 & 0 & \cdots & 0 \\ 1 & \epsilon & \epsilon^2 & \cdots & \epsilon^{s_1} & 0 & \cdots\cdots\cdots & 0 \\ 1 & \epsilon^2 & \epsilon^4 & \cdots\cdots\cdots & \epsilon^{2s_2} & 0 & \cdots & 0 \\ 1 & \epsilon^3 & \epsilon^6 & \cdots\cdots & \epsilon^{3s_3} & 0 & \cdots\cdots & 0 \\ \vdots & & & & & & & \\ 1 & \epsilon^{N-1} & \epsilon^{2(N-1)} & \cdots\cdots & \epsilon^{(N-1)s_{N-1}} & 0 & \cdots & 0 \end{bmatrix} \tag{2}$$

(in the top row, there are s_0+1 non-zero entries), such that $s_j \leq n-1$ for $j = 0,1,\ldots,N-1$.

AMS(MOS) subject classification (1970). Primary 42A20; Secondary 15A18 .

In a more formal style, we define $A = [c_{jk}]_{0\le j,k\le N-1}$ by $c_{jk} = \frac{1}{\sqrt{N}}\,\epsilon^{jk}$ if $j \le s_k$, $= 0$ otherwise.

THEOREM. The estimate (1) is equivalent to

(3) if $A \in \bigcup_{N=2}^{\infty} \mathfrak{A}(N, \sqrt{N})$, the eigenvalues of A^*A are $\le C$.

(The absolute constants C are not necessarily the same in (1) and (3).)

PROOF. We shall write (1) in the form

$$\text{(1')} \qquad \frac{1}{2\pi}\int_0^{2\pi} \max_{0\le s\le n-1} \Big|\sum_{r=0}^{s} a_r e^{rti}\Big|^2 dt \le C \sum_{r=0}^{n-1} |a_r|^2 ,$$

The equivalence of (1) and (1') is not difficult to check, since (1) is obviously equivalent to $\|\max_{k\le n} |s_k(f)|\,\|_2 \le C\|f\|_2$ for every $n = 1,2,\ldots,$ which is the same as

$$\Big\|\max_{k\le n} \Big|\sum_{r=-k}^{k} a_r e^{rti}\Big|\,\Big\|_2^2 \le C \sum_{r=-\infty}^{+\infty} |a_r|^2 , \quad n = 1,2,\ldots .$$

We assume $\sum_{r=0}^{n-1} |a_r|^2 = 1$, and observe that $\varphi(t) = \max_{0\le s\le n-1} \big|\sum_{r=0}^{s} a_r e^{rti}\big|^2$ is the maximum of finitely many trigonometric polynomials $|T_s(t)|^2$, $s = 0,1,\ldots,n-1$ and so φ is a Lipschitz function. We proceed to find an upper bound for the Lipschitz constant M_φ of φ. Clearly,

$$\text{(4)} \qquad M_\varphi \le \max_s \{\text{Lipschitz constant of } |T_s|^2\} \le \max_s \{\|\tfrac{d}{dt}|T_s(t)|^2\|_\infty\} .$$

Since $|T_s(t)|^2 = T_s(t)\,\overline{T_s(t)}$ is a trigonometric polynomial of degree s, it follows from Bernstein's inequality that

$$\text{(5)} \qquad \Big|\frac{d}{dt}\,|T_s(t)|^2\Big| \le s\,\|T_s\|_\infty^2 .$$

Finally, by the Cauchy-Schwarz inequality, and since $\sum_{r=0}^{n-1} |a_r|^2 = 1$,

$$\text{(6)} \qquad \|T_s\|_\infty = \Big\|\sum_0^s a_r e^{rti}\Big\|_\infty \le \sum_0^s |a_r| \le \sqrt{\sum_0^s |a_r|^2}\cdot\sqrt{s} \le \sqrt{s}$$

From (5) and (6) we obtain

$$\Big\|\frac{d}{dt}|T_s(t)|^2\Big\|_\infty \le s^2 \le n^2 ,$$

so that from (4) we conclude that

$$\text{(7)} \qquad M_\varphi \le n^2 .$$

By Jackson's theorem, for some absolute constant C and for every $d = 1,2,\ldots,$ there exists a trigonometric polynomial $U_d(t)$ of degree $\le d$ such that

(8) $$\|\varphi - U_d\|_\infty \leq \frac{CM_\varphi}{d} = \frac{2Cn^2}{d} .$$

We choose $d = n^2$, and let N be any integer $\geq d$. We compare now

$$I(\varphi) = \frac{1}{2\pi}\int_0^{2\pi} \varphi(t)\,dt = \text{the left-hand side of } (1')$$

with the corresponding Riemann sum $R_N(\varphi)$, arising from the partition of $[0,2\pi]$ into N equal parts. Since $N \geq d$, we have that $I(U_d) = R_N(U_d)$, and so we obtain from (8)

$$|I(\varphi) - R_N(\varphi)| = |I(\varphi - U_d) - R_N(\varphi - U_d)| \leq |I(\varphi - U_d)| + |R_N(\varphi - U_d)| \leq$$

$$\leq 2\,\|\varphi - U_d\|_\infty \leq 4C .$$

It follows that the integrals $I(\varphi)$ are bounded by an absolute constant, i.e., (1') holds, if and only if the Riemann sums $R_N(\varphi)$ are bounded by an absolute constant.

Since, for some choice of integers $s_0, s_1, \ldots, s_{N-1}$, such that $s_j < n$ for $j = 0,1,\ldots,N$, we have that

$$R_N(\varphi) = \frac{1}{N}\sum_{\nu=0}^{N-1} \varphi\left(\frac{2\pi\nu}{N}\right) = \sum_{\nu=0}^{N-1} \left|\sum_{r=0}^{s_\nu} a_r \frac{1}{\sqrt{N}} e^{r\frac{2\nu\pi i}{N}}\right|^2 = \|A\vec{a}\|^2$$

where A is defined by (2), $n^2 < N$ and $\vec{a}$ is the N-tuple $[a_0, a_1, \ldots, a_{n-1}, 0, \ldots, 0]$, we conclude that the Riemann sums $R_N(\varphi)$ will be bounded by an absolute constant if and only if the quadratic forms $\|Aa\|^2$ are, on the unit sphere, bounded by an absolute constant, i.e., if and only if all the eigenvalues of A^*A are $\leq C$, for every matrix A with $n^2 < N$.

This proves the equivalence of (1) and (3).

REMARKS. We can rephrase (3) in terms of the norms of operators A, or in terms of the maxima of the corresponding quadratic forms, instead of the eigenvalues of A^*A. With such a rephrasing, it becomes obvious that (3) is a discrete analogue of (1). That becomes particularly clear if we use the terminology of Fourier transform on the group $Z(\text{mod } N)$ and define, in the obvious way, for every function f on that group, the nth partial sum $s_n(f)$ of the "Fourier series" of f. Then (3) translates as follows:

(3') $$\Big\|\sup_{n\leq\sqrt{N}} |s_n(f)|\Big\|_2 \leq C\,\|f\|_2 ,$$

for every N, every function f on $Z(\text{mod } N)$, and some absolute constant C.

When (3) is rephrased as (3'), it even notationally resembles (1), so that our theorem does not appear at all surprising.

We should point out that we still do not know whether the restriction $n < \sqrt{N}$ can be removed.

It should be noted that whereas the implication (1) ⇒ (3) is not obvious, and its proof required application of tools like Bernstein's inequality and Jackson's theorem, the implication (3) ⇒ (1) is trivial: one just has to pass from the Riemann sums to the integral. Even more is true: (1) follows directly not only from (3), but also from the following, apparently weaker, statement

(3") There exists an infinite collection $\{N_j\}$ of positive integers and a function φ, $\varphi(n) \to +\infty$ as $n \to \infty$, $\varphi(n) \leq \sqrt{n}$, with the property that for every $A \in \bigcup_{j=1}^{\infty} \mathfrak{A}(N_j, \varphi(N_j))$, all the eigenvalues of A^*A are $\leq C$.

The situation is currently as follows: we have established in this note that (3) is valid; we know also that from (3) or from (3") one can deduce in a few lines, by a standard argument, that the Fourier series of every square-integrable function converges almost everywhere; however, we do not have yet a direct proof of the elementary statement (3). Until we obtain such a direct proof, we can not be satisfied with our understanding of the convergence of Fourier series.

REFERENCES

1. L. Carleson, On convergence and growth of partial sums of Fourier series, Acta Math. 116(1966), 135-157.

2. R. Hunt, On the convergence of Fourier series, Proceedings of the Conference on Orthogonal Expansions and their Continuous Analogues, Carbondale Press.

3. C. Fefferman, Pointwise convergence of Fourier series, Annals of Mathematics, 98(1973), 551-571.

Department of Mathematics, The Ohio State University, Columbus, OH 43210

Proceedings of Symposia in Pure Mathematics
Volume XXXV, Part 1, 1979

MULTIPLE FOURIER SERIES OF FUNCTIONS OF GENERALIZED BOUNDED VARIATION

Daniel Waterman[1]

ABSTRACT. The status of the localization problem for multiple Fourier series is reviewed. An answer to the problem for the rectangular partial sums of the Fourier series of functions of two variables is presented. The hypothesis is that a function be measurable and, corresponding to each coordinate direction, there is an equivalent function that is of harmonic bounded variation on almost every line in that direction and whose total harmonic variation on those lines is an integrable function of the other variable. This is best possible in the sense that if Λ-variation is substituted for harmonic variation and Λ-bounded variation does not imply harmonic bounded variation, then there is a function of that class for which localization fails for square partial sums. The problems remaining in this area are noted.

The localization principle for one variable asserts that if $f : [-\pi,\pi] \to R$ is integrable and vanishes in an open interval, then the partial sums of the Fourier series of f converge uniformly to zero on any compact subset of the interval. If we consider integrable $f : [-\pi,\pi]^n \to R$, we know the following facts:

1. For $n = 2$, localization may fail for differentiable f [4].

2. For $n = 2$, if f is of bounded variation in the sense of Cesari (BVC), which is equivalent to saying that its gradient is a measure, then not only does localization hold, but the rectangular partial sums converge almost everywhere [1, 13]. Note that BVC contains the Sobolev space W_1^1.

3. For $n > 2$, localization holds for square sums if f is in the Sobolev space W_1^p, $p \geq n-1$ [4], and for $p > n-1$, it holds for rectangular partial sums [8]. When $p < n-1$, there is an $f \in W_1^p$ for which localization by square sums fails, and for $p = n-1$, there is an $f \in W_1^p$ for which localization by rectangular sums fails [4].

The Cesari class, BVC, may be defined alternatively (in R^n) as follows:

AMS(MOS) subject classifications (1970). Primary 26A45; Secondary 26A54, 42A48, 42A92.

[1]Research supported by the National Science Foundation under grant MCS77-00840.

> $f \in BVC$ if f is measurable and, corresponding to each coordinate direction, there is an equivalent function which is of bounded variation on a.e. line in that direction, and whose total variation on those lines is an integrable function of the remaining $(n-1)$ variables.

We shall consider a generalization of this notion. Let $\Lambda = \{\lambda_n\}$ be a non-decreasing sequence of positive real numbers such that $\Sigma\, 1/\lambda_n$ diverges. A function g defined on $[a,b] \subset R$ is said to be of <u>Λ-bounded variation</u> (ΛBV) if $\Sigma |g(a_n)-g(b_n)|/\lambda_n$ converges for every sequence of non-overlapping intervals $[a_n,b_n] \subset [a,b]$. The supremum of such sums is the <u>total Λ-variation</u> of g on $[a,b]$. If $\lambda_n = n$, we say that g is of <u>harmonic bounded variation</u> (HBV). The nature of these functions and the convergence and summability properties of their Fourier series have been studied extensively [9,10,11,14,15,16,17,18,19].

In collaboration with Casper Goffman, we recently obtained results relating these notions to multiple Fourier series [5,6]. We have shown that <u>if the Cesari bounded variation is generalized by replacing ordinary variation by harmonic variation, then, in R^2, the localization principle holds for integrable functions of that class</u>.

We define $V^p_{\Lambda,\alpha}$ to consist of those $f \in L^p$, $p \geq 1$, on an interval in R^n, to which these correspond equivalent f_i, $i = 1,\ldots,n$, such that, on almost every line in the i-th coordinate direction, $f_i \in \Lambda BV$ and V_i, the total Λ-variation of f_i on these lines, is in L^α, $\alpha \geq 1$, as a function of the remaining $(n-1)$ variables. For $n > 2$, we assume further that each f_i, restricted to almost any line in the i-th coordinate direction, has, at each point, a value between the upper and lower limits at that point.

If, for $f \in V^p_{\Lambda,\alpha}$, we choose corresponding functions f_i, $i=1,\ldots,n$, to be right continuous for $-\pi \leq x_i < \pi$ and left continuous at $x_i = \pi$ on almost every line in the i-th coordinate direction, then

$$\|f\|_{\Lambda,p,\alpha} = \|f\|_p + \sum_1^n \|V_i(f_i)\|_\alpha$$

defines a norm on $V^p_{\Lambda,\alpha}$ with its elements considered to be equivalence classes of a.e. equal functions. $V^p_{\Lambda,\alpha}$ is a Banach space under this norm.

Using this fact, we were able to show that our localization theorem is best possible in the following sense:

<u>If ΛBV is not contained in HBV, then the localization property for square partial sums does not hold for the class $V^1_{\Lambda,1}$ in R^2</u>.

In the proof of our localization theorem we made two assumptions:

1. $f_i(x_1,x_2)$ is right continuous as a function of x_i in $[-\pi,\pi)$ and left continuous at $x_i = \pi$ for a.e. x_j, $j \neq i$.

2. $V_i(f_i,[a,b])$, the total Λ-variation of f_i with respect to x_i on the interval $[a,b]$, is a measurable function of x_j, $j \neq i$, for any interval $[a,b] \subset [-\pi,\pi]$.

By quite delicate considerations we were able to show that, for any f in $V^1_{\Lambda,1}(R^2)$ we may choose measurable equivalent functions f_i which satisfy these assumptions. It follows from recently obtained results [11] that these f_i minimize $\|V_i(f_i)\|_1$. There would be no need for such considerations if f were continuous, but it should be noted that there are functions of bounded variation in the sense of Cesari all of whose equivalent functions are nowhere continuous.

In the argument which shows that f_i can be chosen to satisfy the above assumptions, what is important is the almost continuity of f_i in the one other variable x_j. A function is almost continuous at p if there is a set Z of measure zero such that the restriction of the function to the complement of Z is continuous at p. In n dimensions, $n > 2$, we would need almost continuity in the (n-1) other variables. In the case of localization of Fourier series of functions in W_1^p, the requirement $p > n-1$ ensured continuity on almost every (n-1)-dimensional hyperplane perpendicular to a coordinate axis. In the present case it is not clear what conditions are required.

We conclude by describing several problems in this area which remain to be solved.

1. Can the localization result be extended to higher dimensions?

2. In our definition of $V^p_{\Lambda,\alpha}$ we use n equivalent functions. In the case of bounded variation in the sense of Cesari, it was shown by Hughs and Serrin [7,12] that one equivalent function may be used for all coordinate directions. Whether this can be done for $V^p_{\Lambda,\alpha}$ is an interesting problem.

3. Cesari has shown that in the two variable case, BVC implies a.e. convergence of the rectangular partial sums of the Fourier series, while in the case $n=3$, a.e. convergence follows from the additional assumption that the variations of the f_i are in L^p, $p > 1$ [1,2]. Chen has extended Cesari's argument to all $n > 2$ and shown that L^p may be replaced by $L(\log^+ L)^{n-2}$ [3]. We conjecture that $V^1_{H,1}$ suffices for a.e. convergence of double Fourier series. For $n > 2$, we have no conjecture at this time.

4. Is $V^p_{\Lambda,\alpha}$ preserved under rotation of axes? It is well known that BVC is preserved, but the arguments used in that connection have no applicability in the present context. This problem may be very difficult.

REFERENCES

1. L. Cesari, Sulla funzione di due variabli a variazione limitata seconde Tonelli a sulla convergenza della relative serie doppie di Fourier, Rend. Serie Mat R. Univ. Roma 1(1937), 277-294.

2. ________, Sulla funzione di più variabli generalmente a variazone limitata e sulla convergenza della relative serie multiplie di Fourier. Comment. Pontificia Acad. Sci. 7(1939), 171-197.

3. Jau-D. Chen, A theorem of Cesari on multiple Fourier series, Studia Math. 29(1973), 69-80.

4. C. Goffman and F. C. Liu, Localization of square sums of multiple Fourier series, Studia Math. 44(1972), 61-69.

5. ________ and D. Waterman, On localization for double Fourier series, Proc. Natl. Acad. Sci. USA 75(1978), 580-581.

6. ________, The localization principle for double Fourier series, Studia Math., to appear.

7. R. E. Hughs, Functions of BVC type, Proc. A.M.S. 12(1961), 698-701.

8. F. C. Liu, On the localization of rectangular partial sums, Proc. A.M.S. 34(1972), 90-96.

9. C. W. Onneweer and D. Waterman, Fourier series of functions of harmonic bounded fluctuation on groups, J. d'Analyse Math. 27(1974), 79-93.

10. S. J. Perlman, Functions of generalized variation, Fund. Math., to appear.

11. ________ and D. Waterman, Some remarks on functions of Λ-bounded variation, Proc. A.M.S., to appear.

12. J. B. Serrin, On the differentiability of functions of several variables, Arch. Rational Mech. Anal. 7(1961), 359-372.

13. L. Tonelli, Sulla serie doppie di Fourier, Ann. Scuola Norm. Sup. Pisa 6(1937), 315-326.

14. D. Waterman, On convergence of Fourier series of generalized bounded variation, Studia Math. 44(1972), 107-117.

15. ________, On the summability of Fourier series of functions of Λ-bounded variation, Studia Math. 55(1976), 87-95.

16. ________, On Λ-bounded variation, Studia Math. 57(1976), 33-45.

17. ________, Bounded variation and Fourier series, Real Anal. Exch. 3(1977-78), 61-86.

18. ________, Fourier series of functions of Λ-bounded variation, Proc. A.M.S., to appear.

19. ________, Λ-bounded variation: recent results and unsolved problems, Real Anal. Exch., to appear.

SYRACUSE UNIVERSITY

Proceedings of Symposia in Pure Mathematics
Volume XXXV, Part 1, 1979

SOME INEQUALITIES FOR RIESZ POTENTIALS OF TRIGONOMETRIC POLYNOMIALS OF SEVERAL VARIABLES

G. Wilmes[1]

ABSTRACT. Riesz-type inequalities are established enabling one to estimate the norms of Riesz potentials of trigonometric polynomials of several variables by moduli of continuity. Characterizations of the Peetre K-functional related to Riesz potential spaces are considered as applications. Moreover, an inverse theorem of A. Zygmund for the best approximation by trigonometric polynomials is regained and extended.

The main subject of the following considerations are Riesz-type inequalities for Riesz potentials of trigonometric polynomials of several variables. This nomenclature rests upon a result of M. Riesz [9] who proved that

$$\sup_{x\in[-\pi,\pi)} |P_n'(x)| \leq \frac{n}{2} \sup_{x\in[-\pi,\pi)} |P_n(x+\tfrac{\pi}{n}) - P_n(x)| \tag{1}$$

holds for any trigonometric polynomial $P_n(x) := \sum_{k=-n}^{n} c_k e^{ikx}$ with $c_k \in \mathbb{C}$ (the complex plane). There exist various extensions of this inequality which turned out to be quite useful, in particular in connection with certain problems in approximation theory. For example, S.N. Bernstein, S.M. Nikolskii, and S.B. Steckin (1948) obtained ($0<h<2\pi/n$, $k\in\mathbb{P}$, the set of non-negative integers, $1\leq p\leq\infty$)

$$\left\{\int_{-\pi}^{\pi} |P_n^{(k)}(x)|^p dx\right\}^{1/p} \leq \left[\frac{n}{2\sin(hn/2)}\right]^k \left\{\int_{-\pi}^{\pi} |\Delta_h^k P_n(x)|^p dx\right\}^{1/p} \tag{2}$$

as well as an analogous result for entire functions of exponential type (see [11, pp. 215,217] and the literature cited there). Here $\Delta_h^k P_n$ denotes the k-th order difference of P_n with increment h (cf. (8)). For extensions to partial derivatives in the multi-dimensional case see [11, p. 217; 8, p. 66] whereas generalizations in the frame work of general Banach spaces are contained in [8, pp. 56-59].

AMS(MOS) subject classification (1970). 42 A 04, 42 A 08.

[1]Supported by Grant No. II B4 - FA 7109 awarded by the Minister für Wissenschaft und Forschung des Landes Nordrhein-Westfalen.

It was pointed out by R.P. Boas (see [3]) that (2) could be obtained via an idea of P. Civin [6], which enables one to evaluate the norms of convex multipliers. In the following it is indicated how this idea can be carried over in establishing Riesz-type inequalities for Riesz potentials of trigonometric polynomials in the (radial) multi-dimensional case. It should be noted that the treatment is indeed closely related to the corresponding one given in [13] for entire functions of exponential type so that proofs need only be sketched in the following.

Let $\mathbb{R}^N$ be the Euclidean N-space with elements $x := (x_1,\dots,x_n), y, u, \dots$, inner product $xy := x_1y_1+\dots+x_Ny_N$, and norm $|x| := \sqrt{xx}$. Elements of the hypersurface S_{N-1} of the unit sphere in $\mathbb{R}^N$ are denoted by $\omega := (\omega_1,\dots,\omega_N)$, whereas $d\omega$ means the restriction of the Lebesgue measure to S_{N-1}. With

$$Q^N := \{x \in \mathbb{R}^N;\ -\pi \leq x_j < \pi,\ 1 \leq j \leq N\}$$

let $L^p_{2\pi}$ be the space of measurable functions with period 2π in each variable for which

$$\|f\|_p := \{(2\pi)^{-N} \int_{Q^N} |f(x)|^p dx\}^{1/p}, \qquad \|f\|_\infty := \operatorname*{ess\,sup}_{x \in Q^N} |f(x)|,$$

respectively, is finite. Then for $k \in \mathbb{Z}^N$ (the set of integral lattice points in $\mathbb{R}^N$) the k-th Fourier coefficient of $f \in L^1_{2\pi}$ is given by

$$f^\wedge(k) := (2\pi)^{-N} \int_{Q^N} f(x) e^{-ikx} dx.$$

The Riesz potential spaces H^α_p, $1 \leq p \leq \infty$, $\alpha > 0$, are defined by

$$(3) \qquad H^\alpha_p := \{f \in L^p_{2\pi};\ |k|^\alpha f^\wedge(k) = g^\wedge(k) \text{ for some } g =: I^\alpha f \in L^p_{2\pi},\ k \in \mathbb{Z}^N\},$$

equipped with the norm

$$(4) \qquad \|f\|_{p,\alpha} := \|f\|_p + |f|_{p,\alpha}, \qquad |f|_{p,\alpha} := \|I^\alpha f\|_p.$$

As a first step to the inequalities announced for Riesz potentials $I^\alpha P_\rho$ of trigonometric polynomials

$$P_\rho \in T_\rho := \{P_\rho;\ P_\rho(x) := \sum_{|k| \leq \rho} c_k e^{ikx}, \quad c_k := P_\rho^\wedge(k) \in \mathbb{C}\},$$

it is convenient to consider Riesz-type inequalities related to the operators

D_ω^α, $\overline{D}_\omega^\alpha$ which for any $\alpha > 0$, $\omega \in S_{N-1}$ are defined on T_ρ by

$$(5) \qquad (D_\omega^\alpha P_\rho)(x) := \sum_{|k| \leqslant \rho} (ik\omega)^\alpha P_\rho^\wedge(k) e^{ikx},$$

$$(6) \qquad (\overline{D}_\omega^\alpha P_\rho)(x) := \sum_{|k| \leqslant \rho} |k\omega|^\alpha P_\rho^\wedge(k) e^{ikx},$$

the fractional powers of $(ik\omega)^\alpha$ being interpreted as principal values, i.e.

$$(7) \qquad (ik\omega)^\alpha := |k\omega|^\alpha \exp\{i\tfrac{\alpha\pi}{2} \operatorname{sgn} k\omega\}.$$

For $f \in L^p_{2\pi}$, $h \in \mathbb{R}^N$ let us introduce the (radial) difference and modulus of continuity of order $\alpha > 0$ by ($f \in L^p_{2\pi}$, $h \in \mathbb{R}^N$)

$$(8) \qquad \Delta_h^\alpha f(x) := \sum_{j=0}^\infty (-1)^j \binom{\alpha}{j} f(x+jh), \qquad \omega_{\alpha,p}(t,f) := \sup_{0<|h|\leqslant t} \|\Delta_h^\alpha f\|_p.$$

<u>THEOREM 1:</u> Let $0 < r < 2\pi/\rho$, $\alpha > 0$, and $\omega \in S_{N-1}$ be fixed. Then for any $P_\rho \in T_\rho$ there hold the Riesz-type inequalities

$$(9) \qquad \|D_\omega^\alpha P_\rho\|_p \leqslant \left[\frac{\rho}{2\sin(r\rho/2)}\right]^\alpha \|\Delta_{-r\omega}^\alpha P_\rho\|_p \qquad (1 \leqslant p \leqslant \infty),$$

$$(10) \qquad \|\overline{D}_\omega^\alpha P_\rho\|_p \leqslant C_p \left[\frac{\rho}{2\sin(r\rho/2)}\right]^\alpha \|\Delta_{-r\omega}^\alpha P_\rho\|_p \qquad (1 < p < \infty).$$

<u>Proof:</u> Obviously one has ($k \in \mathbb{Z}^N$)

$$(\Delta_h^\alpha t_\rho)^\wedge(k) = t_\rho^\wedge(k) \sum_{j=0}^\infty (-1)^j \binom{\alpha}{j} e^{ijhk} = t_\rho^\wedge(k)(1-e^{ihk})^\alpha.$$

It was shown in [13] that for $|k| \leqslant \rho$, $h := -r\omega$ there holds

$$(ik\omega)^\alpha/(1-e^{ihk})^\alpha = \varphi(k\omega) \exp\{-i\alpha \tfrac{hk}{2}\}$$

with some 2ρ-periodic function $\varphi(t)$ defined by

$$\varphi(t) := (t/2\sin(rt/2))^\alpha \qquad (\rho \leqslant t < \rho).$$

Moreover, φ has an absolutely convergent Fourier series with alternating coefficients, i.e.,

$$\varphi(t) = \sum_{j=-\infty}^\infty a_j e^{i(\pi j t)/\rho}, \qquad (-1)^j a_j \geqslant 0.$$

Thus, setting $y(j) := (\alpha/2)r\omega + (\pi j/\rho)\omega$ and representing φ by its Fourier series, one obtains (cf. [6; 11, p. 215])

$$\|D_\omega^\alpha P_\rho\|_p = \Big\| \sum_{|k|\leq\rho} \sum_{j=-\infty}^{\infty} a_j (1-e^{-irk\omega})^\alpha P_\rho^\wedge(k) e^{ik(x+y(j))} \Big\|_p$$
$$\leq \sum_{j=-\infty}^{\infty} |a_j| \, \|\Delta_{-r\omega}^\alpha P_\rho\|_p = \varphi(\rho) \|\Delta_{-r\omega}^\alpha P_\rho\|_p .$$

This proves (9). Inequality (10) may be derived from (9) by the Marcinkiewicz multiplier criterion (see [13] in connection with [10, p. 263]). □

Concerning the one-dimensional situation, (9) reproduces (2), at the same time generalizing it to derivatives of fractional order. Moreover, an immediate consequence of (9) is the Bernstein inequality

$$\Big\| \sum_{j=-n}^{n} (ij)^\alpha P_n^\wedge(j) e^{ijx} \Big\|_p \leq 2^{1-\alpha} n^\alpha \|P_n\|_p \qquad (0<\alpha\leq 1) \tag{11}$$

which sharpens a result of T. Bang [1] who obtained (11) with a constant $C_\alpha := 2^{1-\alpha}/\Gamma(2-\alpha)$. With exception of the trivial case $p = 2$, however, the best possible constant in (11) seems to be an open problem (for $0 < \alpha < 1$).

In order to derive Riesz-type inequalities for Riesz potentials one needs a connection between the operators $I^\alpha, D_\omega^\alpha$, and $\overline{D}_\omega^\alpha$ (recall (3, 5-6)). This is given via ($P_\rho \in T_\rho$, $x \in \mathbb{R}^N$)

$$I^\alpha P_\rho(x) = \frac{B_\alpha}{\cos(\alpha\pi/2)} \int_{S_{N-1}} D_\omega^\alpha P_\rho(x)\,d\omega \qquad (\alpha \neq 2j+1,\ j \in \mathbb{P}), \tag{12}$$

$$I^\alpha P_\rho(x) = B_\alpha \int_{S_{N-1}} \overline{D}_\omega^\alpha P_\rho(x)\,d\omega \qquad (\alpha > 0) \tag{13}$$

with $B_\alpha := \int_{S_{N-1}} |\omega_1|^\alpha d\omega$. These equalities follow from the fact that $|k|^\alpha$ can be represented by ($k \in \mathbb{Z}^N$)

$$|k|^\alpha = \frac{B_\alpha}{\cos(\alpha\pi/2)} \int_{S_{N-1}} (ik\omega)^\alpha d\omega \qquad (\alpha \neq 2j+1,\ j \in \mathbb{P}), \tag{14}$$

$$|k|^\alpha = B_\alpha \int_{S_{N-1}} |k\omega|^\alpha d\omega \qquad (\alpha > 0). \tag{15}$$

For the details, however, see [13].

<u>THEOREM 2:</u> For any $P_\rho \in T_\rho$, $\rho > 0$, there hold the inequalities

$$\|I^\alpha P_\rho\|_p \leq C_\alpha \rho^\alpha \omega_{\alpha,p}(\rho^{-1}, P_\rho) \qquad (p = 1,\infty; 0<\alpha \neq 2j+1, j \in \mathbb{P}), \tag{16}$$

(17) $$\| I^{\alpha}P_{\rho}\|_{p} \leq C_{\alpha p}\, \rho^{\alpha}\omega_{\alpha,p}(\rho^{-1},P_{\rho}) \qquad (1 < p < \infty, \alpha > 0).$$

Proof: It follows from (9) with $r := 1/\rho$ and (12) that for any $\alpha \neq 2j+1$, $j \in \mathbb{P}$

$$\| I^{\alpha}P_{\rho}\|_{p} = \frac{B_{\alpha}}{|\cos(\alpha\pi/2)|} \Big\| \int_{S_{N-1}} D_{\omega}^{\alpha}P_{\rho}(x)d\omega \Big\|_{p}$$

$$\leq C'_{\alpha}\rho^{\alpha} \int_{S_{N-1}} \| \Delta^{\alpha}_{-(1/\rho)\omega}P_{\rho}\|_{p}d\omega \leq C_{\alpha}\rho^{\alpha}\omega_{\alpha,p}(\rho^{-1},P_{\rho}).$$

Analogously (17) follows from (10) and (13). □

As an application of Thm. 2 let us consider estimates for the modified Peetre K-functional (cf. (3 - 4))

(18) $$K_{\alpha,p}(t,f) := \inf_{g \in H_p^{\alpha}} \{\| f-g\|_{p} + t|g|_{p,\alpha}\} \qquad (\alpha > 0, 1 \leq p \leq \infty)$$

by the modulus of continuity (8). Such relations are often useful to study the rate of convergence by certain linear processes in dependence upon smoothness properties of the associated functions to be approximated. For example, following [12] it is not hard to give an exact description of the approximation by the Riesz means

(19) $$(R,\alpha,\gamma)_{\rho}f(x) := \sum_{|k| \leq \rho} (1 - (|k|/\rho)^{\alpha})^{\gamma} f^{\wedge}(k)e^{ikx} \qquad (f \in L^{p}_{2\pi})$$

if one expresses smoothness via the K-functional (18), namely

(20) $$\| (R,\alpha,\gamma)_{\rho}f - f\|_{p} \sim K_{\alpha,p}(\rho^{-\alpha},f) \qquad (\gamma > (N-1)|\tfrac{1}{p}-\tfrac{1}{2}|).$$

The following characterization of $K_{\alpha,p}$ then delivers direct and inverse theorems for the Riesz means with smoothness expressed by (classical) moduli of continuity.

THEOREM 3: Let $0 < \gamma < \alpha < \beta$. Then for any $f \in L^{p}_{2\pi}$

(21) $$\omega_{\alpha,p}(t,f) \leq C_{\alpha p}\, K_{\alpha,p}(t^{\alpha},f) \leq C'_{\alpha p}\, \omega_{\alpha,p}(t,f) \qquad (1 < p < \infty),$$

(22) $$\omega_{\beta,p}(t,f) \leq C_{\alpha\beta}\, K_{\alpha,p}(t^{\alpha},f) \leq C_{\alpha\gamma}\, \omega_{\gamma,p}(t,f) \qquad (1 \leq p \leq \infty),$$

(23) $$K_{\alpha,p}(t^{\alpha},f) \leq C_{\alpha}\, \omega_{\alpha,p}(t,f) \qquad (p = 1,\infty, \alpha \neq 2j+1, j \in \mathbb{P}).$$

The estimates from above follow immediately from Thm. 2 if one inserts polynomials of best approximation into the definition (18) of $K_{\alpha,p}$. The other inequalities may be proved by means of suitable multiplier criteria, in particular using [5a], but for all details see [13] (compare also [12]).

Note that the restriction to $\alpha \neq 2j+1$, $j \in \mathbb{P}$, in (16) cannot be dropped. Indeed, this would lead via (23) in connection with (20) to contradictions to classical results concerning the approximation by Riesz means (cf. [5,p. 145] for e.g. $\alpha = N = 1$).

Let us finally consider an inverse theorem for the best approximation by trigonometric polynomials in $L^p_{2\pi}$, $1 < p < \infty$,

(24) $$E_{p,\rho}(f) := \inf_{P_\rho \in T_\rho} \|f - P_\rho\|_p =: \|f - P_\rho(f)\| \qquad (f \in L^p_{2\pi}).$$

THEOREM 4: If $f \in L^p_{2\pi}$, $1 < p < \infty$, satisfies for some $\alpha > 0$

(25) $$E_{p,\rho}(f) = O(\rho^{-\alpha}) \qquad (\rho \to \infty),$$

then ($\tilde{p} := \min\{p,2\}$)

(26) $$\omega_{\alpha,p}(t,f) = O(t^\alpha |\log t|^{1/\tilde{p}}).$$

Proof: Note that (25) holds if and only if f belongs to $B^\alpha_{p,\infty}$, where $B^\alpha_{p,q}$, $\alpha > 0$, $1 \leqslant p,q \leqslant \infty$, denote the Besov spaces which may be represented as interpolation spaces between $L^p_{2\pi}$ and H^β_p for some $\beta > \alpha$. Thus with the usual modification for $q = \infty$ (cf. [2,p. 188])

$$B^\alpha_{p,q} := \{f \in L^p_{2\pi}; \|f\|^\alpha_{p,q} < \infty\}, \quad \|f\|^\alpha_{p,q} := \|f\|_p + |f|^\alpha_{p,q},$$

$$|f|^\alpha_{p,q} := \{\int_0^1 [t^{-(\alpha/\beta)} K_{\beta,p}(t,f)]^q \frac{dt}{t}\}^{1/q}.$$

Moreover, by (18), (24 - 25) one has $(t \to 0+)$

$$K_{\alpha,p}(t^\alpha,f) \leqslant \|f - P_{1/t}(f)\|_p + t^\alpha |P_{1/t}(f)|_{p,\alpha} = O(t^\alpha |P_{1/t}(f)|_{p,\alpha}).$$

Thus (26) will be an immediate consequence of Thm. 3 as soon as one has shown the following

THEOREM 5: For any $P_\rho \in \mathcal{T}_\rho$ there holds ($\tilde{p} := \min\{p,2\}$)

$$|P_\rho|_{p,\alpha} \leq C(1+\log^+\rho)^{1/\tilde{p}-1/q} \|P_\rho\|^{\alpha}_{pq} \qquad (1<p<\infty, \tilde{p}\leq q \leq \infty), \tag{27}$$

with $\log^+\rho := \max\{\log\rho, 0\}$.

Proof: For $q=\tilde{p}$ the result follows from the well-known embedding theorem for Besov spaces (cf. [2, p. 152]). Moreover, the Bernstein inequality $|P_\rho|_{p,\beta} \leq C\rho^\beta \|P_\rho\|_p$ (which can e.g. easily be obtained from (17)) implies

$$K_{\beta,p}(t,P_\rho) \leq Ct\rho^\beta \|P_\rho\|_p .$$

Thus, for any $q_o < q$ and $\Theta = (\alpha/\beta)$, $r' := q/q_o$, $r := r'/(r'-1)$, $\gamma := 2\beta/(1-\Theta)$ it follows by Hölder's inequality that $(\rho \geq 1)$

$$\begin{aligned}
(|P_\rho|^{\alpha}_{pq_o})^{q_o} &= \Big(\int_0^{\rho^{-\gamma}} + \int_{\rho^{-\gamma}}^1\Big) t^{-\Theta q_o}[K_{\beta,p}(t,P_\rho)]^{q_o} \frac{dt}{t} \\
&\leq C\rho^{\beta q_o}\|P_\rho\|_p^{q_o} \int_0^{\rho^{-\gamma}} t^{q_o(1-\Theta)-1} dt \\
&\quad + \int_{\rho^{-\gamma}}^1 t^{-1/r}\, t^{-\Theta q_o-(1/r')} [K_{\beta,p}(t,P_\rho)]^{q_o} dt \\
&\leq C'\rho^{\beta q_o}\rho^{-\gamma q_o(1-\Theta)} \|P_\rho\|_p^{q_o} \\
&\quad + \Big(\int_{\rho^{-\gamma}}^1 \frac{dt}{t}\Big)^{1/r}\Big(\int_{\rho^{-\gamma}}^1 [t^{-\Theta}K_{\beta,p}(t,P_\rho)]^{q_o r'} \frac{dt}{t}\Big)^{1/r'} \\
&\leq C''(\log\rho)^{1-(q_o/q)} (|P_\rho|^{\alpha}_{pq})^{q_o} + C'''\rho^{-\beta q_o}\|P_\rho\|_p^{q_o} .
\end{aligned}$$

This completes the proof. □

Counterparts of Thms. 4 - 5 for entire functions of exponential type are found in [7]. Moreover, Thm. 4 extends a result of A. Zygmund (see [14; 11, p. 339]) who proved the result for $\alpha = 1$ in the one-dimensional case. He also gave a counterexample which shows that the exponent $1/\tilde{p}$ in (26) cannot in general be replaced by a smaller one. A different approach to Thm. 4 can be deduced via comparison theorems (see [4]). It is interesting to note that the approach in [14] as well as the comparison theorems in [4] are based upon inequalities of Littlewood-Paley-type. Here such an inequality is actually implicitely contained in (27). Indeed, if one uses Peetre's definition of the

Besov spaces, then (27) for $\tilde{p} = q$ may be interpreted as a Littlewood-Paley-type inequality (cf. [2, pp. 152, 188]).

References

[1] T. Bang, Une inéqualité de Kolmogoroff et les fonctions presque périodiques; Danske Videnskab. Selesk. 19 (1941), Nr. 4.

[2] J. Bergh, J. Löfström, Interpolation Spaces; Springer, Berlin 1976.

[3] R.P. Boas, Comment, in: Mathematical Reviews 9 (1948), 579-580.

[4] J. Boman, H.S. Shapiro, Comparison theorems for a generalized modulus of continuity; Ark. Math. 9 (1971), 91-116.

[5] P.L. Butzer, R.J. Nessel, Fourier Analysis and Approximation, Vol. I: One-Dimensional Theory; Academic Press, New York, and Birkhäuser, Basel 1971.

[5a] P.L. Butzer, U. Westphal, An access to fractional differentiation via fractional difference quotients, in: Fractional Calculus and its Applications; Lecture Notes in Math. 457, Springer, Berlin 1975, 116-145.

[6] P. Civin, Inequalities for trigonometric integrals; Duke Math. J. 8 (1941), 656-665.

[7] R.J. Nessel, G. Wilmes, Nikolskii-type inequalities in connection with regular spectral measures, Acta Math. Acad. Sci. Hungar. 33 (1978), 169-182.

[8] R.J. Nessel, G. Wilmes, Über Ungleichungen vom Bernstein-Nikolskii-Riesz-Typ in Banach Räumen; Forschungsberichte des Landes NRW, Westdeutscher Verlag, Opladen 1978.

[9] M. Riesz, Eine trigonometrische Interpolationsformel und einige Ungleichungen für Polynome; Jber. Deutsch. Math.-Verein. 23 (1914), 354-368.

[10] E.M. Stein, G. Weiss, Introduction to Fourier Analysis on Euclidean Spaces; Princeton Univ. Press, Princeton 1970.

[11] A.F. Timan, Theory of Approximation of Functions of a Real Variable; Macmillan, New York 1963.

[12] W. Trebels, On the approximation behaviour of the Riesz means in $L^p(R^N)$, in: Approximation Theory; Lecture Notes in Math. 556, Springer, Berlin 1976, 428-438.

[13] G. Wilmes, On Riesz-type inequalities and K-functionals related to Riesz potentials in R^N, Numerical Functional Analysis and Optimization 1 (1978), in print.

[14] A. Zygmund, A remark on the integral modulus of continuity; Revista Univ. Nac. Tucuman, A, 7 (1950), 259-269.

Proceedings of Symposia in Pure Mathematics
Volume XXXV, Part 1, 1979

A NOTE ON SOBOLEV SPACES

Björn E.J. Dahlberg

Let W_m^p denote the Sobolev space of functions u on R^n which have all its derivatives of order up to m in L^p. Suppose now that $H \in C_0^\infty(R)$ and $H(0) = 0$. It is well known that $H(u) \in W_1^p$ for all $u \in W_1^p$. It is also known that if $p \geq \frac{n}{m}$ then

(1) $$H(u) \in W_m^p \text{ for all } u \in W_m^p,$$

see Adams [1], where more information about this can be found. Also it is easily seen that if $m = 2$ and $p = 1$ then (1) holds. The purpose of this note is to show that these are essentially the only cases for which (1) hold.

THEOREM 1. Suppose $1 \leq p < \frac{n}{m}$ for $m \geq 3$ and $1 < p < \frac{n}{2}$ for $m = 2$. If $H \in C^\infty(R)$ and $H(u) \in W_m^p$ for all $u \in W_m^p$ then $H(u) \equiv cu$.

PROOF. Let $\{y^j\}_1^\infty$ be a sequence of points such that $|y^j - y^k| \geq 10$ for $j \neq k$. Let $u \in C_0^\infty(R^n)$ be such that $u(x) = x_1$ for $|x| \leq 1$ and the support of u is contained in $\{x: |x| \leq 2\}$. It is easily seen that there are positive numbers A_j and ε_j such that $\varepsilon_j \to 0$, $A_j > 1$ and

(2) $$\begin{cases} \Sigma A_j^p \varepsilon_j^{n-mp} < \infty \quad \text{and} \\ \Sigma A_j^{mp-1} \varepsilon_j^{n-mp} = \infty . \end{cases}$$

Let $v(x) = \Sigma u_j(x)$ where $u_j(x) = A_j \, u\left(\frac{x-y^j}{\varepsilon_j}\right)$. It is easily seen that $v \in C^\infty(R^n)$ and

$$\sum_{|\alpha| \leq m} \int_{R^n} \left|\frac{\partial^\alpha v}{\partial x^\alpha}\right|^p dx \leq C \sum_1^\infty A_j^p \varepsilon_j^{n-mp}$$

so it follows from (2) that $v \in W_m^p$. We shall next prove that H must be a polynomial of degree less than m. Suppose this is not the case. Then $\inf\{|H_r^{(m)}(t)|: a \leq t \leq b\} > 0$ for some $a < b$. Let $S_j = \{x: |x-y^j| < \varepsilon_j$, $a\varepsilon_j < A_j(x_1 - y_1^j) < b\varepsilon_j\}$. Since $A_j/\varepsilon_j \to \infty$ as $j \to \infty$ it follows that the area of S_j is bigger then a constant times ε_j^n/A_j if j is large enough. Also, since the supports of the u_j:s are disjoint it follows that $v = u_j$ in S_j. Hence

$$\int \left|\frac{\partial^m}{\partial x_1^m} H(v)\right|^p dx \geq \text{const} \sum A_j^{pm} \varepsilon_j^{-mp} |S_j| \geq \text{const} \sum A_j^{pm-1} \varepsilon_j^{n-mp},$$

and from (2) it follows that $H(v) \notin W_m^p$. Therefore H must be a polynomial and by considering functions of the form $|x|^{-\theta}\psi(x)$, where $\psi \in C_0^\infty(R^n)$ and $\psi(0) \neq 0$ it is easily seen that H must be linear which yields the theorem.

In Adams [1] it is proved that if $H \in C^\infty(R)$ satisfies $|t^{j-1}D^jH(t)| \leq C_j$ for $t > 0$ and $1 < p \leq \infty$ then

$$H(u) \in W_2^p \quad \text{for all} \quad u \in W_{2,+}^p, \tag{3}$$

where $W_{m,+}^p = \{u \in W_m^p: u \geq 0\}$.

We shall next show that there is essentially no analogue of (3) for $W_{m,+}^p$ for $m \geq 3$ and $1 < p < \frac{n}{m}$. (The case $p \geq \frac{n}{m}$ is of course a consequence of (1).)

THEOREM 2. Suppose $m \geq 3$ and $1 < p < \frac{n}{m}$. If $H \in C^\infty(R)$ and $H(u) \in W_m^p$ for all $u \in W_{m,+}^p$ then $H(t) \equiv ct$ for $t > 0$.

PROOF. As above the result follows if we can show that H equals a polynomial for $t > 0$. Let λ_m be the smallest integer $\geq \frac{m}{2}$. We claim that $D^{\lambda_m}H(t) = 0$ for $t > 0$. Suppose this is not the case and let $\xi > 0$ be such that $D^{\lambda_m}H(\xi) \neq 0$. Let $G(t) = H(t+\xi)$. Since $D^{\lambda_m}G(0) \neq 0$ it follows that there are numbers $c > 0$ and $a > 0$ such that

$$|D^mG(t^2)| \geq ct^{\delta_m} \quad \text{for} \quad 0 < t < a, \tag{4}$$

where $\delta_m = 0$ if m is even and $\delta_m = 1$ if m is odd.

It is easily seen that there are numbers $A_j > 1$, $\varepsilon_j \downarrow 0$ such that

$$\begin{cases} \sum \varepsilon_j < \infty, \ \sum A_j^p \varepsilon_j^{n-mp} < \infty \quad \text{and} \\ \sum A_j^{(mp-1)/2} \varepsilon_j^{n-mp} = \infty. \end{cases} \tag{5}$$

Let $y^j = (0, 5\sum_{k=1}^{j} \varepsilon_k, 0,\ldots,0) \in R^n$. We observe that the balls

$B_j = \{x: |x-y^j| < 2\varepsilon_j\}$ are pairwise disjoint and that there is an $R > 0$ such that $\cup B_j \subset \{x: |x| < R\}$. Let $u_j(x) = A_j\, u\left(\frac{x-y^j}{\varepsilon_j}\right)$, where $u \in C_0^\infty(R^n)$ is non-negative, has its support in $\{x: |x| < 2\}$ and $u(x) = x_1^2$ for $|x| < 1$. Pick a $\varphi \in C_0^\infty(R^n)$ such that $\varphi \geq 0$ and $\varphi(x) = 1$ for $|x| \leq R$. It now follows from (5) that if $v = (\xi + \Sigma u_j)\varphi$ then $v \in W^p_{m,+}$. Arguing as in the proof of Theorem 1 it follows from (4) that $\int_{R^n} \left|\frac{\partial^m}{\partial x_1^m} H(v)\right|^p dx \geq \text{const}\ \Sigma A_j^{(mp-1)/2} \varepsilon_j^{n-mp}$ which taken together with (5) shows that $H(v) \notin W^p_m$. Therefore $D^{\lambda_m} H(t) \equiv 0$ for $t > 0$ and as remarked in the beginning of the proof this yields the theorem.

REFERENCE

[1] D.A. Adams, "On the existence of capacitary strong estimates in R^n", Ark.Mat. 14, 1976, 125-140.

UNIVERSITY OF GÖTEBORG
DEPARTMENT OF MATHEMATICS
FACK
S-402 20 GÖTEBORG
SWEDEN

CHAPTER 2

Hardy spaces and BMO

Proceedings of Symposia in Pure Mathematics
Volume XXXV, Part 1, 1979

SOME PROBLEMS IN THE THEORY OF HARDY SPACES

Guido Weiss[1]

INTRODUCTION. During the past ten years the theory of Hardy spaces has undergone a dramatic development. The classical H^p spaces of holomorphic functions have been given several real variable characterizations. These involve various types of maximal functions and the notion of atoms. These notions, in turn, have paved the way for the development of the theory of Hardy spaces in a very general setting. This development is still in progress. The problems I collected here arise naturally from this development.

I claim no originality concerning the problems that are posed here. Most arose in conversations and collaborations with my colleagues at Washington University: R. R. Coifman, R. Rochberg and M. Taibleson. Others came up in earlier collaborations with E. M. Stein. I am willing, however, to shoulder all blame if some are not well posed or turn out to be of little interest. I can only hope that anyone judging these problems will adopt the basic principle of Antoni Zygmund: "when judging a mathematician's work, one should only integrate his f_+ ."

#1. Suppose $\varphi \in L^1(\mathbb{R})$, $\varphi \geqq 0$ and $\|\varphi\|_1 = 1$. We can then consider the maximal function operator

$$(m_\varphi f)(x) = \sup_{\epsilon>0} \left\{ \left| \int_{-\infty}^{\infty} f(y-t)\varphi_\epsilon(t)dt \right| : |x-y| < \epsilon/2 \right\} ,$$

where $\varphi_\epsilon(t) = \frac{1}{\epsilon}\varphi(\frac{t}{\epsilon})$.

AMS(MOS) subject classifications (1970). Primary 30A78.

[1]Research supported by the National Science Foundation under grant MCS75-02411 A03.

Burkholder, Gundy and Silverstein [1] and C. Fefferman and Stein [10] have characterized real H^p in terms of such functions. When $p = 1$ f belongs to real H^1 if and only if f and $m_\varphi f$ belong to $L^1(\mathbb{R})$ provided φ is a "smooth bump" function (φ Lipschitz of any order α, $0 < \alpha \leqq 1$, and with compact support -- or vanishing at ∞ sufficiently fast -- will do). It is natural to ask if an interesting space is so characterized by a function φ that is not smooth. For example, what functions have integrable Hardy-Littlewood maximal function (this is the maximal function obtained when $\varphi = \chi$, the characteristic function of $(-\frac{1}{2},\frac{1}{2})$)? The following result shows that there is a dramatic difference between the smooth maximal functions and m_χ :

THEOREM. If $m_\chi f \in L^1(\mathbb{R})$, <u>for</u> $f \in L^1(\mathbb{R})$, <u>then</u> $f(x) = 0$ a.e..

PROOF. If f is not 0 a.e. there exists an interval (x_1, x_2) such that $\int_{x_1}^{x_2} f = \alpha \neq 0$. Since $f \in L^1(\mathbb{R})$ there exists $R \gg x_2$ such that

$$\int_R^\infty |f| < \frac{|\alpha|}{4} .$$

Observe that either $\left|\int_{x_1}^R f\right| \geqq \frac{\alpha}{2}$ or $\left|\int_{x_2}^R f\right| \geqq \frac{\alpha}{2}$. Otherwise $\left|\int_{x_1}^{x_2} f\right| = \left|\left(\int_{x_1}^R - \int_{x_2}^R\right) f\right| < \alpha$, which is impossible. Let $k = 1$ or 2 be the index for which $\left|\int_{x_k}^R f\right| \geqq \frac{\alpha}{2}$. Then, for $x \geqq R$ we have

$$(m_\chi f)(x) \geqq \frac{1}{x-x_k} \left|\int_{x_k}^x f\right| \geqq \frac{1}{x-x_k} \left\{\left|\int_{x_k}^R f\right| - \int_R^x |f|\right\} \geqq \frac{1}{x-x_k} \left(\frac{\alpha}{2} - \frac{\alpha}{4}\right) .$$

It is now obvious that $m_\chi f$ cannot belong to $L^1(\mathbb{R})$.

PROBLEM. Are there functions φ that are smoother than χ , yet not Lipschitz of order α , such that the condition f , $mf \in L^1(\mathbb{R})$ characterizes a non-trivial space that is not $H^1(\mathbb{R})$? If such a φ exists, is there an operator, that is analogous to the Hilbert transform, which also characterizes this space?

These questions can be extended to other values of p , $0 < p < 1$. The maximal functions $m_\varphi f$ that characterize H^p require more smoothness conditions as p tends to 0 (this is a reflection of the fact that the duals of H^p are Lipschitz spaces of increasingly higher orders as p tends to 0).

Suppose, then, that φ is sufficiently smooth, so that $m_\varphi f$ characterizes all H^p spaces for $p > p_o$, but not H^{p_o} . What can be said about the space of distributions f with $m_\varphi f \in L^{p_o}$?

The above theorem and these questions are valid in n-dimensions.

#2. The problem we shall pose now is clearly related to the one above. Let us first give some background. We consider $H^1(\mathbb{R}^2)$, the space of all real-valued functions in $L^1(\mathbb{R}^2)$ such that its Riesz transforms belong to $L^1(\mathbb{R}^2)$. These transforms correspond to the multiplier operators

$$(R_1 f)^\wedge(re^{i\theta}) = (\cos\theta)\hat{f}(re^{i\theta}) \quad \text{and} \quad (R_2 f)^\wedge(re^{i\theta}) = (\sin\theta)\hat{f}(re^{i\theta}) \ .$$

Hence, $H^1(\mathbb{R}^2)$ consists of those real $f \in L^1(\mathbb{R}^2)$ such that

$$e^{i\theta}\,\hat{f}(re^{i\theta})$$

is the Fourier transform of a function in $L^1(\mathbb{R}^2)$. J. Garcia-Cuerva [11] has shown that there exists a real $f \notin H^1(\mathbb{R}^2)$ such that $e^{2i\theta}\hat{f}(re^{i\theta})$ is the Fourier transform of a function in $L^1(\mathbb{R}^2)$. Thus, the singular integral operator whose associated multiplier is $e^{2i\theta}$ does not determine $H^1(\mathbb{R}^2)$.

PROBLEM. Which (sets of) singular integral operators determine $H^1(\mathbb{R}^n)$?

If a set $\{S_1, S_2, \ldots, S_k\}$ of singular integral operators does not determine $H^1(\mathbb{R}^n)$, as in the example just given, one could investigate the space $\{f \in L^1(\mathbb{R}^n) : S_j f \in L^1(\mathbb{R}^n) \ , \ j = 1, 2, \ldots, k\}$. Is there a maximal function characterization of it? What "H^1-like" properties does it have? Is it a "natural" space for interpolation?

All these questions can be raised for $H^p(\mathbb{R}^n)$ when $p < 1$. There are some added complications arising from the fact that the Riesz transforms only seem to determine $H^p(\mathbb{R}^n)$ for $(n-1)/n \leqq p \leqq 1$; for lower values of p the operators determined by the "higher gradients" of Calderón and Zygmund appear to be the "right" ones for defining H^p (see [3] and [18] and, also, problem #8 below). There are results of Carleson [4] and Janson [13] in this general area. There is still work to be done.

#3. Let us now turn to the atomic characterization of H^p . On $\mathbb{R}$ a p-atom is a function $a(x)$ satisfying: (1) Supp $a \subset I$ (= a finite interval) and $\|a\|_\infty \leqq |I|^{-1/p}$; (2) for $j = 0, 1, \ldots, [\frac{1}{p} - 1]$ (as usual, $[t]$ denotes the largest integer not exceeding t)

$$\int_{-\infty}^{\infty} a(x)x^j dx = 0 \quad .$$

The space real $H^p(\mathbb{R})$ consists of those linear functionals f on an appropriate Lipschitz space (see [20]) having the form

$$f = \sum_{j=1}^{\infty} \lambda_j a_j \quad , \tag{3}$$

where the a_j's are atoms and $\sum |\lambda_j|^p < \infty$. The "norm" of f is the infimum of all expressions $\left(\sum |\lambda_j|^p\right)^{1/p}$ over all representations (3). It can be shown that, if we increase the number of vanishing moments in (2), we obtain, via representations (3), the same space real $H^p(\mathbb{R})$ (see [20]).

PROBLEM. The definition of a space of linear functionals by means of (3), acting on an appropriate Lipschitz space, makes perfectly good sense if we <u>decrease</u> the number of moments in (2). What are the basic properties of these spaces?

For example, when $p = 1/2$, the space we obtain by using atoms $a(x)$ which only satisfy (2) for $j = 0$ has a dual that can be represented by the Lipschitz functions $\ell(x)$ on $\mathbb{R}$ satisfying $|\ell(x)-\ell(y)| \leqq c|x-y|$ for an appropriate constant c . This fact tends to indicate that these spaces are "natural". The problem we are announcing is really an invitation to study these spaces, their multipliers and other properties. Moreover, we need not restrict ourselves to this one dimensional case; we can consider it whenever the notion of vanishing moments is appropriate in the atomic characterization of H^p (see, also, problem #4 in this connection).

The "right" number of moments that vanish for a p-atom and the smoothness of φ in #1 are probably related. It would be of interest to understand better the nature of this relation.

#4. Atomic H^p spaces can be naturally defined on spaces of homogeneous type (see [9]). The notion of vanishing moments, however, does not make sense in this general setting. For this reason a general theory of H^p spaces on spaces of homogeneous type was developed [9] in which condition (2) in #3 was only assumed for $j = 0$. There exists, however, a condition that replaces (2) that can be extended to much more general settings. In order to describe the situation let us return to $\mathbb{R}$. Suppose $a(x)$ satisfies (1) in #3 and

(2) $$\mathfrak{M}_a(x) = \int_I \frac{a(y)}{1+|x-y|}\,dy \in L^p(\mathbb{R}) \quad .$$

Suppose, first, that $p = 1$. If a has mean 0 then

(3) $$\mathfrak{M}_a(x) = \int_I a(y)\left\{\frac{1}{1+|x-y|} - \frac{1}{1+|x|}\right\} dy \quad .$$

From this it follows easily that $\mathfrak{M}_a(x)$ is integrable (remember that y belongs to the bounded interval I). Conversely, if a does not have mean 0 , $\mathfrak{M}_a$ cannot be integrable: for let $m_I = m_I(a) = \frac{1}{|I|}\int_I a(t)dt$, then

(4) $$\mathfrak{M}_a(x) = \int_I \frac{a(y)-m_I\chi_I(y)}{1+|x-y|}\,dy + m_I\int_I \frac{dy}{1+|x-y|} \quad .$$

The first term in this sum is integrable (by the argument we just gave) while the second is obviously not integrable. Thus, (2), for $p = 1$, is equivalent to $\int a = 0$. In fact, a slight modification of this argument shows that (2), for $\frac{1}{2} < p \leqq 1$, is equivalent to $\int a = 0$ (observe that it follows from (3) that, if $2|y| < |x|$, then $|\mathfrak{M}_a(x)|^p \leqq 2^p|I|^{p-1}(1+|x|)^{-2p}$, which is integrable for $\frac{1}{2} < p$).

Now suppose $p \leqq \frac{1}{2}$. If

(5) $$\int_{-\infty}^{\infty} a(y)dy = 0 = \int_{-\infty}^{\infty} a(y)y\,dy \quad ,$$

then, for $x \gg y$,

$$\mathfrak{M}_a(x) = \int_I a(y)\left\{\frac{1}{1+x-y} - \frac{1}{1+x} - \frac{y}{(1+x)^2}\right\} dy$$
$$= \int_I a(y)\,\frac{y^2}{(1+x)^2(1+x-y)}\,dy \quad .$$

It follows that $|\mathfrak{M}_a(x)| \leqq \frac{c}{(1+x)^{3p}}$. From this, and a similar analysis of the case $x \ll y$, we see that $\mathfrak{M}_a \in L^p(\mathbb{R})$ for $\frac{1}{3} < p$. If (5) does not hold, by subtracting the unique linear polynomial p_I , restricted to I , such that $a - p_I$ has vanishing 0 and first moment and arguing as we did with the expression (4), we obtain the conclusion that (2), for $\frac{1}{3} < p \leqq \frac{1}{2}$, is equivalent to (5).

In general (subtracting the jth partial sum of the Taylor series of $1/(1+|x-y|)$) we see how the moment condition (that at least $\frac{1}{p} - 1$ must vanish) is equivalent to condition (2).

Here is more evidence that expression (2) is the "correct" alternative for the vanishing moments condition. At the other extreme from these classical H^p spaces are those for which only the mean zero condition is the natural one (and, also, these are the Hardy spaces that can be characterized by the "Hardy-Littlewood maximal function"). One such example is furnished by the real line endowed with the dyadic distance $m(x,y)$ which is the length of the smallest dyadic interval containing both x and y. This space is of homogeneous type (it is the real line realization of the two-adic numbers or the two series field; both are local fields). Let us replace condition (2) by

$$\mathfrak{M}_a(x) = \int_I \frac{a(y)}{1+m(x,y)}\, dy \in L^p(\mathbb{R}) , \tag{6}$$

$0 < p \leqq 1$. Since y is restricted to I, $m(x,y) = m(x,0)$ when x is far enough away from y. Thus, if $m_I(a) = 0$,

$$\mathfrak{M}_a(x) = \int_I a(y) \left\{\frac{1}{1+m(x,y)} - \frac{1}{1+m(x,0)}\right\} dy = 0$$

provided $|x|$ is large enough. It follows that (6) is equivalent to $\int a = 0$ for all p such that $0 < p \leqq 1$.

In $\mathbb{R}^n$ the moment condition characterizing the "standard" atomic spaces (see [15]) can be shown to be equivalent to

$$\mathfrak{M}_a(x) = \int_Q \frac{a(y)}{1+|x-y|^n}\, dy \in L^p(\mathbb{R}^n) ,$$

where Q is the ball supporting the atom $a(x)$.

In view of all these facts it is natural to make the following general definition of p-atoms. Suppose (X,m,μ) is a space of homogeneous type with m a measure distance[2]. Then for $0 < p \leqq 1$ we say that $a(y)$, defined on X, is a p-atom provided a is supported on a ball $Q = Q_r(x_0)$, $\|a\|_\infty \leqq \mu(Q)^{-1/p}$ and

$$\mathfrak{M}_a(x) = \int_Q \frac{a(y)}{1+m(x,y)}\, d\mu(y) \in L^p(X) . \tag{7}$$

PROBLEM. With this definition of p-atoms, define the atomic H^p spaces for such spaces of homogeneous type, as was done in [9], and then carry out the program developed in the work just cited.

---

[2] This means that the ball $Q_r(x_0) = \{x \in X: m(x_0,x) < r\}$ has μ-measure of the order of r (see [9]).

Thus, this problem is an invitation to study and develop the theory of atomic H^p spaces based on these atoms. That is, one should examine carefully those spaces of homogeneous type having a measure distance, characterize the duals of the corresponding atomic H^p spaces, develop a theory of interpolation for these spaces, study their molecular structure (as in [20]), their multipliers and appropriately invariant operators (when these notions are meaningful), the relation of these spaces to ones characterized by appropriate maximal functions, and so on. The recent work of Macias and Segovia [16] is probably relevant to several of these questions. Some of the techniques found there might be applicable to obtain a smooth distance function, an appropriate notion of Lipschitz spaces (enabling one to define H^p as linear functionals) and maximal functions that are associated with the atoms being considered.

One might, further, consider other quasi-distances, besides $m(x,y)$, that determine different moment conditions in the classical cases. This, then would be a way of extending problem #3 to spaces of homogeneous type.

#5. The atomic characterization of $H^p(\mathbb{R}^n)$ is obtained by making use of the maximal function characterization of this space [10]. Since the original definition of these spaces involves only the Riesz transforms (and the appropriate maximal functions played no role in the early development of their properties) it is natural to ask whether there is a more direct way of obtaining the atomic decomposition of $H^p(\mathbb{R}^n)$. For example, Coifman and Rochberg [6] obtain a representation of the functions in (weighted) Bergman spaces in terms of a countable number of "special" functions that are <u>molecules</u> (which, in turn, have atomic decompositions; see [20]). Maximal functions play no role in their work. Perhaps their methods extend to other situations.

PROBLEM. Obtain a direct proof of the atomic decomposition of the "classical" H^p spaces (without using maximal functions).

A similar question can be posed for the space BMO. Peter Jones (see [14]) has a direct proof of the fact that $BMO(\mathbb{R}^1)$ consists of the functions of the form $b_1 + \tilde{b}_2$, where $b_1, b_2 \in L^\infty(\mathbb{R}^1)$ and $\tilde{b}_2$ is the Hilbert transform of b_2. His construction involves complex variable techniques (such as the Blaschke product). The problem of extending this to n-dimensions remains open, where the n Riesz transforms play the role of the Hilbert transform.

#6. The following problem has the virtue that it is very simple to pose and is, in some sense, very natural:

PROBLEM. Is there a real variable characterization of those L^∞ functions (say, on $\mathbb{R}$) whose Hilbert transform is bounded?

I confess that I have certain misgivings about posing a problem simply because it is easy to formulate. I do not want it to create an infinitude of investigations about "such and such" functions whose Hilbert transform is "such and such"; where "such and such" could be, say, "continuous" or any other smoothness condition. Despite this wish I will mention a related problem: is there a natural characterization of the space $\{\tilde{f}: f \in L^1(\mathbb{R})\}$?

#7. The following is, by now, an old question that should have been settled long ago. The H^p-space theory introduced by E. M. Stein and myself in [19] dealt with Riesz systems $F(x,y) = (u(x,y), v_1(x,y), \ldots, v_n(x,y))$ defined on

$$\mathbb{R}^{n+1}_+ = \{(x,y): x = (x_1, x_2, \ldots, x_n) \in \mathbb{R}^n , \ y > 0\} ,$$

$n \geqq 2$, satisfying

$$\int_{\mathbb{R}^n} |F(x,y)|^p dx \leqq M < \infty \tag{1}$$

for all $y > 0$ (3). An important property of these systems (crucial for our method) is that $|F(x,y)|^p$ is subharmonic for $p \geqq (n-1)/n$. One can show, by example, that the number $(n-1)/n$ is best possible (in fact, $F(x,y) = \nabla |(x,y)|^{1-n}$, $n \geqq 2$, has the property that $|F(x,y)|^p$ is subharmonic for $p \geqq (n-1)/n$ and is not subharmonic for $p < (n-1)/n$. Because of this fact we were able to carry out our theory only for these values of p $(\geqq (n-1)/n)$. For example, we showed that boundary values

$$\lim_{y \to 0+} F(x,y) = F(x) \tag{2}$$

---

(3) The fact that F is a Riesz system means that u and v_j, , $j = 1, 2, \ldots, n$, are harmonic in $\mathbb{R}^{n+1}_+$ and satisfy $\frac{\partial u}{\partial y} + \sum_1^n \frac{\partial v_j}{\partial x_j} = 0$, $\frac{\partial v_j}{\partial x_k} = \frac{\partial v_k}{\partial x_j}$, $\frac{\partial u}{\partial x_k} = \frac{\partial v_k}{\partial y}$.

exist a.e. in x [(4)] for $p \geqq (n-1)/n$ and in the $L^p(\mathbb{R}^n)$-norm for $p > \frac{n-1}{n}$.

PROBLEM. Do the limits (2), pointwise and in the norm, exist for all Riesz systems in $H^p(\mathbb{R}^{n+1}_+)$ for certain values of $p < \frac{n-1}{n}$? If so, does the rest of the H^p-theory go through for these values (as developed in [19])?

It is known that we do have an H^p theory associated with $\mathbb{R}^{n+1}_+$ if, instead of Riesz systems, we consider the "higher gradient" systems introduced by Calderòn and Zygmund [3] (see #8). There are atomic H^p spaces associated with these systems and, thus, it is natural to ask whether there is a connection between the orders of the gradient involved and the number of vanishing moments that the corresponding atoms must have.

#8. We have just discussed Riesz systems in the (n+1)-dimensional region $\mathbb{R}^{n+1}_+$. Suppose we consider such a system in a domain $D \subset \mathbb{R}^n$, $n \geqq 3$. More precisely, this is a vector-valued function $F = (v_1, \ldots, v_n)$ on D whose components are harmonic and satisfy

$$(1) \qquad \operatorname{Div} F = \sum_{j=1}^{n} \frac{\partial v_j}{\partial x_j} = 0 \quad \text{and} \quad \operatorname{Rot} F = \left(\cdots, \frac{\partial v_j}{\partial x_k} - \frac{\partial v_k}{\partial x_j}, \cdots\right) = (\cdots, 0, \cdots).$$

The important subharmonicity property we discussed in #7, in this n-dimensional setting is that $|F|^p$ is subharmonic for $p \geqq \frac{n-2}{n-1}$. There are several systems of partial differential equations that determine vector-valued harmonic functions F for which there exists a <u>critical value</u> $q < 1$ such that $|F|^p$ is subharmonic for $p \geqq q$. Examples are furnished by the higher gradients of Calderón and Zygmund. If D is simply connected, conditions (1) are equivalent to the existence of a harmonic function h on D such that $\nabla h = F$. The second gradient systems of Calderón and Zygmund are obtained by applying the gradient to each component of F. We then obtain an n^2-dimensional vector, $\nabla^{(2)}h = (h_{11}, h_{12}, \ldots, h_{ij}, \ldots, h_{nn})$. The kth gradient $\nabla^{(k)}h$ is obtained by applying ∇ to each of the components of $\nabla^{(k-1)}h$. It can be shown that the critical value, p_k, for the subharmonicity of $|\nabla^{(k)}h|^p$ tends to 0. In fact, $p_k = (n-2)/(k+n-2)$ (see [3]). The same phenomenon is true for similar systems associated with compact Lie groups (see [7]). I know of no such system, however, for which

---

[(4)] Actually, the a.e. result holds if we let the point (ξ, y) in $\mathbb{R}^{n+1}_+$ tend to $(x,0)$ non-tangentially.

the critical index exceeds $(n-2)/(n-1)$. Perhaps the most general such systems are the ones studied in [8] for which the existence of a critical index $q < 1$ is established.

PROBLEM. Does there exist a general Cauchy-Riemann system, in the sense of [8], for which the critical index exceeds $(n-2)/(n-1)$, where n is the dimension of the domain of the system?

#9. If an operator T maps $H^1(\mathbb{R})$ boundedly into $L^1(\mathbb{R})$, is linear and commutes with translations then it maps into $H^1(\mathbb{R})$. This can be shown by using the Hilbert transform, H, which, because of the hypotheses on T, must commute with T : $HTf \in L^1(\mathbb{R})$ since $HTf = THf \in L^1(\mathbb{R})$ whenever $f \in H^1(\mathbb{R}^1)$ and, thus, $Tf \in H^1(\mathbb{R})$. We are, of course, using the fact that H maps $H^1(\mathbb{R})$ into itself.

This situation is true more generally for $\mathbb{R}^n$ since each Riesz transform maps $H^1(\mathbb{R})$ boundedly into itself (this was first observed by Stein [17]; it can also be shown by observing that the Riesz transforms of atoms are molecules [20]).

Suppose, now, that we consider a "non-isotropic" version of $\mathbb{R}^n$. For example, if $\alpha = (\alpha_1, \alpha_2, \ldots, \alpha_n)$ is an n-tuple of positive integers, not all equal, we can introduce the quasi-distance $d(x,y) = |x-y|^{\alpha} =$ $= |x_1-y_1|^{\alpha_1} + \cdots + |x_n-y_n|^{\alpha_n}$. It is easy to check that $(\mathbb{R}^n, d, \mu)$, where μ is Lebesgue measure, is of homogeneous type. The Heisenberg groups also furnish examples of non-isotropic versions of $\mathbb{R}^n$ (see [9]). In all these situations there is a natural atomic H^1 (or, more generally, H^p) space theory.

PROBLEM. If an operator T maps a non-isotropic $H^1(\mathbb{R}^n)$ space into $L^1(\mathbb{R}^n)$, is linear and commutes with translations, does it map into this non-isotropic $H^1(\mathbb{R}^n)$ space?

Of course, we could have asked if an appropriate set of "Riesz transforms" exists which characterizes $H^1(\mathbb{R}^n)$ (as a non-isotropic atomic space, or a "maximal function space"). It might be true that in certain special cases (such as "parabolic" $H^1(\mathbb{R}^n)$, see [2]) the area function, viewed as a vector-valued singular integral could be used in order to solve this problem.

After this set of problems were compiled, R. R. Coifman and B. Dahlberg solved this problem for the non-isotropic space $(\mathbb{R}^n, d, \mu)$. Their solution appears in these proceedings. This problem remains open, however, for the

Heisenberg group, where the non-isotropic nature varies from point to point and the Coifman-Dahlberg method may not be applicable.

#10. Atoms and molecules have been very useful for obtaining new results in the theory of Hardy spaces. Moreover, they can also be used to obtain new simple proofs of old results (see [9]). For example, the fact that for each $f \in H^1(\mathbb{R})$ there exists $g \in H^1(\mathbb{R})$ such that $|\hat{f}| \leqq |\hat{g}|$ and $\|g\|_{H^1} \leqq c\|g\|_{H^1}$ (c an absolute constant) can be shown by observing that $|f|'$ is a Fourier molecule (see [9]). As a sample of many problems involving the technique of "atomic methods" we pose:

PROBLEM. If $f \in H^1(\mathbb{R}^n)$, $n \geqq 2$, does there exist $g \in H^1(\mathbb{R}^n)$ such that $|\hat{f}| \leqq |\hat{g}|$ and $\|g\|_{H^1} \leqq c\|f\|_{H^1}$?

The classical proof of this result in 1-dimension uses the factorization of holomorphic functions in H^1 into a product of H^2 functions. Atoms can be used to obtain factorization theorems in certain n-dimensional settings (see [6]) which, in turn, give us a result that is analogous to the one we are seeking. The proof given in [9] of the 1-dimensional result, however, does not seem to extend to higher dimensions. Perhaps one can obtain factorization theorems in these cases.

#11. If B is a convex set in $\mathbb{R}^n$, the <u>tube</u>, T_B , <u>over the base</u> B is the domain $T_B = \{x + iy \in \mathbb{C}^n : x \in \mathbb{R}^n , y \in B\}$. If $p > 0$ then F , a holomorphic function on T_B , is said to belong to $H^p_{T_B}$ provided

$$\|F\|_{H^p_{T_B}} = \sup_{y \in B} \left(\int_{\mathbb{R}^n} |F(x+iy)|^p dx \right)^{1/p} < \infty .$$

This definition provides us with another extension to n-dimensions of the theory of Hardy spaces. Some of the theory of H^p spaces over tubes is developed in Chapter III of [18]. One of the natural questions in this theory concerns itself with the existence of boundary values $F(x+iy_0) = \lim_{\substack{y \to y_0 \\ y \in B}} F(x+iy)$.

Even when $n = 2$ and $p = 2$ the situation is not obvious. If we merely ask that the convergence, as $y \to y_0$, is taken in the L^2-norm it is not true that $y \in B$ can approach y_0 unrestrictedly. It is shown that such unrestricted convergence can occur if and only if y_0 is a <u>polygonal boundary point</u> (see pg. 98 of [18]). The situation is known in the case

$n = 3$. Mary Weiss has found necessary and sufficient conditions for L^2-convergence in this case (see page 122, (6.3) in [18]). (The proof has not appeared.)

PROBLEM. Find necessary and sufficient conditions for the unrestricted convergence in L^2 of $F(x+iy)$, as $y \in B$ tends to $y_0 \in \partial B$, when $F \in H^2_{T_B}$, $B \subset \mathbb{R}^n$. Consider other values of p and other types of convergence.

REFERENCES

1. D. Burkholder, R. Gundy and M. Silverstein, A maximal function characterization of the class H^p , Trans. Amer. Math. Soc. 157 (1971), 137-153.

2. A. P. Calderón and A. Torchinsky, Parabolic maximal functions associated with a distribution, Advances in Math. Vol. 16, No. 1 (1975), 1-64.

3. A. P. Calderón and A. Zygmund, On higher gradients of harmonic functions, Studia Math. 24 (1964), 211-226.

4. L. Carleson, Two remarks on H^1 and BMO , Analyse Harmonique d'Orsay 164 (1975), 1-11.

5. R. R. Coifman and R. Rochberg, Another characterization of BMO , to appear in the Proceedings of the A.M.S.

6. R. R. Coifman, R. Rochberg and G. Weiss, Factorization theorems for Hardy spaces in several variables, Ann. of Math 103 (1976), 611-635.

7. R. R. Coifman and G. Weiss, Invariant systems of conjugate harmonic functions associated with compact Lie groups, Studia Math. 44 (1972), 301-308.

8. R. R. Coifman and G. Weiss, On subharmonicity inequalities involving solutions of generalized Cauchy-Riemann equations, Studia Math 36 (1970), 77-83.

9. R. R. Coifman and G. Weiss, Extensions of Hardy spaces and their use in analysis, Bull. of A.M.S. 83 (1977), 569-645.

10. C. Fefferman and E. M. Stein, H^p spaces of several variables, Acta Math. 129 (1972), 137-193.

11. J. Garcia-Cuerva, Weighted H^p-spaces, Dissertationes Math. 162, to appear.

12. J. Garcia-Cuerva, Weighted H^p-spaces, Dissertation, Washington University, 1975.

13. S. Janson, Characterization of H^1 by singular integral transforms on martingales and $\mathbb{R}^n$, Math. Scand. 41 (1977), 140-152.

14. P. Jones, A constructive proof of Fefferman's characterization of $BMO(\mathbb{R}^1)$, to appear.

15. R. Latter, A characterization of $H^p(\mathbb{R}^n)$ in terms of atoms, to appear in Studia Math.

16. R. Macias and C. Segovia, A decomposition into atoms of distributions on spaces of homogeneous type, Preprint.

17. E. M. Stein, Classes H^p , multiplicateurs e fonctions de Littlewood-Paley, C. R. Acad. Sci. Paris Sèr. A-B 263 (1966); ibid. 264 (1967).

18. E. M. Stein and G. Weiss, Introduction to Fourier Analysis on Euclidean Spaces, Princeton Univ. Press (1975), Princeton, N.J.

19. E. M. Stein and G. Weiss, On the theory of harmonic functions of several variables, Acta Math. 103 (1960), 25-62.

20. M. Taibleson and G. Weiss, Molecular characterizations of certain Hardy spaces, (to appear).

DEPARTMENT OF MATHEMATICS, WASHINGTON UNIVERSITY, ST. LOUIS, MISSOURI 63130

Proceedings of Symposia in Pure Mathematics
Volume XXXV, Part 1, 1979

WEAK-TYPE INEQUALITIES FOR H^p AND BMO

Colin Bennett[1] & Robert Sharpley[2]

ABSTRACT. In connection with the Marcinkiewicz interpolation theorem, it appears that the most convenient way of storing the weak-type information for a given operator is in terms of a single inequality called a weak-type inequality. The first part of the paper surveys recent results on weak-type inequalities, together with their applications in harmonic analysis and approximation theory. The second part contains some new results on weak-type inequalities in the H^p-theory. These include a characterization of the Peetre K-functional for (L^p, BMO) in terms of the sharp-function, and a weak-type inequality for the sharp-function which leads to a simple proof of the John-Nirenberg lemma for functions of bounded mean oscillation.

§1. THE HARDY-LITTLEWOOD MAXIMAL OPERATOR AND REARRANGEMENTS.

The Hardy-Littlewood maximal function of a locally integrable function f on R^n is given by

(1.1)
$$(Mf)(x) = \sup_{Q \ni x} \{\frac{1}{|Q|} \int_Q |f(y)|dy\},$$

where the supremum extends over all cubes Q containing x with sides parallel to the coordinate axes. From its origins in function theory [13], it has evolved into an important tool in harmonic analysis [27] and related areas such as probability [23] and ergodic theory [11].

The maximal function takes into account the local, as opposed to the pointwise, behavior of f. It thus provides a representation of the "magnitude" of f amenable to differentiation and integration theory. Quantitative measurement of the magnitude is most naturally made by expressing the function as a member of such function spaces as L^p, L^{pq}, $L^p (\log L)^\alpha$, etc. Hence, most

---

AMS (MOS) subject classifications (1970). Primary: 30A78, 46E30, 46E35. Secondary: 42A40, 47A30.

1) Research supported by National Science and Engineering Research Council (Canada) under Grant A4489.

2) Research supported by National Science Foundation under Grant MCS 77-03666.

applications hinge on the boundedness of the maximal operator M between suitably chosen pairs of rearrangement-invariant spaces. The purpose of this section is to demonstrate that not only are the norms of Mf and f related in various ways but that in fact all such estimates are consequences of a simple relationship between the decreasing rearrangements of the functions themselves (cf. Theorem 1.3).

Indeed, the function

$$f^{**}(t) = \frac{1}{t}\int_0^t f^*(s)ds, \qquad t > 0, \tag{1.2}$$

clearly resembles the maximal function of the decreasing rearrangement f^* of f. Theorem 1.3, to the effect that $(Mf)^* \sim f^{**}$, tells us that f^{**} is also the decreasing rearrangement of the maximal function Mf of f. In particular, for rearrangement-invariant estimates, the maximal function Mf can always be replaced by the simpler function f^{**}, which is itself nothing more than an average of f^*.

The idea that $(Mf)^*$ is dominated by f^{**} goes back in some form to Hardy-Littlewood [13]; cf. also [29, pp. 29-33]. The equivalence of these two functions seems however to have first been pointed out by Herz [14]. In this section we shall present a simple proof of Herz' theorem (Theorem 1.3). The following covering lemma will be needed.

LEMMA 1.1. Let Ω be an open subset of R^n with finite measure. Then there are dyadic cubes Q_j, $j = 1,2,3,\ldots$, with pairwise disjoint interiors, such that

a) $$Q_j \cap \Omega^c \neq \phi, \qquad \text{for each } j;$$

b) $$\Omega \subseteq \bigcup_j Q_j;$$

c) $$|\Omega| \leq \sum_j |Q_j| \leq 2^n|\Omega|.$$

Proof. For each $x\in\Omega$, select a dyadic cube, $Q(x)$ say, of smallest diameter, which contains x and has nonempty intersection with Ω^c. Now subdivide $Q(x)$ into 2^n congruent subcubes and select any one, $\tilde{Q}(x)$ say, which contains x. Clearly $\tilde{Q}(x) \subset \Omega$, and so

$$2^{-n}\,|Q(x)| = |\tilde{Q}(x)| = |\tilde{Q}(x) \cap \Omega| \leq |Q(x) \cap \Omega|. \tag{1.3}$$

Let $K = \{Q(x) : x\in\Omega\}$. Because of the dyadic nature of the cubes and the fact that $|\Omega| < \infty$, each $x\in\Omega$ is contained in a maximal cube, say $\overline{Q}(x)$, from K. Listing the at most countably many cubes in $\{\overline{Q}(x) : x\in\Omega\}$ as $Q_1, Q_2,\ldots$, we see that properties a) and b) are immediate, hence so is the first inequality in c). The remaining inequality follows from the observation that (1.3) is valid for every Q_j : $2^{-n}|Q_j| \leq |Q_j \cap \Omega|$; summing over j, we obtain the desired result.

REMARK 1.2. If Q is a fixed cube containing Ω, and if Ω is open relative to Q, then a similar argument shows that the cubes Q_j can be selected as subcubes of Q (i.e., dyadic with respect to Q) and each Q_j meets $Q\backslash\Omega$.

THEOREM 1.3 (Herz [14]). If f is locally integrable on R^n, then

$$3^{-n}(Mf)^*(t) \leq f^{**}(t) \leq (2^n + 1)(Mf)^*(t), \qquad t > 0. \tag{1.4}$$

Proof. Fix $t > 0$. For the right-hand inequality we can suppose $(Mf)^*(t) < \infty$. The lower semi-continuity of Mf guarantees that the set $\Omega = \{x\varepsilon R^n : (Mf)(x) > (Mf)^*(t)\}$ is open, and the estimate $|\Omega| \leq t$ follows from the equimeasurability of Mf and $(Mf)^*$. Applying Lemma 1.1 to Ω, we obtain a sequence of cubes Q_j, with disjoint interiors, for which properties a), b), and c) of the lemma hold. With $F = (\cup Q_j)^c$, we put $g = \sum_j f\chi_{Q_j}$ and $h = f\chi_F$. The subadditivity of $f \to f^{**}$ gives immediately

$$f^{**}(t) \leq g^{**}(t) + h^{**}(t) \leq t^{-1} \|g\|_1 + \|h\|_\infty. \tag{1.5}$$

But $F \subset \Omega^c$ so

$$\|h\|_\infty \leq \|\chi_F Mf\|_\infty \leq (Mf)^*(t). \tag{1.6}$$

Furthermore, each Q_j meets Ω^c so $|Q_j|^{-1} \int_{Q_j} |f(x)|dx \leq (Mf)^*(t)$. Hence by Lemma 1.1 c),

$$\|g\|_1 = \sum_j \int_{Q_j} |f(x)|dx \leq 2^n|\Omega|(Mf)^*(t) \leq 2^n t(Mf)^*(t).$$

Together with (1.5) and (1.6), this gives the desired result.

In the following argument to establish the left-hand inequality in (1.4) we use the fact that the maximal operator is of weak type (1,1) ($t(Mf)^*(t) \leq 3^n\|f\|_1$) and of strong type (∞,∞) ($\|Mf\|_\infty \leq \|f\|_\infty$), and repeat a standard argument due originally to Calderón [5, Theorem 8]. Thus, we assume $f^{**}(t) < \infty$ and consider the set $E = \{x\varepsilon R^n : |f(x)| > f^*(t)\}$. Again we have $|E| \leq t$. For the functions

$$g(x) = (f(x) - f^*(t)\operatorname{sgn} f(x))\chi_E(x), \qquad h = f-g,$$

we have the estimates

$$\|g\|_1 \leq t(f^{**}(t) - f^*(t)), \qquad \|h\|_\infty \leq f^*(t).$$

Since

$$(Mf)(x) \leq (Mg)(x) + (Mh)(x) \leq (Mg)(x) + \|h\|_\infty,$$

it follows that

$$\begin{aligned}(Mf)^*(t) &\leq (Mg)^*(t) + \|h\|_\infty \leq 3^n t^{-1} \|g\|_1 + \|h\|_\infty \\ &\leq 3^n(f^{**}(t) - f^*(t)) + f^*(t) \leq 3^n f^{**}(t).\end{aligned}$$

This completes the proof.

Herz' theorem, which asserts that $(Mf)^* \sim f^{**}$, enables us to use f^{**} as a "model" for Mf. The Hardy-Littlewood maximal theorem [29, p.32] is an easy consequence since $\|f^{**}\|_p = \|\frac{1}{t}\int_0^t f^*(s)ds\|_p \le c_p\| f^*\|_p$ $(p > 1)$, by virtue of a classical inequality of Hardy [28, p.196].

COROLLARY 1.4 (Hardy-Littlewood). Suppose $1 < p \le \infty$. If $f \in L^p(R^n)$, then $Mf \in L^p(R^n)$ and $\|Mf\|_p \le c_p\|f\|_p$.

The model reveals much more. An interchange in the order of integration gives

$$\int_0^1 f^{**}(t)dt = \int_0^1 \frac{1}{t}\int_0^t f^*(s)ds\, dt = \int_0^1 f^*(s)ds \int_s^1 \frac{dt}{t} = \int_0^1 f^*(s) \log \frac{1}{s}\, ds. \tag{1.7}$$

If we work on the unit circle T, say, (or any fixed ball in R^n) so that the decreasing rearrangements vanish outside of a finite interval (0, 1), then the right-hand side is a norm for the space $L \log^+ L$ (T) [2]. The left-hand side is, by Herz' theorem, equivalent to the L^1-norm of Mf. Thus, on the one hand, we obtain the Hardy-Littlewood result [29, p.32] that $f \in L \log^+ L$ implies $Mf \in L^1$, and on the other, the Stein [26]-Herz [14] converse : $Mf \in L^1$ implies $f \in L \log^+ L$.

COROLLARY 1.5 (Hardy-Littlewood-Stein-Herz). Suppose $f \in L^1(T)$. Then Mf is integrable if and only if $f \in L \log^+ L(T)$.

It should come as no surprise that Corollaries 1.4 and 1.5 fall out so easily when we remark that the Marcinkiewicz interpolation theorem [30, p.112] is lurking in the background. Indeed, we used the usual weak-type hypotheses for M in the proof of Herz' theorem. The fundamental relationship (1.4) is seen therefore as a convenient and concise way of storing the weak-type information for the maximal operator (strictly speaking, this remark applies only to the left-hand inequality in (1.4); the right-hand inequality is a bonus: it gives us the Stein-Herz converse).

In the next section we shall see how this program can be repeated for other basic weak-type operators such as the Hilbert transform and the fractional integrals. This leads to a new way of viewing the Marcinkiewicz interpolation theorem, and to significant extensions (cf. Section 3) of that theorem beyond its traditional domain in harmonic analysis.

§2. WEAK-TYPE INEQUALITIES. The conjugate function, or periodic Hilbert transform, $Hf = \tilde{f}$ of a function $f \in L^1(T)$ is defined [29, p.131] by the principal-value integral

$$(Hf)(e^{it}) \equiv \tilde{f}(e^{it}) = \frac{1}{2\pi}\int_{-\pi}^{\pi} f(e^{ix}) \cot\left(\frac{t-x}{2}\right)dx.$$

Our point of departure is the O'Neil-Weiss [21] inequality

(2.1) $$(Hf)^{**}(t) \le c\left(\frac{1}{t}\int_0^t f^{**}(s)ds + \int_t^1 f^{**}(s)\frac{ds}{s}\right), \qquad 0 < t < 1,$$

which is a fairly easy consequence of the Stein-Weiss [28, p.240] description of $(H\chi)^*$ for characteristic functions χ. Clearly, the inequality has meaning only if f^{**} is integrable, that is, if $f \in L \log L$ (cf. (1.7)). Thus, (2.1) does not describe the action of H on all of L^1 and, in particular, does not explicitly contain the information that H is of weak type (1,1) [29, p.134]. Nevertheless, this additional information can be incorporated into (2.1) by means of an elementary decomposition argument similar to that used in the second half of the proof of Theorem 1.3. In so doing, Bennett-Rudnick [2] established the following inequality

(2.2) $$(Hf)^{*}(t) \le c\left(\frac{1}{t}\int_0^t f^{*}(s)ds + \int_t^1 f^{*}(s)\frac{ds}{s}\right), \qquad 0 < t < 1.$$

valid now for all $f \in L^1$. Note that (2.1) follows, by integration, from (2.2).

This is the exact counterpart, for the Hilbert transform, of the fundamental inequality (1.4) for the maximal operator. As before, the classical estimates are easy consequences:

COROLLARY 2.1 a) (M. Riesz [29, p.253]) $H : L^p \to L^p$, $1 < p < \infty$;
b) (Zygmund [29, p.254]) $H : L \log L \to L^1$;
c) (Zygmund [29, p.254]) $H : L^\infty \to L_{exp}$.

Proof. a) The averaging operator $t^{-1}\int_0^t (\cdot)\, ds$ is bounded on L^p for $p > 1$ (by Hardy's inequality); its adjoint $\int_t^1 (\cdot)ds/s$ is bounded on L^p for $p < \infty$.
b) Integrate each side of (2.1) and use (1.7).
c) If f is bounded, then (2.2) shows that $(Hf)^*(t)$ grows at most logarithmically as $t \to 0$.

What is the interpretation in terms of the Marcinkiewicz interpolation theorem? Certainly H is of weak type (1,1) but on the other hand there seems to be no reasonable way of defining a concept of weak type (∞, ∞) that will be satisfied by the Hilbert transform. And yet precisely this kind of information seems to be encoded in the inequality (2.2) because of the results it produces in Corollary 2.1.

We can better understand what is happening here by considering the Weyl fractional integrals I_λ, $0 < \lambda < 1$ [30, p.135]:

$$(I_\lambda f)(e^{it}) = \frac{1}{\Gamma(\lambda)} \int_0^\infty f(e^{i(t-x)})x^{\lambda-1}dx \quad ^{1)}$$

The operator I_λ is of weak types $(1, (1-\lambda)^{-1})$ and (λ^{-1}, ∞) (cf. [2]). Now we invoke the fundamental contribution of Calderón [5, p.290]: so long as p_0 and p_1 are finite, the pair of weak-type conditions (p_0,q_0) and (p_1,q_1) on an operator T can always be combined and, in fact, are equivalent to a single inequality satisfied by T. In the case of the operator I_λ, the inequality is

(2.3) $$(I_\lambda f)^*(t) \le c\,(t^{\lambda-1} \int_0^t f^*(s)ds + \int_t^1 s^\lambda f^*(s)ds/s), \qquad 0 < t < 1.$$

The point of the exercise is that the Hilbert transform is, in a formal sense, the fractional integral I_0 of order 0. Letting $\lambda \to 0$, we see that the weak-type condition $(1, (1-\lambda)^{-1})$ tends to weak type (1,1), but the condition (λ^{-1},∞) tends to the meaningless weak type (∞,∞). However, the equivalent inequality (2.3) tends to precisely the inequality (2.2)! This suggests that we adopt the Calderón formulation, extended to infinite values of the parameters, as the fundamental notion of weak type.

DEFINITION 2.2 [2]. Suppose $0 < p_0 < p_1 \le \infty$, $0 < q_0, q_1 \le \infty$, with $q_0 \ne q_1$, and let m be the slope of the line segment σ joining the points $(1/p_0, 1/q_0)$ and $(1/p_1, 1/q_1)$ in the plane : $m = (1/q_1 - 1/q_0)/(1/p_1 - 1/p_0)$. Let $S(\sigma)$ be the integral operator defined by

(2.4) $$S(\sigma)f(t) = t^{-1/q_0} \int_0^{t^m} s^{1/p_0} f(s)\frac{ds}{s} + t^{-1/q_1} \int_{t^m}^\infty s^{1/p_1} f(s)\frac{ds}{s}, \quad 0 < t < \infty .$$

We say that an operator T is of weak type $(p_0, q_0; p_1, q_1)$ if

(2.5) $$(Tf)^*(t) \le cS(\sigma)(f^*)(t), \qquad 0 < t < \infty ,$$

for all f for which the right-hand side is finite.

In particular, (2.2) shows that the Hilbert transform is of weak type $(1,1; \infty,\infty)$, and (2.3) that the fractional integrals I_λ are of weak type $(1, 1/(1-\lambda); 1/\lambda,\infty)$. Any inequality of the form (2.5) will be referred to simply as a weak-type inequality.

Once the weak-type inequality (2.5) is established for a given operator T, the interpolation is performed exactly as in Corollary 2.1(a) by means of the Hardy inequalities. Thus, if $0 < \theta < 1$ and

(2.6) $$\frac{1}{p} = \frac{1-\theta}{p_0} + \frac{\theta}{p_1}, \qquad \frac{1}{q} = \frac{1-\theta}{q_0} + \frac{\theta}{q_1},$$

[1)] defined in this way for f with mean value 0 on T, and extended by linearity to all $f \in L^1$.

we apply the Lorentz L^{qr}-norm to each side of (2.5). The operator $S(\sigma)$ is so designed that the right-hand side reduces, via the Hardy inequalities, to the L^{pr}-norm of f. This shows that $T : L^{pr} \to L^{qr}$, for any r, which is precisely Calderón's formulation [5, p.293] (and proof) of the Marcinkiewicz interpolation theorem.

When applied to the Hilbert transform, this result produces the M. Riesz theorem as presented in Corollary 2.1(a). It is natural to ask whether parts (b) and (c), which involve the Zygmund spaces L log L and L exp, can also be derived in this way. While such spaces have traditionally been regarded as Orlicz spaces (thus preventing their incorporation in the Calderón theory), it is nevertheless the case that they can also be regarded as more general types of Lorentz spaces, and can therefore be easily amalgamated with the L^{pq}-spaces. The appropriate framework is furnished by the class of Lorentz-Zygmund spaces $L^{pq}(\log L)^{\alpha}$, introduced by Bennett-Rudnick [2]. A function f (on the circle, say) is in $L^{pq}(\log L)^{\alpha}$, $0 < p, q \le \infty$, $-\infty < \alpha < \infty$, if

(2.7) $$\|f\|_{p,a;\alpha} = \left(\int_0^1 [t^{1/p}(1-\log t)^{\alpha} f^*(t)]^q \frac{dt}{t}\right)^{1/q} < \infty$$

(with the evident modification if $q = \infty$).

Clearly, $L^{pq}(\log L)^{o}$ is the familiar Lorentz space L^{pq}, and it is not hard to show that $L^{pp}(\log L)^{\alpha}$ is the Zygmund space $L^{p}(\log L)^{\alpha}$ when $p < \infty$. The Zygmund space of α-th power exponentially-integrable functions is nothing more than the space $L^{\infty\infty}(\log L)^{-1/\alpha}$. Furthermore, the O'Neil spaces $K^{p}(\log^{+}K)^{\alpha p}$ [18] arise as the Lorentz-Zygmund spaces $L^{p1}(\log L)^{\alpha}$. Complete details are given in [2].

With these foundations in place, it remains to formulate the Marcinkiewicz interpolation theorem in terms of operators of weak type $(p_o,q_o;p_1,q_1)$ acting on Lorentz-Zygmund spaces. The first part gives the "internal" results corresponding to the values $0 < \theta < 1$ in (2.6); the other two parts give the "end-point" results corresponding to $\theta = 0$ and $\theta = 1$.

THEOREM 2.3 [1)] (Bennett-Rudnick [2]). Suppose $0 < p_o < p_1 \le \infty$ and $0 < q_o, q_1 \le \infty$, with $q_o \ne q_1$. Let T be a quasilinear operator of weak type $(p_o,q_o;p_1,q_1)$.

a) If $0 < \theta < 1$ and p, q are given by (2.6), then

$$T : L^{pa}(\log L)^{\alpha} \to L^{qa}(\log L)^{\alpha}$$

---

1) We present only the finite measure space version; the general case is given in [2].

whenever $0 < a \le \infty$ and $-\infty < \alpha < \infty$.

b) If $1 \le a \le b \le \infty$ and $-\infty < \alpha, \beta < \infty$, then

$$T : L^{p_0 a}(\log L)^{\alpha+1} \to L^{q_0 b}(\log L)^{\beta}$$

whenever $\alpha + 1/a = \beta + 1/b > 0$.

c) If $1 \le a \le b \le \infty$ and $-\infty < \alpha, \beta < \infty$, then

$$T : L^{p_1 a}(\log L)^{\alpha+1} \to L^{q_1 b}(\log L)^{\beta}$$

whenever $\alpha + 1/a = \beta + 1/b < 0$.

The essence of the result is that the "index" $\sigma + 1/s$ of the space $L^{rs}(\log L)^{\sigma}$ remains constant when $0 < \theta < 1$ but always decreases by a factor of one in the endpoint cases $\theta = 0$ and $\theta = 1$. This single result directly produces all of the classical rearrangement-invariant estimates for such fundamental operators as the Hilbert transform, the fractional integrals, the maximal operator, and the Fourier transform; complete details can be found in [2]. Furthermore, as DeVore-Riemenschneider-Sharpley [7] have shown, this natural formulation of the Marcinkiewicz theorem lifts effortlessly into a general Banach space context and hence produces further applications in harmonic analysis and approximation theory. These results form the core of the next section.

We conclude our discussion of the rearrangement-invariant case with some remarks on multilinear generalizations of Theorem 2.3. Such results are of importance in dealing with convolution and tensor product operators [19,20,25], for example. Sharpley [25] has developed weak-type inequalities for bilinear (or multilinear) operators T satisfying m individual weak-type estimates. The weak-type inequality has the form

$$T(f,g)^{**}(t) \le c \int_0^\infty \int_0^\infty f^*(r)g^*(s)\, \Psi_\sigma\, (r,s;t) \frac{dr}{r}\frac{ds}{s} \equiv S_\sigma(f^*,g^*)(t) \tag{2.8}$$

where the kernel Ψ_σ is a combination of powers of r, s, and t determined by the m initial estimates.

The "internal" mapping properties of T are obtained from (2.8) exactly as in the linear case (Theorem 2.3a)), namely, by applying appropriate norms to (2.8) and reducing the right-hand side by means of suitable generalizations of the Hardy inequalities (cf. [25]). The analysis of the endpoint cases in Theorem 2.3 b), c) is much more intricate. Nevertheless, it again ultimately depends on what can be regarded as limiting cases of the Hardy inequalities. It would be of some interest to have corresponding inequalities for the multilinear theory.

§3. GENERALIZED WEAK-TYPE INEQUALITIES. The results of the first two sections have been concerned with rearrangements and hence with the magnitude of the function. We now want to consider other characteristics of the function such as its smoothness. DeVore-Riemenschneider-Sharpley [7] made the interesting observation that while magnitude and smoothness are unrelated, the analysis in the two situations is exactly the same. Indeed, for the periodic Hilbert transform H there is the inequality [29, p.121]

$$\frac{\omega(Hf;t)}{t} \le c\left(\frac{1}{t}\int_0^t \frac{\omega(f;s)}{s}\,ds + \int_t^1 \frac{\omega(f;s)}{s}\frac{ds}{s}\right), \qquad 0 < t < 1, \tag{3.1}$$

where $\omega(g;.)$ is the modulus of continuity [29, p.42] of a function g. This inequality has exactly the same structure as (2.1)(or (2.2)) but with f^{**} (or f^*), the measure of magnitude, replaced by $t^{-1}\omega(f;t)$, the measure of smoothness. Proceeding as in Section 2, we obtain the precise analogues of the results in Corollary 2.1. Thus $H : Lip(\alpha,q) \to Lip(\alpha,q)$ if $0 < \alpha < 1$; at the endpoints we find that $H : D \to C$, where D is the Dini class and C the space of continuous functions; and if $f \varepsilon Lip\ 1$, then $\omega(Hf;t)/t$ grows at most logarithmically as $t \to 0$. In fact, by applying the Lorentz-Zygmund norms to (3.1), we obtain the whole spectrum of results corresponding to Theorem 2.3.

The crucial link between (2.1) and (3.1), which allows the abstract theory to unfold, is provided by the Peetre K-functional [4, Chapter 3]. If $f \varepsilon X_0 + X_1$, where (X_0,X_1) is a compatible couple of Banach spaces, then the K-functional $K(f;t) \equiv K(f;t;X_0,X_1)$ is defined by

$$K(f;t) = \inf_{f=f_0+f_1} \left(\|f_0\|_{X_0} + t\|f_1\|_{X_1} \right), \qquad 0 < t < \infty, \tag{3.2}$$

where the infimum is taken over all possible representations $f = f_0 + f_1$ of f with $f_0 \varepsilon X_0$ and $f_1 \varepsilon X_1$.

The point is that $t^{-1}K(f;t;L^1,L^\infty) = f^{**}(t)$ and $t^{-1}K(f;t;C,C^{(1)}) \sim t^{-1}\omega(f;t)$ [4, Chapter 3]. Hence, both (2.1) and (3.1) involve particular kinds of K-functionals related by means of a weak-type inequality.

DEFINITION 3.1 [7]. Let (X_0,X_1), (Y_0,Y_1) be compatible couples of Banach spaces. Suppose $1 \le p_0 \le p_1 \le \infty$, $1 \le q_0, q_1 \le \infty$, $q_0 \ne q_1$, and let σ be the corresponding interpolation segment (cf. Definition 2.2). Let T be a quasi-linear operator carrying $X_0 + X_1$ into $Y_0 + Y_1$. Then T is said to be of (generalized) weak-type σ with respect to (X_0,X_1) and (Y_0,Y_1) if

$$\frac{K(Tf;t;Y_0,Y_1)}{t} \le cS(\sigma)\left[\frac{K(f;(.);X_0,X_1)}{(.)}\right](t), \qquad 0 < t < \infty, \tag{3.3}$$

holds whenever the right-hand side is finite.

Once such an inequality has been established, the mapping properties of T are obtained as before by applying Lorentz-Zygmund norms and obtaining results analogous to Theorem 2.3. We shall not discuss the mapping properties in any great detail here. Instead, we shall try to convey some of the flavor of the theory by pointing out some of the interesting weak-type inequalities and the phenomena they control. For further details, see DeVore-Riemenschneider-Sharpley [7], or the survey article [3].

The smoothness spaces derived from L^p, namely the Besov spaces $B_p^{\lambda,a}$, are defined in terms of the k-th order L^p-modulus of continuity

$$\omega_k(f;t)_p = \sup_{|h|\le t} \|\Delta_h^k f\|_p, \tag{3.4}$$

where Δ_h^k is the k-th power of the difference operator $(\Delta_h f)(x) = f(x+h) - f(x)$. It is a well-known result that $B_p^{\lambda,a}$ is independent of the order k used in its definition, provided only that $k > \lambda$ [4, Chapter 3]. Lying beneath is a weak-type inequality known as Marchaud's inequality [16]. We consider only the simplest case, involving $0 < \lambda \le 1$ and k = 1 or 2. The trivial "direct" estimate

$$\omega_2(f;t)_p \le 2\omega_1(f;t)_p \tag{3.5}$$

follows directly from the definition (3.4). The corresponding "inverse" result is given by Marchaud's inequality:

$$t^{-1}\omega_1(f;t)_p \le c\{\|f\|_p + \int_t^\infty s^{-1}\omega_2(f;s)_p \, ds/s\}. \tag{3.6}$$

Now the Besov space $B_p^{\lambda,a}$ results from applying the Lorentz $L^{1/(1-\lambda),a}$ norm to $t^{-1}\omega_k(f;t)$. Clearly, in view of (3.5) and (3.6), it is immaterial which of k = 1 or k = 2 is used, provided only that the averaging operator $\int_t^\infty (.)ds/s$ remains bounded; as we have noted previously, this is the case when $1/(1-\lambda) < \infty$, that is, when $\lambda < 1$. When $\lambda = 1$, the averaging operator is unbounded and in fact it is well-known that in this case the spaces corresponding to k = 1 and k = 2 no longer coincide (cf. [29, p.47]).

Note that since the K-functionals on the left- and right-hand sides of (3.3) are not necessarily the same, the inequality can store important information even for the identity operator: Marchaud's inequality is an example. Another example arises in the Sobolev-type embedding theorems for Besov spaces. Here we want to compare two k-th order L^p-moduli but this time with k fixed and p varying. The inequality reads

$$\omega_k(f;t)_q \le c\int_0^t s^{-\theta}\omega_r(f;s)_p \, ds/s, \qquad 0 < t < \infty, \tag{3.7}$$

where $\theta = n/p - n/q$ (n the euclidean dimension) and $q > p$ [7]. The usual embeddings of $B_p^{\lambda,a}$ into $B_q^{\mu,b}$, together with the logarithmic end-point estimates, all follow directly by applying Lorentz-Zygmund norms to (3.7). A variant of

(3.7), in which the modulus on the left-hand side is replaced by f^{**}, produces embeddings of Besov spaces into rearrangement-invariant spaces such as L^q, L^{qa}, etc. [7].

The Bernstein-type theorems on the absolute convergence of Fourier series [29, p.243] are also controlled by weak-type inequalities. Such theorems relate the magnitude of the Fourier coefficients $\hat{f}(n)$ to the smoothness of f. The relevant inequality is [7]

$$(\hat{f})^{**}(t) \le c\frac{1}{t}(\|f\|_1 + \int_{1/t}^{1} \frac{\omega(f;s)_p}{s^{1/p}} \frac{ds}{s}), \qquad 1 \le t \le \infty,$$

valid for $f \varepsilon L^p$, $1 \le p \le 2$.

The weak-type inequalities also play an important role in approximation theory. For example, when one of the K-functionals in (3.3) is chosen to measure smoothness, the other can be replaced by the closely related "degree of approximation" functional $E(f;t)$. Thus, in $L^p(T)$ for instance,

$$E(f;t)_p = \inf \{\|f-g\|_p\}$$

where the infimum is taken over all trigonometric polynomials g whose degree does not exceed [t]. The corresponding weak-type inequalities give rise to some of the classical "direct" and "inverse" approximation theorems; see [7] for details.

The moral of the last three sections, if it is not already clear, is that of the numerous phenomena in analysis that can be interpreted in terms of an operator acting between a pair of spaces, a great many are controlled by a weak-type inequality relating an appropriate pair of K-functionals. Furthermore, when it is known that the operator is unbounded at one or other of the "endpoints" (and this, after all, is what the Marcinkiewicz theorem is all about), then it is often possible to anticipate, by including one or the other of the averaging operators $t^{-1}\int_0^t (\cdot)ds$, $\int_t^\infty (\cdot)ds/s$, exactly what form the weak-type inequality must take.

We shall see some good examples of this in the next section when we turn to the theory of H^p-spaces and BMO. The basic characteristic of H^1-functions is not smoothness or magnitude, but something in between (one hesitates to call it analyticity!). Whatever one calls it, it is expressed in terms of the grand maximal function $\mathcal{M}f$, and this (more precisely, $(\mathcal{M}f)^{**}(t)$) is equivalent to $t^{-1}K(f;t;H^1,L^\infty)$ [9]. Similarly, the basic characteristic of BMO-functions is oscillation which is expressed in terms of the sharp-function $f^\#$. We shall show (in Section 6) that $f^{\#*}(t)$ is equivalent to $t^{-1}K(f;t;L^1,BMO)$. These key descriptions of the K-functionals lead to some interesting weak-type inequalities which play a fundamental role in the theory. For example, the weak-type inequality (4.15) relating magnitude (f^{**}) and oscillation ($f^{\#*}$) will lead us to a simple proof of the John-Nirenberg lemma and other results.

§4. INTERPOLATION BETWEEN L^1 AND BMO. The space BMO of functions of bounded mean oscillation was devised by John-Nirenberg [15] for the purpose of studying regularity properties of solutions of elliptic partial differential equations. BMO is the Banach space (of equivalence classes modulo the constants) of all locally integrable functions f on R^n for which

$$\|f\|_{BMO} = \sup_Q Q(|f - Q(f)|)$$

is finite, where $Q(f) = |Q|^{-1}\int_Q f(y)dy$ and the supremum is taken over all cubes Q with sides parallel to the coordinate axes.

The BMO-norm measures the oscillation of f on all of R^n. The local oscillation is expressed by the "sharp-function"

$$f^{\#}(x) = \sup_{Q\ni x} Q(|f - Q(f)|) \tag{4.1}$$

which was introduced by Fefferman-Stein [10]; the supremum is now taken over only those cubes Q which contain the point x. Clearly, $f \in BMO$ if and only if $f^{\#} \in L^\infty$, and

$$\|f\|_{BMO} = \|f^{\#}\|_\infty. \tag{4.2}$$

The space BMO could just as well be defined in terms of the quantity

$$f^b(x) = \sup_{Q\ni x}\{\inf_c Q(|f-c|)\}; \tag{4.3}$$

indeed, it is easily verified that

$$f^b(x) \le f^{\#}(x) \le 2f^b(x). \tag{4.4}$$

Note that as an immediate consequence of (1.1) and (4.1) we have

$$f^{\#}(x) \le 2Mf(x) \le 2\|f\|_\infty. \tag{4.5}$$

Together with (4.2), this shows that L^∞ is continuously embedded in BMO.

One reason for the importance of BMO is that it arises as (essentially) the range of certain singular integral operators, such as the Hilbert or Riesz transforms, acting on L^∞. Consequently, the interpolation properties of BMO are also of much interest. Now while BMO contains L^∞, the fundamental John-Nirenberg lemma (of which we give an alternate proof in Corollary 4.6) shows that it is only "slightly" larger than L^∞. Hence, we might expect the (θ,q)-interpolation spaces for the pairs (L^1,L^∞) and (L^1,BMO) to be the same. This result, namely

$$(L^1,L^\infty)_{\theta,q} = L^{pq} = (L^1,BMO)_{\theta,q}, \qquad \theta = 1 - 1/p, \tag{4.6}$$

valid for $0 < \theta < 1$ and $0 < q \le \infty$, is due to Hanks [12].

While the first equivalence in (4.6) is, of course, well-known [4, p.186], it will nevertheless be instructive to examine its proof. Recall [4, p.167] that f belongs to the interpolation space $(X_o,X_1)_{\theta,q}$ if $(\int_o^\infty[t^{-\theta}K(f;t)]^q dt/t)^{1/q}$

is finite. Since $K(f;t;L^1,L^\infty) = tf^{**}(t)$ [4, p.184], we see that the norms in $(L^1,L^\infty)_{\theta,q}$ and L^{pq} are

$$\left(\int_0^\infty [t^{1/p}f^{**}(t)]^q dt/t\right)^{1/q}, \qquad \left(\int_0^\infty [t^{1/p}f^{*}(t)]^q dt/t\right)^{1/q}, \qquad \theta = 1-1/p,$$

respectively. That the first dominates the second is clear from the "direct" inequality

$$f^* \le f^{**} .$$

But in the other direction these quantities are related by the "inverse" weak-type inequality

$$f^{**}(t) \le \frac{1}{t}\int_0^t f^*(s)ds .$$

Hence, by applying L^{pq}-norms, we see that the two spaces coincide whenever the averaging operator is bounded on L^{pq}, that is, whenever $p > 1$.

While this is an admittedly trivial example of a weak-type inequality, it does point out a direction to be followed in establishing the more complex second equivalence in (4.6). Indeed, the inclusion $(L^1,L^\infty)_{\theta,q} \subseteq (L^1,\mathrm{BMO})_{\theta,q}$ follows immediately from the <u>direct inequality</u>

$$K(f;t;L^1,\mathrm{BMO}) \le 2K(f;t;L^1,L^\infty), \tag{4.7}$$

which itself follows from the definition of the K-functional and the obvious estimate $\|f\|_{\mathrm{BMO}} \le 2\|f\|_\infty$. By analogy with the previous situation, what is needed to establish the reverse inclusion is a weak-type inequality, <u>inverse</u> to (4.7). This we shall do in Corollary 4.4. In the process, we shall establish a weak-type inequality which is inverse to the direct inequality $(f^\#)^* \le 2(Mf)^*$, embodied in (4.5). This is the content of the next theorem.

<u>THEOREM 4.1.</u> If $f\epsilon L^1 + L^\infty$, then

$$(Mf)^*(t) \le c\int_t^\infty (f^\#)^*(s)\,\frac{ds}{s} + (Mf)^*\,(+\infty), \qquad 0 < t < \infty . \tag{4.8}$$

Proof. The open set $\Omega = \{f^\# > (f^\#)^*(2t)\} \cup \{Mf > (Mf)^*(2t)\}$ has measure

$$|\Omega| \le 4t. \tag{4.9}$$

Applying Lemma 1.1, we obtain a covering of Ω by cubes $Q_j, j = 1,2,\ldots,$ with pairwise disjoint interiors, such that each Q_j has nonempty intersection with Ω^c, and $\sum_j|Q_j| \le 2^n|\Omega|$. If $F = (\bigcup_j Q_j)^c$, define

$$g = \sum_j [f-Q_j(f)]\chi_{Q_j}, \qquad h = \sum_j Q_j(f)\chi_{Q_j} + f\chi_F .$$

Then $f = g + h$, and so

$$(Mf)^*(t) \le (Mg)^*(t) + \|h\|_\infty \le 3^n t^{-1}\|g\|_1 + \|h\|_\infty , \tag{4.10}$$

since M is of weak-type (1,1). For any point $x_j \varepsilon Q_j \cap \Omega^c$, we have $Q_j(|f-Q_j(f)|) \leq f^{\#}(x_j) \leq (f^{\#})^*(2t)$. Hence, by (4.9) and Lemma 1.1(c),

$$\|g\|_1 = \sum_j |Q_j| Q_j(|f-Q_j(f)|) \tag{4.11}$$

$$\leq \sum_j |Q_j| (f^{\#})^*(2t) \leq 4 \cdot 2^n t (f^{\#})^*(2t).$$

On the other hand, $Q_j(|f|) \leq Mf(x_j) \leq (Mf)^*(2t)$, and so

$$\|h\|_\infty = \max \{\|f\chi_F\|_\infty, \sup_j Q_j(|f|)\} \tag{4.12}$$

$$\leq \max \{\|(Mf)\chi_F\|_\infty, (Mf)^*(2t)\} \leq (Mf)^*(2t).$$

Combining inequalities (4.10), (4.11), and (4.12), we obtain

$$(Mf)^*(t) \leq 4 \cdot 6^n (f^{\#})^*(2t) + (Mf)^*(2t). \tag{4.13}$$

Iteration of (4.13) N times, with $2^k t$ replacing $2^{k-1}t$, $k = 1,2,\ldots,N$, gives

$$(Mf)^*(t) \leq 4 \cdot 6^n \sum_{k=1}^{N} (f^{\#})^*(2^k t) + (Mf)^*(2^N t).$$

But since $(f^{\#})^*$ is nonincreasing, the sum can be estimated in terms of an integral:

$$(Mf)^*(t) \leq 8 \cdot 6^n \int_t^{2^N t} (f^{\#})^*(s)\frac{ds}{s} + (Mf)^*(2^N t). \tag{4.14}$$

Inequality (4.8) now follows by letting N tend to infinity.

<u>COROLLARY 4.2.</u> If $f \varepsilon L^1 + L^\infty$, then

$$f^{**}(t) \leq c\{\int_t^\infty (f^{\#})^*(s)\frac{ds}{s} + f^{**}(+\infty)\}, \qquad 0 < t < \infty. \tag{4.15}$$

Moreover, if $f^{**}(+\infty) = 0$ (in particular, if $f \varepsilon L^r$ for any $r < \infty$), then there are constants c_1 and c_2 independent of f such that

$$c_1\|f\|_{pq} \leq \|f^{\#}\|_{pq} \leq c_2\|f\|_{pq}, \qquad 1 < p < \infty, \quad 0 < q < \infty. \tag{4.16}$$

Proof. The inequality (4.15) follows directly from (4.8) and Theorem 1.3. We introduce the notation P and P' for the averaging operators

$$P(f^*)(t) = \frac{1}{t}\int_0^t f^*(s)ds \tag{4.17}$$

and

$$P'(f^*)(t) = \int_t^\infty f^*(s)\frac{ds}{s}. \tag{4.18}$$

From (4.5) and (1.4) we have $(f^{\#})^*(t) \leq cf^{**}(t)$. Hence, applying L^{pq}-norms and using Hardy's inequality for P, we obtain $\|f^{\#}\|_{pq} \leq c\|f\|_{pq}$, provided $1 < p \leq \infty$. In the other direction, we note that $f^{**}(+\infty) = 0$, by hypothesis. Hence, by (4.15), $f^*(t) \leq f^{**}(t) \leq cP'(f^{\#})^*(t)$. Applying L^{pq}-norms and using

Hardy's inequality for P', we now obtain $\|f\|_{pq} \le c\|f^{\#}\|_{pq}$, provided $1 \le p < \infty$. This completes the proof.

It will be shown in §6 that $K(f;t;L^1,BMO) \sim t(f^{\#})^*(t)$. All that is needed at the present time is the "easy" half of this estimate:

LEMMA 4.3. If $f \in L^1 + BMO$, then

(4.19) $$t(f^{\#})^*(t) \le cK(f;t;L^1,BMO), \qquad 0 < t < \infty .$$

Proof. If $f = g + h$, with $g \in L^1$ and $h \in BMO$, then $f^{\#} \le g^{\#} + h^{\#} \le g^{\#} + \|h\|_{BMO}$, so $t(f^{\#})^*(t) \le t(g^{\#})^*(t) + t\|h\|_{BMO}$. Using (4.5) and the weak (1,1) estimate for M, we obtain $t(f^{\#})^*(t) \le 2\cdot 3^n\|g\|_1 + t\|h\|_{BMO}$. Taking the infimum over all decompositions $f = g + h$, we get the desired inequality (4.19).

Now we are in a position to establish the weak-type inequality which is inverse to (4.7), and hence to identify the interpolation spaces $(L^1,BMO)_{\theta,q}$.

COROLLARY 4.4. The identity operator is of generalized weak-type $[1,1;\infty,\infty)$ with respect to the pairs (L^1,BMO) and (L^1,L^∞), that is,

(4.20) $$\frac{K(f;t;L^1,L^\infty)}{t} \le c\left(\int_t^\infty \frac{K(f,s;L^1,BMO)}{s}\frac{ds}{s} + \lim_{N\to\infty}\frac{K(f;N;L^1,L^\infty)}{N}\right).$$

Consequently,

(4.21) $$(L^1,L^\infty)_{\theta,q} = (L^1,BMO)_{\theta,q}, \qquad 0 < \theta < 1, \quad 0 < q \le \infty .$$

Proof. Inequality (4.20) follows from (4.15), (4.19) and the fact that $K(f;t;L^1,L^\infty) = tf^{**}(t)$. One of the inclusions in (4.21) is given by (4.7) so we need only show that $(L^1,BMO)_{\theta,q} \subseteq (L^1,L^\infty)_{\theta,q}$. First suppose $f \in L^1 \cap BMO$. Then $f^{**}(+\infty) = 0$ and so the last term in (4.20) vanishes. Applying the L^{pq}-norm (with $\theta = 1-1/p$) to (4.20), and using Hardy's inequality for the operator P', we obtain

$$\|f\|_{(L^1,L^\infty)_{\theta,q}} \le c\|f\|_{(L^1,BMO)_{\theta,q}}, \qquad f \in L^1 \cap BMO.$$

But $L^1 \cap BMO$ is dense in the interpolation space $(L^1,BMO)_{\theta,q}$ [4, Chapter 3] for $0 < q < \infty$, so the inequality persists for all $f \in (L^1,BMO)_{\theta,q}$. This establishes (4.21) for $q < \infty$. The remaining case $q = \infty$ is then settled by using the reiteration theorem [4, p.177].

Now let us show how the John-Nirenberg lemma can be derived from the weak-type inequality (4.15). Actually, what we need is a "local" version of (4.15) relating to a fixed cube Q. At the same time, however, we may as well establish a more general inequality which will also give us basic information concerning the $L^{p,\lambda}$ spaces [22]. For a fixed cube Q, and $0 \le \alpha \le 1$, let

$$f_Q^{\#,\alpha}(x) = \sup_{\substack{Q' \subseteq Q \\ x \varepsilon \bar{Q}'}} \{|Q'|^{-\alpha/n} Q'(|f-Q'(f)|)\}\chi_Q(x),$$

and

$$(M_Q f)(x) = \sup_{\substack{Q' \subseteq Q \\ x \varepsilon \bar{Q}'}} \{Q'(|f|)\}\chi_Q(x).$$

The local version of (4.15) is as follows.

LEMMA 4.5. Suppose Q is a fixed cube in R^n, possibly R^n itself, and let $0 \le \alpha \le 1$. Then there is a constant c, depending only on the dimension n, such that for each locally integrable function f on R^n,

$$(4.22) \qquad (f\chi_Q)^{**}(t) \le c\left\{\int_t^{|Q|} (f_Q^{\#,\alpha})^*(s) s^{\alpha/n} \frac{ds}{s} + Q(|f|)\right\}, \qquad 0 < t \le \frac{|Q|}{2}.$$

Before proceeding with the proof of Lemma 4.5, let us examine some of its consequences. Note that (4.15) is in fact the special case of (4.22) corresponding to $Q = R^n$ and $\alpha = 0$. In the resulting inequalities (4.16), which are due originally to Fefferman-Stein [10], the non-trivial assertion is that $f^{\#} \varepsilon L^p$ implies $f \varepsilon L^p$, for $1 < p < \infty$. The John-Nirenberg lemma can be regarded as the limiting case $p = \infty$ of this result: it asserts that if $f^{\#} \varepsilon L^{\infty}$ (i.e., $f \varepsilon$ BMO), then f is "locally" exponentially integrable.

COROLLARY 4.6 (John-Nirenberg [15]). Suppose f is locally integrable on R^n and let Q be a fixed cube in R^n. Then

$$(4.23) \qquad [(f-Q(f))\chi_Q]^{**}(t) \le c\int_t^{|Q|} (f_Q^{\#,0})^*(s) \frac{ds}{s}, \qquad 0 < t < \frac{|Q|}{2}.$$

If $f \varepsilon$ BMO, then

$$K(Q) = \|f_Q^{\#,0}\|_\infty = \sup_{Q' \subseteq Q} Q'(|f-Q'(f)|) < \infty,$$

and

$$(4.24) \qquad [(f-Q(f))\chi_Q]^*(t) \le cK(Q)\log^+\left(\frac{2|Q|}{t}\right), \qquad 0 < t < \infty.$$

In terms of the distribution function, (4.24) asserts

$$(4.25) \qquad |\{x \varepsilon Q : |f(x) - Q(f)| > \lambda\}| \le 2|Q|\exp\left(\frac{-\lambda}{cK(Q)}\right), \qquad 0 < \lambda < \infty.$$

Proof. The inequality (4.23) follows by applying (4.22) to the function $f - Q(f)$. The resulting constant term $Q(|f - Q(f)|)$ does not exceed $(f_Q^{\#,0})^*(|Q|-)$. Indeed, $Q(|f - Q(f)|) \le f_Q^{\#,0}(x)$, for every $x \varepsilon Q$, that is, on a set of measure $|Q|$. Since a function and its decreasing rearrangement are equidistributed, the result follows. But now $(f_Q^{\#,0})^*(|Q| -)$ is dominated by a constant multiple of the integral in (4.22), since $t \le |Q|/2$. This establishes (4.23). The inequality (4.24) is an easy consequence since $g^* \le g^{**}$ and $[(f - Q(f))\chi_Q]^*(t) = 0$ for $t > |Q|$. Finally, (4.25) is an equivalent restatement of (4.24) since the decreasing rearrangement and the distribution function are mutually inverse.

The space $L^{p,\lambda}$ (cf. [22]) consists of those locally integrable functions f on R^n for which the norm (modulo constants)

$$\|f\|_{L^{p,\lambda}} = \sup_Q \{|Q|^{1-\lambda/n} Q(|f - Q(f)|^p)\}^{1/p}, \qquad 1 \le p < \infty ,$$

is finite. Clearly, $L^{1,n}$ = BMO and so, by analogy, the space $L^{p,n}$, $1 \le p \le \infty$, is often denoted by BMO(p).

<u>COROLLARY 4.7 [15].</u> For $1 \le p < \infty$, BMO(p) = BMO, with equivalent norms.

Proof. The inclusion BMO(p) $\subseteq$ BMO results directly from (4.24) and the fact that $\int_2^\infty (\log u)^p du/u^2 < \infty$. The opposite containment is an immediate consequence of Hölder's inequality.

<u>COROLLARY 4.8</u> (Campanato [6], Meyers [17]). Suppose $1 \le p < \infty$ and $n < \lambda \le n + p$. Then

$$L^{p,\lambda} = \mathrm{Lip}(\alpha), \qquad \alpha = (\lambda - n)/p,$$

with equivalent norms.

Proof. Note that $0 < \alpha \le 1$. It is easy to see that $\mathrm{Lip}(\alpha) \subseteq L^{p,\lambda}$: if $f \varepsilon \mathrm{Lip}(\alpha)$, then for any $x \varepsilon Q$,

$$|f(x) - Q(f)| \le \frac{1}{|Q|} \left|\int_Q [f(x) - f(y)]dy\right| \le c\|f\|_{\mathrm{Lip}(\alpha)} |Q|^{\alpha/n} ,$$

from which the finiteness of the $L^{p,\lambda}$-norm follows. In the other direction, Holder's inequality gives

$$(4.26) \qquad f_Q^{\#,\alpha}(x) \le \sup_{\substack{Q' \subseteq Q \\ x \varepsilon Q'}} \{|Q'|^{-\alpha/n} Q'(|f - Q'(f)|^p)^{1/p}\} \le \|f\|_{L^{p,\lambda}}, \quad x \varepsilon Q ,$$

that is, $f_Q^{\#,\alpha}$ is bounded on Q by $\|f\|_{L^{p,\lambda}}$. Using this estimate in (4.22) (applied to $f - f_Q$ rather than f, and with $t \to 0$), we find that

$$\|(f - f_Q)\chi_Q\|_\infty \le c\|f\|_{L^{p,\lambda}} |Q|^{\alpha/n}$$

Hence

$$|f(x) - f(y)| \le 2c\|f\|_{L^{p,\lambda}} |Q|^{\alpha/n}, \qquad x,y\in Q,$$

and so $f \in \mathrm{Lip}(\alpha)$. Note that if $K_\alpha(Q)$ denotes the supremum of the left-hand side of (4.26), then the same argument produces the last estimate with $\|f\|_{L^{p,\lambda}}$ replaced by $K_\alpha(Q)$. Hence, if $K_\alpha(Q) \to 0$ as $|Q|\to 0$, we see that $f \in \mathrm{lip}\,(\alpha)$.

As a final application of (4.22), we complement Corollary 4.7 by showing that $\mathrm{BMO}(p) = \mathrm{BMO}$, for $0 < p < 1$ (cf. [12]). Let

$$f_Q^p(x) = \sup_{\substack{Q'\subseteq Q \\ x\in \bar{Q}'}} \{\inf_{c_{Q'}} Q'(|f - c_{Q'}|^p)^{1/p}\}\chi_Q(x). \tag{4.27}$$

The space BMO(p) consists of those f for which $\|f^p_{R^n}\|_\infty < \infty$.

<u>COROLLARY 4.9.</u> If $0 < p < 1$, then

$$\mathrm{BMO}(p) = \mathrm{BMO}, \tag{4.28}$$

and

$$\frac{K(f;t;L^p,L^\infty)}{t} \le c\{\int_t^\infty [\frac{K(f;s;L^p,\mathrm{BMO})}{s}]^p \frac{ds}{s} + (|f|^p)^{**}(+\infty)\}^{1/p}. \tag{4.29}$$

Proof. Hölder's inequality gives

$$(|f|^p)_Q^{\#,0}(x) \le 2[f_Q^p(x)]^p, \tag{4.30}$$

since $\big||f(x)|^p - |c|^p\big| \le |f(x) - c|^p$ for $0 < p < 1$. Using (4.30) in the inequality (4.23) (applied to $|f|^p$ instead of f) we obtain

$$(|f|^p\chi_Q)^{**}(t) \le c\{\int_t^{|Q|} (f_Q^p)^*(s)^p \frac{ds}{s} + Q(|f|^p)\}, \qquad 0 < t < \frac{|Q|}{2}. \tag{4.31}$$

Now replace f by $f-c_Q$ and estimate f_Q^p from above by $f^p_{R^n}$ to get

$$[(f-c_Q)\chi_Q]^*(t)^p \le c\ \|f\|^p_{\mathrm{BMO}(p)} \log^+(\frac{2|Q|}{t}).$$

Taking p-th roots and integrating from 0 to $|Q|$ we find that $\|f\|_{\mathrm{BMO}} \le c\|f\|_{\mathrm{BMO}(p)}$. The reverse inequality follows immediately from Hölder's inequality. This establishes (4.28).

The inequality (4.31) with $Q = R^n$ gives

$$\{\frac{1}{t}\int_0^t f^*(s)^p ds\} \le c\{\int_t^\infty (f^p_{R^n})^*(s)^p \frac{ds}{s} + (|f|^{p})^{**}(+\infty)\}, \qquad 0 < t < \infty. \tag{4.32}$$

But $K(f;t;L^p,L^\infty) \sim \{\int_0^{t^p} f^*(s)^p ds\}^{1/p}$, so taking p-th roots in (4.32) and changing variables $t \to t^p$ we get

$$\frac{K(f;t;L^p,L^\infty)}{t} \le c\{\int_{t^p}^\infty (f^p_{R^n})^*(s)^p \frac{ds}{s} + (|f|^p)^{**}(+\infty)\}^{1/p}.$$

This, together with the estimate

$$(4.33)\qquad (f^p_{R^n})^*(t^p) \le c\,\frac{K(f;t;L^p,BMO)}{t},$$

establishes (4.29). The proof of (4.33) is exactly the same as that of (4.19), but uses the equivalence of BMO and BMO(p), $0 < p < 1$, as well as the simple fact that $g^p_{R^n}(x) \le M(|g|^p)^{1/p}(x)$.

We close this section with a proof of Lemma 4.5. Since it closely follows the proof of Theorem 4.1, we shall be brief.

Proof of Lemma 4.5. If $0 < t \le |Q|/2$, then the set

$$\Omega_Q = \{x : f_Q^{\#,\alpha}(x) > (f_Q^{\#,\alpha})^*(2t) \text{ or } (M_Qf)(x) > (M_Qf)^*(2t)\}$$

has measure not exceeding $4t$. Since $\Omega_Q \subseteq Q$, the remark following Lemma 1.1 shows that there is a covering $\{Q_j\}$ of Ω_Q by cubes Q_j, with pairwise disjoint interiors, such that each Q_j is contained in Q and has nonempty intersection with $Q-\Omega_Q$, and $\Sigma|Q_j| \le 2^n\,|\Omega_Q|$. If $f = g + h$, then

$$(4.34)\qquad (f\chi_Q)^{**}(t) \le t^{-1}\|g\chi_Q\|_1 + \|h\chi_Q\|_\infty$$

and

$$(4.35)\qquad (M_Qf)^*(t) \le 3^nt^{-1}\|g\chi_Q\|_1 + \|h\chi_Q\|_\infty ,$$

since M is of weak type (1,1). With $F = Q-(\cup Q_j)$ and

$$g = \sum_j [f-Q_j(f)]\chi_{Q_j}, \qquad h = \sum_j Q_j(f)\chi_{Q_j} + f\chi_F ,$$

we have

$$(4.36)\qquad \|g\|_1 = \sum_j |Q_j|^{1+\alpha/n}\{|Q_j|^{-\alpha/n}Q_j(|f-Q_j(f)|)\}$$

$$\le (\sum_j|Q_j|)^{1+\alpha/n}(f_Q^{\#,\alpha})^*(t) \le ct^{1+\alpha/n}(f_Q^{\#,\alpha})^*(2t),$$

and

$$(4.37)\qquad \|h\|_\infty \le \max\{\sup_j Q_j(|f|),\ \|f\chi_F\|_\infty\} \le (M_Qf)^*(2t).$$

Using (4.36) and (4.37) in the inequalities (4.34) and (4.35), we obtain

$$(4.38)\qquad (f\chi_Q)^{**}(t) \le ct^{\alpha/n}(f_Q^{\#,\alpha})^*(2t) + (M_Qf)^*(2t)$$

and

$$(4.39)\qquad (M_Qf)^*(t) \le ct^{\alpha/n}(f_Q^{\#,\alpha})^*(2t) + (M_Qf)^*(2t).$$

Let $N \ge 1$ be the unique integer satisfying $2^Nt \le |Q| < 2^{N+1}t$. Using (4.39) to estimate the right-hand side of (4.38), and iterating this process (N-1) times, we find

$$(f\chi_Q)^{**}(t) \le \sum_{k=1}^{N} (2^k t)^{\alpha/n} (f_Q^{\#,\alpha})^*(2^k t) + (M_Q f)^*(2^N t). \tag{4.40}$$

The sum can be estimated by the integral $\int_t^{2^N t} s^{\alpha/n} (f_Q^{\#,\alpha})^*(s)\, \frac{ds}{s}$, and the constant term by

$$(M_Q f)^*(2^N t) \le [M(f\chi_Q)]^*(2^N t) \le 3^n (2^N t)^{-1} \| f\chi_Q \|_1 \le 2.3^n Q(|f|).$$

This establishes (4.22) and hence completes the proof.

§5. INTERPOLATION FOR (H^1, L^∞) AND (H^1, BMO).

Rivière-Sagher [24] showed that the (θ,q)-interpolation spaces for the pair (H^1, L^∞) are the same as those for (L^1, L^∞), namely

$$(H^1, L^\infty)_{\theta,q} = L^{pq}, \qquad 0 < \theta < 1, \quad 0 < q \le \infty, \tag{5.1}$$

where $\theta = 1-1/p$. Subsequently, Fefferman-Rivière-Sagher [9] used the newly-developed methods of the Fefferman-Stein real-variable H^p-theory [10] to determine the K-functional for H^p and L^∞. Thus

$$K(f;t;H^p, L^\infty) \sim \{\int_0^{t^p} (Mf)^*(s)^p ds\}^{1/p},$$

where Mf is the grand maximal function of f. It follows directly from the weak-type (1,1) and strong-type (∞,∞) properties of M that

$$\frac{K(f;t;H^1,L^\infty)}{t} \le c\, \frac{1}{t} \int_0^t f^{**}(s)ds \le c\, \frac{1}{t} \int_0^t \frac{K(f;s;L^1,L^\infty)}{s} \frac{ds}{s}. \tag{5.2}$$

This inequality asserts that the identity operator is of generalized weak type $(1,1;\infty,\infty)$ for the pairs (L^1, L^∞) and (H^1, L^∞). This weak-type inequality is the inverse of the direct estimate

$$\frac{K(f;t;L^1,L^\infty)}{t} \le c\, \frac{K(f;t;H^1,L^\infty)}{t}, \tag{5.3}$$

which holds because $\|f\|_{L^1} \le \|f\|_{H^1}$. Note that (5.3) and (5.2) (via Hardy's inequality) imply (5.1).

In this section, we present a proof of (5.2) using only the weak-type (1,1) and strong-type (2,2) properties of the Riesz transforms. This gives a good illustration, as does the Rivière-Sagher result, of the fact that the (θ,q) interpolation spaces, for $0 < \theta < 1$, can often be determined without specific knowledge of the K-functional. Such knowledge is usually required, however, in dealing with the situation at the endpoints $\theta = 0$ or $\theta = 1$. Furthermore, once (5.2) has been established, we can combine it with the weak-type inequalities from the previous section and thereby determine the (θ,q) interpolation spaces $(0 < \theta < 1)$ for H^1 and BMO (which, once again, are the Lorentz spaces L^{pq} [12]). Finally, we present a basic inequality for the Hilbert transform (Corollary 5.5) which, besides storing information relative to H^1 and BMO, implies the rearrangement-invariant inequality (2.1) of O'Neil-Weiss.

DEFINITION 5.1 [10]. The Hardy space $H^1(R^n)$ consists of those L^1-functions f whose Riesz transforms

$$R_j f(x) = c_n \int_{R^n} f(x-y) \frac{y_j}{|y|^{n+1}}\, dy, \qquad j = 1,2,\dots,n,$$

also belong to $L^1(R^n)$. H^1 is a Banach space when given the norm

$$\|f\|_{H^1} = \|f\|_{L^1} + \sum_{j=1}^{n} \|R_j f\|_{L^1} .$$

The right-hand side of (5.2) is finite for all t if and only if $\int_0^1 f^{**}(t)\, dt < \infty$. Now it is not hard to show (cf. [2, Corollary 10.2], for example) that this condition is equivalent to membership of f in the space $L \log L + L^\infty$, that is, f is representable as a sum $f = g + h$, with $h \in L^\infty$ and

$$\int_{R^n} |g(x)| \log\, (2 + |g(x)|)dx < \infty .$$

Indeed, the functional $\int_0^1 f^{**}(t)dt$ defines an equivalent norm on $L \log L + L^\infty$. The elementary argument used in the proof of this equivalence shows, in particular, that

$$\int_\Omega |g(x)| \log (2 + |g(x)|)dx \le \|g\|_{L^1} \log (2 + \|g\|_{L^1}) + \int_0^1 g^{**}(t)dt, \tag{5.4}$$

for any locally integrable function g whose support Ω has measure at most one. We shall need to use this inequality in the proof of the next theorem.

THEOREM 5.2. If $f \in L \log L + L^\infty$, then

$$t^{-1} K(f;t;H^1,L^\infty) \le ct^{-1} \int_0^t f^{**}(s)ds, \qquad 0 < t < \infty. \tag{5.5}$$

Proof. Since the Riesz transforms commute with dilations, it is enough to establish (5.5) for $t = 1$. Also, by homogeneity, we can assume

$$\int_0^1 f^{**}(s)ds \le 1. \tag{5.6}$$

Hence, it will suffice to find a constant c, depending only on the dimension n, such that

$$K(f;1;H^1;L^\infty) \le c, \tag{5.7}$$

for all f satisfying (5.6). This will be done by producing $g \in H^1$ and $h \in L^\infty$ with $f = g + h$ and

(5.8) $$\|g\|_{H^1} + \|h\|_{L^\infty} \le c.$$

Now the open set $\Omega = \{(Mf)(x) > (Mf)^*(1)\}$ has measure at most one. Let $\{Q_k\}$ be a Whitney decomposition [27, Chapter 1] of Ω : the cubes Q_k therefore have union Ω, have pairwise disjoint interiors, and satisfy $(\beta_n \cdot Q_k) \cap \Omega^c \ne \phi$, where $\beta_n = 10n^{1/2}$ (here $\beta_n \cdot Q$ denotes the cube concentric with Q, having β_n-times the diameter of Q). If $F = \Omega^c$, define

$$g = \sum_k [f - Q_k(f)]\chi_{Q_k}, \qquad h = \sum Q_k(f)\chi_{Q_k} + f\chi_F.$$

Since $\beta_n \cdot Q_k$ meets F, we have

$$|Q_k(f)| \le Q_k(|f|) \le \beta_n^n(\beta_n \cdot Q_k)(|f|) \le \beta_n^n\ (Mf)^*(1).$$

Hence, by (1.4) and (5.6),

(5.9) $$|Q_k(f)| \le (3\beta_n)^n \int_0^1 f^*(s)ds \le (3\beta_n)^n \int_0^1 f^{**}(s)ds \le (3\beta_n)^n .$$

Similarly,

(5.10) $$\|f\chi_F\|_\infty \le \|(Mf)\chi_F\|_\infty \le (Mf)^*(1) \le 3^n,$$

and so (5.9) and (5.10) combine to give

(5.11) $$\|h\|_\infty \le (3\beta_n)^n.$$

To estimate the H^1-norm of g, we first note that

(5.12) $$g^{**}(s) \le cf^{**}(s), \qquad 0 < s < 1.$$

Indeed, if E is any subset of Ω with $|E| = t$, then

$$\int_E |g| \le \sum_k \int_{E\cap Q_k} |f - Q_k(f)| \le \sum_k \int_{E\cap Q_k} |f| + \sum_k |E\cap Q_k| \beta_n^n(\beta_n \cdot Q_k)(|f|).$$

Since each cube $(\beta_n \cdot Q_k)$ meets Ω^c, we have, using (1.4), $(\beta_n \cdot Q_k)(|f|) \le (Mf)^*(1) \le 3^n f^{**}(1)$. Hence

$$\int_E |g| \le \int_E |f| + |E|(3\beta_n)^n f^{**}(1) \le \int_0^t f^*(s)ds + t(3\beta_n)^n f^{**}(t).$$

Taking the supremum over all sets E of measure t, we obtain (5.12) with $c = (3\beta_n)^n + 1$.

Next we use the following classical estimate for the Riesz transforms of a function b with mean value zero and support in a cube Q:

(5.13) $$\|R_j b\|_{L^1} \le c\left(\int_Q |b(x)| \log(2 + |b(x)|)dx + |Q|\right), \qquad j = 1,2,\dots,n.$$

The proof uses a standard Marcinkiewicz-type decomposition argument (cf. [8, vol. II, p.166]) involving the weak-type (1,1) and strong-type (2,2) properties of the Riesz transforms. We omit the details.

Applying (5.13) to the functions $g\chi_{Q_k}$, we find

$$\|g\|_{H^1} \le \|g\|_{L^1} + \sum_{j=1}^{n} \sum_k \|R_j(g\chi_{Q_k})\|_{L^1}$$

$$\le \|g\|_{L^1} + c \sum_k \{\int_{Q_k} |g(x)|\log(2 + |g(x)|)dx + |Q_k|\}$$

$$= \|g\|_{L^1} + c\{\int_{\Omega} |g(x)|\log(2 + |g(x)|)\, dx + |\Omega|\}.$$

Since $|\Omega| \le 1$, we can apply (5.4) to obtain

$$\|g\|_{H_1} \le c\{\|g\|_{L^1}\log(2 + \|g\|_{L^1}) + \int_0^1 g^{**}(t)dt\ .$$

Hence, using the estimate $\|g\|_{L^1} = \int_0^1 g^*(t)dt \le \int_0^1 g^{**}(t)dt$, we see from (5.12) and (5.6) that $\|g\|_{H^1}$ is bounded from above by a constant (depending only on n). This completes the proof.

For the corresponding function spaces on the unit circle, Zygmund's theorem (Corollary 2.1(b)) asserts that $L \log L \subseteq H^1$. This result is decidedly false in R^n because, for example, $H^1(R^n)$ contains no positive functions. There is however the following natural analogue in R^n of Zygmund's theorem, which follows directly from (5.5) with t = 1.

<u>COROLLARY 5.3.</u> $\qquad L \log L + L^\infty \subseteq H^1 + L^\infty\ .$

As we remarked above, an immediate consequence of the previous theorem and Corollary 4.4 is that

$$(5.14) \qquad (H^1, L^\infty)_{\theta,q} = L^{pq} = (L^1, BMO)_{\theta,q}, \qquad 0 < \theta < 1, \quad 0 < q \le \infty,$$

where $\theta = 1-1/p$. But the embeddings $L^\infty \subseteq BMO$ and $H^1 \subseteq L^1$ imply

$$(H^1, L^\infty)_{\theta,q} \subseteq (H^1, BMO)_{\theta,q} \subseteq (L^1, BMO)_{\theta,q}.$$

Together with (5.14), this produces the following result, due to Hanks [12].

<u>COROLLARY 5.4.</u> If $\theta = 1-1/p$, then

$$(H^1, BMO)_{\theta,q} = L^{pq}, \qquad 0 < \theta < 1, \quad 0 < q \le \infty\ .$$

For any operator T carrying H^1 into L^1 and L^∞ into BMO, it follows easily (cf. [4, p.180]) from the definition of the K-functional that

$$K(Tf;t;L^1,BMO) \le cK(f;t;H^1,L^\infty).$$

Hence, using the Fefferman-Rivière-Sagher result and Lemma 4.3, we obtain, for the Riesz transforms in particular, the following interesting inequality.

COROLLARY 5.5. If $f \in H^1 + L^\infty$, then the Riesz transforms satisfy

$$[(R_j f)^{\#}]^*(t) \le c(Mf)^{**}(t), \qquad 0 < t < \infty, \quad j = 1,2,\dots,n, \tag{5.15}$$

where M is the grand maximal function.

Let f be a simple function so, in particular, $(R_j f)^{**}(+\infty) = 0$. Applying (4.15) and (5.15) to $R_j f$ we obtain

$$\int_0^1 (R_j f)^*(s)ds \le c\int_1^\infty [(R_j f)^{\#}]^*(s)\frac{ds}{s} \le c\int_1^\infty (Mf)^{**}(s)\frac{ds}{s}.$$

Hence (5.5) and a change in the order of integration gives

$$\int_0^1 (R_j f)^*(s)ds \le c\{\int_0^1 f^{**}(u)du + \int_1^\infty f^{**}(u)\frac{du}{u}\}. \tag{5.16}$$

This, in the notation of [2], simply asserts the boundedness (on the simple functions) of R_j from the space $L \log L + L^{\infty 1}$ into $L^1 + L^\infty$. Since the simple functions are dense in $L \log L + L^{\infty 1}$, the inequality (5.16) persists for all f. A simple dilation argument and the fact that the Riesz transforms commute with dilations now gives the basic rearrangement-invariant inequality for the Riesz transforms, due to O'Neil-Weiss [21] (which we originally discussed in (2.1) in the context of the unit circle).

COROLLARY 5.6. If $f \in L \log L + L^{\infty 1}$, then the Riesz transforms satisfy

$$(R_j f)^{**}(t) \le c\{\frac{1}{t}\int_0^t f^{**}(s)ds + \int_t^\infty f^{**}(s)\frac{ds}{s}\}, \qquad 0 < t < \infty,\ j=1,2,\dots,n. \tag{5.17}$$

The fundamental inequality (5.15) thus stores not only the information $R_j : H^1 \to L^1$ and $R_j : L^\infty \to BMO$, but also, via (5.17), the rearrangement-invariant behavior of the Riesz transforms. Since a potentially "deeper" inequality could result from encoding the information $R_j : H^1 \to H^1$ and $R_j : BMO \to BMO$, it would seem to be a problem of some interest to describe in concrete terms the K-functional for the pair (H^1, BMO).

§6. THE K-FUNCTIONAL FOR L^1 AND BMO. In this final section, we show that the K-functional for L^1 and BMO can be identified with $t(f^{\#})^*(t)$.

THEOREM 6.1. There are constants c_1 and c_2, depending only on the dimension n, such that for any $f \in (L^1 + BMO)(R^n)$,

(6.1) $$c_1 t(f^{\#})^*(t) \le K(f;t;L^1,BMO) \le c_2 t(f^{\#})^*(t), \qquad 0 < t < \infty .$$

Proof. The first inequality in (6.1) was established in Lemma 4.3. In order to prove the second inequality, we must exhibit $g \in L^1$ and $h \in BMO$ such that $f = g + h$ and

(6.2) $$\|g\|_{L^1} + t\|h\|_{BMO} \le c_2 t(f^{\#})^*(t).$$

With t fixed, the open set $\Omega = \{x : f^{\#}(x) > (f^{\#})^*(t)\}$ has measure not exceeding t. Let $F = \Omega^c$ and let $\{Q_j\}$ be a Whitney covering of Ω (cf. [27]). Thus, the cubes Q_j are dyadic and satisfy

(6.3) $$\Omega = \bigcup_j Q_j ;$$

(6.4) $$|Q_j \cap Q_k| = 0, \qquad \text{unless } j = k;$$

(6.5) $$\text{diam } (Q_j) \le \text{dist } (Q_j,F) \le 4 \text{ diam } (Q_j), \qquad j = 1,2,\ldots$$

If $\alpha\cdot Q$ ($\alpha > 0$) denotes the cube concentric with Q but having α-times the diameter, then (6.5) shows that with $\beta = \beta(n) = 10n^{1/2}$,

(6.6) $$(\beta\cdot Q_j) \cap F \ne \phi, \qquad j = 1,2,\ldots$$

We shall denote by $\overline{Q}_j$ the cube $\beta\cdot Q_j$.

The decomposition to be used in (6.2) is given by

(6.7) $$g(x) = \sum_j (f(x) - Q_j(f))\chi_{Q_j}(x)$$

and

(6.8) $$h(x) = \sum_j Q_j(f)\chi_{Q_j}(x) + f(x)\chi_F(x).$$

For any cube Q (so $\overline{Q} = \beta\cdot Q$), it is clear that

(6.9) $$Q(|f - Q(f)|) \le 2\beta^n \,\overline{Q}(|f - \overline{Q}(f)|).$$

Now, for every Whitney cube Q_j, we see from (6.6) that $\overline{Q}_j$ meets F, on which the oscillation is small. Hence $\overline{Q}_j(|f - \overline{Q}_j(f)|) \le (f^{\#})^*(t)$, for every j. Combining this with (6.9), and using the properties (6.3) and (6.4) of the Whitney decomposition, we obtain the following estimate for the L^1-norm of g:

(6.10) $$\|g\|_{L^1} = \sum_j |Q_j| Q_j(|f - Q_j(f)|) \le 2\beta^n \sum_j |Q_j|(f^{\#})^*(t) \le 2\beta^n t(f^{\#})^*(t) .$$

Before estimating $\|h\|_{BMO}$ we shall need some further properties of the

Whitney cubes Q_j. For a fixed index j_o, let $J_o = \{j : Q_j \cap Q_{jo} \neq \phi\}$. Thus J_o is the set of indices corresponding to Q_{j_o} and every cube Q_j that "touches" Q_{jo}. The following estimate, which is a consequence of (6.5) and the dyadic nature of the cubes, shows that all of the cubes Q_j, $j \varepsilon J_o$, have approximately the same size:

$$\frac{1}{4} \operatorname{diam}(Q_{j_o}) \leq \operatorname{diam}(Q_j) \leq 4 \operatorname{diam}(Q_{j_o}), \qquad j \varepsilon J_o . \tag{6.11}$$

In particular,

$$\frac{3}{2} \cdot Q_{j_o} \subseteq \bigcup_{j\varepsilon J_o} Q_j \subseteq 9 . Q_{j_o} \tag{6.12}$$

The required estimate for h in (6.2) is that $\|h\|_{BMO} \leq c\,(f^{\#})^*(t)$. For this, it will suffice to show that there is a constant c, depending only on the dimension n, such that

$$Q(|h - Q(h)|) \leq c(f^{\#})^*(t), \tag{6.13}$$

for all cubes Q in R^n. In fact, if we can find any constant $\alpha = \alpha_Q$ for which

$$A(Q) \equiv Q(|h - \alpha|) \leq c(f^{\#})^*(t),$$

then (6.13) will follow, by way of (4.5).

Now fix Q and let $K = \{k : Q_k \cap Q \neq \phi\}$. On each Q_k, the function h is constant and equal to $Q_k(f)$. Hence

$$A(Q) = \sum_{k\varepsilon k} \frac{|Q_k \cap Q|}{|Q|\ |Q_k|} \left|\int_{Q_k} (f(x) - \alpha)dx\right| + \frac{1}{|Q|} \int_{Q\cap F} |f(x) - \alpha|\, dx. \tag{6.14}$$

There are three cases to consider in estimating A(Q):

<u>Case 1.</u> Suppose $K = \phi$. Then $Q \subseteq F$ and we select $\alpha = Q(f)$. Hence

$$A(Q) \leq Q(|f - Q(f)|) \leq (f^{\#})^*(t),$$

since Q contains a point of F.

<u>Case 2.</u> Suppose, for some $j_o \varepsilon K$, that

$$\operatorname{diam}(Q) < \frac{1}{4} \operatorname{diam}(Q_{j_o}). \tag{6.15}$$

This, and the fact that $Q_{j_o} \cap Q \neq \phi$, implies that $Q \subseteq (3/2)\cdot Q_{j_o}$ and hence, by (6.12), that $Q \cap F = \phi$. We claim that $K \subseteq J_o$, that is, any Whitney cube Q_k touching Q must also touch Q_{jo}. The point is that, by (6.15), Q is in the interior of $(3/2)\cdot Q_{jo}$ at a positive distance from the boundary. Thus any Q_k, $k \varepsilon K$, since it touches Q, will intersect $(3/2)\cdot Q_{jo}$ in a set of positive measure. The first relation in (6.12) therefore shows that Q_k intersects some

Q_j, $j\varepsilon J_o$, in a set of positive measure. But the Whitney cubes have mutually disjoint interiors, so this implies $Q_k = Q_j$, that is, $k\varepsilon J_o$. Hence $K \subseteq J_o$.

Let $\tilde{Q}$ denote the cube $(\beta \cdot Q_{j_o}) = (10n^{1/2} \cdot Q_{j_o})$, and select $\alpha = \tilde{Q}(f)$. The second term on the right of (6.14) vanishes because $Q \cap F = \phi$. In the first term we use the trivial estimate $|Q_k \cap Q| \le |Q|$, replace the index set K by the larger J_o, and use the first part of (6.11) to get

$$A(Q) \le \sum_{j\varepsilon J_o} \frac{1}{|Q_j|} \int_{Q_j} |f(x) - \tilde{Q}(f)|dx \le \frac{4^n}{|Q_{j_o}|} \sum_{j\varepsilon J_o} \int_{Q_j} |f(x) - \tilde{Q}(f)|dx.$$

Now $|Q_{j_o}| = \beta^{-n}|\tilde{Q}|$ and the second relation in (6.12) shows that the cubes Q_j, $j\varepsilon J_o$, are (disjoint) subsets of $\tilde{Q}$. Hence

$$A(Q) \le (4\beta)^n \, \tilde{Q}(|f - \tilde{Q}(f)|) \le (4\beta)^n (f^{\#})^*(t),$$

since $\tilde{Q}$ contains a point of F.

<u>Case 3.</u> Here $K \ne \phi$ and for all $k\varepsilon K$,

$$\frac{1}{4} \operatorname{diam}(Q_k) \le \operatorname{diam}(Q). \tag{6.16}$$

An immediate consequence is that each Q_k, $k\varepsilon K$, is contained in $9 \cdot Q$:

$$\bigcup_{k\varepsilon K} Q_k \subseteq 9 \cdot Q . \tag{6.17}$$

Hence, by (6.6), the cube $\tilde{Q} = (9\beta) \cdot Q$ meets F : $\tilde{Q} \cap F \ne \phi$. Furthermore, by (6.17), the sets Q_k, $k\varepsilon K$, and $Q \cap F$ are disjoint subsets of $\tilde{Q}$. Returning to (6.14) we thus have

$$\begin{aligned} A(Q) &\le \frac{1}{|Q|} \{\sum_{k\varepsilon K} \int_{Q_k} |f(x) - \alpha|dx + \int_{Q\cap F} |f(x) - \alpha|\} \\ &\le \frac{1}{|Q|} \int_{\tilde{Q}} |f(x) - \alpha|dx = \frac{(9\beta)^n}{|\tilde{Q}|} \int_{\tilde{Q}} |f(x) - \alpha|dx. \end{aligned}$$

Choosing $\alpha = \tilde{Q}(f)$, we obtain finally

$$A(Q) \le (9\beta)^n \, \tilde{Q}(|f - \tilde{Q}(f)|) \le (9\beta)^n (f^{\#})^*(t),$$

since $\tilde{Q}$ contains a point of F.

Collecting the results from all three cases, we see that $A(Q) \le (9\beta)^n (f^{\#})^*(t)$, for any cube $Q \subseteq R^n$, and so

$$\|h\|_{BMO} \le 2(9\beta)^n (f^{\#})^*(t).$$

This, together with (6.10), establishes the second inequality in (6.1) with $c_2 = 2\beta^n + 2(9\beta)^n$. Since $\beta = 10n^{1/2}$ depends only on n, the proof is complete.

REMARK 6.2. The key element of the proof of Theorem 6.1 is the construction of a conditional expectation of f (namely the function h) which lies in BMO. The Whitney covering, in which the cubes are arranged in "geometric progression", seems to be essential here. Since arbitrary conditional expectations do not preserve BMO-functions, this construction may be of independent interest.

REMARK 6.3. With only slight modification to the proof of Theorem 6.1, it is possible to show that the K-functional for the pair (L^p, BMO), where $0 < p < 1$, is equivalent to the functional $t(f^p_{R^n})^*(t^p)$, where $f^p_{R^n}$ is the L^p-analogue of the sharp-function defined by (4.27).

REFERENCES

1. R.A. Adams, "Sobolev Spaces", Academic Press, New York, 1975.

2. C. Bennett and K. Rudnick, "On Lorentz-Zygmund spaces", to appear in Dissertationes Math., Vol. 175.

3. C. Bennett and R. Sharpley, "Weak type inequalities in analysis", Proc. Conf. Oberwolfach, Linear Operators and Approximation, August 1977. ISNM, Vol. 40, Birkhauser, Basel, 1978, 151-162.

4. P.L. Butzer and H. Berens, "Semigroups of Operators and Approximation", Springer, Berlin, 1967. MR 37 #5588.

5. A.P. Calderón, "Spaces between L^1 and L^∞ and the theorem of Marcinkiewicz" Studia Math 26 (1966), 273-299. MR 34 #3295.

6. S. Campanato, "Proprietà di hölderiantà di alcune classi di funzioni", Ann. Scuola Norm. Sup. Pisa 17 (1963), 175-188. MR 27 #6119.

7. R. DeVore, S. Riemenschneider, and R. Sharpley, "Weak interpolation in Banach spaces", to appear. J. Functional Anal.

8. R.E. Edwards, "Fourier Series", Holt, Rinehart, and Winston, New York, 1967. MR 36 #5588.

9. C. Fefferman, N.M. Rivière, and Y. Sagher, "Interpolation between H^p spaces: The real method", Trans. Amer. Math. Soc. 191 (1974), 75-81. MR 52 #8909.

10. C. Fefferman and E.M. Stein, "H^p spaces of several variables", Acta Math. 129 (1972), 137-193.

11. A.Garsia, "Topics in Almost Everywhere Convergence", Markham, Chicago. 1970. MR 41 #5869.

12. R. Hanks, "Interpolation by the real method between BMO, $L^\alpha (0 < \alpha < \infty)$, and $H^\alpha (0 < \alpha < \infty)$", Indiana Univ. Math. J. 26 (1977), 679-689.

13. G.H. Hardy and J.E. Littlewood, "A maximal function theorem with function theoretic applications", Acta Math. 54 (1930), 81-116.

14. C. Herz, "The Hardy-Littlewood maximal theorem", Symposium on Harmonic Analysis, Univ. of Warwick, 1968.

15. F. John and L. Nirenberg, "On functions of bounded mean oscillation", Comm. Pure Appl. Math 14 (1961), 415-426. MR 24 #A1348.

16. G.G. Lorentz, "Approximation of Functions", Holt, Rinehart, and Winston, New York, 1966. MR 35 #4642.

17. N.G. Meyers, "Mean oscillation over cubes and Hölder continuity", Proc. Amer. Math. Soc. 15 (1964), 717-721. MR 29 #5969.

18. R. O'Neil, "Les fonctions conjugées et les intégrales fractionnaires de la classe $L(\log^+ L)^s$", C.R. Acad. Sci. Paris 263 (1966), 463-466. MR 35 #717.

19. R. O'Neil, "Convolution operators and L(p,q) spaces", Duke Math J. 30 (1963), 129-142. MR 26 #4193.

20. R. O'Neil, "Integral transforms and tensor products on Orlicz and L(p,q) spaces", J. Analyse Math. 21 (1968), 1-276.

21. R. O'Neil and G. Weiss, "The Hilbert transform and rearrangement of functions", Studia Math. 23 (1963), 189-198. MR 28 #3298.

22. J. Peetre, "On the theory of $L^{p,\lambda}$ spaces", J. Functional Anal. 4 (1969), 71-87. MR 29 #3300.

23. K.E. Petersen, "Brownian Motion, Hardy Spaces and Bounded Mean Oscillation", Cambridge Univ. Press, Cambridge, 1977.

24. N.M. Rivière and Y. Sagher, "Interpolation between L^∞ and H^1, the real method", J. Functional Anal. 14 (1973), 401-409. MR 50 #14204.

25. R. Sharpley, "Multilinear weak-type interpolation of m n-tuples with applications", Studia Math. 60 (1977), 179-194.

26. E.M. Stein, "Note on the class L log L", Studia Math. 32 (1969), 305-310. MR 40 #799.

27. E.M. Stein, "Singular Integrals and Differentiability Properties of Functions", Princeton Univ. Press, Princeton, N.J., 1970. MR 44 #7280.

28. E.M. Stein and G. Weiss, "Introduction to Fourier Analysis on Euclidean Spaces", Princeton univ. Press, Princeton, N.J. 1971. MR 46 #4102.

29. A. Zygmund, "Trigonometric Series, Vol. I", 2nd rev. ed., Cambridge Univ. Press, New York, 1959. MR 21 #6498.

30. A. Zygmund, "Trigonometric Series, Vol. II", 2nd ed. rev., Cambridge Univ. Press, New York, 1959. MR 21 #6498.

Proceedings of Symposia in Pure Mathematics
Volume XXXV, Part 1, 1979

SINGULAR INTEGRAL CHARACTERIZATIONS OF NONISOTROPIC H^p SPACES AND THE F. AND M. RIESZ THEOREM

R. R. Coifman[1] and Björn Dahlberg

We start by describing a version of the F. and M. Riesz theorem valid in $\mathbb{R}^n$:

Let $\rho(t)$ be a nonnegative increasing function on $[0,\infty)$ and

$\Gamma_\rho = \left\{\xi \in \mathbb{R}^n\colon \sum_{i=2}^{n} |\xi_i| < \rho(\xi_1)\ ,\ \xi_1 \geq 0\right\}$. Then any finite Borel measure whose Fourier transform is supported in Γ_ρ is necessarily absolutely continuous.

We observe that the case $\rho(\xi_1) \leq C\ \xi_1$ is well known see [5] and is obtained by showing that the measure is actually in $H^1(T_\Gamma)$ (where T_Γ is a tube domain over a cone containing Γ_ρ).

It turns out that the result stated above can easily be proved by a reduction to one dimension and that the measure has to belong to a certain "maximal" H^1 space. Actually the following more precise version is true:

THEOREM I. Let $\rho_i(t)$ be nonnegative functions increasing to ∞ , $i = 2,3,\ldots,n$ and

$$\Gamma_\rho = \{\xi \in \mathbb{R}^n\colon |\xi_i| \leq \rho_i(\xi_1) \quad i = 2,\ldots,n \quad \xi_1 \geq 0\}$$

Then there exists a test function $\Phi \in \mathcal{S}$ with $\int_{\mathbb{R}^n} \Phi\, dx = 1$ and a constant

---

AMS(MOS) subject classifications (1970). Primary 30A78.

[1]Research supported by the National Science Foundation under grant MCS75-02411 A03.

C_p such that

$$0 < p < \infty \quad , \quad \|\sup_{\lambda>0} |\Phi_\lambda * f|\,\|_p \le C_p \limsup_{\lambda\to\infty} \|\Phi_\lambda * f\|_p$$

for all tempered distributions f whose Fourier transform is supported in Γ_ρ, where

$$\Phi_\lambda(x) = \lambda \prod_{i=2}^{n} \rho_i(\lambda)\Phi(\lambda x_1, \rho_1(\lambda)x_2, \ldots, \rho_n(\lambda)x_n) \quad .$$

The F. and M. Riesz theorem mentioned above is an immediate corollary. In fact, for $p = 1$ and $f = \nu$, a finite Borel measure, the right hand side is necessarily bounded; thus, $f^*(x) = \sup_\lambda |\Phi_\lambda * \nu|$ is integrable. Since $\Phi_\lambda * \nu$ converges to ν weakly (as $\lambda \to \infty$) and is dominated by f^*, it follows that ν is absolutely continuous. (We have $\nu(0) \le \int_0 f^* dx$ for each open set 0).

Another immediate corollary is the possibility of characterizing H^p spaces of the kind studied by A. P. Calderón and A. Torchinsky [1], see article by G. Weiss on related questions [6]. These spaces are defined as spaces of tempered distributions f for which $\sup_\lambda |\Phi_\lambda * f| \in L^p$ where Φ_λ is as above with $\rho_i(\lambda) = \lambda^{\alpha_i}$ $\alpha_i > 0$.

It turns out that the space is independent of the choice of the test function Φ (provided $\int \Phi\, dx \ne 0$ and the α_i's are fixed). Various other norms or definitions are given in [1], [3].

Using theorem I we can now prove

THEOREM II. There exist $2n$ singular integral operators K_i such that, for $f \in \mathcal{S}$ and $0 < p < \infty$,

$$C_p \|\sup_\lambda |\Phi_\lambda * f|\,\|_p \le \sum_{i=1}^{2n} \|K_i(f)\|_p \le C'_p \|\sup_\lambda |\Phi_\lambda * f|\,\|_p$$

(where $C_p > 0$, $C'_p > 0$ depend only on p and Φ).

The proof follows an idea of L. Carleson [2]. We can find a C^∞ partition of unity on Σ_{n-1} consisting of $2n$ functions $\omega_i(\xi)$ having the following properties:

$$\text{for } i = 1,\ldots,n \qquad e_i \in \text{support } \omega_i \subseteq \{\xi \in \Sigma_{n-1}\colon \xi_i > 0\}$$

$$\text{for } i = n+j \quad j = 1,\ldots,n \qquad -e_i \in \text{supp } \omega_{n+j} \subseteq \{\xi \in \Sigma_{n-1}\colon \xi_i < 0\}$$

(here $\{e_i\}$ denotes the standard orthonormal basis of $\mathbb{R}^n$).

If we now extend ω_i to $\mathbb{R}^n$ by homogeneity i.e. by requiring: $\omega_i(\xi_1,\ldots,\xi_n) = \omega(\lambda\xi_1,\lambda^{\alpha_2^i}\xi_2,\ldots,\lambda^{\alpha_n}\xi_n)$ we can define

$$K_i(f) = (\omega_i(\xi)\hat{f}(\xi))^{\vee}$$

As proved in [1] K_i is bounded on H^p . By construction $\omega_i(\xi)\hat{f}(\xi)$ is supported in a region of the type Γ_ρ with $\rho_i(t) = C_i t^{\beta_i}$ ($\beta_i = \alpha_i$ or $\frac{1}{\alpha_i}$) from Theorem I we get

$$\|\sup_\lambda |\Phi_\lambda * K_i(f)|\,\|_p \leq C_p \|K_i(f)\|_p \qquad f \in \mathcal{S} .$$

Moreover since

$$f = \sum_{i=1}^{2n} K_i(f)$$

we obtain the left inequality in the theorem.

We now sketch a proof of theorem I. We use the notation $x = (x_1,x')$ with $x' = (x_2,\ldots,x_n)$ and observe that since $\hat{f}$ is supported in $\xi_1 \geq 0$ it is of analytic type in the x_1 variable.

We let φ denote a test function on $\mathbb{R}^1$ whose Fourier transform $\hat{\varphi}(\xi)$ equals 1 in $|\xi| \leq 1/2$ and vanishes for $|\xi| \geq 1$. The main point of the proof is the following obvious identity:

$$(1,1) \qquad \hat{\varphi}\left(\frac{\xi_1}{\lambda}\right) \hat{f}(\xi_1,\xi_2,\ldots,\xi_n) = \hat{\varphi}\left(\frac{\xi_1}{\lambda}\right) \hat{\varphi}\left(\frac{\xi_2}{2\rho_2(\lambda)}\right),\cdots,\hat{\varphi}\left(\frac{\xi_n}{2\rho_n(\lambda)}\right)\hat{f}(\xi_1,\ldots,\xi_n)$$

valid whenever $\hat{f}$ is supported in Γ_ρ .

We now define Φ_λ by $\hat{\Phi}_\lambda(\xi) = \hat{\varphi}\left(\frac{\xi_1}{\lambda}\right) \hat{\varphi}\left(\frac{\xi_2}{2\rho_2(\lambda)}\right),\cdots,\hat{\varphi}\left(\frac{\xi_n}{2\rho_n(\lambda)}\right)$ and $F_\lambda(x) = \Phi_\lambda * f$.

(1,2) Clearly $F_\mu(x) = \Phi_\mu * F_\lambda$ if $2\rho_i(\mu) < \rho_i(\lambda)$, $(\rho_1(\lambda) = \lambda)$.

We fix λ large and observe that for almost all x' the distribution $F_\lambda(x_1,x')$ is in $H^p(\mathbb{R}^1)$, being holomorphic in the x_1 variable.[2]

---

[2] One sees easily that $F_\lambda(x_1,x')$ has a holomorphic extension to $F_\lambda(z,x')$ in $\mathcal{I}m\, z > 0$ which is of polynomial growth in z and has boundary values in $L^p(\mathbb{R})$ and thus belongs to $H^p(\mathbb{R})$.

Using the one dimensional Fefferman-Stein theory we obtain

$$\|\sup_{\mu} |\varphi_\mu * F_\lambda(\cdot, x')| \|_p^p \leq C_p \int_{-\infty}^{\infty} |F_\lambda(x_1, x')|^p \, dx_1 \quad ,$$

where $\varphi_\mu = \mu \varphi(\mu x)$.

It now follows from (1,1) and (1,2) that

$$\int_{-\infty}^{\infty} \sup_{\mu \leq N} |\Phi_\mu * F_\lambda(x_1, x')|^p \, dx_1 = \int \sup_{\mu \leq N} |F_\mu(x_1, x')|^p \, dx_1$$

$$\leq C_p \int_{-\infty}^{\infty} |F_\lambda(x_1, x')|^p \, dx_1 \qquad \text{for } \lambda \text{ sufficiently large.}$$

Integrating in x' we obtain, for λ large,

$$\int_{R^n} \sup_{\mu \leq N} |F_\mu(x)|^p \, dx \leq C_p \int |F_\lambda(x)|^p \, dx \quad ;$$

letting, first, $\lambda \to \infty$ and, then, $N \to \infty$, we obtain the desired result.

REFERENCES

1. A. P. Calderón and A. Torchinsky, Parabolic maximal functions associated with a distribution, Adv. in Math., Vol. 16, April 1975.

2. L. Carleson, Two remarks on H^1 and B.M.O., Analyse Harmonique d'Orsay 164 (1975).

3. R. R. Coifman and G. Weiss, Extensions of Hardy spaces and their use in analysis, Bull. A.M.S., Vol. 83, No. 4, July 1977.

4. C. Fefferman and E. M. Stein, H^p spaces of several variables, Acta Math. 129 (1972), 137-193.

5. E. M. Stein and G. Weiss, Introduction to Fourier analysis on Euclidean spaces, Princeton Univ. Press, Princeton, N.J., 1971.

6. G. Weiss, Some problems in the theory of Hardy spaces, these Proceedings.

DEPARTMENT OF MATHEMATICS, WASHINGTON UNIVERSITY, ST. LOUIS, MISSOURI 63130

Proceedings of Symposia in Pure Mathematics
Volume XXXV, Part 1, 1979

A MAXIMAL THEORY FOR GENERALIZED HARDY SPACES

by
Roberto A. Macías and Carlos Segovia

ABSTRACT. A maximal function is introduced for distributions acting on certain spaces of Lipschitz functions defined on spaces of homogeneous type. A decomposition into atoms for distributions whose maximal functions belong to L^p, $p \leq 1$, is obtained.

0. INTRODUCTION. Our purpose is to construct a maximal theory for H^p spaces defined on spaces of homogeneous type. With this aim, we introduce a class of spaces of homogeneous type that we call normal spaces of order α and we define a space E^α of Lipschitz functions that plays the role of the C^∞ functions with bounded support on R^n. By a distribution on E^α we mean a continuous linear functional on E^α. A maximal function is associated to a distribution which is similar to the maximal functions considered, for instance, in [2] and [5]. We state Calderón-Zygmund type lemmas for distributions and functions generalizing lemma A of [6]. Using these lemmas we prove that a distribution whose maximal function belongs to L^p, $p \leq 1$, can be decomposed into a series of multiples of p-atoms. This decomposition allows us to obtain that the atomic H^p spaces as defined in [4] and [9] can be identified with the space of all the distributions on E^α whose maximal functions belong to L^p. These results were announced in [10]. Detailed proves are included in [11] and [12]. Decompositions of this type were obtained in [3] for the real line, in [8] for R^n and in [1] for diagonalizable parabolic metrics.

1. PRELIMINARIES. A quasi-distance on a set X is a non-negative function $d(x,y)$ defined on $X \times X$ satisfying: $d(x,y) = 0$ if and only if $x = y$, $d(x,y) = d(y,x)$ and there exists a finite constant K such that $d(x,y) \leq K(d(x,z) + d(z,y))$. Two quasi-distances $d_1(x,y)$ and $d_2(x,y)$ will be considered as equivalent if there

AMS(MOS) subject classifications (1970). Primary 30A78, 43A85, 44A25; Secondary 26A16, 26A69.

exists $0 < c \leq 1$ such that $c \cdot d_1(x,y) \leq d_2(x,y) \leq c^{-1} \cdot d_1(x,y)$. Given a quasi-distance $d(x,y)$ on X, the family $U_r = \{(x,y) : d(x,y) < r\}$, $r > 0$, of subsets of $X \times X$ is a basis for a uniformity on X. We shall say that this is the uniformity generated by $d(x,y)$. The topology induced on X by this uniformity shall be called the d-topology. The balls $B(x,r) = \{y : d(x,y) < r\}$, $r > 0$, form a basis of neighbourhoods of the point x. The uniformities generated by equivalent quasi-distances, as well as the induced topologies, are coincident. In the case when $d(x,y)$ is a distance we have that $|d(x,z) - d(y,z)| \leq d(x,y)$. For the general case of a quasi-distance, the following result holds.

(1.1) THEOREM. Let $d(x,y)$ be a quasi-distance on X. There exists a quasi-distance $d'(x,y)$ equivalent to $d(x,y)$, a finite constant C and a number $0 < \alpha < 1$ such that for every x, y and z in X and $r > 0$

$$(1.2) \qquad |d'(x,z) - d'(y,z)| \leq C \cdot r^{1-\alpha} d'(x,y)^{\alpha}$$

holds, provided that both $d'(x,z)$ and $d'(y,z)$ be less than r.

PROOF. Let $b = 3K^2$. The family of sets $V_n = \{(x,y) : d(x,y) < b^{-n}\}$, n an integer, is a countable basis for the uniformity generated by $d(x,y)$ satisfying $V_n o V_n o V_n \subset V_{n-1}$ (where "o" stands for the composition of relations). By the metrization theorem for uniform spaces (see [7], Chapter 6), there is a distance $\rho(x,y)$ such that $V_n \subset \{(x,y) : \rho(x,y) < 2^{-n}\} \subset V_{n-1}$. This implies that $2^{-1} \cdot \rho(x,y) \leq d(x,y)^{\alpha} \leq 4 \cdot \rho(x,y)$, where α is given by $b^{\alpha} = 2$. Defining $d'(x,y) = \rho(x,y)^{1/\alpha}$, we get that $d'(x,y)$ is a quasi-distance which is equivalent to $d(x,y)$. Moreover, from the triangular property of the distance $\rho(x,y)$ and the mean value theorem, we get that $d'(x,y)$ satisfies (1.2). ///

A first consequence of theorem (1.1) is that the family of all the balls with center at x, defined with the quasi-distance $d'(x,y)$, is a basis of open neighbourhoods of the point x.

Let $d(x,y)$ be a quasi-distance on X and let μ be a non-negative measure defined on a σ-algebra of subsets of X which contains the balls and the open subsets of X. We shall say that (X,d,μ) is a space of homogeneous type if there exists a finite constant A such that for every $x \in X$ and $r > 0$

$$0 < \mu(B(x,2r)) \leq A \cdot \mu(B(x,r)) .$$

The set M of all points x such that $\mu(\{x\}) > 0$ is countable and it can be shown (see [11]) that if $x \in M$, there exists $r > 0$ such that $B(x,r) = \{x\}$.

Let $\phi(x)$ be a function which is integrable on bounded measurable subsets of

(X,d,μ). Denote by $m_B(\phi)$ the average of $\phi(x)$ on B, i.e. $m_B(\phi) = \mu(B)^{-1} \int_B \phi(x)\,d\mu(x)$.

Let $0 < \beta < \infty$ and $1 \leq q \leq \infty$. We say that $\phi(x)$ belongs to $\mathrm{Lip}(\beta,q)$ if there exists a finite constant C such that

$$(1.3) \qquad (\mu(B)^{-1} \int_B |\phi(x) - m_B(\phi)|^q \,d\mu(x))^{1/q} \leq C \cdot \mu(B)^{\beta}$$

holds for every ball B. The (β,q)-norm of $\phi(x)$ is defined as the least constant C satisfying the condition (1.3) above and shall be denoted by $\|\phi\|_{\beta,q}$. We say that a function $\phi(x)$ belongs to $\mathrm{Lip}(\beta)$, $\beta > 0$, if there is a finite constant C such that for every x and y in X, we have $|\phi(x) - \phi(y)| \leq C \cdot d(x,y)^{\beta}$. The least constant C satisfying this condition will be denoted by $\|\phi\|_{\beta}$. We shall say that a measurable function belongs to BMO (bounded mean oscillation) if it satisfies (1.3) for $\beta = 0$ and $q = 1$. It will be convenient for our purposes to define Lip(0) as the set of all functions $\phi(x)$ belonging to BMO such that for every ball B and $\varepsilon > 0$ there exists a continuous function $\psi(x)$ satisfying $\int_B |\phi(x)-\psi(x)|\,d\mu(x) < \varepsilon$. For any function $\phi(x)$ in Lip(0), $\|\phi\|_o$ stands for the least constant for which (1.3) holds with $\beta = 0$ and $q = 1$. We observe that if we change the quasi-distance of the space of homogeneous type by an equivalent quasi-distance, the sets of functions $\mathrm{Lip}(\beta,q)$ and $\mathrm{Lip}(\beta)$ remain unchanged. Moreover, the corresponding norms turn out to be equivalent.

Another consequence of theorem (1.1) and the former observation is the existence of an abundant family of non-constant functions with bounded support belonging to $\mathrm{Lip}(\beta)$. This can be seen by taking any function $h(t)$, $t \geq 0$, infinitely differentiable with bounded support and defining $\phi(x) = h(d'(x,z)/r)$ for any $z \in X$ and $r > 0$.

Let us assume that (X,d,μ) is a space of homogeneous type with the property that the balls are open sets. By theorem (1.1), this assumption does not imply a loss of generality. Then, the function defined by

$$(1.4) \qquad \delta(x,y) = \inf\{\mu(B) : B \text{ is a ball containing } x \text{ and } y\}, \text{ if } x \neq y \text{ and } \delta(x,x) = 0$$

is a quasi-distance. It is not necessarily true that $\delta(x,y)$ is equivalent to $d(x,y)$. However, it has the properties of the δ-topology coinciding with the d-topology and the μ-measure of a ball, defined with the quasi-distance δ, being comparable to its radius. More precisely, there exist finite and positive constants A_1, A_2, K_1 and K_2 such that for every x in X and $r > 0$

$$(1.5) \qquad A_1\, r \leq \mu(B_\delta(x,r)) \quad \text{if } r \leq K_1 \mu(X),$$

$$(1.6) \qquad B_\delta(x,r) = X \quad \text{if } r > K_1 \mu(X),$$

$$(1.7) \qquad A_2\, r \geq B_\delta(x,r)) \quad \text{if } r \geq K_2 \mu(\{x\}) \text{ and}$$

(1.8) $B_\delta(x,r) = \{x\}$ if $r < K_2\mu(\{x\})$.

It is easy to verify that these relations between $\delta(x,y)$ and the measure μ imply that (X,δ,μ) is a space of homogeneous type. A space of homogeneous type such that its quasi-distance and its measure satisfy (1.5) through (1.8) shall be called a normal space of homogeneous type or, more briefly, a normal space. Given a space of homogeneous type (X,d,μ), by means of theorem (1.1) and (1.4), one can always find a quasi-distance $\delta(x,y)$ such that (X,δ,μ) is a normal space of homogeneous type. The usefulness of this change of quasi-distances is illustrated by the following theorem.

(1.9) THEOREM. Let (X,d,μ) be a space of homogeneous type such that the balls are open sets. Let $\delta(x,y)$ be the quasi-distance associated with this space in (1.4). Then, a function belongs to the space $\mathrm{Lip}(\beta,q)$ on (X,d,μ), $0 < \beta < \infty$, $1 \leq q \leq \infty$, if and only if it coincides almost everywhere with a function belonging to the space $\mathrm{Lip}(\beta)$ on (X,δ,μ). Moreover, the norms on the corresponding spaces are equivalent.

For a proof of this theorem see [11]. In particular, this theorem shows that for any given $\beta > 0$ the sets of functions $\mathrm{Lip}(\beta,q)$ are the same for every $1 \leq q \leq \infty$.

Next we shall review the definition of the atomic (or generalized) Hardy spaces on spaces of homogeneous type (see [4]). Let (X,d,μ) be a space of homogeneous type. We shall say that an integrable function $a(x)$ is a p-atom, $0 < p \leq 1$, on (X,d,μ) if $\int a(x)d\mu(x) = 0$ and there is a ball B containing the support of $a(x)$ such that $\|a\|_\infty \leq \mu(B)^{-1/p}$. If $\mu(X)$ is finite we shall consider the function $a(x) \equiv \mu(X)^{-1}$ as a p-atom. It follows directly from the definitions that if $a(x)$ is a p-atom, then $L_a(\phi) = \int a(x)\phi(x)d\mu(x)$ defines a linear functional on $\mathrm{Lip}(1/p-1,1)$ for $0<p<1$ and on $\mathrm{Lip}(0)$ for $p=1$. Endowing these spaces with the norms $\|\phi\|_{(1/p-1,1)}$ and $\|\phi\|_o$, when $\mu(X)=\infty$, and $\|\phi\|_{(1/p-1,1)} + \mu(X)^{-1}\int|\phi(x)|d\mu(x)$ and $\|\phi\|_o + \mu(X)^{-1}\int|\phi(x)|d\mu(x)$, when $\mu(X) < \infty$, it turns out that L_a is bounded and $\|L_a\| \leq 1$. To simplify the notation, we shall denote the functional L_a simply as a. Let us consider a numerical sequence $\{\lambda_i\}$ such that $\Sigma_i|\lambda_i|^p < \infty$ and a sequence $\{a_i(x)\}$ of p-atoms, then the series $\Sigma_i \lambda_i a_i$ converges strongly to a linear functional f such that $\|f\| \leq (\Sigma_i|\lambda_i|^p)^{1/p}$. The atomic Hardy space $H^p(X,d,\mu)$ is defined as the set of all the linear functionals (on $\mathrm{Lip}(1/p-1,1)$ or $\mathrm{Lip}(0)$) that can be represented as a series $\Sigma_i \lambda_i a_i$ of the type just described. For an element f of $H^p(X,d,\mu)$, let $\|f\|_{H^p}$ be the infimun of $(\Sigma_i|\lambda_i|^p)^{1/p}$ over all the possible representations of f as $\Sigma_i \lambda_i a_i$. The space $H^p(X,d,\mu)$, $0<p\leq 1$, endowed with the distance $\|f-g\|^p_{H^p}$ becomes a complete topological vector space. The dual space of H^p can be identified with $\mathrm{Lip}(1/p-1,1)$ if $0<p<1$ and with $\mathrm{Lip}(0)$ for $p=1$.

We recall that if $\delta(x,y)$ is the quasi-distance associated with the space of homogeneous type (X,d,μ) in (1.4), then (X,δ,μ) is a normal space of homogeneous type and the functions of $\mathrm{Lip}(\beta,q)$ defined with the quasi-distance $d(x,y)$ coincide almost everywhere with the functions of $\mathrm{Lip}(\beta)$ defined with the quasi-distance $\delta(x,y)$. Moreover, it can be verified that a function $a(x)$ is a p-atom on (X,d,μ) if and only if it is a fixed multiple of a p-atom on (X,δ,μ). This implies that the space $H^p(X,d,\mu)$ coincides with $H^p(X,\delta,\mu)$, being their norms equivalent.

By theorem (1.1) we can assume, without loss of generality, that the quasi-distance $\delta(x,y)$ satisfies (1.2). A normal space of homogeneous type satisfying (1.2) for some $\alpha > 0$ shall be called a normal space of homogeneous type of order α or, shortly, a normal space of order α. As we noted before, (1.2) implies the existence of non-constant functions with bounded support belonging to $\mathrm{Lip}(\alpha)$ on the space (X,δ,μ). Consequently, since the dual space of H^p can be identified with the space $\mathrm{Lip}(1/p-1)$, $0 < p < 1$, we get that the H^p spaces are non-trivial, at least for $(1+\alpha)^{-1} < p \leq 1$.

2. MAXIMAL FUNCTIONS. Let (X,d,μ) be a normal space of homogeneous type of order α. Let us take a point x_o in X and let n be a positive integer. We shall denote by E_n^α the set of all functions with supports contained in the closure of $B(x_o,n)$ and belonging to $\mathrm{Lip}(\beta)$ for all $0<\beta<\alpha$. We shall also denote by $E_n{}^\alpha$ the topological vector space (E_n^α, Z_n^α), where Z_n^α is the topology for E_n^α defined by the family of norms $\{\|\cdot\|_\beta : 0 < \beta < \alpha\}$ and $\|\cdot\|_\infty$. This topological vector space is a Frechet space and the topology Z_{n+1}^α restricted to E_n^α coincides with Z_n^α. We denote by E^α the strict inductive limit of the sequence $\{E_n^\alpha\}_{n=1}^\infty$. The set E^α consists of all functions with bounded supports belonging to $\mathrm{Lip}(\beta)$ for every $0 < \beta < \alpha$. The definition of E^α does not depend on the x_o chosen in the definition of the spaces E_n^α. We shall say that a linear functional f on E^α is a distribution on E^α if it is continuous.

For γ, $0 < \gamma < \alpha$, and x in X, we introduce a class $T_\gamma(x)$ which will allow us to define maximal functions of distributions on E^α. We shall say that a function ψ belonging to E^α is in $T_\gamma(x)$ if there exists r such that $r \geq K_2\mu(\{x\})$, the support of ψ is contained in $B(x,r)$, $r\|\psi\|_\infty \leq 1$ and $r^{1+\gamma}\|\psi\|_\gamma \leq 1$. Let f be a distribution on E^α and $0 < \gamma < \alpha$. We define the γ-maximal function $f_\gamma^*(x)$ of f as $f_\gamma^*(x) = \sup\{|\langle f,\psi\rangle| : \psi \in T_\gamma(x)\}$. In the sequel, we shall sometimes denote $f_\gamma^*(x)$ by $f^*(x)$, when no confusion arises. The following lemma will be needed. For a proof see [12].

(2.1) LEMMA. (Partition of the unity). Let (X,d,μ) be a normal space of order α and Ω an open subset of finite measure, strictly contained in X. There exists a natural number M, which depends on the space only, a sequence $\{x_n\}$ of points of Ω and a sequence of functions $\{\phi_n\}$, $\phi_n \in \mathrm{Lip}(\alpha)$, such that if $r_n = (10K^2)^{-1}\,\mathrm{dist}(x_n, C\Omega)$ then

(2.2) the balls $B(x_n, r_n/4K)$ are pairwise disjoint,

(2.3) $\cup_n B(x_n, r_n) = \Omega$,

(2.4) for every n, the number of balls $B(x_j, 5Kr_j)$ whose intersections with $B(x_n, 5Kr_n)$ are non-empty is at most M,

(2.5) the support of $\phi_n(x)$ is contained in $B(x_n, 2r_n)$,

(2.6) for every x in $B(x_n, r_n)$, $\phi_n(x) \geq M^{-1}$,

(2.7) there exists a constant c depending on the space only such that $\| \phi_n \|_\alpha \leq c \cdot r_n^{-\alpha}$ and

(2.8) $\Sigma_n \phi_n(x) = \chi_\Omega(x)$.

Next, we state some results that generalize lemma A of [6]. The proofs of these results can be found in [12].

(2.9) LEMMA (Calderón-Zygmund type lemma for distributions). Let f be a distribution on E^α such that for some γ, $0 < \gamma < \alpha$, and some p, $(1+\gamma)^{-1} < p \leq 1$, its γ-maximal function belongs to $L^p(X, d\mu)$. Let $t^p > \int f^*_\gamma(x)^p d\mu(x)/\mu(X)$ and $\Omega = \{x : f^*_\gamma(x) > t\}$. This set Ω is open, strictly contained in X and $\mu(\Omega) < \infty$. Let $\{\phi_n(x)\}$ be the partition of the unity associated with Ω in lemma (2.1). If we define

$$S_n(\psi)(x) = \phi_n(x)[\int \phi_n(z)d\mu(z)]^{-1} \cdot \int [\psi(x) - \psi(z)]\,\phi_n(z)d\mu(z)$$

then, $\langle b_n, \psi\rangle = \langle f, S_n(\psi)\rangle$ defines a distribution b_n on E^α and

(2.10) $(b_n)^*_\gamma(x) \leq c \cdot t\,[r_n/(d(x,x_n) + r_n)]^{1+\gamma}$ if $x \notin B(x_n, 4Kr_n)$ and

(2.11) $(b_n)^*_\gamma(x) \leq c \cdot f^*_\gamma(x)$ if $x \in B(x_n, 4Kr_n)$.

Moreover, $\Sigma_n b_n$ converges strongly in the dual space of E^α to a distribution b satisfying

(2.12) $b^*_\gamma(x) \leq c \cdot t\,\Sigma_n [r_n/(d(x,x_n) + r_n)]^{1+\gamma} + c \cdot f^*_\gamma(x)\chi_\Omega(x)$ and

(2.13) $\int b^*_\gamma(x)^p d\mu(x) \leq c \cdot \int_\Omega f^*_\gamma(x)^p d\mu(x)$.

The distribution $g = f - b$ satisfies.

(2.14) $g^*_\gamma(x) \leq c \cdot t\,\Sigma_n [r_n/(d(x,x_n) + r_n)]^{1+\gamma} + c \cdot f^*_\gamma(x)\chi_\Omega(x)$.

From (2.14) we get that $g^*_\gamma(x)$ belongs to $L^2(X, d\mu)$. It can be shown that for such a distribution g, there exists a function g(x) belonging to the closure of E^{α} in $L^2(X,d\mu)$ such that $\langle g,\psi \rangle = \int g(x)\psi(x)\,d\mu(x)$ and $|g(x)| \leq c \cdot g^*_{\gamma}(x)$. Moreover, by (2.13), we see that for any given $\varepsilon > 0$ if t is chosen big enough, the distribution f - g satisfies $\int (f-g)^*_{\gamma}(x)^p d\mu(x) = \int b^*_{\gamma}(x)^p d\mu(x) < \varepsilon$. This means that the distribution f can be approximated by a distribution which coincides with a function in $L^2(X,d\mu)$.

For a distribution f induced by a function, we have the following refinement of lemma (2.9).

(2.15) LEMMA (Calderón-Zygmund type lemma for functions). Let f(x) be a function in $L^r(X,d\mu)$, $1 \leq r < \infty$, and $0 < \gamma < \alpha$, $(1+\gamma)^{-1} < p \leq 1$. Assume that the γ-maximal function $f^*_{\gamma}(x)$ of the distribution f induced by f(x) belongs to $L^p(X,d\mu)$ and $|f(x)| \leq c \cdot f^*_{\gamma}(x)$ almost everywhere on X. Letting t, b and g have the same meaning as in lemma (2.9), we have

(2.16) if $m_n = [\int \phi_n(z)\,d\mu(z)]^{-1} \cdot \int f(y)\phi_n(y)\,d\mu(y)$, then $|m_n| \leq c \cdot t$,

(2.17) the distribution induced by the function $b_n(x) = [f(x) - m_n]\,\phi_n(x)$ coincides with b_n,

(2.18) the series $\Sigma_n b_n(x)$ converges, for every $x \in X$ and in $L^r(X,d\mu)$, to a function b(x) which induces a distribution coinciding with b, and

(2.19) the function $g(x) = f(x) - b(x)$ can be expressed as $g(x) = f(x)\chi_{C\Omega}(x) + \Sigma m_n\phi_n(x)$. Moreover, $|g(x)| \leq c \cdot t$ and the distribution induced by g(x) coincides with g.

3. ATOMIC DECOMPOSITION. In this paragraph we shall show that a distribution f such that its γ-maximal function $f^*_{\gamma}(x)$ belongs to $L^p(X,d\mu)$ can be decomposed into a series of multiples of p-atoms. The bulk of the proof is concentrated in lemma (3.1). The proofs shall be given for the case $\mu(X) = \infty$. The case of $\mu(X) < \infty$ requires minor modifications and can be found in [12].

(3.1) LEMMA. Let h(x) be a function in $L^2(X,d\mu)$ such that $|h(x)| \leq 1$, assume that for some γ, $0 < \gamma < \alpha$, and some q, $(1+\gamma)^{-1} < q < 1$, the γ-maximal function $h^*_{\gamma}(x)$ of the distribution h induced by h(x) belongs to $L^q(X,d\mu)$. Then, for every p, $q < p \leq 1$, there exists a numerical sequence, $\{\lambda_k\}$, and a sequence of p-atoms, $\{a_k(x)\}$, such that for every $\psi(x)$ in E^{α}

$$\langle h,\psi \rangle = \Sigma_k \lambda_k \int a_k(x)\psi(x)\,d\mu(x) \quad \text{and} \quad \Sigma_k |\lambda_k|^p \leq c \int h^*_{\gamma}(x)^q d\mu(x)$$

hold, with c a finite constant independent of h(x).

PROOF. Given s, $0 < s < 1$, we define $\{H_k(x)\}_{k=0}^{\infty}$ inductively as $H_o(x) = h(x)$

and for $k \geq 1$, $H_k(x)$ is the function $g(x)$ given in (2.19) for $f(x) = H_{k-1}(x)$ and $t = s^k$. With the same notations of lemma (2.15), we have that the sequence just defined satisfies

(3.2) $H_k(x) = H_{k-1}(x) - \Sigma_n b_{k,n}(x)$,

(3.3) $|H_k(x)| \leq c \cdot s^k$ and

(3.4) $H_k^*(x) \leq h_\gamma^*(x) + 3^{1+\gamma} c \cdot \Sigma_{i=1}^k s^i \Sigma_n [r_{i,n}/(d(x,x_{i,n}) + r_{i,n})]^{1+\gamma}$.

Properties (3.2) and (3.3) follow immediately from the definition of $H_k(x)$ and (2.19). We shall obtain (3.4) by induction on k. For $k = 0$, (3.4) is obvious. Assume that (3.4) holds for $k - 1$, then, if $x \notin E_k = \{y : H_{k-1}^*(y) > s^k\}$, by (3.2) and (2.10) we obtain

$$H_k^*(x) \leq H_{k-1}^*(x) + c \cdot s^k \Sigma_n [r_{k,n}/(d(x,x_{k,n}) + r_{k,n})]^{1+\gamma}$$

which by the inductive hypothesis implies (3.4). If $x \in E_k$, from (3.3) we get

$$H_k^*(x) \leq c \cdot s^k \leq 3^{1+\gamma} c \cdot s^k \Sigma_n [r_{k,n}/(d(x,x_{k,n}) + r_{k,n})]^{1+\gamma} .$$

Next, we shall estimate the measure of E_k. By definition of E_k and (3.4) we have

$$s^{kq}\mu(E_k) \leq \int H_{k-1}^*(x)^q d\mu(x) \leq \int h^*(x)^q d\mu(x) +$$

$$3^{(1+\gamma)q} c^q \Sigma_{i=1}^{k-1} s^{iq} \Sigma_n \int [r_{i,n}/(d(x,x_{i,n}) + r_{i,n})]^{(1+\gamma)q} d\mu(x),$$

therefore, recalling that by hypothesis $(1 + \gamma)^{-1} < q < 1$, we obtain

$$s^{kq}\mu(E_k) \leq C \cdot [\int h^*(x)^q d\mu(x) + \Sigma_{i=1}^{k-1} s^{iq}\mu(E_i)] ,$$

where the constant C depends on γ, q and the space only. If we denote $\gamma_o = \int h^*(x)^q d\mu(x)$ and $\gamma_i = s^{iq}\mu(E_i)$ for $i \geq 1$, the estimatives above can be written as $\gamma_k \leq C \Sigma_{i=1}^{k-1} \gamma_i$. It can be easily seen that this inequality for γ_k implies that $\gamma_i \leq \gamma_o (C + 2)^i$ for $i \geq 0$. Consequently, we get

$$s^{kq}\mu(E_k) \leq (C + 2)^k \int h^*(x)^q d\mu(x) .$$

Now, since by (3.2), $h(x) = H_k(x) + \Sigma_{i=1}^k \Sigma_n b_{i,n}(x)$ and by (3.3), $|H_k(x)| \leq c \cdot s^k$, $0 < s < 1$, it follows that $h = \Sigma_i \Sigma_n b_{i,n}$. From (2.17) we have that the support of $b_{i,n}(x)$ is contained in the ball $B(x_{i,n}, 2r_{i,n})$ and $|b_{i,n}(x)| \leq 2c \cdot s^{i-1}$. Therefore, defining $\lambda_{i,n} = 2c \cdot s^{i-1} \cdot \mu(B(x_{i,n}, 2r_{i,n}))^{1/p}$, it turns out that the functions $e_{i,n}(x) = \lambda_{i,n}^{-1} b_{i,n}(x)$ are p-atoms and $h = \Sigma_i \Sigma_n \lambda_{i,n} e_{i,n}$. By the estimate for $\mu(E_k)$ obtained above, we have

$$\Sigma_{i=1}^\infty \Sigma_n |\lambda_{i,n}|^p \leq \Sigma_{i=1}^\infty 2^p c^p M s^{(i-1)p} \mu(E_i) \leq$$

$$2^p c^p M s^{-p} [\Sigma_{i=1}^\infty s^{(p-q)i} (C + 2)^i] \int h_\gamma^*(x)^q d\mu(x) \leq C' \int h_\gamma^*(x)^q d\mu(x),$$

if s is chosen in such a way that $s^{p-q}(C + 2) < 1$. ///

(3.5) THEOREM. Let f be a distribution on E^α such that for some γ, $0<\gamma<\alpha$, and some p, $(1+\gamma)^{-1}<p\leq 1$, its γ-maximal function $f^*_\gamma(x)$ belongs to $L^p(X,d\mu)$. Then, there exists a sequence of p-atoms, $\{a_n(x)\}$, and a numerical sequence, $\{\lambda_n\}$, such that $f = \Sigma_n \lambda_n a_n$ strongly in the dual space of E^α. Moreover, there exist two positive and finite constants c'_p and c''_p independent of f, such that

$$c'_p \int f^*_\gamma(x)^p d\mu(x) \leq \Sigma_n |\lambda_n|^p \leq c''_p \int f^*_\gamma(x)^p d\mu(x).$$

PROOF. Let $B_k(x)$ and $G_k(x)$ be the functions $b(x)$ and $g(x)$ associated with $f(x)$ in lemma (2.15) for $t = 2^k$, k integer, and denote $\Omega_k = \{x: f^*_\gamma(x) > 2^k\}$. Since $f(x) = B_k(x) + G_k(x)$ for every integer k, we define $h_k(x) = G_{k+1}(x) - G_k(x) = B_k(x) - B_{k+1}(x)$. The γ-maximal function $h^*_k(x)$ satisfies

$$h^*_k(x) \leq c\cdot 2^k \quad \text{and} \tag{3.6}$$

$$h^*_k(x) \leq C\cdot 2^k \, \Sigma_{j=k}^{k+1} \Sigma_i \, [\, r_{j,i}/(d(x,x_{j,i}) + r_{j,i})\,]^{1+\gamma}. \tag{3.7}$$

For $x \in \Omega_k$, these estimates follow from (2.19). For $x \notin \Omega_k$, the estimates are a consequence of (2.12). Taking q, $(1+\gamma)^{-1} < q < p \leq 1$, from (3.7) we get

$$\int h^*_k(x)^q d\mu(x) \leq C\cdot 2^{kq}\, \mu(\Omega_k). \tag{3.8}$$

On the other hand, it can be shown that $\Sigma_{-\infty}^{\infty} h_k$ converges strongly to f. By (3.6) and (3.8), we have that $c^{-1}2^{-k}h_k$ satisfies the hypotheses of lemma (3.1). Therefore, there exists a numerical sequence $\{\lambda_{k,i}\}$ and a sequence $\{a_{k,i}\}$ of p-atoms such that $c^{-1}2^{-k}h_k = \Sigma_i \lambda_{k,i} a_{k,i}$ and

$$\Sigma_i |\lambda_{k,i}|^p \leq C\cdot 2^{-kq} \int h^*_k(x)^q d\mu(x) \leq C\cdot \mu(\Omega_k).$$

Then, defining $\rho_{k,i} = c\cdot 2^k \lambda_{k,i}$, we get that $f = \Sigma_k \Sigma_i \rho_{k,i} a_{k,i}$ and

$$\Sigma_k \Sigma_i |\rho_{k,i}|^p \leq C\cdot \Sigma_{-\infty}^{\infty} 2^{kp}\mu(\Omega_k) \leq C\cdot \int f^*_\gamma(x)^p d\mu(x). ///$$

By applying theorem (3.5) it can be proved that the atomic H^p spaces defined in the first paragraph are essentially the same as the spaces of distributions on E^α with maximal functions in L^p. More precisely.

(3.9) THEOREM. Let $0<\gamma<\alpha$ and $(1+\gamma)^{-1} < p \leq 1$. The linear transformation that maps an element f of H^p into its restriction $\tilde{f}$ to E^α is one to one and onto the space of distributions on E^α such that their γ-maximal functions belong to $L^p(X,d\mu)$. Moreover, there exist two positive and finite constants c_1 and c_2 such that

$$c_1 \|f\|_{H^p} \leq [\, \int \tilde{f}^*_\gamma(x)^p d\mu(x)\,]^{1/p} \leq c_2 \|f\|_{H^p}.$$

For a proof of this theorem see [12].

REFERENCES

1. A. P. Calderón, *An atomic decomposition of distributions in parabolic H^p spaces*, Advances in Math. 25, Nº 3 (1977), 216 - 225.

2. A. P. Calderón and A. Torchinsky, *Parabolic maximal functions associated with a distribution*, Advances in Math. 16 (1975), 1 - 64.

3. R. R. Coifman, *A real variable characterization of H^p*, Studia Math. 51 (1974), 267 - 272.

4. R. R. Coifman and G. Weiss, *Extensions of Hardy spaces and their use in analysis*, Bull. Amer. Math. Soc., 83 (1977), 569 - 645.

5. C. Fefferman and E. M. Stein, *H^p spaces of several variables*, Acta Math. 129 (1972), 137 - 193.

6. C. Fefferman, N. M. Rivière and Y. Sagher, *Interpolation between H^p spaces: the real method*, Trans. Amer. Math. Soc., 191 (1974), 75 - 81.

7. J. L. Kelley, *General Topology*, D. van Norstrand, New York, 1955.

8. R. H. Latter, *A characterization of $H^p(\mathbb{R}^n)$ in terms of atoms*, Studia Math. 62 (1977), 92 - 101.

9. R. A. Macías, *Interpolation theorems on generalized Hardy spaces*, Doctoral Dissertation, Washington Univ., St. Louis, 1974.

10. R. A. Macías and C. Segovia, *On the decomposition into atoms of distributions on Lipschitz spaces*, Atas do XI Colóquio Brasileiro de Matemática, July 1977, Poços de Caldas.

11. ———, *Lipschitz functions on spaces of homogeneous type*, Advances in Math. (to appear).

12. ———, *A decomposition into atoms of distributions on spaces of homogeneous type*, Advances in Math. (to appear).

DEPARTAMENTO DE MATEMÁTICA, UNIVERSIDADE ESTADUAL DE CAMPINAS, 13.100, CAMPINAS, SP, BRAZIL.

Proceedings of Symposia in Pure Mathematics
Volume XXXV, Part 1, 1979

LOCAL HARDY SPACES

David Goldberg*

Abstract. A new type of Hardy space is described. The main advantage of this space over the classical one is that pseudo-differential operators are bounded on it for $0 < p < \infty$. This local Hardy space also has applications to the classical theory.

1.

The classical Hardy space (as considered by Hardy, Littlewood and M. Riesz) is the set of holomorphic $f(z)$ in the unit disc $\{z \in C: |z| < 1\}$ normed by $\sup_{0<r<1} (\int |f(re^{i\theta})|^p \, d\theta)^{1/p}$. If $f = u + iv$, it will be convenient to identify $f(z)$ with $u(\theta) = \lim_{r\to 1} u(re^{i\theta})$. To recover f, use $u(re^{i\theta}) = P_r * u = \frac{1}{2\pi} \int \frac{1-r}{1-2r\cos(\theta-t)+r^2} f(\theta)d\theta$, and then solve the Cauchy-Riemann equations to get v, hence f. The symbol H^p will denote this set of "boundary values".

In this set-up, the theorem of M. Riesz on L^p boundedness of the Hilbert transform says that $H^p = L^p$ for $p > 1$, and the F. and M. Riesz theorem that $H^1 \subset L^1$. When $p < 1$, H^p is a space of tempered distributions (although this is elementary, it was not explicitly pointed out until recently.)

As an example, let $f(z) = \frac{z+1}{z-1}$, which is a classical H^p function for any $p < 1$. Then $f = u + iv$ with $u(re^{i\theta}) = P_r(\theta)$, so $\frac{z+1}{z-1}$ is identified with $u(\theta) = \delta(\theta)$ (the Dirac delta function at 0). To recover f, write $u(re^{i\theta}) = P_r * \delta = P_r$.

The phenomena to be discussed here are peculiar to the half-plane version of Hardy spaces, that is (in the classical set-up) the norm is now $\sup_{y>0} (\int |f(x + iy)|^p dx)^{1/p}$, where f is holomorphic in the half plane $\{z = x + iy : y>0 \}$. Since f is L^p bounded (on the lines $x + iy$, $-\infty<x<\infty$) if and only if u and v are, it is not surprising that H^1 is exactly all $f \in L^1$ with $Hf \in L^1$. Recall that $Hf = \lim_{\varepsilon\to 0} \frac{1}{\pi} \int_{|y|>\varepsilon} \frac{f(x-y)}{y} dy$ (or $(\hat{Hf}) = \text{sgn}\hat{f}$) is the Hilbert transform which maps the boundary values of u to those of v.

AMS (MOS) subject classifications (1970). Primary 42A40, 42A68, 30A78, 35S05

*Research partially supported by a National Science Foundation graduate fellowship at Princeton University.

2.

One reason for the interest in H^p when $p < 1$ is that it is a space which has many properties of L^p when $p > 1$. Here are two examples.

Theorem 1: If $m(x)$ satisties $|m^{(k)}(x)| \le C_k|x|^{-k}$ and $(T_m f)^\wedge = m\hat{f}$ (defined on Schwartz space $\mathcal{S}$) then $||T_m f||_{H^p} \le C||f||_{H^p}$ if $o < p < \infty$.

Theorem 2: If $o < p < 2$, $\int \frac{|\hat{f}(x)|^p}{|x|^{2-p}} \le C||f||^p_{H^p}$

The first theorem reduces to (a special case of) the Marcinkiewicz interpolation theorem when $p > 1$, the second may be thought of as a companion to the Hausdorff-Young theorem (which estimates the L^p norm of $\hat{f}$ for $p > 2$).

For simplicity, only $p = 1$ will be considered from this point on. The key fact which enables both of these theorems to work is that when $f \varepsilon H^1$ then $\int f = o$. Here is a quick proof. If $f \varepsilon L^1$ and $Hf \varepsilon L^1$ then $(Hf)^\wedge = \text{sgn} x \hat{f}$ is continuous (being the Fourier transform of an L^1 function) so $\hat{f}(o) = o$, but $\hat{f}(o) = \int f$. Since the Hilbert transform satisfies the hypotheses of Theorem 1 the conclusion can not possibly follow unless $\int f = o$. In Theorem 2, if $\hat{f}(o) \neq o$, the integral will not even be finite, much less satisfy the estimate.

Now in certain problems motivated by partial differential equations, it is convenient to measure smoothing properties in terms of L^p, Lipschitz and Sobolev spaces. In addition, since the study of a differential equation on an open set can be reduced to the study of pseudo-differential operators on the boundary (which will be some manifold), it is crucial that pseudo-differential operators are bounded on the above spaces and that they are well-defined on manifolds. Since H^p has many properties of L^p, it is tempting to place it together with the above mentioned spaces. But unfortunately, the very property which enabled Theorems 1 and 2 to hold ($f \varepsilon H^1 \Rightarrow \int f = o$) also causes H^1 to fail to behave properly on manifolds and with respect to pseudo-differential operators. To see this, note that if $\Psi \varepsilon \mathcal{S}$ then $f \to \Psi f$ is not bounded on H^1 (since $\int \Psi f \neq o$). This map is a pseudo-differential operator and also is a "patching" map of the kind necessary for working on a manifold.

3.

There is a local space (call it h^p) which both extends many L^p theorems to $p < 1$ and also is suitable for working with manifolds and pseudo-differential operators. It has the property that $\mathcal{S} \subset h^p$, thus Theorems 1 and 2 will not hold as stated, but will hold in a suitably modified form.

Theorem 1': If $m(x)$ satisfies $|m^{(k)}(x)| \leq C_k|x|^{-k}$ and is smooth at o, then $||T_m f||_{h^p} \leq C\,||f||_{h^p}$.

Theorem 2': If $o < p < 2$, $\int \frac{|\hat{f}(x)|^p}{(1+|x|)^{2-p}} \leq C||f||^p_{h^p}$

It is reasonable to ask how "natural" h^p is. To test this, consider its dual $(h^p)^*$. Now whereas $(H^p)^* = \Lambda_\alpha$ with a norm $||f||_{\Lambda_\alpha} = \sup_{x,y} \frac{|f(x)-f(y)|}{|x-y|^\alpha}$ (for $o < \alpha < 1$ $\alpha = 1-p^{-1}$), the dual of h^p is Λ_α with more familiar norm $\sup_{x,y} \frac{|f(x)-f(y)|}{|x-y|^\alpha} + ||f||_\infty$. One way to summarize briefly the distinction between H^p and h^p is that h^p is adapted to partial differential equations (its dual is the Lipschitz space considered in that theory), whereas H^p belongs to Fourier analysis proper, (it and its dual are invariant under appropriate dilations).

The entries in the following table compare the definitions of h^1 against those for H^1.

H^1	h^1
1. $f\varepsilon L^1$ $Hf\varepsilon L^1$ $Hf = \int f(x-y)\frac{1}{y}\,dy$	$f\varepsilon L^1$ $hf\varepsilon L^1$ $hf = \int f(x-y)k(y)dy$ $k(y) = 1/y$ $\|y\|<1$, $k(y) = 0, \|y\|>2$
2. $f(x) = \lim_{y\to o} F(x+iy)$, F holomorphic in $\{x+iy: y>o\}$ $\sup_{y>o} \int \|F(x+iy)\|dx < \infty$	$f(x) = \lim_{y\to o} F(x+iy)$, F holomorphic in $\{x+iy: o < y < 1\}$ $\sup_{o<y<1} \int \|F(x+iy)\|dx < \infty$
3. $\sup_{y>o} \|P_y * f\| \varepsilon L^1$ $P_y = \frac{y}{x^2+y^2}$ the Poisson kernel for $\{x+iy: y>o\}$	$\sup_{y>o} \|K_y * f\| \varepsilon L^1$ $K_y = \frac{\cosh(1-2y)\pi x}{\cosh \pi x}$, the Poisson kernel for $\{x+iy: o<y<1\}$
4. $\sup_{t>o} \|\phi_t * f\| \varepsilon L^1$ $\phi \varepsilon \mathcal{S}$, $\int\phi \neq o$.	$\sup_{o<t<1} \|\phi_t * f\| \varepsilon L^1$ $\phi\varepsilon\mathcal{S}$, $\int\phi \neq o$

The fourth condition makes it obvious that $\mathcal{S} \subset h^1$. Note the connection between Theorem 1' and condition 1: both consider smooth multipliers, which are in fact pseudo-differential operators.

4.

One way to think of h^p is as the extension of L^p to $p < 1$, and as such it is very natural in connection with pseudo-differential operators and manifolds. However, h^p is also a useful tool in classical Fourier analysis. To illustrate this consider the following;

DeLeeuw's Theorem: If $(Tf)^\wedge = m\hat{f}$ is a bounded multiplier on $L^p(R)$ then $(Sf)^\wedge(k) = m(k)\,\hat{f}(k)$ $(k\in Z)$ is bounded on $L^p(T)$.

Here is a brief sketch of the proof. It can be assumed that $f(x) = \Sigma a_k\, \phi^{2\pi ixk}$, so consider $f_\varepsilon(x) = \Phi_\varepsilon(x)\, f(x)$ $(\Phi\varepsilon \mathcal{S}$ $\Phi_\varepsilon(x) = \varepsilon^{-n}\Phi(\frac{x}{\varepsilon}))$ as a function on R. Then by hypothesis $||Tf_\varepsilon||_{L^p(R)} \le C||f_\varepsilon||_{L^p(R)}$. As $\varepsilon\to\infty$

$|| f_\varepsilon||_{L^p(R)} \to ||f||_{L^p(T)}$ and $||Tf_\varepsilon||_{L^p(R)} \to ||Tf||_{L^p(T)}$ which gives the theorem.

To extend this theorem to H^1 involves the following difficulty: $\Phi_\varepsilon f$ passes out of the class H^1. However, the theory of h^p spaces is perfectly adapted to solve this problem via

Lemma: If $\phi\varepsilon\mathcal{S}$, $\int\phi = 1$ and $\hat{\phi}^{(k)}(o) = o$ for $k \ge 1$, then $||f-\phi*f||_{H^1} \le C||f||_{h^1}$.

Proof: $H(f-\phi*f) = hf$ (take the Fourier transform of both sides), so $||f-\phi*f||_{H^1} = ||f-\phi*f||_{L^1} + ||H(f-\phi*f)||_{L^1} \le ||f||_{L^1} + ||hf||_{L^1} = ||f||_{h^1}$.

Now the extension of deLeeuw's theorem to H^1 is simply three easy steps:

I. Since $f\varepsilon h^1 \Rightarrow \Phi f\varepsilon h^1$ the L^p proof works for h^1

II. Let $\tilde{T}f = T(f-\phi*f)$. Then by the lemma $||\tilde{T}f||_{H^1(R)} \le C||f-\phi*f||_{H^1} \le C||f||_{h^1(R)} \le C||f||_{H^1(R)}$ so from I, $||\tilde{S}f||_{H^1(T)} \le C||f||_{H^1(T)}$.

III. $(S-\tilde{S})(f) = \hat{f}(o)$ but $|\hat{f}(o)|_{H^1(T)} = |\hat{f}(o)| \le \int|f| \le ||f||_{H^1}$

so since $S-\tilde{S}$ and $\tilde{S}$ are bounded on $H^1(T)$, then S is, which proves the theorem.

A proof without the use of h^p (but much longer) has also been obtained by Yves Meyer. A detailed account of these results (together with extensions to $n > 1$ and $p < 1$) with complete references will appear in Duke Math J. 46 (1979).

DEPAUL UNIVERSITY
2323 N. SEMINARY
CHICAGO, ILLINOIS 60614

Proceedings of Symposia in Pure Mathematics
Volume XXXV, Part 1, 1979

DISTRIBUTIONS WITH STRONG MAXIMAL FUNCTIONS IN $L^p(R^n)$

W. R. MADYCH

ABSTRACT. We indicate an H^p theory of distributions whose strong maximal functions are in L^p. Analogues of several classical theorems are obtained.

In [2], Fefferman and Stein developed a real variable theory of certain H^p spaces which were introduced in [8] and were originally defined in terms of harmonic functions on the upper half space of R^{n+1} and their gradients. They showed that these spaces can be defined in terms of certain maximal functions which are closely related to the theory of differentiating integrals with respect to the one parameter family of balls or (equivalently) cubes. Subsequently a theory of H^p spaces based on more general one parameter Vitali families was developed in [1].

In this note we indicate an H^p theory based on maximal functions which are related to the differentiability of integrals with respect to the n parameter family of rectangles with sides parallel to the coordinate axis, see [3].

Γ denotes the cone $\{t = (t_1,\ldots,t_n) : t_j > 0,\ j = 1,\ldots,n\}$ and if ϕ is in $\mathcal{S} = \mathcal{S}(R^n)$ then ϕ_t is defined by $\phi_t(x_1,\ldots,x_n) = (\prod_{i=1}^n t_i)^{-1}\phi(\frac{x_1}{t_1},\ldots,\frac{x_n}{t_n})$. x and y always denote elements of R^n, t denotes an element of Γ, and $\underline{dt} = (\prod_{i=1}^n t_i)^{-1}dt$ where dt is Lebesque measure. Otherwise all notation is standard.

Given a tempered distribution f on R^n and a ϕ in $\mathcal{S}$ we define the maximal functions f_ϕ and $f_{\phi,\lambda}$, $\lambda > 0$, as follows:

$$f_\phi(x) = \sup_{y,t}\{|f*\phi_t(y)| : |x_i-y_i| < t_i,\ i = 1,\ldots,n\},$$

$$f_{\phi,\lambda}(x) = \sup_{y,t} |f*\phi_t(y)| \prod_{i=1}^n (1 + |x_i-y_i|/t_i)^{-\lambda}.$$

The space H^p, $0 < p < \infty$, is defined as the class of all those distributions f in $\mathcal{S}'$ for which f_ϕ is in L^p where $\phi(x) = \exp(-|x|^2)$. The quasi-norm

AMS(MOS) subject classifications (1970). 46F05, 26A69.

in H^p is taken to be $\|f\|_{H^p} = \|f_\phi\|_p$.

If $1 < p < \infty$, it is well known that $\| \ \|_{H^p}$ is equivalent to the usual L^p norm and $H^p = L^p$. What follows is mainly of interest in the case $0 < p \leq 1$.

Theorem 1: If λ is greater than p^{-1} and f_ϕ is in L^p, $0 < p < \infty$, then $f_{\phi,\lambda}$ is also in L^p and $\| f_{\phi,\lambda}\|_p \leq C\| f_\phi\|_p$, where C is a constant which depends on p and λ but not on f.

As a consequence of Theorem 1 it is possible to characterize H^p in terms of a "grand" maximal function as in [2]. Furthermore, we have

Corollary 1: If f is in H^p and ϕ is in $\mathcal{S}$ then

$$\| f*\phi_t \|_\infty \leq C\|f\|_{H^p}\left(\prod_{i=1}^n t_i\right)^{-\frac{1}{p}}$$

where C is a constant independent of f and t.

Let $X_n = \{\xi \in R^n : \xi_i = 0 \text{ for some } i = 1,\dots,n\}$ and $\mathcal{S}_o = \{\phi \in \mathcal{S} : D^\nu\phi(\xi) = 0$ for all multi-indices ν and all ξ in $X_n\}$. If A is a class of tempered distributions then $\hat{A} = \{\hat{f} : f \in A\}$ where $\hat{f}$ denotes the Fourier transform of f. $\mathcal{D}_o = \{f \in C^\infty : f$ has compact support in $R^n \setminus X_n\}$.

Corollary 2: $\hat{\mathcal{D}}_o$ and $\hat{\mathcal{S}}_o$ are dense in H^p.

Corollary 3: H^p is complete.

Corollary 4: If f is in H^p and $\phi \in \mathcal{S}$ then $\lim_{t\to 0} f*\phi_t(x)$ exists for almost all x and is in L^p.

Given positive numbers p,q and a real number α, $B_\alpha^{p,q}$ is the collection of all f's in $\mathcal{S}'$ for which

(i) $f*\psi_t$ is in L^p for every t in Γ and

(ii) $B_\alpha^{p,q}(f) = \{\int_\Gamma ((\prod_{i=1}^n t_i)^{-\alpha}\| f*\psi_t\|_p)^q \underline{dt}\}^{\frac{1}{q}}$ is finite.

Here ψ denotes a function in $\mathcal{S}$ whose Fourier transform is non-negative and supported in $\{\xi: 1 \leq |\xi_i| \leq 2,\ i = 1,\dots,n\}$, and such that $\int_\Gamma \hat{\psi}(t\xi)\underline{dt} = 1$ for ξ in $R^n \setminus X_n$. Note that $t = (t_1\xi_1,\dots,t_n\xi_n)$. In the case that p and/or q are equal to ∞, (i) and (ii) have the usual meaning.

If both p and q are greater than or equal to one then $B_\alpha^{p,q}(f)$ is a norm on $(\mathcal{S}'/\mathfrak{h})^\wedge \cap B_\alpha^{p,q}$. Namely, $B_\alpha^{p,q}$ may be considered as a subspace of $(\mathcal{S}'/\mathfrak{h})^\wedge$ where $\mathfrak{h}$ is the class of distributions with support in X_n. We will not go into further technical details here. Equivalent norms on $B_\alpha^{p,q}$ have been considered in [4].

Theorem 2: If $0 < p < 1$, then $H^p \subset B_\alpha^{1,1}$, where $\alpha = 1 - \frac{1}{p}$. Furthermore $B_\alpha^{1,1}(f) \leq C\| f \|_{H^p}$, where $\alpha = 1 - \frac{1}{p}$ and C is a constant independent of f.

Since $(\mathcal{S}'/\mathfrak{h})^\wedge$ is the dual of $\hat{\mathcal{S}}_o$, the expression $\langle g,f\rangle$ has the usual meaning for $g \in (\mathcal{S}'/\mathcal{n})^\wedge$ and $f \in \hat{\mathcal{S}}_o$. As a consequence of the above theorem we have $|\langle g,f\rangle| \leq C\|f\|_{H^p} B_\alpha^{\infty,\infty}(g)$ where $\alpha = \frac{1}{p} - 1$. Hence, if we use the standard notation $(H^p)^*$ to denote the dual of H^p, we have

<u>Corollary 5</u>: <u>If $0 < p < 1$ then $(H^p)^* = B_\alpha^{\infty,\infty}$, where $\alpha = \frac{1}{p} - 1$. This duality is to be interpreted in the following sense</u>:

(i) <u>Given g in $B_\alpha^{\infty,\infty}$, the linear functional $\ell(f) = \langle g,f\rangle$, defined initially on $\hat{\mathcal{S}}_o$ may be extended continuously to all of H^p.</u>

(ii) <u>Given a bounded linear functional ℓ on H^p, there is a unique g in $B_\alpha^{\infty,\infty}$ such that $\ell(f) = \langle g,f\rangle$ for all f in $\hat{\mathcal{S}}_o$.</u>

REFERENCES

[1] A. P. Calderon and A. Torchinsky, "Parabolic maximal functions associated with a distribution", <u>Advances in Math</u>. 16(1975), 1-64.

[2] C. Fefferman and E. M. Stein, "H^p spaces of several variables", <u>Acta Math</u>. 129(1972), 137-193.

[3] B. Jessen, J. Marcinkiewicz, and A. Zygumnd, "Note on the differentiability of multiple integrals", <u>Fund. Math</u>. 25(1936), 217-234.

[4] W. R. Madych, "Absolute Summability of Fourier Transforms on R^n", <u>Indiana Univ. Math. J</u>. 25(1976), 467-479.

[5] N. M. Riviere, "The Fourier method in approximation theory," unpublished lecture notes.

[6] E. M. Stein, <u>Singular integrals and differentiability properties of functions</u>, Princeton, 1971.

[7] E. M. Stein and G. Weiss, "<u>Introduction to Fourier analysis on Euclidean spaces</u>, Princeton, 1970.

[8] E. M. Stein and G. Weiss, "On the theory of harmonic function of several variables I. The theory of H^p spaces", <u>Acta Math</u>. 103(1960),

[9] A. Zygmund, <u>Trigonometric Series</u>, 2nd ed., Cambridge, 1968.

DEPARTMENT OF MATHEMATICS, IOWA STATE UNIVERSITY, AMES, IOWA 50011

Proceedings of Symposia in Pure Mathematics
Volume XXXV, Part 1, 1979

WEIGHTED HARDY SPACES

José García - Cuerva

This paper reviews the results presented in [9], where a theory of weighted Hardy spaces is developed for the upper half plane and some applications are found.

For a weight w in the class A_∞ (defined in [3] and [12]), and for $0<p<\infty$, let $H^p(w)$ be the space of all functions $F(x+it)$ analytic on the upper half plane $\mathbb{R}^2_+ = \{x+it : t>0\}$ for which:

$$\|F\|_{H^p(w)} \equiv \sup_{t>0} \{\int_{-\infty}^{\infty} |F(x+it)|^p w(x)\,dx\}^{1/p} < \infty$$

The classical results about the behavior at the boundary of an H^p function, extend to this situation in the following way:

THEOREM 1. *Let* $F \varepsilon H^p(w)$. *Then:*

(i) $\lim_{t\to 0} F(x+it)$ *exists for a.e.* $x\varepsilon\mathbb{R}$ *and the function* $F(x) \equiv \lim_{t\to 0} F(x+it)$ *is in* $L^p(w)$. *Actually* F *has non-tangential limits at a.e.* $x\varepsilon\mathbb{R}$

(ii) $\lim_{t\to 0} \int_{-\infty}^{\infty} |F(x+it) - F(x)|^p w(x)\,dx = 0$ *and consequently*

$$\lim_{t\to 0} \int_{-\infty}^{\infty} |F(x+it)|^p w(x)\,dx = \int_{-\infty}^{\infty} |F(x)|^p w(x)\,dx$$

(iii) $\|F\|_{H^p(w)} \leqslant (\mathrm{const})\{\int_{-\infty}^{\infty} |F(x)|^p w(x)\,dx\}^{1/p}$

Here is an outline of the proof: One shows that the non-tangential maximal function $m_F(x) \equiv \sup\{|F(y+it)| : |x-y| < t\}$ is in $L^p(w)$. From this and a result of Calderón (Theorem 3.19 in [14]), (i) follows. Then (ii) is immediate. The way to prove that $m_F \varepsilon L^p(w)$ is to consider the subharmonic function $s(z) = |F(z)|^\varepsilon$ where $\varepsilon>0$ is small. This function is in $L^q(w(x)dx)$ uniformly in $t>0$ where $q = p/\varepsilon$. If ε is so small that w is in the class A_q (defined in [3] and [12]), then s has a least harmonic majorant which is the Poisson integral of some $s_0 \varepsilon L^q(w)$. Once this is proved we have: $m_F \leqslant (\mathrm{const})\, s_0^{*1/\varepsilon}$, where s_0^* is the Hardy-Littlewood maximal function of s_0 , and therefore $\int_{-\infty}^{\infty} m_F(x)^p w(x)\,dx \leqslant (\mathrm{const}) \int_{-\infty}^{\infty} |s_0(x)|^q w(x)\,dx < \infty$.
Finally (iii) follows from the observation that $s_0(x) = |F(x)|^\varepsilon$. The crucial

---

AMS (MOS) subject classifications (1970). Primary 30A78, 42A40, 42A18.

fact is the existence of s_0. This can be proved as in the case $w \equiv 1$ by using the weak-* compactness of the set $\{s(\cdot+it) : t>0\}$ in $L^{q'}(w)$ where $q' = q/(q-1)$. The only difference with the case $w \equiv 1$ is that the behavior of s for $|x| \longrightarrow \infty$ or $t \longrightarrow \infty$ may be a litte worse. However the fact that $w \varepsilon A_q$ allows us to obtain estimates which are still good enough to prove the majorization by using a maximum principle for subharmonic functions. One can also prove by using harmonic majorants the following useful result:

THEOREM 2. *Let* w_1 *and* w_2 *be* A_∞ *weights. Suppose that* $F \varepsilon H^{p_1}(w_1)$ *and its boundary function belongs to* $L^{p_2}(w_2)$ *with* $p_2 > p_1$ *then* $F \varepsilon H^{p_2}(w_2)$.

For a weight w, we will denote by $q_0 \equiv q_0(w)$, the critical exponent for w, that is, the infimun of all the q's such that w is in the class A_q. If $F \varepsilon H^p(w)$ with $p > q_0$, then $F(x+it) = (F * P_t)(x)$ and $F(x) = f(x) + i\tilde{f}(x)$ where f is real and $\tilde{f}$ is its Hilbert transform. The correspondence $F \longleftrightarrow f$ is an equivalence of metric linear spaces between $H^p(w)$ and $\operatorname{Re} L^p(w)$. The situation is quite different for $p \leqslant q_0$. We will concentrate our attention on these spaces, in particular on those for which $p \leqslant 1$.

It is easy to find nice dense subclasses of $H^p(w)$. A typical result is this:

THEOREM 3. *For any positive integer* N, *let* S_N^p *be the subspace formed by the functions* $F \varepsilon H^p(w)$ *such that:*

(i) F *is continuous on* $\overline{\mathbb{R}^2_+} = \{(x,t) : t \geqslant 0\}$

(ii) $F(x+it) = O(|x|^{-N})$ *as* $|x| \longrightarrow \infty$ *uniformly in* t *for each strip* $0 \leqslant t \leqslant t_0$.

Then S_N^p *is dense in* $H^p(w)$

COROLLARY. *For each* $q > 0$ $H^p(w) \cap H^q(w)$ *is dense in* $H^p(w)$.

This corollary for $q > q_0$, allows us to see $H^p(w)$ as the completion under a certain norm or quasi-norm of a space of nice real functions defined on the boundary. For example for $0 < p \leqslant 1$ one can use the quasi-norm $\|f\|^p_{L^p(w)} + \|\tilde{f}\|^p_{L^p(w)}$ for nice f on the boundary.

Once the space $H^p(w)$ has been realized as a space of functions on the boundary, it seems natural to try to find equivalent quasi-norms which may provide new useful ways to look at $H^p(w)$. Let us see how the results in [1] and [7] on maximal function characterizations and those in [2] on atomic decomposition, extend to the weighted case:

Let ϕ be such that $|\phi(x)| \leqslant (\text{const})/(1+|x|)^\alpha$ where $\alpha > 1$. For f a nice function on $\mathbb{R}$, let $f(x,t) = (f * \phi_t)(x)$ where for $t > 0$, $\phi_t(x) = t^{-1}\phi(t^{-1}x)$. Then we can define the following maximal functions:

(i) The non-tangential maximal function:

$$\phi^*_\nabla(f)(x) = \sup\{|f(y,t)| : |x-y| < t\}$$

(ii) In general, for $N \geqslant 1$, the non-tangential maximal function of amplitude N

$$\phi^*_{\nabla,N}(f)(x) = \sup\{|f(y,t)| : |x-y| < Nt\}$$

(iii) For $M \geqslant 1$, the tangential maximal function with exponent M:

$$\phi^{**}_M(f)(x) = \sup\{|f(y,t)|(t/(|x-y|+t))^M : (y,t)\in\mathbb{R}^2_+\}$$

The gauge (defined for f nice):

$$(1) \qquad f \longmapsto \|P^*_\nabla(f+i\tilde{f})\|^p_{L^p(\omega)}$$

defines $H^p(\omega)$. That is, the space of functions on which (1) is finite, endowed with the quasi-norm given by (1), becomes, once completed, an equivalent copy of $H^p(\omega)$. We will find some other gauges which are equivalent to (1) (two gauges are said to be equivalent if they dominate each other i.e.: if they bound each other up to multiplicative constants). Actually we will start with the gauge

$$(2) \qquad f \longmapsto \|P^*_\nabla(f)\|^p_{L^p(\omega)}$$

obviously dominated by (1). We will eventually get to the fact that (1) and (2) are equivalent, which is the extension to our context of the Burkholder-Gundy and Silverstein theorem ([1]).

In general for any ϕ as above if ω is in the class A_q, we have:

$$(3) \qquad \|\phi^*_{\nabla,N}(f)\|^p_{L^p(\omega)} \leqslant (\text{const})N^q\, \|\phi^*_\nabla(f)\|^p_{L^p(\omega)}$$

By putting together these estimates for $N = 2^k$ $k = 0,1,\ldots$, we can see that if $M > q_0/p$, then:

$$(4) \qquad \|\phi^{**}_M(f)\|^p_{L^p(\omega)} \leqslant (\text{const})\, \|\phi^*_\nabla(f)\|^p_{L^p(\omega)}$$

As in [7], by considering $\sigma(x) = \int_1^\infty \psi(s)P_s(x)\,ds$ for an appropriate ψ, we can find a σ in the Schwartz class S for which:

$$(5) \qquad \sigma^*_\nabla(f)(x) \leqslant (\text{const})P^{**}_M(f)(x)$$

Now it is not difficult to pass from σ to any smooth approximation to the identity just as it is done in [7]. In particular we found it useful to consider the following maximal operator:

$$S^*_M(f)(x) = \sup\{|\textstyle\int_{-\infty}^\infty f(t)\phi(t)\,dt| / (\int_{-\infty}^\infty |\phi(u)|\,du + |I_\phi|^{M+1} \int_{-\infty}^\infty |\phi^{(M+1)}(u)|\,du)\}$$

where the supremum is taken over all the C^∞ functions ϕ with compact support and such that $\operatorname{dist}(x,I_\phi) < |I_\phi|$, I_ϕ being the smallest closed interval containing the support of ϕ. If $M > q_0/p$ we have:

(6) $$S_M^*(f)(x) \leq (\text{const})\sigma_M^{**}(f)(x)$$

The maximal function $S_M^*(f)$ allows us to estimate averages of f against a very large class of smooth functions. This is the key to the atomic decomposition which we shall now describe.

Let w be a weight with critical exponent q_0. For $0 < p \leq 1$ and r such that $r > p$ and $w \in A_r$, a (p,r)-atom with respect to w will be a real-valued function a, supported in a compact interval I and satisfying:

(i) $$(w(I)^{-1}\int_I |a(x)|^r w(x)\,dx)^{1/r} \leq w(I)^{-1/p}$$

if $r < \infty$ or $\|a\|_\infty \leq w(I)^{-1/p}$ if $r = \infty$.

(ii) $$\int_{-\infty}^{\infty} a(x)x^k\,dx = 0 \quad \text{for } k = 0,1,\dots,[q_0/p]-1$$

where $[q_0/p]$ stands for the biggest integer $\leq q_0/p$.

It is easily seen that for a (p,r)-atom a:

(7) $$\|\tilde{a}\|^p_{L^p(w)} \leq (\text{const}) \quad \text{and}$$

(8) $$\|P_\nabla^*(a+i\tilde{a})\|^p_{L^p(w)} \leq (\text{const})$$

with constants which don't depent on a. Thus atoms are simple examples of functions in $H^p(w)$. Actually every function in $H^p(w)$ can be decomposed into atoms. In fact we have:

THEOREM 4. *Let* f *be a nice function (say* $f \in \operatorname{Re} L^q(w)$ *with* $q > q_0$*) such that* $S_M^*(f) \in L^p(w)$ *where* $M > q_0/p$. *Then there exist:*

(i) *A sequence* (a_k) *of* (p,∞)*-atoms with respect to* w, *and*

(ii) *A sequence* (λ_k) *of real mumbers satisfying:*

$$\sum_k |\lambda_k|^p \leq (\text{const})\ \|S_M^*(f)\|^p_{L^p(w)}$$

such that $f(x) = \sum_k \lambda_k a_k(x)$ *a.e. and also in* $L^q(w)$.

The proof is as in [2].

For $f \in \operatorname{Re} L^q(w)$ let $N_{p,r}(f) = \inf\{\sum_k |\lambda_k|^p : f(x) = \sum_k \lambda_k a_k(x)$ a,e. and in $L^q(w)\}$. Theorem 4 says simply that $N_{p,\infty}(f) \leq (\text{const})\|S_M^*(f)\|^p_{L^p(w)}$. On $\operatorname{Re} L^q(w)$ we have the following chain of gauges, where $\prec$ means "is dominated by":

(9) $$N_{p,r}(f) \prec N_{p,\infty}(f) \prec \|S_M^*(f)\|^p_{L^p(w)} \prec \|\sigma_M^{**}(f)\|^p_{L^p(w)} \prec$$

$$\prec \|\sigma^*_\nabla(f)\|^p_{L^p(w)} \prec \|P^{**}_M(f)\|^p_{L^p(w)} \prec \|P^*_\nabla(f)\|^p_{L^p(w)} \prec$$

$$\prec \|P^*_\nabla(f+i\tilde f)\|^p_{L^p(w)} \quad ; \quad w\varepsilon A_r \ , \ r<\infty.$$

But it follows immediately from (8) that $\|P^*_\nabla(f+i\tilde f)\|^p_{L^p(w)} \leqslant (\text{const})N_{p,r}(f)$ which closes the chain of gauges and shows that all of them are equivalent. The conclusion is that the completion of the space determined on $\operatorname{Re} L^q(w)$ by any of the equivalent gauges in (9), is an equivalent copy of $H^p(w)$.

The atomic characterization of $H^p(w)$ leads easily to a simple description of its dual $(H^p(w))^*$, contained in the following statement:

THEOREM 5. *Let* $\Lambda\varepsilon(H^p(w))^*$. *Suppose that* $w\varepsilon A_r$, $p<r<\infty$. *Then, there is a real valued function* ℓ, *such that:*

(i) ℓ/w *is locally in* $L^{r'}(w)$.

(ii) *There is a constant* $C \leqslant (\text{const})\,\|\Lambda\|$ *such that for every compact interval* I:

$$(10) \qquad (w(I)^{-1}\int_I |(\ell(x)-P_I(\ell)(x))/w(x)|^{r'} w(x)dx)^{1/r'} \leqslant Cw(I)^{(1/p)-1}$$

where $P_I(\ell)$ *is the unique polynomial of degree* $\leqslant [q_0/p]-1$ *such that* $\int_I(\ell(x)-P_I(\ell)(x))x^k dx = 0$ *for* $k=0,\dots,[q_0/p]-1$.

(iii) *For every* $f\varepsilon L^r(w)$ *supported in a compact interval and such that* $\int f(x)x^k dx = 0$ *for* $k=0,\dots,[q_0/p]-1$ *if we call* $F(x+it) = (P_t*(f+i\tilde f))(x)$ *then:*

$$(11) \qquad \Lambda(F) = \int_{-\infty}^{\infty} f(x)\ell(x)dx$$

Of course the function ℓ *is not unique. It is only determined up to sum of a polynomial of degree* $\leqslant [q_0/p]-1$.

Conversely given ℓ *satisfying* (i) *and* (10) *with a constant* C, *the functional defined by* (11) *extends in a unique way to an element* $\Lambda\varepsilon(H^p(w))^*$ *and* $\|\Lambda\| \leqslant (\text{const})\, C$.

We obtain as a byproduct the equivalence of the conditions (10) for different values of r'. In particular for $1\leqslant r'<q_0'$. If $q_0=1$ and $p<1$, the range will also include $q_0'=\infty$ i.e. it will be $1\leqslant r'\leqslant\infty$ (for $r'=\infty$ the left hand side of (10) will simply be $\|(\ell-P_I(\ell))/w|_\infty)$. For $w(x)\equiv 1$ and $p=1$, this equivalence is the well-known theorem of John and Nirenberg ([11]) about the characterization of B.M.O., which is $(H^1)^*$.

There is another kind of weighted H^p which we will consider now. For $(1/2)<p\leqslant 1$, $r\geqslant 1$ and $r>p$, a (p,r)-atom of homogeneous type (h.t.) with respect to w will be a real valued function a, supported in a compact interval I, and satisfying:

(i) $$(w(I)^{-1}\int_I |a(x)|^r w(x)\,dx)^{1/r} \leqslant w(I)^{-1/p} \quad \text{if} \quad r<\infty$$

$$\text{or} \quad \|a\|_\infty \leqslant w(I)^{-1/p} \quad \text{if} \quad r=\infty$$

(ii) $$\int_{-\infty}^{\infty} a(x)w(x)\,dx = 0$$

This is the natural notion of atom for the space of homogeneous type associated to w (see [4] and [5]).

We define $h_r^1(w)$ as the space of all real-valued functions $f(x)$ for which there is a decomposition $f(x)=\sum_k \lambda_k a_k(x)$ a.e., with $\sum_k |\lambda_k|<\infty$ and each a_k a $(1,r)$-atom of h.t. with respect to w. On $h_r^1(w)$ we consider the norm $N_{1,r}(f) = \inf\{\sum_k |\lambda_k|\}$ where the infimum is taken over all the decompositions of f into $(1,r)$-atoms of h.t. with respect to w. It can be proved that all the spaces $h_r^1(w)$ coincide as sets and all the norms $N_{1,r}$ are equivalent. We can simply write $h^1(w)$. We also have the following:

THEOREM 6. *If* w *is an* A_∞ *weight, then the mapping* $f(x) \longmapsto f(x)w(x)$ *is an equivalence between* $h^1(w)$ *and* $\mathrm{Re}\,H^1(\mathbb{R})$ *(The* H^1 *corresponding to* $w\equiv 1$, *viewed as a space of real boundary functions).*

The proof is based upon the fact that an A_∞ weight satisfies a reverse Hölder's inequality with respect to Lebesgue measure (see [3]).

All this can be extended to $p<1$. For $\alpha>0$ let $L_\alpha(w)$ be the space of equivalent classes $[\ell]$ modulo constants of functions ℓ for which there is a constant C such that for any x, y and any interval I containing x and y: $|\ell(x)-\ell(y)| \leqslant Cw(I)^\alpha$. The infimum of all these constants C gives a norm on $L_\alpha(w)$ which, endowed with it, becomes a Banach space.

If $\alpha=(1/p)-1$ and a is a (p,r)-atom of h.t. with respect to w, then the mapping $[\ell] \longmapsto \int_{-\infty}^{\infty} \ell(x)a(x)w(x)\,dx$ is a bounded linear functional L_a on $L_\alpha(w)$ with norm dominated by a constant independent of a.

$h_r^p(w)$ will be the subspace of $(L_{(1/p)-1}(w))^*$ formed by those functionals L for which there is a decomposition $L=\sum_k \lambda_k L_{a_k}$ converging in the topology of $(L_{(1/p)-1}(w))^*$, such that $\sum_k |\lambda_k|^p<\infty$ and the a_k's are (p,r)-atoms of h.t. with respect to w. For $L \in h_r^p(w)$, let $N_{p,r}(L) = \inf\{\sum_k |\lambda_k|^p\}$ where the inf is taken over all decompositions as above. $N_{p,r}$ is a quasi-norm on $h_r^p(w)$. It can be seen that for p fixed, the spaces $h_r^p(w)$ obtained for different r's, coincide as sets, and the corresponding quasi-norms $N_{p,r}$ are equivalent.

The main result which relates the "atomic" or "homogeneous type" h^p spaces and the "analytic type" H^p spaces is the following:

THEOREM 7. *Let* w *be a weight with critical exponent* q_0. *Let* $(q_0/(1+q_0)) < p \leqslant 1$. *Then there is an equivalence between* $h^p(w)$ *and* $H^p(w^{1-p})$ *which for* (p,r)*-atoms of h.t. with respect to* w, *is given by:*

$$a(x) \longmapsto P_t * (aw + i(aw)^\sim)(x)$$

Theorem 7 includes theorem 6 as a particular case. Notice, however, that for $p=1$ we simply get the ordinary H^1 space and not a weighted H^1 space.

Weighted Hardy spaces appear naturally when one views radial functions in $H^p(\mathbb{R}^n)$ for $n>1$, as functions on the real line. For example if we denote by $H^1_{rad}(\mathbb{R}^n)$ the subspace of $H^1(\mathbb{R}^n)$ formed by the radial functions and by $h^1_{even}(|r|^{n-1})$ the subspace of $h^1(|r|^{n-1})$ formed by the even functions, we have the following:

THEOREM 8. *The mapping* $f(r)\longmapsto F(x)=f(|x|)$ *is an equivalence between the Banach spaces* $h^1_{even}(|r|^{n-1})$ *and* $H^1_{rad}(\mathbb{R}^n)$.

A simple geometric proof can be given by looking at atoms. Combining theorems 8 and 6 one gets:

THEOREM 9. *The mapping* $F(x)=f(|x|)\longmapsto f(|r|)|r|^{n-1}$ *is an equivalence between* $H^1_{rad}(\mathbb{R}^n)$ *and* $H^1_{even}(\mathbb{R})$.

Actually it is the same to consider even functions or functions living in $[0,\infty[$. If we denote by $H^1_+(\mathbb{R})$ the subspace of $\operatorname{Re}H^1(\mathbb{R})$ formed by the functions living in $[0,\infty[$, we have:

THEOREM 10. *The mapping*

$$F(x)=f(|x|)\longmapsto f(r)r^{n-1}\chi_{[0,\infty[}(r)$$

is an equivalence between $H^1_{rad}(\mathbb{R}^n)$ *and* $H^1_+(\mathbb{R})$.

These results suggest to investigate the relation between the Riesz transforms of a radial function $F(x)=f(|x|)$ in $\mathbb{R}^n$, and the Hilbert transform of the function $f(|r|)|r|^{n-1}$ viewed either as an even function or as a function living in $[0,\infty[$. This was done in [9], and we will now describe briefly the results obtained there.

The size of the Riesz transforms of order 1, R_jF, $j=1,\dots,n$, of a radial function $F(x)=f(|x|)$ in $L^1(\mathbb{R}^n)$, can be obtained by looking at a unique radial function $RF(x)=g(|x|)$. In particular $F\in H^1(\mathbb{R}^n)$ if and only if $g(r)r^{n-1}$ is in $L^1([0,\infty[)$. But for $r>0$, we have, assuming, as we may, that $\int_0^\infty f(s)s^{n-1}ds=0$:

$$g(r)r^{n-1}=c\{(f(s)s^{n-1})^\sim(r)-(f(s)s^{n-1})^\sim(-r)+ \tag{12}$$
$$+c'A(f(s)s^{n-1})(r)+T(f(s)s^{n-1})(r)\}$$

where $f(s)s^{n-1}$ is regarded as a function living in $[0,\infty[$; $A(f(s)s^{n-1})(r)= = r^{-1}\int_0^r f(s)s^{n-1}ds$ and T is an operator bounded in $L^1([0,\infty[)$. A does not map $L^1([0,\infty[)$ into itself. However it can be seen very easily by looling at atoms, that it does map $H^1_+(\mathbb{R})$ boundedly into $L^1([0,\infty[)$. This provides a different proof of a part of theorem 10, namely of the fact that if $f(r)r^{n-1}$ is in $H^1_+(\mathbb{R})$, then $F\in H^1(\mathbb{R}^n)$. This proof is independent of the geometry of $\mathbb{R}^n$ and can be applied to the weighted spaces studied in [13]. In order to prove the converse it follows easily from (12) that we just need to see that for

$F \epsilon H^1(\mathbb{R}^n)$ we have $A(f(s)s^{n-1}) \epsilon L^1([0,\infty[)$. In [9] we prove this fact by obtaining a decomposition similar to (12) for the Riesz transforms of order 2 of F. The decomposition turns out to be simpler: the Hilbert transform doesn't appear and all we get is the average operator A and an operator bounded in $L^1([0,\infty[)$. This is quite interesting. We not only get that if $F \epsilon H^1(\mathbb{R}^n)$, then $A(f(s)s^{n-1}) \epsilon L^1([0,\infty[)$; we are also able to prove the following:

THEOREM 11. *The system of Riesz transforms of order 2 does not characterize* $H^1(\mathbb{R}^n)$.

In fact we show that there is a radial function F in $L^1(\mathbb{R}^n)$ but not in $H^1(\mathbb{R}^n)$, whose Riesz transforms of order 2 are all in $L^1(\mathbb{R}^n)$. In order to prove this, one just needs to see that there is a function h in $\mathrm{Re}\, L^1([0,\infty[)$ but not in $\mathrm{Re}\, H^1(\mathbb{R})$ such that $A(h) \epsilon L^1([0,\infty[)$. This is achieved by considering for $k = 1,2,\ldots$ the function $h_k(r) = k\chi_{[k,k+k^{-1}]}(r) - k^{-1}\chi_{[k+k^{-1},2k+k^{-1}]}(r)$. It can be seen that $\int_0^\infty |A(h_k)(r)|dr \leqslant 1$, while $\int_{k/2}^k |\tilde{h}_k(r)|dr \geqslant (\mathrm{const})(\log k)$. From these estimates the existence of h can be derived by a Functional Analysis argument or h can be constructed directly by putting together some of the h_k's .

Theorem 11 came as a surprise. The opposite was generally believed to hold (see for example the conjecture made by C. Fefferman in [6]). A similar phenomenon occurs in the context of local fields (see [8]). A generalization of theorem 11 can be seen in [10] .

REFERENCES

1. D.L. Burkholder, R.F. Gundy and M.L. Silverstein, *A maximal function characterization of the class* H^p, Trans. Amer. Math. Soc. 157 (1971), 137-153, M.R. 43 # 527.

2. R.R. Coifman , *A real variable characterization of* H^p, Studia Math. 51 (1974), 269-274, M.R. 50 # 10784.

3. R.R. Coifman and C. Feffermann, *Weighted norm inequalities for maximal functions and singular integrals*, Studia Math. 51 (1974), 241-250, M.R. 50 # 10670 .

4. R.R. Coifman and G. Weiss, *Analyse harmonique non-commutative sur certains espaces homogenes*, Lecture Notes in Math., vol. 242, Springer-Verlag, Berlin and New York, 1971.

5. ______, *Extensions of Hardy spaces and their use in Analysis*, Bull. Amer. Math. Soc., 83 (1977), 569-643.

6. C. Fefferman, *Harmonic Analysis and* H^p *spaces*, (Studies in Math., vol. 13, M. Ash Ed.) Math. Assoc. of Amer., 1976, 38-75.

7. C. Fefferman and E.M. Stein, H^p *spaces of several variables*, Acta Math. 129 (1972), 137-193.

8. A. Gandulfo, J. García-Cuerva and M. Taibleson, *Conjugate system characterizations of* H^1 *: counterexamples for the euclidean plane and local fields*, Bull. Amer. Math. Soc. 82 (1976), 83-85, M.R. 52 # 14820.

9. J. García-Cuerva, *Weighted* H^p *spaces*, Dissertationes Mathematicae 162, to appear.

10. S. Janson, *Characterizations of* H^1 *by singular integral transforms on martingales and* $\mathbb{R}^n$. Math. Scand. 41 (1977), 140-152.

11. F. John and L. Nirenberg, *On functions of bounded mean oscillation*, Comm. Pure Appl. Math. 14 (1961), 415-426, M.R. 24 # A1348.

12. B. Muckenhoupt, *Weighted norm inequalities for the Hardy maximal function*, Trans. Amer. Math. Soc. 165 (1972), 207-226, M.R. 45 # 2461.
13. B. Muckenhoupt and E.M. Stein, *Classical expansions and their relations to conjugate harmonic functions*, Trans. Amer. Math. Soc., 118 (1965), 17-92.
14. E.M. Stein and G. Weiss, *Introduction to Fourier Analysis on Euclidean spaces*, Princeton U. Press, 1971, M.R. 46 # 4102.

DEPARTMENT OF MATHEMATICS, WASHINGTON UNIVERSITY, ST. LOUIS, Mo 63130

Current address: Sección de Matemáticas, Facultad de Ciencias, Universidad de Salamanca, Spain.

Proceedings of Symposia in Pure Mathematics
Volume XXXV, Part 1, 1979

Weighted Hardy Spaces on Lipschitz Domains

by

Carlos E. Kenig

Introduction: The main object of this paper is to describe a recently developed theory of weighted H^p spaces on Lipschitz domains in the complex plane. In the first section we introduce some basic notation and preliminary material. In the second section we describe the basic results obtained in [7] , where detailed proofs can be found. The third section contains some further remarks and applications. An outline of their proofs is given.

Acknowledgements: Most of these results are part of the author's doctoral dissertation directed by Professor A. P. Calderón at the University of Chicago. I would like to thank him for his guidance and encouragement. It is a pleasure to thank the Victor J. Andrew Foundation for their support during my graduate studies.

Section 1

We begin by fixing some notation that will be used throughout the whole paper. Λ will denote a curve in the complex plane, given parametrically by $z(t) = t + i\eta(t)$, where η is a real valued function in the class Lip 1 , with Lipschitz constant M . Thus, η has a derivative almost everywhere, which is in L^∞ , and such that $\|\eta'\|_\infty = M$. $\mathcal{O}$ will denote the open set above the curve, i.e., $\mathcal{O} = \{z \in \mathbb{C}, z = x + iy, y > \eta(x)\}$. Arc length measure on Λ will be denoted by ds , and if $E \subset \Lambda$, $s(E)$ will denote its arc length measure. Let ν be any measure on Λ that is absolutely continuous with respect to arc length ds . We will identify such measures with their density with respect to ds , $\frac{d\nu}{ds}$, and also call this density ν .

We will also consider two such domains $\mathcal{O}_1$ and $\mathcal{O}_2$ simultaneously. A conformal mapping from $\mathcal{O}_2$ onto $\mathcal{O}_1$ that maps ∞ to ∞ will be denoted $\sigma_{1,2}$. It extends as a homeomorphism of $\overline{\mathcal{O}_2}$ onto $\overline{\mathcal{O}_1}$. If ν is any measure on $\partial \mathcal{O}_1 = \Lambda_1$, then $\sigma_{1,2}(\nu)$ will denote the measure on $\partial \mathcal{O}_2 = \Lambda_2$ such that $\sigma_{1,2}(\nu)(F) = \nu(\sigma_{1,2}(F))$, for $F \subset \partial \mathcal{O}_2$.

For $0 < \alpha < \frac{\pi}{2}$, and $z = x + iy$ a complex number, $\Gamma_\alpha(z)$ will denote the open angle with axis in the vertical direction, vertex z , opening α and pointing upwards, i.e. $\Gamma_\alpha(z) = \{w = a + ib \in \mathbb{C} , b > y ,$ and $|x - a| < (\tan\alpha\}) (b - y)\}$. We remark that if $0 < \alpha < [\arctan 1/M]$, then, for some $\varepsilon > 0$, $\Gamma_{\alpha+\varepsilon}(z) \subset \mathcal{O}$ for all $z \in \Lambda$.

The letter C will denote a constant, which need not be the same in different occurences.

We now recall the definition of the classes A_p of Muckenhoupt (see [8]) .

<u>Definition 1.1</u>: If $1 < p < \infty$, and ν is a measure on Λ which is absolutely continuous with respect to ds , and such that its density, also called ν , is nonnegative and locally integrable with respect to ds , we say that $\nu \in A_p$ on Λ if there exists a constant $C_p > 0$ such that for all intervals $I \subset \Lambda$,

$$\left(\frac{1}{s(I)} \int_I \nu \, ds\right) \cdot \left(\frac{1}{s(I)} \int_I \nu^{-(1/p-1)} \, ds\right)^{p-1} \leq C_p$$

<u>Definition 1.2</u>: $\nu \in A_\infty$ on Λ if there exist $C > 0$, $\delta > 0$ such that given any interval I in Λ , and any measurable subset $E \subset I$, then

$$\frac{\nu(E)}{\nu(I)} \leq C \left(\frac{s(E)}{s(I)}\right)^\delta .$$

Lemma 1.3 : $A_\infty = \bigcup_{p>1} A_p$.

This lemma is due to Muckenhoupt ([8]) .

Since the region $\mathcal{O}$ is simply connected, it is conformally equivalent to $\mathbb{R}^2_+$. The conformal mapping will be a basic technical tool in the sequel. So let $\Phi : \mathbb{R}^2_+ \longrightarrow \mathcal{O}$ be conformal, and such that $\Phi(\infty) = \infty$. $\psi : \mathcal{O} \longrightarrow \mathbb{R}^2_+$ will denote its inverse. Their main quantitative properties are:

Theorem 1.4: i) $|\arg \Phi'(z)| \leq \arctan M \quad \forall z \in \mathbb{R}^2_+$

ii) $|\Phi'(x)| \in A_2$ on $\mathbb{R}$

iii) $\frac{\Phi'(z)}{(i+z)^2} \in H^1(\mathbb{R}, dx)$, the classical Hardy space.

Remark: Since $|\Phi'(x)| \in A_2$ on $\mathbb{R}$, it belongs to A_p on $\mathbb{R}$, for any $p \geq 2$. However, if $p < 2$, $|\Phi'(x)|$ need not belong to A_p . More precisely, given $1 < p < 2$, there exists a domain $\mathcal{O}$ such that $|\Phi'(x)| \notin A_p$ on $\mathbb{R}$. More generally,

Lemma 1.5: Let $\nu \in A_2$ on $\mathbb{R}$. Then, there exists a Lipschitz domain $\mathcal{O}$ such that $\nu(x) \approx |\Phi'(x)|$.

We now wish to describe how additional smoothness conditions on η affect the A_p properties of $|\Phi'(x)|$.

Lemma 1.6: Given any p , $1 < p < 2$, there exists an $\eta \in C^1(\mathbb{R})$, $\eta' \in L^\infty(\mathbb{R})$ such that, for the corresponding domain $\mathcal{O}$, $|\Phi'(x)| \notin A_p$. However, if $\eta \in C^1(\mathbb{R})$, $\eta' \in L^\infty(\mathbb{R})$, and is uniformly continuous, then $|\Phi'(x)| \in A_p$ for all p , $1 < p < \infty$.

Section 2: Let ν be a measure on Λ , absolutely continuous with respect to arc length ds , and in A_∞ . For $0 < p < \infty$, consider the class $H^p(\mathcal{O}, d\nu)$ of functions F , analytic in $\mathcal{O}$, and such that $M_\alpha F(z) = \sup_{\zeta \in \Gamma_\alpha(z)} |F(\zeta)| \in L^p(\Lambda, d\nu)$, where $0 < \alpha < \arctan 1/M$.

These classes are well defined ; i.e., if $0 < \beta < \arctan 1/M$, and we consider $M_\beta F$ instead, we obtain the same space, with comparable norm. More precisely,

Lemma 2.1: $\nu\ \{z \in \Lambda,\ M_\beta F(z) > \lambda\} \leq C\ \nu\{z \in \Lambda,\ M_\alpha F(z) > \lambda\}$, where C depends only on ν, $M\ \alpha$ and β.

Using a subharmonicity argument, we can show that M_0 also defines the same classes with comparable norm.

Suppose we have two such domains $\mathcal{O}_1$ and $\mathcal{O}_2$, and consider $\sigma_{1,2} : \mathcal{O}_2 \longrightarrow \mathcal{O}_1$. We want to analyze its effects on our spaces. To do so, we first analyze its effects on measures.

Lemma 2.2: If ν is any measure in A_∞ on Λ_1, then $\sigma_{1,2}(\nu)$ is in A_∞ on Λ_2.

In particular, $\omega(z(t)) = |\psi'(z(t))|$ is in A_∞ on Λ.

We are now ready to state one of our main results:

Theorem 2.3: $F \in H^p(\mathcal{O}_1, d\nu)$ iff $F \circ \sigma_{1,2} \in H^p(\mathcal{O}_2, d\sigma_{1,2}(\nu))$.

We also seek alternate characterizations for $H^p(\mathcal{O}, d\nu)$. We first need some definitions.

Definition 2.4: Assume $\nu \in A_\infty$ on Λ. Then, $AE(\nu) = \{$ G analytic and never 0 on $\mathcal{O}$, with a non-tangential limit almost everyhwere $G(z)$, $z \in \Lambda$, such that $|G(z)| \approx \nu(z)$, and such that there exists $m \geq 0$ with $\dfrac{G \circ \Phi(w)\ \ \Phi'(w)}{(i + w)^m} \in H^1(\mathbb{R}, dx)\}$.

Definition 2.5: Assume $\nu \in A_\infty$ on Λ_1, and $\mu \in A_\infty$ on Λ_2, $\sigma_{1,2}\ \ \mathcal{O}_2 \longrightarrow \mathcal{O}_1$, and $\Phi_1 : \mathbb{R}^2_+ \longrightarrow \mathcal{O}_1$, $\Phi_2 : \mathbb{R}^2_+ \longrightarrow \mathcal{O}_2$ the respective conformal mappings. Then $AE(\nu, \mu) = \{G$, analytic and never 0 on $\mathcal{O}_2$, having a non-tangential limit $G(z)$ a.e. on Λ_2, such that $|G(z)| \approx \dfrac{d\sigma_{1,2}(\nu)}{d\mu}$, and such that there exists $m \geq 0$, and $H \in AE(\mu)$ such that $\dfrac{G \circ \Phi_2(w)\ \ H \circ \Phi_2(w)\ \Phi_2'(w)}{(i + w)^m} \in H^1(\mathbb{R}^2_+, dx)\}$.

Lemma 2.6: Assume $\nu \in A_\infty$ on Λ_1, and $\mu \in A_\infty$ on Λ_2. Then, $AE(\nu) \neq \phi$, $AE(\nu, \mu) \neq \phi$.

Our further characterizations are:

Theorem 2.7: Let $\nu \in A_\infty$ on Λ. For an F analytic in $\mathcal{O}$, the following are equivalent:

i) $F \in H^p(\mathcal{O}, d\nu)$

ii) $\sup_{h>0} \left(\int_\Lambda |F(z + ih)|^p \, d\nu \right)^{1/p} < +\infty$

iii) For any $G \in AE(\nu)$, $F [G]^{1/p} \in H^p(\mathcal{O}, ds)$.

Corollary 2.8: Assume $\mathcal{O}_1$ and $\mathcal{O}_2$ are two Lipschitz domains, $\sigma_{1,2} : \mathcal{O}_2 \longrightarrow \mathcal{O}_1$ and ν is a measure in A_∞ on Λ_1, μ a measure in A_∞ on Λ_2. Assume $G \in AE(\nu, \mu)$. Then, $F \in H^p(\mathcal{O}_1, d\nu)$ iff $[F \circ \sigma_{1,2}] [G]^{1/p} \in H^p(\mathcal{O}_2, d\mu)$.

Two specific instances of corollary 2.8 deserve to be stated separately.

Corollary 2.9: Assume that ν is in A_∞ on Λ. Then, there exists a function G, analytic and never 0 on $\mathbb{R}^2_+$, with non-tangential limit $G(x)$ a.e., such that $|G(x)| \approx \Phi(\nu)$, and such that $F \in H^p(\mathcal{O}, d\nu)$ iff $(F \circ \Phi) [G]^{1/p} \in H^p(\mathbb{R}^2_+, dx)$.

Corollary 2.10: Assume that ν and μ are in A_∞ on Λ, then there exists a function G, analytic and never 0 in $\mathcal{O}$, with non-tangential limit almost everywhere on Λ, such that, for $z \in \Lambda$, $|G(z)| \approx \frac{d\nu}{d\mu}$, and such that $F \in H^p(\mathcal{O}, d\nu)$ iff $F [G]^{1/p} \in H^p(\mathcal{O}, d\mu)$.

An immediate application of Corollary 2.9, and of the classical factorization theorems for ordinary H^p spaces on the upper-half plane (see [4]) gives:

Theorem 2.11: Assume that $F \in H^p(\mathcal{O}, d\nu)$. Then, $F = BH$, where $|B(z)| \leq 1$ for $z \in \mathcal{O}$, $|B(z)| = 1$ for a.e. (ds) $z \in \Lambda$ and H

is in $H^p(\mathcal{O}, d\nu)$ and has no zeros in $\mathcal{O}$.

Theorem 2.12: Assume that $F \in H^p(\mathcal{O}, d\nu)$. Then, $F = F_1 F_2$, where $F_i \in H^{2p}(\mathcal{O}, d\nu)$.

We recall that $\omega(z(t)) = |\phi'(z(t))|$ is in A_∞ on Λ, and $\Phi(\omega) = dx$. We then obtain the following application to ordinary H^p spaces on the upper half plane: For $y > 0$, let $C_y = \psi\{z + iy, z \in \Lambda\}$. Let ds_y be arc length on C_y. Then,

Corollary 2.13: $F \in H^p(\mathbb{R}^2_+, dx)$ iff $\sup_{y>0} \int_{C_y} |F(z)|^p \, ds_y < +\infty$.

Our last characterization of $H^p(\mathcal{O}, d\nu)$ involves area integrals and Littlewood-Paley g functions.

$$\text{Let } S_\alpha F(z) = \left(\iint_{\Gamma_\alpha(z)} |F'(x+iy)|^2 \, dx\,dy \right)^{1/2}, \text{ and}$$

$$gF(z) = \left(\int_{\operatorname{Im} z}^{+\infty} (y - \operatorname{Im} z) \, |F'(\operatorname{Re} z + iy)|^2 \, dy \right)^{1/2}.$$

Then,

Theorem 2.14: For $F \in H^p(\mathcal{O}, d\nu)$, $0 < p < +\infty$, we have

$$\|S_\alpha F\|_{L^p(\Lambda, d\nu)} \approx \| gF \|_{L^p(\Lambda, d\nu)} \approx \| F \|_{H^p(\mathcal{O}, d\nu)}$$

We also study conjugate harmonic functions, boundary values and the Dirichlet problem in $\mathcal{O}$, with boundary data in $L^p(\Lambda, d\nu)$. We prove the analogue of the Burkholder-Gundy-Silverstein theorem ([1]) :

Theorem 2.15: If u is harmonic, real valued in $\mathcal{O}$, then $u = \operatorname{Re} F$, $F \in H^p(\mathcal{O}, d\nu)$ iff $M_\alpha u \in L^p(\Lambda, d\nu)$.

We show the existence of a number p_0, depending on Λ and ν, such that if $p > p_0 \geq 1$, $f \in L^p(\Lambda, d\nu)$, there exists u harmonic in $\mathcal{O}$, $u \longrightarrow f$ non-tangentially, and such that $M_\alpha u \in L^p(\Lambda, d\nu)$. If $1 < p \leq p_0$, there are f in $L^p(\Lambda, d\nu)$ for which no such u can be found.

If $d\nu = ds$, then $1 \leq p_0 < 2$, and examples (coming from lemma 1.5) are produced to show that if $1 < p < 2$, there exists a Lipschitz domain $\mathcal{O}$ such that $p \leq p_0$. This is an essential difference with $\mathbb{R}^2_+$. In the case of bounded Lipschitz domains, this follows from more general (n-dimensional) results of B.E.J. Dahlberg ([3]). However, Dahlberg ([3]), and E.B. Fabes, M. Jodeit Jr. and N. M. Rivière ([5]) have shown that for a bounded C^1 domain, $p_0 = 1$. We give examples, coming from lemma 1.6, to show that this is no longer the case for unbounded domains. In this case, to get $p_0 = 1$, uniform continuity of η' is needed.

Section 3:

Our first remark, which was observed jointly with E.B. Fabes, is about regularity for the Dirichlet problem.

Definition 3.1: $L^p_1(\Lambda) = \{f \in L^p(\Lambda, ds)$ such that $\frac{df}{ds} \in L^p(\Lambda, ds)\}$.

Theorem 3.2: Assume $f \in L^2_1(\Lambda)$. Then, if u is the harmonic function in $\mathcal{O}$ such that $u \longrightarrow f$ non-tangentially, and $\|M_\alpha u\|_{L^2(\Lambda,ds)} \leq C\|f\|_{L^2(\Lambda,ds)}$, we have, $\|M_\alpha \nabla u\|_{L^2(\Lambda,ds)} \leq C\|f\|_{L^2_1(\Lambda)}$.

Proof: Let $g(x) = f \circ \Phi(x)$. Then, $g \in L^2(\mathbb{R}, |\Phi'(x)|\,dx)$, and $\frac{dg}{dx}$ exists and belongs to $L^2(\mathbb{R}, \frac{ds}{|\Phi'(x)|})$. Let v be the Poisson extension of g to $\mathbb{R}^2_+$. $|\Phi'(x)| \in A_2$, and hence $\frac{1}{|\Phi'(x)|} \in A_2$; thus the Hilbert ransform H, and the Hardy-Littlemood maximal operator are bounded in $L^2(\mathbb{R}, \frac{dx}{|\Phi'(x)|})$. It follows then that

$$(*)\ \|M_\alpha(\nabla v)\|_{L^2(\mathbb{R}, \frac{dx}{|\Phi'(x)|})} \leq C\left\|\frac{dg}{dx}\right\|_{L^2(\mathbb{R}, \frac{dx}{|\Phi'(x)|})} \leq C\|f\|_{L^2_1(\Lambda)}.$$

Consider now $u = v \circ \psi$. Let $F = \frac{\partial u}{\partial x} + i\frac{\partial u}{\partial y}$. F is analytic in $\mathcal{O}$. Denote $H = \frac{\partial v}{\partial x} + i\frac{\partial v}{\partial y}$, H is analytic in $\mathbb{R}^2_+$ and $|F| \approx |H \circ \psi|\,|\psi'|$ by the Cauchy-Riemann equations.

Note that $\Phi(|\psi'(z)|\, ds) = \frac{dx}{|\Phi'(x)|}$. Therefore, by theorems 2.7 and 2.3 $F \in H^2(\mathcal{O}, ds)$ iff $H \circ \psi \in H^2(\mathcal{O}, |\psi'(z)|\, ds)$ iff $H \in H^2(\mathbb{R}^2_+, \frac{dx}{|\Phi'(x)|})$. But this follows from (*), and hence $F \in H^2(\mathcal{O}, ds)$, thus, our estimate follows.

This theorem is sharp: if $p > 2$, it can be shown that there exists a Lipschitz domain $\mathcal{O}$ and a function $f \in L^p_1$ such that the above estimate cannot hold. On the other hand, if $p < 2$, the Dirichlet problem with boundary data in $L^p(\Lambda, ds)$ cannot be solved in a general Lipschitz domain.

A similar technique applies to the Neumann problem. Here, the range is $1 < p \leq 2$, which is also sharp. In [5], it is shown that on a bounded C^1 domain we have regularity in the Dirichlet problem for $1 < p < \infty$, in any dimension. What happens on a Lipschitz domain for $n \geq 3$ is still an open question.

Using theorems 2.3 and 2.7, it is possible to interpolate analytic families of linear operators, with change of measure. The technique consists in proving first the result for p large, and then, using multilinear operators and corollary 2.9, to pass to arbitrary p. We now give a typical such application.

Lemma 3.3 : Let $\mathcal{A} = \{ F$ analytic in $\mathcal{O}$, such that $F \circ \psi \in \mathcal{S}(\overline{\mathbb{R}^2_+}) \}$. Then, $\mathcal{A}$ is dense in $H^p(\mathcal{O}, d\nu)$, $0 < p < \infty$, for any $\nu \in A_\infty$.

Let $S = \{z \in \mathbb{C}, 0 \leq \operatorname{Re} z \leq 1\}$, and suppose to each z in S, there is assigned a linear operator T_z on $\mathcal{A}$, into measurable functions on a measure space $(N, d\omega)$. Assume this is an analytic family, with admissible growth (see [10], chapter V, section 4).

Theorem 3.4 : Assume for each F in $\mathcal{A}$,

$$\|T_{iy} F\|_{L^{q_0}(N, d\omega)} \leq M_0(y) \|F\|_{H^{p_0}(\mathcal{O}, d\nu_0)}, \quad \text{and}$$

$$\|T_{1+iy} F\|_{L^{q_1}(N, d\omega)} \leq M_1(y) \|F\|_{H^{p_1}(\mathcal{O}, d\nu_1)}, \quad \text{where}$$

$1 \leq q_j \leq \infty$, $0 < p_j < +\infty$, $\nu_j \in A_\infty$, and $M_j(y)$ are independent of F, and satisfy $\sup_{-\infty < y < +\infty} e^{-b|y|} \log M_j(y) < +\infty$, for some $b < \pi$. Then, if $0 \leq t \leq 1$, there exists a constant M_t such that for all $F \in \mathcal{A}$,

$$\|T_t F\|_{L^{q_t}(N, d\omega)} \leq M_t \|F\|_{H^{p_t}(\mathcal{O}, d\nu_t)},$$

where $\frac{1}{p_t} = \frac{(1-t)}{p_0} + \frac{t}{p_1}$, $\frac{1}{q_t} = \frac{(1-t)}{q_0} + \frac{t}{q_1}$, $\nu_t = \nu_0^{(1-t)} \cdot \nu_1^t$.

Similar results for weighted H^p spaces on $\mathbb{R}_+^{n+1}$, with n arbitrary, have been recently obtained by J. O. Strömberg and A. Torchinsky (personal communication).

Using theorem 2.3 and theorems of J. Garcia Cuerva ([6]) on atomic decompositions of weighted H^p spaces on $\mathbb{R}_+^2$ and their duality properties, we can immediately obtain corresponding results for our spaces. For simplicity, we formulate them only for the spaces $H^1(\mathcal{O}, ds)$ and $H^1(\mathcal{O}, d\omega)$ $(\omega = |\psi'|)$.

An atom a on Λ is a real valued function, supported on an interval $I \subset \Lambda$, and such that $\|a\|_\infty \leq \frac{1}{s(I)}$, and $\int a d\omega = 0$. Let $A(z)$ be the function analytic in $\mathcal{O}$, such that the real part of its boundary values coincides with a A is called an analytic atom.

A homogeneous atom b on Λ is a real valued function, supported on an interval $I \subset \Lambda$, and such that $\|b\|_\infty \leq \frac{1}{\omega(I)}$, $\int b d\omega = 0$. The corresponding analytic function B is called a homogeneous analytic atom. Then, we have

<u>Theorem 3.5</u>: Assume F in analytic in $\mathcal{O}$. Define $N(F)$ as the infimum

of all the sums $\sum_{j=1}^{\infty} |\lambda_j|$, corresponding to all the decompositions $F(z) = \sum_{j=1}^{\infty} \lambda_j A_j(z)$, where A_j is an analytic atom, and the convergence is uniform on compact sets. Then, F is in $H^1(\mathcal{O}, ds)$ if and only if $N(F) < +\infty$. In this case, $N(F) \approx \|F\|_{H^1(\mathcal{O}, ds)}$.

Moreover, if we consider $H^1(\mathcal{O}, ds)^*$, i.e., the set of all continuous linear functionals $\ell : H^1(\mathcal{O}, ds) \longrightarrow \mathbb{R}$, we have:

Theorem 3.6: Assume $\ell \in H^1(\mathcal{O}, ds)^*$. Then, there exists a locally integrable function (with respect to $d\omega$) $\ell(z)$ on Λ, which is unique modulo constants, such that

$$\frac{1}{s(I)} \int_I |\ell(z) = m_I(\ell)| \, d\omega \leq C ,$$

where C is independent of the intend $I \subset \Lambda$, and $m_I \ell = \frac{1}{\omega(I)} \int_I d\omega$, and such that if a is an atom, and A the corresponding analytic atom, then

$$\ell(A) = \int_\Lambda a \cdot \ell d\omega .$$

Moreover, if for such a function ℓ we deine a linear function on A by the above formula, it has a unique extension to a bounded linear functional on $H^1(\mathcal{O}, ds)$.

If we want to prove analogous results for $H^1(\mathcal{O}, d\omega)$, we notice that it corresponds to $H^1(\mathbb{R}^2_+, dx)$. Thus, if we insert homogeneous everywhere in theorem 3.5, it is valid for $H^1(\mathcal{O}, d\omega)$. If in theorem 3.6, we replace the condition on ℓ by $\frac{1}{\omega(I)} \int_I |\ell - m_I \ell| \, d\omega \leq C$, and write homogeneous everywhere, we get the analogous result for $H^1(\mathcal{O}, d\omega)$. We note that, since $\omega \in A_\infty$, the last condition is equivalent to $\frac{1}{s(I)} \int_I |\ell - m_I \ell| \, ds$, the ordinary BMO space on Λ (Muckenhoupt and Wheeden [9]).

This last theorem leads to the following remark: Let

$$Cf(z) = \frac{1}{2\pi i} \int_\Lambda \frac{f(z)}{z - \zeta} \, d\zeta \quad , \quad z \in \Lambda .$$

A. P. Calderón ([2]) showed that there exists a constant $a > 0$ such that if $\|\eta'\|_\infty < a$, then C is a bounded operator in $L^p(\Lambda, ds)$, for $1 < p < \infty$. Assume this is the case. A natural conjecture is that f is the real part of the boundary values of an $H^1(\mathcal{O}, ds)$ function if and only if f and Cf belong to $L^1(\Lambda, ds)$. However, this is false. It can be shown, using theorem 3.5, that there are functions f such that $f, Cf \in L^1(\Lambda, ds)$, and yet f is not the real part of the boundary values of any function in $H^1(\mathcal{O}, ds)$. On the other hand, if f is the real part of the boundary values of a function in $H^1(\mathcal{O}, d\omega)$, then f/ω and $C(f/\omega)$ are in $L^1(\Lambda, ds)$.

Bibliography:

[1] D. L. Burkholder, R. F. Gundy and M. L. Silverstein, A maximal function characterization of the class H^p, Trans. Amer. Math. Soc. 157 (1971), p.p. 137 - 153.

[2] A. P. Calderón, On the Cauchy integral on Lipschitz curves, and related operators, Proc. Nat. Acad. of Sciences 74, n°.4 (1977), p. p. 1324 - 1327.

[3] B. E. J. Dahlberg, On the Poisson integral for Lipschitz and C^1 domains, to appear, Studia Math.

[4] P. L. Duren, Theory of H^p spaces, Academic Press, New York and London, 1970.

[5] E. B. Fabes, M. Jodeit, Jr., and N. M. Rivière, Potential Techniques for boundary value problems on C^1 domains, to appear, Acta Mathematica.

[6] J. Garcia Cuerva, Weighted H^p spaces, PhD dissertation, Washington

University, St. Louis, Mo., 1975.

[7] C. E. Kenig, Weighted H^p spaces on Lipschitz domains, (to appear).

[8] B. Muckenhoupt, The equivalence of two conditions for weight functions, Studia Math. 49 (1974), p.p. 101 - 106.

[9] B. Muckenhoupt and R. Wheeden, Weighted bounded mean oscillation and the Hilbert transform, Studia Math. 54 (1976), p.p. 221 - 237.

[10] E. M. Stein and G. Weiss, Instroduction to Fourier Analysis on Euclidean space, Princeton Univ. Press, Princeton, N.J., 1971.

Department of Mathematics
Princeton University
Princeton, N.J. 08540

Proceedings of Symposia in Pure Mathematics
Volume XXXV, Part 1, 1979

THE ATOMIC DECOMPOSITION OF HARDY SPACES

by

Robert H. Latter[1]

In this note we will survey some results relating to the atomic characterization of certain Hardy spaces. To fix the ideas involved we begin by discussing the usual H^p spaces of the upper half-plane, $\mathbb{R}^2_+ = \{(x,y) : y > 0\}$. Define $H^p(\mathbb{R}^2_+)$ $(0 < p < \infty)$ to be the space of all functions F, holomorphic in $\mathbb{R}^2_+$, for which

$$\sup_{y>0} \left[\int_{\mathbb{R}} |F(x+iy)|^p dx\right]^{\frac{1}{p}} = \|F\|_{H^p} < \infty.$$

If $F = u+iv \in H^p$, then u has boundary values $f = \lim_{y\to 0} u(x+iy)$ in the sense of distributions. We denote by Re H^p the space of such boundary distributions with the norm $\|f\|_{H^p} = \|F\|_{H^p}$. If $p > 1$, it follows from the theorem of M. Riesz that Re H^p is essentially isometrically isomorphic to the real-valued functions in $L^p(\mathbb{R})$. Thus, in what follows, we will assume $0 < p \leq 1$ unless otherwise indicated. The following theorem gives a purely real variables characterization of Re H^p.

THEOREM (Burkholder, Gundy, Silverstein [1]). *$f \in$ Re H^p if and only if f is the boundary distribution of a harmonic function u which satisfies $u^*(x) = \sup_{|x-y|<t} |u(y,t)| \in L^p$. Moreover, $\|f\|_{H^p} \sim \|u^*\|_{L^p}$.*

The following theorem shows that the Poisson kernel does not enjoy a special role in the maximal function characterization of Re H^p.

THEOREM (Fefferman, Stein [9]). *Let $\varphi \in S$ satisfy $\hat{\varphi}(0) \neq 0$. Then $f \in$ Re H^p if and only if $\varphi^* f(x) = \sup_{|x-y|<t} |\varphi_t * f(y)| \in L^p$ where $\varphi_t(x) = \frac{1}{t}\varphi(\frac{x}{t})$. Moreover, $\|f\|_{H^p} \sim \|\varphi^* f\|_{L^p}$.*

If A is a class of suitably normalized functions $\varphi \in S$ we may define a "grand maximal function" $f^*(x) = \sup_{\varphi\in A} \varphi^* f(x)$ which also characterizes Re H^p as above.

We now turn to the atomic characterization of Re H^p. We call a

[1]Research supported in part by NSF Grant No. MCS78-02128.

function a on $\mathbb{R}$ a p-*atom* if a is supported on an interval I and satisfies

(i) $|a| \leq |I|^{-1/p}$, and

(ii) $\int_{\mathbb{R}} a(x)x^j dx = 0$ if $j \leq \frac{1}{p} - 1$.

THEOREM (Coifman [5]). *If* $f \in \operatorname{Re} H^p$, *then there exist* $\lambda_i \in \mathbb{C}$ *and* p-*atoms* a_i *such that*

(1) $$\sum_{i=1}^{\infty} |\lambda_i|^p \leq C\|f\|_{H^p}^p$$

and

(2) $$f = \sum_{i=1}^{\infty} \lambda_i a_i.$$

This theorem has an easy converse which is a consequence of the fact that $\|\varphi^* f\|_{L^p} \leq C$ if a is a p-atom and φ is a "bump function." We will now outline a proof of this theorem in case $p = 1$ which extends easily to $0 < p < 1$ and $n > 1$, and which may be adapted to prove some of the theorems which will be stated below.

Put $\Omega_k = \{f^* > 2^k\}$, $k = 0, \pm 1, \ldots$. Let $f = g_k + b_k$ where $|g_k| \leq C2^k$ and $b_k = \sum_{i=1}^{\infty} (f-m_i^k)\varphi_i^k$ be the Calderón-Zygmund decomposition of f given by Fefferman, Rivière and Sahger [8]. Here $m_i^k = \frac{1}{\int \varphi_i^k} \int f\varphi_i^k$ and $\{\varphi_i^k\}$ is a Whitney partition of unity on Ω_k. Because $\operatorname{supp} b_k \subset \Omega_k$ we may write

$$f = \sum_{k=-\infty}^{\infty} g_{k+1} - g_k = \sum_{k=-\infty}^{\infty} b_k - b_{k+1}.$$

Now let $m_{ij}^{k+1} = \frac{1}{\int \varphi_j^{k+1}} \int_{\mathbb{R}} (f-m_j^{k+1})\varphi_i^k \varphi_j^{k+1}$. Notice that $\sum_{i=1}^{\infty} m_{ij}^{k+1} = 0$ and $|m_{ij}^{k+1}| \leq C2^k$. Thus

$$\begin{aligned} b_k - b_{k+1} &= \sum_{i=1}^{\infty} (f-m_i^k)\varphi_i^k - \sum_{j=1}^{\infty} (f-m_j^{k+1})\varphi_j^{k+1} \\ &= \sum_{i=1}^{\infty} [(f-m_i^k)\varphi_i^k - \sum_{j=1}^{\infty} (f-m_j^{k+1})\varphi_j^{k+1}\varphi_i^k] \\ &= \sum_{i=1}^{\infty} \{(f-m_i^k)\varphi_i^k - \sum_{j=1}^{\infty} [(f-m_j^{k+1})\varphi_i^k - m_{ij}^{k+1}]\varphi_j^{k+1}\} \\ &= \sum_{i=1}^{\infty} \beta_i^k \end{aligned}$$

where $\beta_i^k = C2^k|I_i^k|a_i^k$, a_i^k is an atom, and $\sum_{k=-\infty}^{\infty} \sum_{i=1}^{\infty} 2^k |I_i^k| \leq \int_0^{\infty} |f^* > \lambda| d\lambda \leq C\|f\|_{H^1}$.

More details of this proof may be found, for example, in [13].

We now turn to a discussion of the parabolic H^p spaces of Calderón and Torchinsky ([3] and [4]). Let $A_t = t^P$ $(0 < t < \infty)$ be a group of linear transformations on $\mathbb{R}^n$ with infinitesimal generator P satisfying $(Px,x) \geq (x,x)$ where (,) is the usual inner product on $\mathbb{R}^n$. Let $\rho(x)$ denote the corresponding homogeneous norm: $\rho(A_t x) = t\rho(x)$. If $\varphi \in S$ and $\hat{\varphi}(0) \neq 0$, define for $t > 0$, $\varphi_t(x) = t^{-\gamma}\varphi(A_t^{-1}x)$ where $\gamma = \operatorname{tr} P$. Let

$$F(x,t) = (f*\varphi_t)(x)$$

and

$$\varphi^* f(x) = \sup_{\rho(y)<t} |F(x+y,t)|.$$

We say $f \in H^p$, $0 < p < \infty$, if $\varphi^* f \in L^p$. Also, $\|f\|_{H^p} = \|\varphi^* f\|_{L^p}$. Calderón and Torchinsky have shown that the definition of H^p is independent of $\varphi \in S$ $(\hat{\varphi}(0) \neq 0)$ and, moreover, H^p is characterized by a grand maximal function $f^* = \sup_{\varphi \in A} \varphi^* f$ being in L^p.

A p-*atom* a in this context is a function supported in a ball $B_r(x) = \{y:\rho(x-y)<r\}$ which satisfies

(i) $\|a\|_{L^\infty} \leq |B_r(x)|^{-1/p}$ and

(ii) $\int_{\mathbb{R}^n} a(x)x^\alpha \, dx = 0$ if $|\alpha| \leq \gamma(\frac{1}{p} - 1)$.

The atomic H^p space is defined by

$$H^p_{at} = \{f = \sum_{i=1}^\infty \lambda_i a_i : \Sigma|\lambda_i|^p < \infty, \ a_i \text{ p-atoms}\}$$

$$\|f\|^p_{H^p_{at}} = \inf \sum_{i=1}^\infty |\lambda_i|^p$$

where the infimum is taken over all decompositions of f as a sum of atoms. We then have the following theorem.

THEOREM ([13]). $H^p = H^p_{at}$ $(0 < p \leq 1)$. *Moreover,* $\|\cdot\|_{H^p} \sim \|\cdot\|_{H^p_{at}}$.

The proof outlined above for the case m = 1 extends easily to prove this theorem. Moreover, because this proof depends in no essential way on the group structure of $\mathbb{R}^n$ it may be used, along with results of D. Geller [11], to obtain an atomic characterization of the Hardy spaces on the Heisenberg group. In case P is diagonalizable this theorem is due to A. P. Calderón [2] and Latter [12].

Finally we will outline some of our results on the unit ball $B = \{z:|z|<1\} \subset \mathbb{C}^n$. In this situation there are three natural ways to define a Hardy space. First there is a maximal function H^p space: Let $\Sigma = \partial B = \{z \in \mathbb{C}^n : |z|=1\}$. The Poisson-Szegö kernel for B is defined by

$$P_{rz}(\zeta) = c\left(\frac{1-r^2}{|1-rz\cdot\bar{\zeta}|^2}\right)^n$$

$(z,\zeta \in B;\ z\cdot\bar{\zeta} = z_1\bar{\zeta}_1 + \dots + z_n\bar{\zeta}_n)$. If f is a distribution on Σ we define

$$P^*f(z) = \sup_{|1-w\cdot\bar{z}|<1-r} |\int fP_{rw}|.$$

Let

$$H^p_{max} = \{f : P^*f \in L^p\},$$

$$\|f\|_{H^p_{max}} = \|P^*f\|_{L^p}.$$

Next we define the atomic Hardy spaces. In order to avoid the technicalities involved in defining the analogue of vanishing moments in this case we will restrict our attention to $1 \geq p > \frac{2n}{2n+1}$. Then a *p-atom* a is either the constant function $\sigma(\Sigma)^{-1/p}$ or a function supported on a ball $B_s(z) = \{\zeta \in \Sigma : |1-z\cdot\bar{\zeta}|<s\}$, which satisfies

(i) $|a| \leq \sigma(B_s(z))^{-1/p}$ and

(ii) $\int_\Sigma a(z)d\sigma(z) = 0$.

The atomic H^p space is then defined exactly as in the case of the parabolic H^p spaces.

THEOREM ([10]). $H^p_{max} = H^p_{at}$ $(0 < p \leq 1)$. *Moreover,* $\|\cdot\|_{H^p_{max}} \sim \|\cdot\|_{H^p_{at}}$.

This theorem is proved as above. The major difficulty occurs in showing that the mean values m_i^k are properly bounded. This is done by defining an appropriate grand maximal function f^* and showing $P^*f \in L^p \Longrightarrow f^* \in L^p$.

The classical definition of Hardy space in B is in terms of holomorphic functions. $\mathcal{H}^p$ is the space of functions F, holomorphic in B, which satisfy the growth condition

$$\sup_{0<r<1} \|F(rz)\|_{L^p(d\sigma)} = \|F\|_{\mathcal{H}^p} < \infty.$$

Such functions have boundary values $f = \lim_{r\to 1} F(rz)$ in the sense of distributions. We will identify $\mathcal{H}^p$ with its boundary values. Then $\mathcal{H}^2 \subset L^2$. Let $P : L^2 \to \mathcal{H}^2$ be the orthogonal projection.

THEOREM ([10]). *Let* $0 < p \leq 1$. *Then* P *extends to a bounded projection on* H^p_{max} *with range equal to* $\mathcal{H}^p : PH^p_{max} = \mathcal{H}^p$. *In other words, every* $f \in \mathcal{H}^p$ *has a decomposition* $f = \sum_{i=1}^{\infty} \lambda_i A_i$ *where* $\Sigma|\lambda_i|^p \leq C\|f\|^p_{\mathcal{H}^p}$ *and* $A_i = Pa_i$, a_i *a p-atom, and conversely.*

This theorem and the corollary below extend to $p < 1$ results of Coifman, Rochberg, and Weiss [6] for $p = 1$.

COROLLARY. *Let* $0 < p \leq 1$. *If* $f \in H^p$, *then there are* $g_i, h_i \in H^{2p}$ *such that* $f = \sum_{i=1}^{\infty} g_i h_i$ *and* $\sum_{i=1}^{\infty} \|g_i\|_{H^{2p}} \|h_i\|_{H^{2p}} \leq C\|f\|_{H^p}$.

REFERENCES

[1] D. L. Burkholder, R. F. Gundy, M. L. Silverstein, A maximal function characterization of the class H^p, Trans. A.M.S. 157(1971), 137-153.

[2] A. P. Calderón, An atomic decomposition of distributions in parabolic H^p spaces, Adv. in Math. 25(1977), 216-225.

[3] A. P. Calderón, A. Torchinsky, Parabolic maximal functions associated with a distribution, Adv. in Math. 16(1975), 1-63.

[4] A. P. Calderón, A. Torchinsky, Parabolic maximal functions associated with a distribution II, Adv. in Math. 24(1977), 101-171.

[5] R. R. Coifman, A real variable characterization of H^p, Studia Math. 51(1974), 269-274.

[6] R. R. Coifman, R. Rochberg, G. Weiss, Factorization theorems for Hardy spaces in several complex variables, Ann. of Math. 103(1976), 611-635.

[7] R. R. Coifman, G. Weiss, Extensions of Hardy spaces and their use in analysis, Bull. AMS 83(1977), 569-645.

[8] C. Fefferman, N. Rivière, Y. Sahger, Interpolation of H^p spaces, the real method, Trans. AMS.

[9] C. Fefferman, E. M. Stein, H^p spaces of several variables, Acta Math. 129(1972), 137-193.

[10] J. B. Garnett, R. H. Latter, The atomic decomposition for Hardy spaces in several complex variables, to appear in Duke Math. J.

[11] D. Geller, Fourier analysis on the Heisenberg group: Schwartz space, to appear.

[12] R. H. Latter, A decomposition of $H^p(\mathbb{R}^n)$ in terms of atoms, Studia Math. 62(1977), 92-101.

[13] R. H. Latter, A. Uchiyama, The atomic decomposition for parabolic H^p spaces, to appear.

[14] M. Taibleson, G. Weiss, The molecular characterization of certain Hardy spaces, to appear.

Proceedings of Symposia in Pure Mathematics
Volume XXXV, Part 1, 1979

THE MOLECULAR CHARACTERIZATION OF HARDY SPACES

Mitchell H. Taibleson[1] and Guido Weiss[1]

ABSTRACT. The molecular-decomposition characterization of the real Hardy spaces on $\mathbb{R}^n$ is outlined. Extensions to Hardy spaces on the half-spaces $\mathbb{R}^{n+1}_+$ and the unit disk in $\mathbb{R}^2$ is described. Applications to the holomorphic Hardy (Bergman) spaces on the half-plane and disk in $\mathbb{C}$ and the harmonic Hardy spaces on the Euclidean half-spaces $\mathbb{R}^{n+1}_+$ are given.

The results we describe here are extensions of results on the atomic characterization of $H^p(\mathbb{R}^n)$ as given by Coifman and Weiss [2] and Latter [3].

ATOMS. We say that $a(x)$ is a (p,q,s)-atom on $\mathbb{R}^n$, $0 \le 1 \le q$, $p < q$, $s \ge [n(\frac{1}{p} - 1)]$ ($[\cdot]$ is the greatest integer function) if $a(x)$ is supported on a ball B, $\|a\|_q |B|^{\frac{1}{p}-\frac{1}{q}} \le 1$ and $\int a(x)x^\nu dx = 0$ if $|\nu| \le s$ (where, as usual $\nu = (\nu_1, \ldots, \nu_n)$ is an n-tuple of non-negative integers, $|\nu| = \nu_1 + \cdots + \nu_n$, and $x_1^{\nu_1} \cdots x_n^{\nu_n}$. The space $H^{p,q,s}(\mathbb{R}^n)$ is just the collection, $f = \Sigma \lambda_k a_k$ where each a_k is a (p,q,s)-atom and $\Sigma |\lambda_k|^p < \infty$. If $p = 1$ convergence is in L^1 and if $p < 1$ it is as a continuous linear functional on an appropriate space of smooth functions (of the Morrey-Campanato type). The "norm": $\inf \Sigma |\lambda_k|^p$, over all such representations, defines a metric with respect to which $H^{p,q,s}$ is a Frechét space.

The principal result for $H^{p,q,s}$ is that whatever is the (admissible) choice of q and s we obtain the same space and the corresponding norms are mutually equivalent and so we may simply refer to any one of these spaces as $H^p_{at}(\mathbb{R}^n)$. This result appears in [4], §2 and Appendix D. In Latter's result,

---

AMS(MOS) subject classifications (1970). Primary 30A78, 42A18, 42A40.

[1]Research supported by the National Science Foundation under grant MCS75-02411 A03.

referred to above, he showed that spaces defined by $(p,\infty,[n(\frac{1}{p}-1)])$-atoms agree with the classical H^p spaces (maximal, conjugate or Littlewood-Paley norms) and so with an ambiguity with respect to the norm used, we may refer to any such space as $H^p(\mathbb{R}^n)$.

The novelty in this result is that we may take more than the minimum number of moments and still obtain the same space. One important consequence is that we can extend R. Macias' result on interpolation of an analytic family of operators mapping $H^p(\mathbb{R}^n)$ to L^q (see [2, Th. E]), for all $p > 0$. This result is also an important tool for obtaining the multiplier theorems mentioned below.

Another corollary to the equivalence of the $H^{p,q,s}$ spaces for various admissible choices of q and s is that we are able then to identify the duals of these spaces as equivalence classes of functions, and the spaces have equivalent norms. The situation for $p = 1$, $q = \infty$, $s \geq 0$ will illustrate this point. Let $\mathcal{P}_s$ be the class of polynomials of degree at most s , so $\mathcal{P}_0 = \mathbb{R}$. Let $BMO(s) = \{g: \sup_{B \in \mathbb{R}^n} \inf_{P \in \mathcal{P}_s} \left[\frac{1}{|B|}\int_B |g(x)-P(x)|\,dx\right] < \infty\}.$

If $g \in BMO(s)$ we let the indicated quantity be the norm of g , and we may view $BMO(s)$ as a space of functions or as a space of equivalence classes, identifying those functions that differ by a polynomial in $\mathcal{P}_s$. As function spaces, the larger the value of s , the larger is the class of functions permitted and $BMO(0)$ is, of course, the usual BMO space. But we can identify the dual of $H^{p,q,s}$ by a natural pairing and the dual of $H^{1,\infty,s}$ is $BMO(s)$ (as a space of equivalence classes). We can then show that $\tilde{g} \in BMO(s)$, as a space of functions, iff there is a $g \in BMO$ such that $g - \tilde{g} \in \mathcal{P}_s$ and the norms are equivalent. We get similar results for other values of p and q .

One of the major reasons for using the "atomic" characterization of H^p is that atoms are so simple that many "deep" results are reduced to elementary calculations on atoms or their Fourier transforms. (If you like to work in the Fourier transform domain you will probably think $(p,2,s)$-atoms are the simplest; but if you like to work in the space domain, (p,∞,s)-atoms will be your favorite). Many such applications are given in [2]. A problem arises, however, because for quite a few applications, atoms are much too simple.

For example, for all but the most trivial multiplier and integral operators, either atoms are not sent into atoms, or if they are, the size of the support of the image supplies no useful information. For another example, there are versions of Hardy spaces where members are holomorphic or harmonic functions. If one looks for a characterization of such functions in terms of simple elements, those simple elements cannot be atoms since there is no such object as a holomorphic or harmonic atom, since atoms have compact support.

One solution to this problem is to construct a more general class of functions to use as building blocks that are "simple enough". Coifman and Weiss [2] introduced a candidate for such a function, and termed it a "molecule". We generalize that concept here.

MOLECULES. We say that $M(x)$ is a (p,q,s,ϵ)-molecule on $\mathbb{R}^n$ centered at x_0 if p, q and s are related as before (for atoms) and $\epsilon > \max(s/n, \frac{1}{p} - 1)$. Furthermore we require that $\int M(x)x^\nu dx = 0$, $0 \le |\nu| \le s$, and, if $\alpha = 1 - \frac{1}{p} + \epsilon$, $\beta = 1 - \frac{1}{q} + \epsilon$, then

$$\|M(\cdot)\|_q^{\alpha/\beta} \, \|M(\cdot)|\cdot - x_0|^{n\beta}\|_q^{1-\alpha/\beta} = \mathfrak{N}(M) < \infty .$$

The following should make clear just what a typical molecule looks like. If M is a $(1,\infty,0,\epsilon)$-molecule then $\int M dx = 0$, M is bounded, as is $M(x)|x|^{n(1+\epsilon)}$. Take, for instance, $P(x,y) = \frac{1}{C_n} \frac{y}{(|x|^2+y^2)^{\frac{n+1}{2}}}$, $x \in \mathbb{R}^n$, $y > 0$, the Poisson kernel for $\mathbb{R}^{n+1}_+$. Then, as functions of $x \in \mathbb{R}^n$, $\frac{\partial}{\partial y} P(x,y) = \frac{1}{C_n} \frac{|x|^2 - ny}{(|x|^2+y^2)^{\frac{n+3}{2}}}$ is a $(1,\infty,0,\epsilon)$-molecule on $\mathbb{R}^n$ centered at $x = 0$ for all ϵ, $0 < \epsilon < 1$, and $\frac{\partial}{\partial x_k} P(x,y) = \frac{-(n+1)}{C_n} \frac{x_k y}{(|x|^2+y^2)^{\frac{n+3}{2}}}$ is a $(1,\infty,0,\epsilon)$-molecule centered at $x = 0$ for all ϵ, $0 < \epsilon < 2$.

The main result for molecules is:

THEOREM. If $a(x)$ is a (p,q,s)-atom then $a(x)$ is a (p,q,s,ϵ)-molecule (centered at the center of the ball on which it is supported) for all $\epsilon > \max(s/n, \frac{1}{p} - 1)$ and $\mathfrak{N}(a) \le C$, C independent of $a(x)$. If $M(x)$ is a (p,q,s,ϵ)-molecule then $M = \Sigma \lambda_k a_k$, where each a_k is either a (p,q,s)-atom or a (p,∞,s)-atom and $[\Sigma |\lambda_k|^p]^{1/p} \le C\,\mathfrak{N}(M)$, C independent of M.

From this theorem we see that $H^p(\mathbb{R}^n) = H^p_{at}(\mathbb{R}^n)$ could just as well have been defined in terms of decompositions into molecules. The proof appears in [4, §2].

Applications to multiplier transforms and integral operators are, conceptually, quite straight-forward, although the details are somewhat messy. One simply shows that atoms are mapped into molecules if the operator is reasonable. For fractional integration operators, Calderón-Zygmund operators and similar operators with "smooth" kernels, the argument is direct and elementary estimates are all that is needed. For multiplier transforms we use $(p,2,s)$-atoms, with s large, and exploit the Plancherel relations. Observe that the Fourier transform of an atom is C^∞ and vanishes at the origin at an order that depends on s. Let a be the atom, m the multiplier and $\hat{M} = m\hat{a}$. The moment condition for M can be determined from the behaviour of $\hat{M}$ at the origin. Since $\|M\|_2 = \|m\hat{a}\|_2$ and $\|M(\cdot)|\cdot - x_0|^{n\beta}\|_2$ can be estimated by computing the L^2 norm of the derivatives of $\hat{M} = m\hat{a}$ (if $n\beta$ is an integer), we only need Hörmander-like conditions on the multiplier to deal with both the size and the moment condition of the transformed atom. Examples and details are given in §4 and Appendix E of [4].

APPLICATIONS TO SPACES OF HOLOMORPHIC AND HARMONIC FUNCTIONS. Let us now turn to some applications of these ideas to spaces of holomorphic and harmonic functions that are being studied by Coifman and Rochberg [1]. (A summary appears in these proceedings.) Perhaps the simplest example to consider is the Bergman space A_1 on R^2_+, the set of functions, $F(z)$, holomorphic on $R^2_+ = \{z = x + iy: y > 0\}$ such that

$$A_1(F) = \int_{R^2_+} |F(z)|\,d\mu(z) < \infty, \quad d\mu(z) = dxdy.$$

Coifman and Rochberg start with the observation that for some constant c, $B_0(z,\zeta) = c(z-\bar{\zeta})^{-2}$ is a reproducing kernel for $L^2(R^2_+, d\mu)$. Let $B_r(z,\zeta) = (B_0(z,\zeta))^{r+1}$ and $M^{(r)}_\zeta(z) = B_r(z,\zeta)/B_{r-1}(\zeta,\zeta)$. (The functions $B_r(z,\zeta)$ are Bergman kernels. They are reproducing kernels for various classes of functions. There are analogues for all Siegal domains of type 2. The particular case, $D = \{z \in C: |z| < 1\}$ is discussed below.) The kernel is reproducing in the sense that $F(z) = \int_{R^2_+} F(\zeta)B_0(z,\zeta)d\mu(\zeta)$. From this starting point they show that $F \in A_1$ iff there is a complex sequence $\{c_k\}$ and a sequence $\{\zeta_k\} \subset R^2_+$ such that $F(z) = \Sigma\, c_k M^{(r)}_{\zeta_k}(z)$, and $\Sigma\, |c_k| < \infty$ $(r > 0)$.

Furthermore, $\inf \Sigma\, |c_k|$, over all such representations, is equivalent to $A_1(F)$. A straight-forward calculation shows that $M^{(r)}_\zeta(z) = c|\zeta|^{2r}/(z-\bar{\zeta})^{2(r+1)}$ is a $(1,q,0,\varepsilon)$-molecule entered at ζ, for all q , $1 < q \leq \infty$, $\epsilon > r$, with a molecular norm that is uniformly bounded in $\zeta \in \mathbb{R}^n_+$. By a molecule for $\mathbb{R}^2_+$ we mean a molecule for $\mathbb{R}^2$ that is supported on $\mathbb{R}^2_+$. There is a fully developed atomic and molecular theory for $\mathbb{R}^2_+$ and it is given a more complete description in [4]. It follows from all of this that $A_1 \subset H^1(\mathbb{R}^2_+)$, and with a little more work it can be shown that A_1 consists exactly of the holomorphic functions in $H^1(\mathbb{R}^2_+)$.

There is a similar situation for $D = \{z \in \mathbb{C}: |z| \leq 1\}$. One lets $B_0(z,\zeta) = C(1-z\bar{\zeta})^{-2}$ and $M_\zeta^{(r)} = (1-|\zeta|^2)^{2r}/(1-z\bar{\zeta})^{2(r+1)}$. We obtain the representation for $F \in A_1$ in the form $F(z) = F(0) + \Sigma\, C_k M_{\zeta_k}^{(r)}(z)$.

There are several extensions to be noted:

For $0 < p \leq 1$ one uses the observation that $\{[B_0(\zeta,\zeta)]^{1/p}/[B_r(\zeta,\zeta)]\}B_r(z,\zeta)$ is a (p,q,s,ϵ)-molecule centered at ζ provided $r > \epsilon$ and p , q , s and ϵ are related as above (with $n = 2$).

The extension to the weighted spaces adds a considerable order of difficulty. For such spaces the size of the atom or molecule (and the maximal function) is taken with respect to the weighted measure $\nu_\alpha(z)d\mu(z)$, $\nu_\alpha(z) = B_0(z,z)^{-\alpha}$. From the point of view of the "real theory" there are two ways that we can require that moments up to order s vanish. Thus, $\int F(z)x^{\nu_1}y^{\nu_2}d\mu(z) = 0$ or $\int F(z)x^{\nu_1}y^{\nu_2}\nu_\alpha(z)d\mu(z) = 0$, for $0 \leq \nu_1 + \nu_2 \leq s$. For holomorphic functions these two notions agree. For the real theory they are distinct, require separate developments and lead to quite different descriptions of the dual spaces.

The most delicate situation is for the case where the moments vanish with respect to $d\mu(z)$ but the size is measured with respect to the weighted measure. One constructs a family of Euclidean balls whose measures grow regularly with respect to the weighted measure restricted to the domain. The geometry for $\mathbb{R}^2_+$ is not too complicated; this cannot be said for D . The methods used involve: weighted norm inequalities, the theory of spaces of homogeneous type and some delicate estimates for Gram-Schmidt polynomials and their coefficients, where we need to show that certain estimates hold uniformly over a family of regions that vary continuously with respect to a parameter in a compact Hausdorff space. This specific case (for D , where size is measured with respect to a weighted measure and moments vanish with respect to Euclidean measure is given in [4] in §3 (where the results are stated with a sketch of the proofs) and in Appendicies A, B and C (where the details are given).

We note that there is a similar development for harmonic functions on the half-space $\mathbb{R}^{n+1}_+ = \{(x,y): x \in \mathbb{R}^n , y > 0\}$. Let

$P(x,y) = \frac{1}{C_n} \frac{y}{(|x|^2+y^2)^{\frac{n+1}{2}}}$ be the Poisson kernel for $\mathbb{R}^{n+1}_+$. As above, one starts with the observation that if $h(x,y)$ is a "nice enough" harmonic function then $\frac{(-2)^k}{\Gamma(k)} \frac{\partial^k}{\partial y^k} P(x-\xi, y+\eta)\eta^{k-1}$ is a reproducing kernel. (This follows from the Poisson integral representation modified by integration by parts!)

This fact is used together with the observation that if we multiply this kernel by the correct constant we get a family of molecules with uniformly bounded norm:

$$M^{(k)}_{(\xi,\eta)}(x,y) = C_k \frac{\eta^{k-1}}{\eta^{(n+1)(1/p-1)}} \frac{\partial^k}{\partial \eta^k} P(x-\xi, y+\eta)$$

is a (p,q,s,ϵ)-molecule on $\mathbb{R}^{n+1}_+$ centered at (ξ,η) (or at $(\xi,0)$), of uniformly bounded norm if $k-1 > (n+1)\epsilon$ and p, q, s and ϵ are related as above (for $\mathbb{R}^{n+1}$ molecules).

Coifman and Rochberg show that if $\int_{\mathbb{R}^{n+1}_+} |h(x,y)|^p dxdy < \infty$, h harmonic on $\mathbb{R}^{n+1}_+$, then h can be decomposed into harmonic molecules $M^{(k)}_{(\xi,\eta)}(x,y)$ as was done above for holomorphic functions, and, thus such a function is in $H^p(\mathbb{R}^{n+1}_+)$; it turns out that this class consists exactly of the harmonic functions in $H^p(\mathbb{R}^{n+1}_+)$.

These applications have illustrated some of the uses of the molecular theory for Hardy spaces.

We close with an observation that follows directly from the considerations above for harmonic functions. Suppose $F(x,y)$ is harmonic on $\mathbb{R}^{n+1}_+$ and $A_1(F) = \int_0^\infty \int_{\mathbb{R}^n} |F(x,y)| dxdy < \infty$. Let $G(x,y) = \begin{cases} F(x,y), & y > 0 \\ 0, & y \leq 0 \end{cases}$.

Then $G \in H^1(\mathbb{R}^{n+1})$ and $\|G\|_{H^1} \leq CA_1(F)$. To see this we note that F has a molecular development, relative to $\mathbb{R}^{n+1}_+$, and so it has an atomic development, with atoms supported on the half-space. But such atoms are also atoms associated with $\mathbb{R}^{n+1}$, so $G \in H^1(\mathbb{R}^{n+1})$ in the atomic sense; hence, $G \in H^1(\mathbb{R}^{n+1})$ in the classical sense (with respect to Riesz transforms, maximal functions, Littlewood-Paley, or what have you). The norm boundedness is obvious.

REFERENCES

1. R. R. Coifman and R. Rochberg, Bergman spaces of holomorphic functions, to appear.

2. R. R. Coifman and G. Weiss, Extensions of Hardy spaces and their use in analysis, Bull. Amer. Math. Soc. 83 (1977), 569-645.

3. R. H. Latter, A decomposition of $H^p(\mathbb{R}^n)$ in terms of atoms, Studia Math., to appear in vol. 62 (1978).

4. M. H. Taibleson and G. Weiss, The molecular characterization of certain Hardy spaces, to appear.

DEPARTMENT OF MATHEMATICS, WASHINGTON UNIVERSITY, ST. LOUIS, MISSOURI 63130

Proceedings of Symposia in Pure Mathematics
Volume XXXV, Part 1, 1979

A CHARACTERIZATION OF $H^1(\Sigma_{n-1})$

Fulvio Ricci and Guido Weiss[1]

1. Let B_n be the unit ball in R^n , $n \geq 2$, and Σ_{n-1} be its boundary.

We define the Hardy space $H^1(B_n)$ as the Banach space of vector valued functions $F: B_n \to R^n$ satisfying the condition

$$\sup_{r<1} \int_{\Sigma_{n-1}} |F(rx')|\,dx' \equiv \|F\|_{H'} < \infty \tag{1}$$

(where dx' denotes the normalized rotation invariant measure on Σ_{n-1}) and the system of generalized Cauchy-Riemann equations

$$\begin{cases} \dfrac{\partial F_i}{\partial x_j} = \dfrac{\partial F_j}{\partial x_i} & i,j = 1,\ldots,n \\ \displaystyle\sum_{j=1}^{n} \frac{\partial F_1}{\partial x_j} = 0 & . \end{cases} \tag{2}$$

A function F which satisfies (2) is called a Riesz system; condition (2) is equivalent to $F = \nabla h$, for a harmonic function h .

When $n = 2$, $H^1(B_2)$ is the classical Hardy space on the unit disc.

The following basic properties of the classical H^1 space on the disc hold for $H^1(B_n)$:

AMS(MOS) subject classifications (1970). Primary 30A67.

[1]Research supported in part by the National Science Foundation under grant MCS75-02411 A03.

i) any $F \in H^1(B_n)$ has non-tangential boundary values $F(x')$ at almost any point x' of Σ_{n-1} ;

ii) $F_r(x') = F(rx')$ tends to $F(x')$ in the L^1-norm as r tends to 1 ;

iii) $F(x)$ is the Poisson integral of $F(x')$.

A Riesz system F is completely determined by the harmonic real valued function $f(x) = x \cdot F(x)$, which equals $r\frac{\partial h}{\partial r}$, if $F = \nabla h$. In fact, this gives a $1:1$ correspondence between Riesz systems in B_n and harmonic functions $f(x)$ such that $f(0) = 0$ and have real values.

If $F \in H^1(B_n)$, then $\lim_{r\to 1} f(rx') = f(x') \in L^1(\Sigma_{n-1})$ exists a.e. and in the L'-norm.

To see how we can reconstruct $F(x)$ from the real valued function $f(x')$ on Σ_{n-1}, we proceed as follows: since $f(x')$ has mean value 0 on Σ_{n-1}, we obtain the following expression for the function h (which, without loss of generality, we can assume satisfies $h(0) = 0$):

$$h(x) = \int_{\Sigma_{n-1}} f(y')dy' \int_0^1 \left\{\frac{1-\rho^2|x|^2}{|y'-\rho x|^n} - 1\right\} \frac{d\rho}{\rho} \quad .$$

Differentiating we have (3) :

$$F(x) = \int_{\Sigma_{n-1}} f(y')dy' \left[\frac{1-|x|^2}{|y'-x|^n} \frac{x'}{|x|} + n\int_0^1 \frac{1-\rho^2|x|^2}{|y'-\rho x|^{n+2}} d\rho \frac{y'-(y'\cdot x')x'}{|x|} \right]$$

for $x \in R^n \backslash \{0\}$, x' always denotes $\frac{x}{|x|}$.

If we are given $f \in L^1(\Sigma_{n-1})$, with $\int_{\Sigma_{n-1}} f(x')dx' = 0$, standard techniques show that the system defined by (3) is in $H^1(B_n)$ if and only if the principal value operator H with kernel

$$\text{(4)} \qquad H(x',y') = \int_0^1 \frac{1-\rho^2}{|y'-\rho x'|^{n+2}} d\rho \; (y'-(y'\cdot x')x')$$

maps f into an integrable vector valued function on Σ_{n-1} . $Hf(x')$ is the "angular part" of $F(x')$ and can be regarded as a generalization of the conjugate function operator on Σ_1 .

DEFINITION. $H^1(\Sigma_{n-1})$ is the space of real valued integrable functions f on Σ_{n-1} which have mean value zero and are such that $Hf \in L^1(\Sigma_{n-1}, R^n)$.

The theory of Fefferman and Stein [3] can be applied to show that there is a maximal function characterization of $H^1(\Sigma_{n-1})$ by means of smooth approximate identities. Also, $H^1(\Sigma_{n-1})$ is the atomic space on Σ_{n-1} with respect to the measure dx' and to Euclidean distance (as defined in [2]; the proof follows the same lines as in [5]). Observe that since we only deal with functions with mean value zero, the "unnatural" constant atom $a(x) = 1$ does not appear.

If we define in the same way $H^p(B_n)$ and $H^p(\Sigma_{n-1})$ for $p > 1$, we have the following theorem:

THEOREM 1 (Korányi-Vági [4]). If $1 < p < \infty$, $H^p(\Sigma_{n-1})$ consists of all functions in $L^p(\Sigma_{n-1})$ with mean value zero.

2. Calderón and Zygmund [1] proved that if f is an odd function in $L^1(\Sigma_{n-1})$, then $\frac{f(x')}{|x|^n}$ is a singular integral kernel whose associated operator K_f is bounded on $L^p(R^n)$ for $1 < p < \infty$.

A standard method for studying even kernels uses the Riesz transforms to reduce the problem to odd kernels (see [6]). It is natural to ask what is the largest class of even kernels for which this argument can be applied.

We shall prove that this class is $H^1(\Sigma_{n-1})$ [2].

The Riesz transforms R_j on R^n are defined by the multipliers $-i\,\frac{t_j}{|t|}$, $t \in R^n$. Let us take a function $f \in C^\infty(\Sigma_{n-1})$; the composition R_jK_f is well defined as a translation and dilation invariant operator on $L^p(R^n)$, $1 < p < \infty$, and it consists of a multiple of the identity plus a singular integral operator corresponding to a kernel that we call

$$\frac{\tilde{R}_j f(x')}{|x|^n} .$$

Define the vector valued operator $\tilde{R}$ on $C^\infty(\Sigma_{n-1})$ as

$$\tilde{R}f = (\tilde{R}_1 f, \ldots, \tilde{R}_n f)$$

and the operator S as the boundary values of the right hand side in (3).

---

[2] The fact that kernels in $H^1(\Sigma_{n-1})$ give bounded operators on $L^p(R^n)$, for $1 < p < \infty$, has been proved by W. Connett, using the atomic characterization of $H^1(\Sigma_{n-1})$.

We have the following theorem:

THEOREM 2. The space $H^1(\Sigma_{n-1})$ is characterized by the property $\tilde{R}f \in L^1(\Sigma_{n-1}, R^n)$; more precisely, the operator $\tilde{R}$ extends to a bounded operator from $H^1(\Sigma_{n-1})$ to $L^1(\Sigma_{n-1}, R^n)$, and there is a linear combination of $\tilde{R}$ and S which is bounded from $L^1(\Sigma_{n-1})$ to $L^1(\Sigma_{n-1}, R^n)$.

PROOF. Since we know that S is bounded from $H^1(\Sigma_{n-1})$ to $L^1(\Sigma_{n-1}, R^n)$, we only have to prove the last assertion.

Take a spherical harmonic P_k of degree $k \geq 1$ on Σ_{n-1} . Then

(5) $$SP_k(x') = \frac{1}{k} \nabla P_k(x')$$

(∇ denotes the gradient of the harmonic polynomial P_k in R^n).

In order to compute $\tilde{R}\, P_k$, we have to consider the kernel $K_{P_k} = \dfrac{P_k(x')}{|x|^n}$.

We have

$$(R_j P_k)\hat{}(t) = -i^{-k+1} \pi^{n/2} \frac{\Gamma(k/2)}{\Gamma(\frac{n+k}{2})} \frac{t_j}{|t|} P_k(t')$$

(this formula and the following one can be found in [6], pp. 160-162); therefore

(6) $$R_j K_{P_k} = -\pi \frac{\Gamma(\frac{k}{2})\Gamma(\frac{n+k-1}{2})}{\Gamma(\frac{k+1}{2})\Gamma(\frac{n+k}{2})} \frac{\partial}{\partial x_j} \frac{P_k(x')}{|x|^{n-1}}$$

If we call $\beta_{n,k}$ the constant in (6), we obtain

(7) $$\tilde{R}P_k(x') = \beta_{n,k} \nabla \frac{P_k(x')}{|x|^{n-1}} (x')$$

$$= \beta_{n,k}(\nabla P_k(x') - (n+k-1)P_k(x')x') \; .$$

We can expand $\beta_{n,k}$ as

(8) $$\beta_{n,k} = \frac{c_1}{k} + \frac{c_2}{k^2} + \cdots + \frac{c_s}{k^s} + 0\left(\frac{1}{k^{s+1}}\right)$$

where $s = [\frac{n+1}{2}]$ and the c_j's are constants.

From (5), (7) and (8) we have

$$(c_1 S-\tilde{R})P_k(x') = (a_0 + \frac{a_1}{k} + \cdots + \frac{a_{s-1}}{k^{s-1}} + 0\,(\frac{1}{k^s}))P_k(x')x'$$

$$+ (\frac{b_2}{k^2} + \cdots + \frac{b_s}{k^s} + 0\,(\frac{1}{k^{s+1}}))\nabla P_k(x')$$

The following considerations show that $c_1 S-\tilde{R}$ extends to a bounded operator from $L^1(\Sigma_{n-1})$ to $L^1(\Sigma_{n-1},R^n)$:

i) the operator Φ_1 defined by the zonal kernel

$$\Phi_1(x',y') = \sum_{k=1}^{\infty} \frac{1}{k} Z_k(x',y') = \int_0^1 \left(\frac{1-\rho^2}{|y'-\rho x'|^n} -1\right) \frac{d\rho}{\rho}$$

maps $L^1(\Sigma_{n-1})$ into $L^p(\Sigma_{n-1})$, for $p < \frac{n-1}{n-2}$ (here Z_k denotes the zonal harmonic of degree k); the same holds for the operators Φ_ℓ with kernels

$$\Phi_\ell(x',y') = \sum_{k=1}^{\infty} \frac{1}{k^\ell} Z_k(x',y') \quad ,$$

ℓ being a positive integer, since they are iterates of the operator Φ_1 ;

ii) since we have chosen s large enough, any kernel of the form

$$\sum_{k=1}^{\infty} 0\,(\frac{1}{k^s})\, Z_k(x',y')$$

defines an operator mapping $L^1(\Sigma_{n-1})$ into $L^2(\Sigma_{n-1})$;

iii) the operators which map P_k into $\frac{1}{k^\ell}\nabla P_k$ can be expressed as $S\Phi_{\ell-1}$, for $\ell \geq 2$; therefore by theorem 1 they map $L^1(\Sigma_{n-1})$ into $L^p(\Sigma_{n-1},R^n)$, for $p < \frac{n-1}{n-2}$; by the same argument, an operator which maps P_k into $0\,(\frac{1}{k^{s+1}})\nabla P_k$ maps $L^1(\Sigma_{n-1})$ into $L^2(\Sigma_{n-1},R^n)$.

This completes the proof.

REMARKS. i) Actually, both S and $\tilde{R}$ map $H^1(\Sigma_{n-1})$ into $H^1(\Sigma_{n-1},R^n)$.

ii) Since we are on a manifold of dimension $n-1$ we would like to have $H^1(\Sigma_{n-1})$ characterized by the L^1-boundedness of $n-1$ real valued operators; as a matter of fact, if $n-1$ components of either S or $\tilde{R}$ are in $L^1(\Sigma_{n-1})$, then the other is also in $L^1(\Sigma_{n-1})$.

The argument we mentioned at the beginning of this section gives the following result:

COROLLARY. If $f(x')$ is an even function in $H^1(\Sigma_{n-1})$, the singular integral operator K_f is bounded on $L^p(R^n)$, for $1 < p < \infty$.

REFERENCES

1. A. P. Calderón and A. Zygmund, On Singular Integrals, Amer. J. Math. 78 (1956), 289-309.
2. R. R. Coifman and G. Weiss, Extensions of Hardy Spaces and Their Use in Analysis, Bull. Amer. Math. Soc. 83 (1977), 569-645.
3. C. Fefferman and E. M. Stein, H^p Spaces of Several Variables, Acta Math. 129 (1972), 137-193.
4. A. Korányi and S. Vági, Singular Integrals on Homogeneous Spaces and Some Problems of Classical Analysis, Annali della Sc. Norm. Sup. Pisa xxv (1971), 575-648.
5. R. Latter, A Characterization of $H^p(R^n)$ in Terms of Atoms, to appear in St. Math. 1978.
6. E. M. Stein and G. Weiss, Introduction to Fourier Analysis on Euclidean Spaces, Princeton, 1971.

SCUOLA NORMALE SUPERIORE di PISA, PISA, ITALY

DEPARTMENT OF MATHEMATICS, WASHINGTON UNIVERSITY, ST. LOUIS, MISSOURI 63130

Proceedings of Symposia in Pure Mathematics
Volume XXXV, Part 1, 1979

TWO CONSTRUCTIONS IN BMO

John B. Garnett

We discuss two theorems on functions of bounded mean oscillation, each proved by geometric construction.

§1. Norm BMO in the usual way,

$$\|\varphi\|_* = \sup_Q \frac{1}{|Q|} \int |\varphi - \varphi_Q| dx$$

where Q is a cube in R^d, $|Q|$ is its measure, and φ_Q is the average of φ over Q. The first theorem estimates the distance.

$$\inf_{g \in L^\infty} \|\varphi - g\|_*$$

by the constant in the John-Nirenberg inequality

$$\sup \frac{|\{x \in Q : |\varphi(x) - \varphi_Q| > \lambda\}|}{|Q|} < e^{-A\lambda} \tag{1}$$

for λ large. Define

$$A(\varphi) = \sup\{A : (1) \text{ holds for all } \lambda \geq \lambda_0(A,\varphi)\} .$$

By the John-Nirenberg theorem [10], $\varphi \in$ BMO if and only if $A(\varphi) > 0$, and in fact

$$A(\varphi) \geq C/\|\varphi\|_* .$$

It is clear that

$$A(\varphi) = +\infty \text{ if } \varphi \in L^\infty$$

and that

$$A(\log|x|) = 1 .$$

<u>THEOREM 1</u>: <u>There are constants</u> c_1 <u>and</u> c_2, <u>depending only on the dimension, such that</u>

$$c_1/A(\varphi) \le \inf_{g\in L^\infty} \|\varphi - g\|_* \le c_2/A(\varphi).$$

The left inequality is trivial, and so the right inequality is the real theorem, proved jointly with my student P. Jones [7].

By Fefferman's theorem [6],

$$\varphi = g_0 + \sum_{j=1}^{d} R_j g_j \tag{2}$$

where R_j is the Riesz transform, $R_j g = \frac{-x_j}{|x|}\hat{g}$, and where $g_0, g_1, \ldots, g_d \in L^\infty$ with

$$\|\varphi\|_* \sim \sum_0^d \|g_j\|_\infty .$$

Thus Theorem 1 can be reformulated as

$$1/A(\varphi) \sim \inf\left\{\sum_1^d \|g_j\|_\infty : (2) \text{ holds for some } g_0 \in L^\infty\right\}.$$

A consequence is a higher dimensional Helson-Szegö theorem. Let $w \ge 0$ be a weight function on R^d. The Riesz transforms are bounded on $L^2(wdx)$ if and only if

$$\sup_Q \left(\frac{1}{|Q|}\int_Q wdx\right)\left(\frac{1}{|Q|}\int_R\int_Q \frac{1}{w}\, dx\right) < \infty . \tag{A_2}$$

See Hunt, Muckenhoupt and Wheeden [9] and Coifman and Fefferman [2]. When $d = 1$ there is an earlier necessary and sufficient condition, due to Helson and Szegö

$$\log w = g_0 + Hg_1,\ \|g_1\|_\infty < \pi/2,\ g_0 \in L^\infty,$$

where H denotes the Hilbert transform or conjugate function. These two results can be connected to Theorem 1 via

LEMMA 1: *If* $\varphi = \log w$, *then*

$$A(\varphi) > 1 \Longleftrightarrow w \text{ has } (A_2).$$

This lemma is easily derived from Jensen's inequality and the fact that (A_2) implies (A_p) for some $p < 2$. For $d = 1$ the lemma means that Theorem 1 follows from the equivalence between (A_2) and the Helson-Szegö condition. See [9] and [5]. For $d > 1$ we obtain a new Helson-Szegö result.

COROLLARY 1: *There are constants* $B_1(d)$ *and* $B_2(d)$ *such that if* (2) *holds with* $\sum_1^d \|g_j\|_\infty < B_1$, *then* e^φ *satisfies* (A_2). *Conversely, if* e^φ *has* (A_2), *then* (2) *holds for* φ *with* $\sum_1^d \|g_j\|_\infty < B_2$.

My teacher P. Jones has observed that $B_1 \neq B_2$ in the case $d > 1$, no matter how we norm the vector $(\|g_1\|_\infty, \|g_2\|_\infty, \ldots, \|g_d\|_\infty)$. Perhaps we must accept Lemma 1 as the sharp higher dimensional Helson-Szegö theorem.

§2. To give a rough (and slightly inaccurate) proof of Theorem 1, take $d = 1$. Let $A < A(\varphi)$ and fix $\lambda > \lambda_0(A,\varphi)$, $\lambda > 10\|\Phi\|_*$. We use (1) to build a function f such that

$$|\varphi - f| < C\lambda$$
$$\|f\|_* \leq C/A$$

The only interesting case is when $1/A$ is small compared to $\|\varphi\|_*$.

The function f is a sum of little functions called adapted functions. Let I be a dyadic interval with triple $\tilde{I}$. Say that $a(x)$ is adapted to I if $a(x)$ is Lipschitz and

(a) $|a(x)| \leq 1$

(b) $a(x) = 0$, $x \notin \tilde{I}$

(c) $|\nabla a(x)| \leq 1/|I|$, a.e.

LEMMA 2: *If* $\{I_j\}$ *is a sequence of dyadic intervals such that*

$$\sum_{I_j \subset I} |I_j|/|I| \leq K \tag{3}$$

for every interval I, *and if* $a_j(x)$ *is adapted to* I_j, *then*

$$\|\sum a_j\|_* \le cK.$$

The proof is not difficult. Use (c) to control $\sum_{I_j \subset I} a_j(x)$ on an interval I.

Change scale so that $A = \log 4$. Fix a dyadic interval I and suppose $\varphi_I = 0$. Let λ be an integer, $\lambda > \lambda_0(A, \varphi)$, $\lambda > 10\|\varphi\|_*$. Let $R = \{R_1, R_2, \dots\}$ consist of the maximal dyadic subintervals of I for which $\varphi_{R_K} > \lambda$. By (1),

$$\sum |R_k|/|I| < c\, 4^{-\lambda}$$

and we can select dyadic intervals $\{I_j\}$ contained in I such that (3) holds with fixed K and such that each R_k is covered by λ intervals I_j. This means there are functions $a_j(x)$ adapted to I_j so that $f_1 = \sum a_j$ satisfies

$$0 \le f_1 \le \lambda$$

$$f_1 = \lambda \text{ on } R_k$$

$$\|f_1\|_* \le C$$

and f_1 has support in $\tilde{I}$.

Repeating this process with $\varphi - \lambda$ on each R_k and continuing, we obtain $f_1, f_2, \dots$ so that

$$\varphi - (f_1 + f_2 + \cdots) < C\lambda$$

and

$$\|f_1 + f_2 + \cdots\|_* \le C.$$

Some technical difficulties involving large positive and negative values of $\varphi(x)$ must be confronted to get the full result

$$-C\lambda < \varphi - (f_1 + f_2 + \cdots) < C\lambda$$

and we refer to [7] for the rest of the story.

Adapted functions can be eliminated from the proof by using the fact that

$$\|\log M(\chi_E)\|_* \leq C$$

where M is the Hardy-Littlewood maximal function. See [4] or [3]. Setting $E = \bigcup R_k$, one can then simply take

$$f_1 = (\alpha + \beta \log M(\chi_E))^+$$

in the above argument. See [11], where this method is used to obtain Theorem 1 on spaces of homogeneous type. However, the basic difficulties concerning changes of sign remain the same, and adapted functions do provide certain conceptual advantages.

The construction is similar to Carleson's proof [1] that

(4) $$\varphi(t) = \int \frac{1}{y} K\left(\frac{x-t}{y}\right) d\mu(x,y) + \text{bounded}$$

where K is a kernel like the Poisson kernel and where μ is a Carleson measure:

(5) $$|\mu|(I \times (0,|I|)) \leq M|I|.$$

Condition (3) is suggestive of Carleson measures and the proof of Theorem 1 amounts to obtaining (4) with the constant M in (5) as small as possible.

§3. To state the second theorem we keep $d = 1$, although the result holds for $d > 1$. Let $u(x,y)$ be the Poisson integral of a BMO function φ. Then $y|\nabla u|^2 dxdy$ is a Carleson measure

(6) $$\iint_{I\times(0,|I|]} y|\nabla u|^2 dxdy \leq C|I| .$$

For some applications it would be more desirable if $|\nabla u| dxdy$ were a Carleson measure. That is unfortunately false [12]. As a compromise we have

THEOREM 2: If $\varphi \in$ BMO and if $\varepsilon > 0$, then there is $\psi(x,y) \in C^\infty(y > 0)$ such that

(7) $$|u(x,y) - \psi(x,y)| < \varepsilon$$

<u>and</u>

$$(8)\qquad \iint\limits_{I\times(0,|I|]} |\nabla u|\,dxdy \leq (C(\varepsilon,\|\varphi\|_*))\,|I|$$

<u>for all intervals</u> I.

Here is an application of Theorem 2 not an obvious direct consequence of (6). Fix $h > 0$ and let $N_\varepsilon^h(x)$ be the number of ε-oscillations of $u(x,y)$ on the segment $(0,h]$. That is, $N_\varepsilon^h u(x) > n$ if there are $0 < y_0 < y_1 < \cdots < y_n \leq h$ such that

$$|u(x,y_{j+1}) - u(x,y_j)| \geq \varepsilon.$$

By Fatou's theorem $N_\varepsilon^h(x) < \infty$ almost everywhere. By Theorem 2,

$$\frac{1}{h}\int_{x_0}^{x_0+h} N_\varepsilon^h u(x)\,dx \leq C'(\varepsilon,\|\varphi\|_*)$$

for every x_0.

Theorem 2 was suggested by work of Varopoulos [13], [14]. The construction resembles that used in the Corona Theorem, which in turn can be derived from Theorem 2. The function $\psi(x,y)$ cannot in general be harmonic.

We briefly describe the proof. By (6) we can take $\psi = u$ when $y|\nabla u|$ is large. On the place where $y|\nabla u|$ is small it is possible to compare $u(x,y)$ to a dyadic martingale, by replacing $u(x,y)$ with its average over certain horizontal slits. A simple stopping time procedure then yields a piecewise constant solution of (7) and (8) on $\{y|\nabla u|$ small$\}$. It is not difficult to mollify this solution into a C^∞ function. The details will appear elsewhere.

University of California, Los Angeles
University de Paris-Sed, Orsay

REFERENCES

1. L. Carleson, "Two Remarks on H^1 and BMO", Advances in Math. 22(1976) 269-277.

2. R. R. Coifman and C. Fefferman, "Weighted Norm Inequalities for Maximal Functions and Singular Integrals," Studia Math. 51(1974) 241-250.

3. R. R. Coifman and R. Rochberg, "Another Characterization of BMO," preprint.

4. A. Cordoba and C. Fefferman, "Weighted Norm Inequalities for the Hilbert Transform," Studia Math. 62, (1976) pp. 97-101.

5. C. Fefferman, "Recent Progress in Classical Fourier Analysis," Proc. Int. Cong. Math. Vancouver, 1974, I, pp. 95-118.

6. C. Fefferman and E. M. Stein, "H^p Spaces of Several Variables," Acta Math. 129 (1972) pp. 137-193.

7. J. B. Garnett and P. W. Jones, "The Distance in BMO to L^∞" Annals of Math. (to appear).

8. H. Helson and G. Szegö, "A Problem in Prediction Theory", Ann. Math. Pure. Appl. 51 (1960) pp. 107-138.

9. R. A. Hunt, B. Muckenhoupt and R. L. Wheeden, "Weighted Norm Inequalities for the Conjugate Function and Hilbert Transform," Trans. Amer. Math. Soc. 176 (1973), pp. 227-251.

10. F. John and L. Nirenberg, "On Functions of Bounded Norm Oscillation," Comm. Pure Appl. Math., 14 (1961), pp. 415-426.

11. P. W. Jones, "Constructions with Functions of Bounded Mean Oscillation," Thesis UCLA 1978.

12. W. Rudin, "The Radial Variation of Analytic Functions," Duke Math. J. 22 (1955) pp. 235-242.

13. N. Th. Varopoulos, "A Remark on BMO and Bounded Harmonic Functions," Pac. Jour. Math. 72 (1978).

14. N. Th. Varopoulos, "BMO Functions and the $\bar{\partial}$-equation," Pac. Jour. Math. 71 (1977) pp. 221-273.

Proceedings of Symposia in Pure Mathematics
Volume XXXV, Part 1, 1979

INVARIANT SUBSPACES AND SUBNORMAL OPERATORS

James E. Brennan

1. INTRODUCTION. It has long been asked if every bounded linear operator T on an infinite dimensional Hilbert space H has a nontrivial closed invariant subspace. If T is normal the answer is yes and it has been known for many years. But, until recently, the question remained open for operators obtained by restricting a normal operator to one of its closed invariant subspaces. Such operators are called *subnormal* and were first studied by Halmos in [10]. Since then they have become the subject of considerable investigation.

In order to study the invariant subspace problem it can be assumed that T has a cyclic vector, i.e., for some $x \in H$ the linear span of $x, Tx, T^2x, \ldots$ is dense in H. If, in addition, T is subnormal it can be shown that there is a positive measure μ carried on the spectrum of T such that T is unitarily equivalent to multiplication by the complex identity function Z in $H^2(d\mu)$, where $H^2(d\mu)$ is the closure of the polynomials in $L^2(d\mu)$. Furthermore, if T is normal then $H^2(d\mu) = L^2(d\mu)$. This suggests that the methods of classical function theory can be applied to the investigation of subnormal operators. Bram [2] and Wermer [15] were among the first to actually do so and since then the idea has been used repeatedly. Wermer, in particular, showed that if T is subnormal and if $\operatorname{spec}(T)$ has two-dimensional Lebesgue measure zero then T has a nontrivial invariant subspace. His argument was based on an old result of Hartogs and Rosenthal [11] concerning the uniform closure of the rational functions on certain compact subsets of the plane.

The purpose of this note is to describe some new results in this area and to indicate some problems that have not yet been solved.

2. POINT EVALUATIONS AND INVARIANT SUBSPACES. There are a number of ways in which one might try to establish the existence of invariant subspaces for a subnormal operator. As a rule, however, these subspaces generally arise in connection with the zero sets of functions in $H^2(d\mu)$. For example, if $H^2(d\mu) = L^2(d\mu)$ and if X is any subset of the support of μ with $0 < \mu(X) < ||\mu||$ then $S = \{f \in H^2(d\mu): f = 0 \text{ a.e. on } X\}$ is a nontrivial closed subspace invariant under multiplication by Z. On the other hand, if

$H^2(d\mu) \neq L^2(d\mu)$ it often happens that there is a point ξ such that the map $Q \to Q(\xi)$ can be extended from the polynomials to a bounded linear functional on $H^2(d\mu)$, i.e.,

$$|Q(\xi)| \leq K||Q||_{L^2(d\mu)}$$

for every polynomial Q and some absolute constant K. Such a point ξ is called a *bounded point evaluation* for $H^2(d\mu)$ or, more precisely, $H^2(d\mu)$ is said to have a bounded point evaluation at ξ. In this case if we let S be the closure in $H^2(d\mu)$ of the set of polynomials vanishing at ξ we again obtain a nontrivial invariant subspace, since $(Z - \xi) \in S$ and $1 \notin S$.

Therefore, if we could prove that either $H^2(d\mu)$ has a bounded point evaluation or $H^2(d\mu) = L^2(d\mu)$ the existence of invariant subspaces for subnormal operators would follow. Although that program has been carried out for a wide class of measures (cf. [3]), it is still not known if the dichotomy persists in general. Recently, however, Brown [4] was able to answer the invariant subspace question by considering a third alternative. In particular, he proved the following theorem:

THEOREM 1. *If* $H^2(d\mu) \neq L^2(d\mu)$ *and if* $H^2(d\mu)$ *has no bounded point evaluations then there exists a point* ξ *and two functions* $f, g \in H^2(d\mu)$ *so that*

$$Q(\xi) = \int Qf\bar{g}\, d\mu \tag{2.1}$$

for every polynomial Q.

The proof makes use of Sarason's results [14] on the weak-* density of polynomials. By letting S be the closed linear span of $(Z - \xi)f$, $(Z - \xi)^2 f, \ldots$ we obtain a subspace which is evidently invariant under multiplication by Z and, whenever (2.1) is satisfied, it is nontrivial.

The question remains, however, as to whether the hypotheses of Theorem 1 can ever be realized. That is, are there any measures μ for which $H^2(d\mu)$ has no bounded point evaluations and $H^2(d\mu) \neq L^2(d\mu)$? The next theorem is based on results of the author [3] and unpublished work of S. V. Hruščev and it answers this question for most measures which are absolutely continuous with respect to area. Since the proof is rooted in ideas from what is generally regarded as harmonic analysis we shall give a brief sketch of the details.

THEOREM 2. *Let* $\omega > 0$ *have compact support and let* dA *be two-dimensional Lebesgue measure. If*

$$\int \omega(\log^+ \omega)^2\, dA < \infty \tag{2.2}$$

then either $H^2(\omega dA)$ *has a bounded point evaluation or* $H^2(\omega dA) = L^2(\omega dA)$.

PROOF. Let us suppose that $H^2(\omega dA)$ has no bounded point evaluations and let g be any function in $L^2(\omega dA)$ which is orthogonal to the polynomials, i.e., $\int Qg\omega\, dA = 0$ for every polynomial Q. In order to prove that $H^2(\omega dA) = L^2(\omega dA)$ it is sufficient to show that $k = g\omega = 0$ a.e. We shall accomplish this by proving that the Cauchy transform

$$\hat{k}(Z) = \int \frac{k(\xi)}{\xi - Z}\, dA_\xi$$

vanishes almost everywhere and, because $\frac{\partial}{\partial \bar{Z}}(\hat{k}) = -\pi k$, the desired result will follow.

For each $\lambda > 0$ let $E_\lambda = \{Z: |\hat{k}(Z)| < \lambda\}$. Under the assumption that

$$\int \frac{|k(Z)|\log^+|k(Z)|}{|Z - \xi_o|}\, dA_Z < \infty \tag{2.3}$$

we shall construct a family of probability measures ν_δ in such a way that

(i) ν_δ is concentrated on $E_\lambda \cap \{|Z - \xi_o| < \delta\}$;

(ii) $\lim_{\delta\to 0} \int \hat{k}\, d\nu_\delta = \hat{k}(\xi_o)$.

Since each ν_δ is carried by the set where $|\hat{k}| < \lambda$ it follows that $|\hat{k}(\xi_o)| \le \lambda$ and since λ is arbitrary we conclude that $\hat{k}(\xi_o) = 0$. On the other hand, by virtue of (2.2) and our choice of g, $\int |k| \log^+ |k|\, dA < \infty$ and consequently (2.3) is satisfied a.e. Hence $\hat{k} = 0$ a.e.

The idea for constructing the measures ν_δ goes back to Carleson [6] and has been used by the author on several occasions (cf. [3] and [12]). The first step is to verify that, for any point ξ_o, almost every circle $|Z - \xi_o| = r$ meets E_λ in a set of positive linear measure. It is here that our assumption concerning the absence of bounded point evaluations is used. If we suppose that $|\hat{k}|$ is bounded below on $|Z - \xi_o| = r$ for a set of r's of positive measure it can be shown that

$$|Q(\xi_o)| \le K \int |Q|\,|g|\,\omega\, dA$$

for every polynomial Q and some fixed numerical constant K, which is a contradiction. The specifics of the argument can be found in [3].

For convenience we may assume that $\xi_o = 0$. As an additional aid let χ denote the characteristic function of E_λ and let $\ell(r)$ be the linear measure of $E_\lambda \cap (|Z| = r)$. Setting

$$\sigma(B) = \int_B \frac{\chi(Z)}{\ell(|Z|)}\, dA$$

for every Lebesgue measurable set B, we obtain a measure on E_λ and $\sigma(|Z| < \delta) = \delta$. We define ν_δ to be the restriction of $(1/\delta)\sigma$ to the disk $|Z| < \delta$.

By interchanging the order of integration,

$$\int \hat{k}(\zeta)\, d\nu_\delta(\zeta) = \int_{|Z|<2\delta} + \int_{|Z|\geq 2\delta} \left(\int \frac{d\nu_\delta(\zeta)}{Z - \zeta} \right) k(Z)\, dA_Z$$

and it is an easy matter to check that

$$\lim_{\delta\to 0} \int_{|Z|\geq 2\delta} \left(\int \frac{d\nu_\delta(\zeta)}{Z - \zeta} \right) k(Z)\, dA_Z = \hat{k}(0).$$

The difficulty in verifying property (ii) comes in proving that the integral over $|Z| < 2\delta$ tends to zero as $\delta \to 0$.

The argument that we shall present is due to S. V. Hruščev. In order to simplify the exposition we shall use the notation

$$U_\delta(Z) = \int \frac{d\nu_\delta(\zeta)}{|Z - \zeta|} .$$

The crucial step is to obtain an estimate for $m(U_\delta > y)$, the measure of the set where $U_\delta > y$. In particular, it follows from a well-known theorem of J. E. Littlewood (cf. [8, p.368]) that

$$m(U_\delta > y) \leq \frac{||U_\delta||_q^q}{y^q} \leq K \left(\sup_Z \int \frac{d\nu_\delta(\zeta)}{|Z - \zeta|^{2-p}} \right)^{p-1} \cdot y^{-q},$$

where p, q are conjugate indices and the constant K remains bounded as $p \to 1$. From this it can easily be deduced that

$$m(U_\delta > y) \leq \text{const. } \delta^2 e^{-Ky\delta} \tag{2.4}$$

provided $y > 0$ and $y\delta > 1$. Here again the constants are absolute. This is precisely what is needed to complete the proof of the theorem.

It remains only to verify that

$$\lim_{\delta\to 0} \int_{|Z|<2\delta} U_\delta |k|\, dA = 0$$

To this end we decompose $|Z| < 2\delta$ into the union of two disjoint sets X and Y as follows:

(iii) $Y = \{U_\delta \leq 1/\delta\} \cap \{|Z| < 2\delta\}$

(iv) $X = \{U_\delta > 1/\delta\} \cap \{|Z| < 2\delta\}$.

Integrating over Y we find that

$$\int_Y U_\delta |k|\, dA \leq \int_{|Z|<2\delta} \frac{|k|}{\delta}\, dA \leq \int_{|Z|<2\delta} \frac{|k|}{|Z|}\, dA$$

and, since (2.3) is satisfied with $\xi_o = 0$, the right hand integral tends to zero as $\delta \to 0$. To see what happens on the complementary set X we first fix a constant $a > 0$, which we shall specify more precisely in a moment. On the sets $\{U_\delta \leq a \log^+|k|\}$ and $\{U_\delta > a \log^+|k|\}$ we have respectively

(v) $U_\delta|k| \le a|k|\log^+|k|$

(vi) $U_\delta|k| < U_\delta e^{(1/a)U_\delta}$

and consequently

$$\int_X U_\delta|k|\,dA \le a\int_X |k|\log^+|k|\,dA + \int_X U_\delta e^{(1/a)U_\delta}\,dA.$$

If we now let $X_n = \{Z \in X: n/\delta < U_\delta(Z) \le (n+1)/\delta\}$ then $X = \cup_{n=1}^{\infty} X_n$ and from (2.4) it follows that $m(X_n) \le C\delta^2 e^{-Kn}$. Hence,

$$\int_X U_\delta e^{(1/a)U_\delta}\,dA \le \text{const.} \sum_{n\ge 2} e^{[(1/a\delta) - K]n}$$

and if $1/a\delta = K/2$ the right side is $O(\delta)$. With the constant a chosen in this way we therefore have

$$\int_X U_\delta|k|\,dA \le \frac{K}{2\delta}\int_{|Z|<2\delta} |k|\log^+|k|\,dA + O(\delta)$$

and according to (2.3) the last integral tends to zero as $\delta \to 0$. QED

REMARK. In [3] the author has given a somewhat different proof that $\hat{k} = 0$ a.e., assuming that almost every circle $|Z - \xi_o| = r$ meets E_λ in a set of positive linear measure. The argument there makes use of the coarea formula of Federer [9] and the theory of sets of finite perimeter introduced by DeGiorgi in [7]. The key point is that $\nabla\hat{k} \in L^1$, a fact guaranteed by the Calderón-Zygmund theory of singular integrals [5], since $\int|k|\log^+|k|\,dA < \infty$.

3. POINT EVALUATIONS AND ANALYTIC CONTINUATION. Under the assumptions of Theorem 2 we have shown that if $H^2(\omega dA) \ne L^2(\omega dA)$ then $H^2(\omega dA)$ has a bounded point evaluation at some point ξ_o which, incidentally, we can take to be in the support of ω. By a further appeal to the ideas of Carleson used in that endeavor it can also be shown that there is a neighborhood U of ξ_o and a fixed constant K so that

$$|Q(\xi)| \le K||Q||_{L^2(\omega dA)}$$

for every polynomial Q and every $\xi \in U$. Thus, we have the following generalization of Theorem 2.

THEOREM 3. *Let $\omega \ge 0$ have compact support and suppose that $\int\omega(\log^+\omega)^2\,dA < \infty$. If $H^2(\omega dA) \ne L^2(\omega dA)$ then there exists a point ξ_o in the support of ω and a neighborhood U of ξ_o such that every $f \in H^2(\omega dA)$ extends analytically to U.*

The continuation phenomenon has been studied for many years in connection with the completeness problem for domains. Here one is given a simply connected domain Ω and a weight $\omega > 0$ on Ω. In accordance with previous notation we let $H^2(\Omega, \omega dA)$ stand for the closure of the polynomials in $L^2(\Omega, \omega dA)$ and we let $L^2_a(\Omega, \omega dA)$ denote the L^2 functions which are analytic in Ω. Because $\omega > 0$, $H^2 \subset L^2_a$. The problem is to decide when $H^2 = L^2_a$.

For weights ω which depend only on Green's function near the boundary and for which $\omega \in L^\infty(\Omega, dA)$ it is now possible to give a characterization of completeness in terms of analytic continuation. Based on the discussion of the preceding section, it can be shown that $H^2(\Omega, \omega dA) \neq L^2_a(\Omega, \omega dA)$ if and only if every $f \in H^2(\Omega, \omega dA)$ extends analytically across a fixed piece of $\partial\Omega$ (cf. [3]). In case $\omega \equiv 1$ this was conjectured by Mergeljan [13] in 1955 and under various other stringent hypotheses the continuation phenomenon has been observed by Keldyš, Šaginjan, Havin, Beurling and others. For additional information and background material see [3] and [12] and the articles cited there.

4. SOME OPEN QUESTIONS.

(1) Given an arbitrary measure μ is it possible to give a complete characterization of the invariant subspaces in $H^2(d\mu)$ as was done by Beurling [1] for the classical Hardy space $H^2(d\theta)$?

(2) If $H^2(d\mu)$ has no bounded point evaluation must $H^2(d\mu) = L^2(d\mu)$?

(3) Can $H^2(d\mu)$ have just one bounded point evaluation?

(4) If $H^2(d\mu) \neq L^2(d\mu)$ is this accompanied by analytic continuation?

(5) What are the answers to these questions if $d\mu = \omega dA$ with $\omega \in L^1$?

REFERENCES

[1] A. Beurling, On two problems concerning linear transformations in Hilbert space, Acta Math. 81(1948), 239-255.

[2] J. Bram, Subnormal operators, Duke Math Journal. 22(1955), 75-94.

[3] J. Brennan, Point evaluations, invariant subspaces and approximation in the mean by polynomials, J. Functional Analysis (to appear).

[4] S. Brown, Some invariant subspaces for subnormal operators, Thesis, Univ.of California, Santa Barbara (1978).

[5] A. P. Calderón and A. Zygmund, On the existence of certain singular integrals, Acta Math. 88(1952), 85-139.

[6] L. Carleson, Mergeljan's theorem on uniform polynomial approximation, Math. Scand. 15(1965), 167-175.

[7] E. DeGiorgi, Su una teoria generale della misura (r - 1)-dimensionale in uno spazio ad r dimensioni, Ann. Mat. Pura Appl. 36(1954), 191-213.

[8] J. Deny, Sur la convergence de certaines intégrales de la théorie du potentiel, Arch. Math. 5(1954), 367-370.

[9] H. Federer, Curvature measures, Trans. Amer. Math. Soc. 93(1959), 418-491.

[10] P. Halmos, Normal dilations and extensions of operators, Summa Brasil. Math., fasc. 9, vol. 2(1950).

[11] F. Hartogs and A. Rosenthal, Über Folgen analytischer Funktionen, Math. Ann. 104(1931), 606-610.

[12] M. S. Melnikov and S. O. Sinanjan, Questions in the theory of approximation of functions of one complex variable, Contemporary Problems of Mathematics, vol. 4, Itogi Nauki i Tekhniki, VINITI, Moscow(1975), 143-245; J. Soviet Math. 5(1976), 688-752.

[13] S. N. Mergeljan, General metric criteria for the completeness of systems of polynomials, Dokl. Akad. Nauk SSSR 105(1955), 901-904.

[14] D. Sarason, Weak-star density of polynomials, J. Reine Angew. Math. 252 (1972), 1-15.

[15] J. Wermer, Report on subnormal operators, Report of an International Conference on Operator Theory and Group Representations, New York (1955).

DEPARTMENT OF MATHEMATICS
UNIVERSITY OF KENTUCKY
LEXINGTON, KENTUCKY 40506

CHAPTER 3

Harmonic functions, potential theory and theory of functions of one complex variable

Proceedings of Symposia in Pure Mathematics
Volume XXXV, Part 1, 1979

HARMONIC FUNCTIONS IN LIPSCHITZ DOMAINS

Björn E.J. Dahlberg

1. NONTANGENTIAL LIMITS. If u is positive and harmonic in a sufficiently smooth domain then it is a classical fact, which goes back to Fatou that u has nontangential limits outside a set of vanishing surface area. This result can be extended in several ways and we will in this note describe some of the results that extend Fatou's theorem to more general domains. In this connection Lipschitz domains will appear naturally which depends on the following geometrical fact.

LEMMA 1. Suppose Γ is a cone with vertex at the origin. Then there is a $\delta > 0$ which only depends on Γ such that for any set $E \subset R^n$ with diameter less than δ there is a halfspace $H = \{x \in R^n: \langle x-x_0, e\rangle > 0\}$ such that $\bar{E} \subset H$ and $H \cap (\bigcup_{P \in E} \Gamma + P)$ is a Lipschitz domain.

We remark that with a cone we mean a circular, possibly truncated cone which is convex and open.

A function $\varphi: R^n \to R$ such that

$$(1.1) \qquad |\varphi(x) - \varphi(x')| \leq M|x-x'|$$

is called a <u>Lipschitz function</u>. A bounded domain $D \subset R^n$ is called a <u>Lipschitz domain</u> if ∂D can be covered by finitely many right circular cylinders L whose bases are at a positive distance from ∂D such that to each cylinder L there is a Lipschitz function $\varphi: R^{n-1} \to R$ and a coordinatesystem (x,y), $x \in R^{n-1}$, $y \in R$ such that the y-axis is parallell to the axis of symmetry of L and $L \cap D = L \cap \{(x,y): y > \varphi(x)\}$ and $L \cap \partial D = L \cap \{(x,y): y = \varphi(x)\}$.

If the functions φ all can be chosen to be continuously differentiable then D is called a C^1-domain. If in addition the gradients $\nabla\varphi$ all satisfy $|\nabla\varphi(x) - \nabla\varphi(x')| \leq C|x-x'|^\alpha$ then D is called a $C^{1,\alpha}$-domain.

A cone Γ with vertex at a point $P \in \partial D$, where D is a domain, is called a <u>nontangential cone</u> if there is a cone Γ' such that

(1.2) $\bar{\Gamma} - \{P\} \subset \Gamma' \subset D$.

Suppose now that u is a function in a domain D such that for all $P \in E$, where $E \subset \partial D$, there is a nontangential cone $\gamma(P)$ with vertex at P such that $u(Q)$ does not have a limit as $Q \to P$, $Q \in \gamma(P)$. We now make the following observation. Let Σ be a countable dense subset of $S = \{P \in R^n : |P| = 1\}$ and let $\mathcal{F}$ be the countable family of cones with vertex at 0 which have their axis of symmetry determined by a point in Σ and both the height and opening angle are rational numbers. To each $P \in E$ there is now a cone $\Gamma_p \in \mathcal{F}$ such that $\overline{\gamma(P)} - \{P\} \subset \Gamma(P) \subset D$, where $\Gamma(P) = \Gamma_p + P$. We next decompose E into an at most countable family of subsets $\{F\}$ such that all the cones Γ_p, $P \in F$ correspond to the same cone in $\mathcal{F}$. From Lemma 1 follows that each F can be decomposed into an at most countable family of subsets $\{G\}$ such that to each G there is a halfspace H such that $\bar{G} \subset H$ and $H \cap \bigcup_{P \in G} [\Gamma(P)]$ is a Lipschitz domain. If now the behaviour of u in Lipschitz domains were known then we would be able to control the size of the (hopefully) exceptional set E.

This program can be carried out for the case when a) u is positive and b) u is a Cauchyintegral $\int \frac{d\mu(t)}{z-t}$ (Calderon [3]).

We recall that a function u is said to have the nontangential limit L at a point $P \in \partial D$ if

$$u(Q) \to L,\ Q \to P \text{ and } Q \in \Gamma$$

for all nontangential cones Γ with vertex at P.

Hunt and Wheeden [17] proved that positive harmonic functions in Lipschitz domains had nontangential limits outside an exceptional set of vanishing harmonic measure. This result was an extension to R^n of the classical fact (Priwalow [20]) that positive harmonic functions in domains bounded by rectifiable Jordan curves had nontangential limits outside a set of vanishing 1-dimensional Hausdorff measure. We recall that harmonic measure is defined in the following way. For f continuous on ∂D let Hf be the solution of the Dirichlet problem with boundary values f. If $P \in D$ then it follows from the maximum principle that $f \to Hf(P)$ is a positive linear functional of norm 1 on $C(\partial D)$. Hence it can be represented by a probability measure ω_p, the harmonic measure evaluated at P, and we have

$$(1.2) \qquad Hf(P) = \int_{\partial D} f(Q)\omega_p(dQ) .$$

We remark that from Harnack's inequality it follows that the measures ω_p are mutually absolutely continuous and consequently we write $\omega(E) = 0$ if

$\omega_p(E) = 0$ for some P. For more information about harmonic measures we refer to Helms [16].

We denote by σ the surface measure of ∂D if D is a Lipschitz domain.

In Dahlberg [7] it was proved that if $D \subset R^n$ is a Lipschitz domain and $E \subset \partial D$ then $\omega(E) = 0$ if and only if $\sigma(E) = 0$.

If we combine these results we obtain:

THEOREM 1. Suppose $D \subset R^n$ is a domain and $E \subset \partial D$ is such that to each $P \in E$ there is a cone $\gamma(P)$ with vertex at P such that $\gamma(P) \subset R^n - D$. Let u be harmonic in D and suppose that to each $P \in E$ there is a cone $\Gamma(P)$ with vertex at P such that $\Gamma(P) \subset D$ and $\inf u(Q)$: $Q \in \Gamma(P)\} > -\infty$. Then u has a finite nontangential limit at all points $P \in E$ except for a set of vanishing (n-1)-dimensional Hausdorff-measure.

PROOF. By considering the sets $E_j = \{P \in E: \inf_{\Gamma(P)} u > j\}$, $j \in Z$, separately we see that it is sufficient to treat the case when $u > 0$ on $\bigcup_{P\in E} \Gamma(P)$. Using the geometric construction we indicated above we see that there is no loss in generality in assuming that there is a Lipschitz domain $D_1 \subset R^n - D$ such that $E \subset \partial D_1$. By repeating the same construction from the inside we see that we may also assume that there is a Lipschitz domain $D_2 \subset D$ such that $E \subset \partial D_2$. We now recall that Lipschitz functions are differentiable almost everywhere (see Stein [21]) which means that all Lipschitz domains have tangent planes a.e. on the boundary. Also if E_1 denotes the set of all $P \in E$ for which the tangent planes of ∂D_1 and ∂D_2 coinside then $\sigma(E - E_1) = 0$. This means that if $P \in E_1$ and γ is a cone with vertex at P and nontangential with respect to D then γ is nontangential with respect to D_2 provided the diameter of γ is sufficiently small. Hence the theorem follows from the results in Hunt and Wheeden [17] and Dahlberg [7].

Using this type of reasoning one has similar results for Cauchy integrals, see Havin [15] and Zygmund [26].

When D is a halfspace the result of Theorem 1 is contained in Carleson [4] and his result is an extension of Calderon [2], where the case of the two-sided boundedness was settled. A treatment of the situation for $n = 2$ can be found in Priwalow [20] and Zygmund [25].

2. HARMONIC MEASURE. We will in this section indicate some properties of the harmonic measure for Lipschitz domains.

The Green function G of a domain $D \subset R^n$ is given by $G(P,Q) = c_n|P-Q|^{2-n} - h_Q(P)$, where h is chosen so $\Delta_p G(P,Q) = -\delta_Q$, δ_Q is the point measure at Q and $G(P,Q) = 0$ for $P \in \partial D$. (If $n = 2$ then $|P-Q|^{2-n}$ is replaced by $-\log|P-Q|$.)

If D is a $C^{1,\alpha}$ domain, $0 < \alpha < 1$, then it is well known that harmonic measure and surface measure are equivalent and $d\omega_p(Q) = (\partial/\partial n_Q)G(P,Q)$, where $\partial/\partial n$ denotes differentiation in the direction of the unit inward normal. If k_p denotes the density of the harmonic measure (i.e. $k_p(Q) = (\partial/\partial n_Q)G(P,Q)$ then $|k_p(Q) - k_p(Q')| \leq C|P-Q|^\alpha$ (see Widman [24]). However, no regularity results of this sort can be expected for the case of a general Lipschitz domain. For, let $D_\beta = \{z \in C: |\arg z| < \beta/2\}$, where $0 < \beta < 2\pi$. Then $f(z) = z^{\pi/\beta}$ maps D_β onto D_π, from which it follows that the density of the harmonic measure of D_β is comparable to $|z|^{-1+\pi/\beta}$ as $z \to 0$. Letting $\beta \to 2\pi$ we see that we can't expect the density of the harmonic measure to be better than $L^{2+\varepsilon}(\sigma)$ for some $\varepsilon > 0$.

THEOREM 2. Let $D \subset R^n$ be a Lipschitz domain. Then surface measure and harmonic measure are equivalent and if $P \in D$ then $d\omega_p = k d\sigma$ for some $k \in L^{2+\varepsilon}(\sigma)$, $\varepsilon = \varepsilon(P) > 0$. Also there is a constant $C = C(P,D)$ such that

$$(2.1)\qquad \left(\frac{1}{\sigma(A)}\int_A k^2 d\sigma\right)^{1/2} \leq C\,\frac{1}{\sigma(A)}\int_A k\,d\sigma$$

for all sets A of the form $\{Q \in \partial D: |P_0-Q| < r\}$ for some $P_0 \in \partial D$.

If the condition (2.1) holds with the exponent 2 replaced by a $q \in (1,\infty)$ we say that k satisfies a B_q-condition.

One very interesting example of functions satisfying a B_q-condition are Jacobians of quasiconformal mappings, see Gehring [13]. In Coifman and Fefferman [5] and Gehring [13] it is shown that if k satisfies a B_q-condition then k also satisfies a $B_{q+\varepsilon}$-condition for some $\varepsilon > 0$. Therefore (2.1) gives that $k \in L^{2+\varepsilon}(\sigma)$ for some $\varepsilon > 0$.

A proof of Theorem 2 can be found in Dahlberg [7].

We remark that if D is assumed to be C^1, then k satisfies condition B_q for all $q < \infty$.

One important consequence of (2.1), which we shall now outline is that we can solve the Dirichlet problem for D with boundary values in $L^p(\sigma)$, $p \geq 2$.

Suppose now that D is a Lipschitz domain which is starshaped with respect to a ball $B \subset D$ (i.e. whenever $P \in D$ and $Q \in B$ then the line segment between P and Q is contained in D). We remark that if $\delta > 0$ is chosen sufficiently small then the halfspace H can be chosen so that the domain constructed in Lemma 1 has this property.

It now follows from Hunt-Wheeden [17] that for $P \in \partial D$ we have

(2.2) $$\sup_{\Gamma(P)} |Hf(Q)| \leq Cmf(P),$$

where $\Gamma(P)$ is the set of all $Q \in D$ which are on a line through P and B, $mf(P) = \sup_{P \in A} (\frac{1}{\omega(A)} \int_A f d\omega)$, A as in Theorem 2 and ω is the harmonic measure of D evaluated at the center of B. If $d\omega = kd\sigma$ and k satisfies B_q then Hölder's inequality then $(\frac{1}{p}+\frac{1}{q} = 1)$

$$\frac{1}{\omega(A)} \int_A |f| d\omega \leq (\frac{1}{\sigma(A)} \int_A |f|^p d\sigma)^{1/p} (\frac{1}{\sigma(A)} \int_A k^q d\sigma)^{1/q} (\frac{\sigma(A)}{\omega(A)})$$
$$\leq B_q (\frac{1}{\sigma(A)} \int_A |f|^p d\sigma)^{1/p}$$

which yields

(2.3) $$mf(P) \leq C(M|f|^p)^{1/p},$$

where $Mh(P) = \sup_{P \in A} \frac{1}{\sigma(A)} \int_A |h| d\sigma$ is the Hardy-Littlewood maximal function. Since M is of weak type $(1,1)$ it follows that m is weak type (p,p). Since k satisfies the $B_{2+\varepsilon}$-condition for some $\varepsilon > 0$ it follows that mf is weak type $(2-\delta, 2-\delta)$ for some $\delta \in (0,1)$. By Marcinkiewicz interpolation theorem it follows that if $2 \leq p \leq \infty$ then

(2.4) $$\|mf\|_{L^p(\sigma)} \leq C_p \|f\|_{L^p(\sigma)} .$$

Suppose now D is a Lipschitz domain and to each $P \in \partial D$ there is associated a cone $\Gamma(P)$ with vertex at P. We say that $\{\Gamma(P)\}$ is a <u>regular family of cones</u> if we can partition E into finitely many subsets $\{E\}$ such that to each E there are cones Γ_i, $i = 1,2,3$, with vertex at 0 such that for each $P \in E$ we have

$$\Gamma_1 + P \subset \Gamma(P) \subset \gamma + P \subset \Gamma_3 + P \subset D,$$

where $\gamma = \bar{\Gamma}_2 - \{0\}$.

Using (2,2), (2,4) and a patching argument the following result can be shown for a general Lipschitz domain (Dahlberg [8]).

THEOREM 3. Suppose $\{\Gamma(P)\}$ is a regular family of cones. If $2 \leq p \leq \infty$ and $f \in L^p(\sigma)$ then $f \in L^1(\omega)$ and

(2.5) $$\|f^*\|_{L^p(\sigma)} \leq C_p \|f\|_{L^p(\sigma)},$$

where $f^*(P) = \sup\{|Hf(Q)|: Q \in \Gamma(P)\}$.

We remark if D is assumed to be C^1 then it follows from (2.3) and the fact that k satisfies condition B_q for all $q < \infty$ that (2.5) holds for $1 < p \leq \infty$. A new proof of the C^1-result has recently been given by Fabes et al [11] using Calderons result [3] that the Cauchy integral preserves $L^p(\sigma)$, $1 < p < \infty$ and integral equation techniques.

We define the area integral of u as

$$Au(P) = \Big(\int_{\Gamma(P)} |\nabla u(Q)|^2 d(Q)^{2-n} dm(Q)\Big)^{1/2},$$

where ∇u is the gradient of u, $d(Q) = \mathrm{dist}\{Q,\partial D\}$ and dm denotes integration with respect to the Lebesgue measure. We define the nontangential maximal function of u as

$$Nu(P) = \sup\{|u(Q)|: Q \in \Gamma(P)\}.$$

Let Φ be continuous, nondecreasing and satisfy $\Phi(0) = 0$ and $\Phi'(2t) \leq$ $\leq C\phi'(t)$. In the case when D is a halfspace and μ is an A_∞-measure on ∂D then for suitably normalized harmonic functions u it holds that

$$\int_{\partial D} \Phi(A(u))d\mu \sim \int_{\partial D} \phi(N(u))d\mu,$$

Fefferman and Stein [12], Burkholder and Gundy [1], Gundy and Wheeden [14].

For Lipschitz domains the analoguous result is (Dahlberg [9]):

THEOREM 4. Let $D \subset R^n$ be a Lipschitz domain and $P_0 \in D$ a fixed point. Assume that $\{\Gamma(P)\}$ is a regular family of cones. If μ is a measure on ∂D which satisfies the A_∞-condition then for all harmonic functions u in D which vanish at P_0

$$\int_{\partial D} \Phi(A)d\mu \sim \int_{\partial D} \phi(N)d\mu,$$

where $A = A(u)$ and $N = N(u)$.

It is easily seen that if u is harmonic in D and $u(P_0) = 0$ then

$$\int_{\partial D} A^2(u)d\sigma \sim \int_{\partial D} |\nabla u|^2 d(Q)dm(Q).$$

As a consequence of Theorem 3 we therefore have that

$$\int_{\partial D} u^2 d\sigma \sim \int |\nabla u|^2 d(Q)dm(Q).$$

We shall next describe the boundary behaviour of subharmonic functions. Let $\{P_j\}_1^\infty$ be a countable dense set in a bounded domain $D \subset R^n$ and let $u(P) = -\sum_{j=1}^{\infty} 2^{-j}|P-P_j|^{2-n}$. Then u is ≤ 0 and subharmonic in D. Since $u(P_j) = -\infty$ for all j it follows that u can't have finite nontangential limits at any boundary point.

The classical result in this direction is due to Littlewood [19], namely that if u is subharmonic in a ball and u has a positive harmonic majorant then u has radial limits a.e.

If $D \subset R^n$ is a domain and $e: \partial D \to \{P: |P| = 1\} = S$ is a field of directions we say that e is nontangential if to each $P \in \partial D$ there is a cone Γ with vertex at P such that $\Gamma \subset D$ and $e(P)$ is along the axis of symmetry of Γ.

THEOREM 5. (Dahlberg [10].) Let u be subharmonic in a Lipschitz domain D. Suppose that $e: \partial D \to S$ is nontangential and is Lipschitz continuous. If u has a positive harmonic majorant then

$$\lim_{t \downarrow 0} u(P + te(P)) \text{ exists a.e. on } \partial D .$$

We remark that the Lipschitz continuity of e can't be replaced by Hölder continuity with exponent α less than 1, for counterexamples see Dahlberg [10].

3. PROBLEMS. We conclude by posing some problems.

PROBLEM 1. It is well known that (2.1) implies that $\log k$ is in BMO. Suppose that D is a bounded C^1-domain. Is $\log k$ in VMO? One reason that suggests this is that k satisfies B_q for all $q < \infty$.

PROBLEM 2. One general problem to investigate is how much of the H^p-theory of smooth domains (as presented in say Coifman and Weiss [6]) can be carried over to the case of Lipschitz domains. For the case $n = 2$ this has been studied by Kenig [18].

We would like to point out a special problem. Suppose $F = (u_1,\dots,u_n)$ is the gradient of a harmonic function and $\sup_{0<\varepsilon<1} \int_{\partial D_\varepsilon} |F| d\sigma < \infty$, where $\{D_\varepsilon\}$ is a decreasing one-parameter family of domains with parallell boundaries and $D_0 = D$ is a Lipschitz domain. If $p = \frac{n-2}{n-1}$ and $v = |F|^p$ then v is subharmonic (Stein and Weiss [22, p. 234] and has boundary values in $L^r(\sigma)$, where $r = \frac{n-1}{n-2}$. If $n = 3$ then $r = 2$ and consequently theorem 3 gives that the maximal function of v is in $L^2(\sigma)$, i.e. the maximal function of F is in

$L^1(\sigma)$. An interesting question is if this is true for $n \geq 4$.

PROBLEM 3. Let $f \in L^p(\sigma)$, where σ is the surface measure of the boundary of a bounded Lipschitz domain $D \subset R^n$. The Neumann problem is to find the harmonic function which has normal derivative f. (A necessary condition for this is that $\int_{\partial D} f d\sigma = 0$.) In Fabes et al [11] it is solved for the case when D is C^1 but it has not been solved for the general Lipschitz domain (except when $n = 2$ it can be shown to be solvable when $1 < p \leq 2+\varepsilon$, which is sharp.)

PROBLEM 4. If

$$L = \Sigma \frac{\partial}{\partial x_j} (a_{jk} \frac{\partial}{\partial x_k})$$

is a uniformly elliptic, selfadjoint operator with L^∞-coefficients then the Dirichlet problem can be solved for $f \in C(\partial\Omega)$. As in (1.2) this defines a harmonic measure but the relation between this harmonic measure and surface measure is not known in general, not even for balls. For properties of solutions of quasilinear equations in nonselfadjoint form see Widman [23].

REFERENCES.

[1] D.L. Burkholder and R.F. Gundy, "Distribution function inequalities for the area integral", Studia Math. 44 (1972), 527-544.

[2] A.P. Calderon, "On the behaviour of harmonic functions at the boundary", Trans. AMS 68 (1950), 47-54.

[3] A.P. Calderon, "Cauchy integrals on Lipschitz curves and related operators", Proc. Nat. Acad. Sci 74 (1977).

[4] L. Carleson, "On the existence of boundary values for harmonic functions of several variables", Ark. Mat. 4 (1962), 393-399.

[5] R.R. Coifman and C. Fefferman, "Weighted norm inequalities for maximal functions and singular integrals", Studia Math. 51 (1974), 24-250.

[6] R.R. Coifman and G. Weiss, "Extensions of Hardy spaces and their use in analysis", Bull. AMS 83 (1977), 569-645.

[7] B.E.J. Dahlberg, "Estimates of harmonic measure", Arch. Rat. Mech. Anal. 65 (1977), 275-288.

[8] B.E.J. Dahlberg, "On the Poisson integral for Lipschitz and C^1-domains", to appear in Studia Math.

[9] B.E.J. Dahlberg, "Weighted norm inequalities for the Lusin area integral and the nontangential maximal functions for functions harmonic in a Lipschitz domain", to appear in Studia Math.

[10] B.E.J. Dahlberg, "On the existence of radial boundary values for functions subharmonic in a Lipschitz domain", Ind. U. Math. J. 27 (1978), 515-526.

[11] E. Fabes, J. Jodeit and N. Rivière, "Potential techniques for boundary value problems on C^1-domains", preprint.

[12] C. Fefferman and E.M. Stein, "H^p-spaces of several variables", Acta Math. 129 (1972), 137-193.

[13] F.W. Gehring, "The L^p-integrability of the partial derivatives of a quasiconformal mapping, Acta Math. 130 (1973), 265-277.

[14] R.F. Gundy and R.L. Wheeden, "Weighted integral inequalities for the nontangential maximal function, Lusin area integral and Walsh-Paley series", Studia Math. 49 (1974), 107-124.

[15] V.P. Havin, "Boundary properties of integrals of Cauchy type and of conjugate harmonic functions in regions with rectifiable boundary", Mat. Sb(N.S.) 68 (110) 1965, 499-517 (Russian).

[16] L.L. Helms, "Introduction to potential theory", Wiley-Interscience, New York, 1969.

[17] R. Hunt and R. Wheeden, "On the boundary values of harmonic functions", Trans. AMS, 132 (1968), 307-322.

[18] C. Kenig, Thesis , University of Chicago, 1978.

[19] J.E. Littlewood, "On functions subharmonic in a circle (II)", Proc. London Math. Soc. (2) 28 (1928), 383-394.

[20] I.I. Priwalow, "Randeigenschaften analytischer Funktionen", Deutscher Verlag der Wissenschaften, Berlin, 1956.

[21] E.M. Stein, "Singular integrals and differentiability properties of functions", Princeton University Press, Princeton 1970.

[22] E.M. Stein and G. Weiss, "Introduction to Fourier Analysis on Euclidean spaces", Princeton University Press, Princeton, 1971.

[23] K.-O. Widman, "On the boundary behaviour of solutions to a class of elliptic partial differential equations", Ark. Mat. 6 (1967), 485-533.

[24] K.-O. Widman, "Inequalities for the Green function and boundary continuity of the gradient of solutions of elliptic equations", Math. Scand. 21 (1967), 17-37.

[25] A. Zygmund, "Trigonometric Series", 2nd ed., Cambridge Univ. Press, New York 1959.

[26] A. Zygmund, "Integrales Singulières", Springer Lecture Notes 204 (1971).

UNIVERSITY OF GÖTEBORG
DEPARTMENT OF MATHEMATICS
FACK
S-402 20 GÖTEBORG
SWEDEN

Proceedings of Symposia in Pure Mathematics
Volume XXXV, Part 1, 1979

A SURVEY OF HARMONIC FUNCTIONS ON SYMMETRIC SPACES

Adam Korányi[1]

Introduction

Riemannian symmetric spaces include $\mathbb{R}^n$ as an example. This shows that the subject indicated in the title is enormously vast; we are going to limit ourselves here to a homogeneous space point of view, which means by and large that we are interested only in harmonic functions defined on the entire space under consideration.

This still includes the ordinary theory of harmonic functions on the unit disc $U \subset \mathbb{C}$, since U is a globally symmetric space under the Poincaré metric, and the harmonic functions with respect to this metric are the same as the ordinary ones.

On our class of spaces, even on $\mathbb{R}^n$, we will consider several different notions of harmonicity, and we will also consider eigenfunctions of the Laplacian and its generalizations for eigenvalues different from zero. In fact, in certain cases we are forced to do this even in order to be able to handle the case of eigenvalue zero (cf. Example C in §1). The study of general eigenspaces is of great importance for group representation theory too; here we will be able only to allude to these questions and refer for more information to Helgason's expository article [He 6].

The types of questions we are mainly interested in include the generalization of the Poisson integral for U. (This means, first of all, finding the proper analogue of the circle for each type of space and for each notion of harmonicity.) Once a Poisson representation for some class is known, we will ask for a method of reconstructing from the function the object of which it is the Poisson integral: Most interesting in this direction are the pointwise convergence questions which have given rise to the highly interesting investigations of Marcinkiewicz and Zygmund on the polydisc [Z, Ch. XVII], and have led to a number of other important problems and results (cf. [St. 1]). We will also discuss other questions such as the relationship between weak and

AMS(MOS) subject classifications (1970). Primary 43A85, 22E30; Secondary 31B30

[1]Partially supported by the National Science Foundation under grant MCS 78-02916.

strong harmonicity and certain notions of ultrastrong harmonicity arising naturally in the study of Poisson integrals.

We shall not discuss, besides mentioning it here briefly, the problem of various equivalent characterizations of harmonic H^p-spaces where results of A. Debiard [D] on the complex ball and Gundy-Stein [G-S] on the polydisc suggest that the natural setting of the problem is on Riemannian symmetric spaces; for local analogues of these results we refer to [K-P] (case of rank one) and [M-M] where the breakthrough in the direction of higher rank was made. We shall also leave aside the results, belonging to more general settings but closely related to our subject, in [O-S] about non-Riemannian symmetric spaces and in [R] about still wider classes of groups.

In §1 we give the basic definitions and discuss a few fundamental examples. Then we discuss $\mathbb{R}^n$ in detail; in §2 we even restrict ourselves to the case of a finite group acting on $\mathbb{R}^n$, and in §3 we deal with more general compact groups acting. (Our principal notion of harmonicity depends on such a group.) In §4 finally we discuss the case of a symmetric space of non-compact type (i.e., the case of a noncompact semisimple Lie group).

Looking at the three cases in this order is more than a heuristic device. As will be seen, the nicest complete results about $\mathbb{R}^n$ hold when the group acting on it is a finite group generated by reflections, or the maximal compact subgroup of a semisimple Lie group (case of a "Cartan motion group"). It is a much used basic fact about symmetric spaces of non-compact type that to each one of them there is associated a Cartan motion group, and to this a finite group generated by reflections, the Weyl group. In studying a problem about symmetric spaces it is not only instructive, but in many cases even necessary to first solve the two analogous more elementary problems on $\mathbb{R}^n$.

In this survey a number of unsolved problems will be stated. Some of these, especially those about $\mathbb{R}^n$, may not even be very difficult. They seem not to have been studied because they arise from a point of view inspired by the semisimple theory that is relatively new.

The author is grateful to L. Flatto for some interesting discussions about the subject of §2 and to K. Gross for a careful reading of the preliminary version of this article and for a number of useful suggestions.

§1. Strongly and Weakly Harmonic Functions

We deal with several notions of harmonicity. The most important one is defined as follows.

Let X be a connected manifold on which a Lie group G acts transitively. Let o be an arbitrarily chosen base point in X and let K be its stabilizer in G; we assume that K is compact. (When convenient, we identify X with the factor space G/K under the map $g \cdot o \to gK$.)

Let $D(X,G)$ be the algebra of linear differential operators on X which commute with the G-action (i.e., operators D such that $D(F\circ g) = (DF)\circ g$ for all smooth F and all $g\varepsilon G$). Let $D_+(X,G)$ be the subalgebra of those D that annihilate constants (i.e., have zero constant term).

Following Godement [Go] we say that F is harmonic, or, for emphasis, strongly harmonic on X (with respect to G!) if $DF = 0$ for all $D\varepsilon D_+(X,G)$. It is known [Go], [He 1] that F is strongly harmonic if and only if it is locally integrable and

$$F(g\cdot o) = \int_K F(gk\cdot x)\,dk$$

for all $g\varepsilon G$, $x\varepsilon X$. Here dk is Haar measure. (The theorem is true with the obvious modification for F defined only on an open subset of X; but, apart from this, local questions about strongly harmonic functions have not been studied at all.)

A space X as described always has a K-invariant Riemannian metric (it is even unique up to constant if the action of K on the tangent space at o is irreducible, which happens in most of the interesting cases). With respect to this metric there is a Laplace-Beltrami operator Δ on X. We say that F is weakly harmonic if $\Delta F = 0$. Since Δ is clearly G-invariant, strong harmonicity implies weak harmonicity.

These are the two main notions of harmonicity, but later we will also have to deal with some further still stronger ones.

We will also consider eigenfunctions of Δ with eigenvalue different from zero. Similarly, we will consider joint eigenfunctions of $D_+(X,G)$ (or what is the same, of $D(X,G)$), i.e., functions F on X such that

$$DF = \lambda(D)F \tag{1.1}$$

for all $D\varepsilon D(X,G)$ with some $\lambda(D)\varepsilon\mathbf{C}$. In this case, obviously, $D\to\lambda(D)$ is a character (i.e., a homomorphism into $\mathbf{C}$) of the algebra $D(X,G)$.

For this situation we have an immediate extension of Godement's theorem ([He 1, Ch. X] with a somewhat different proof): One has to remark first that, given a homomorphism λ, if the system (1.1) has a non-zero solution, then it has a unique solution ϕ_λ such that ϕ_λ is K-invariant and $\phi_\lambda(o) = 1$. In fact, the existence is clear by translating if necessary, and averaging on K. The uniqueness follows since ϕ_λ, being an eigenfunction of Δ is analytic and all its derivatives at o are determined by (1.1) and the condition of K-invariance (for details see [Go] or [Ko 1, Thm. 1.1.]). Now the theorem is the following:

F is a solution of (1.1) if and only if it is locally integrable, and

$$\int_K F(gk\cdot x)\,dk = \phi_\lambda(x)F(g\cdot o) \tag{1.2}$$

for all g,x.

To prove this: if F is a solution of (1.1), the left-hand side of (1.2) as a function of x is a K-invariant solution of (1.1), hence a multiple of $\phi_\lambda(x)$: Now set $x = o$. Conversely, by a regularization argument one may assume that F is smooth, then apply $D \varepsilon D(X,G)$ to both sides of (1.2) as functions of x, and then set $x = o$.

Before illustrating these notions on examples, we describe the basic method, due to Furstenberg [Fu 2], for constructing joint eigenfunctions of $D(X,G)$.

Let B be a homogeneous space of G on which the subgroup K also acts transitively. Let us consider functions $\mu : X \times B \to \mathbb{C}$ satisfying the cocycle identity

$$\mu(g\cdot x,b) = \mu(g\cdot o,b)\mu(x,g^{-1}\cdot b) \quad . \tag{1.3}$$

Applying $D \varepsilon D(X,G)$ to both sides as functions of x and setting $x = o$ one sees that $x \to \mu(x,b)$ is an eigenfunction with eigenvalue $D\mu(o,b)$ (which is clearly independent of b and is our $\lambda(D)$; note that in this case $\phi_\lambda(x) = \int_B \mu(x,b)db$ with the K-invariant measure db.)

By superpositions ("continuous linear combinations") of such cocycles it seems that one can get all solutions of (1.2). An _a priori_ proof of this, however, is known only for a certain special class of cases, where methods of positivity can be used [Fu 2].

Now, to illustrate our definitions and to anticipate some of the problems discussed later we describe some examples.

Example A. Let $X = \mathbb{R}^n$ with its usual metric, let K be a closed subgroup of $O(n)$, and let G be the group generated by the rotations in K and all translations of $\mathbb{R}^n$. So G is a semi-direct product, $G = \mathbb{R}^n \times_s K$.

The weakly harmonic functions are the usual harmonic functions, $\Delta F = 0$, but the class of strong harmonics depends, of course, on the choice of K. If $K = \{e\}$, $D(X,G)$ consists of all linear differential operators with constant coefficients, and the only strongly harmonic functions are the constants. If $K = SO(n)$, $D(X,G)$ consists of the polynomials in Δ, so weak and strong harmonicity coincide and Godement's Theorem specializes to the classical mean value theorem. If $n = 2m$ and $\mathbb{R}^n = \mathbb{C}^m$, one can take $K = T$ acting by $z \to e^{i\theta}z$ or $K = T^m$ acting on each coordinate separately: The strongly harmonic functions will be the pluriharmonics or n-harmonics, respectively. Other interesting choices of K will be discussed in §§2, 3.

Example B. If X is the unit disc U in $\mathbb{C}$ and G the group of all fractional linear transformations preserving U, we have the simplest case of a symmetric space of non-compact type (cf. §4). The Poincaré metric is G-invariant, and G is the connected component of the group of all isometries.

The Laplace-Beltrami operator is

$$\Delta = (1-|z|^2)^2 \frac{\partial^2}{\partial z \partial \bar{z}}$$

and every G-invariant operator is a polynomial in Δ. So weak and strong harmonicity coincide, and both coincide with ordinary harmonicity in the plane. (But the last statement fails obviously for eigenfunctions of Δ with non-zero eigenvalues!)

Let $P_b(z) = P(z,b)$ ($|z| < 1$, $|b| = 1$) be the ordinary Poisson kernel of U. Then of course $\Delta P_b = 0$, and more generally, the powers are eigenfunctions of Δ:

$$\Delta P_b^{\frac{1}{2}+\alpha} = (\alpha^2 - \tfrac{1}{4}) P_b^{\frac{1}{2}+\alpha} . \tag{1.4}$$

(This can be seen almost without any computation: P_b is characterized up to constant as the positive harmonic function with zero boundary values everywhere except at b. Hence a fractional linear transformation T carrying U to the upper halfplane H and b to ∞ must carry P_b into the function y. The Poincaré metric of H is $ds^2 = (2y)^{-2}(dx^2+dy^2)$, hence $\Delta_H = 4y^2 \frac{\partial^2}{\partial z \partial \bar{z}}$, hence $\Delta_H\, y^{\frac{1}{2}+\alpha} = (\alpha^2 - \frac{1}{4}) y^{\frac{1}{2}+\alpha}$. Since T is an isometry, (1.4) follows.)

There are well-known classical results about the Poisson representation of harmonic functions that are positive or belong to one of the Hardy classes h^p ($p \geq 1$). There is also the classical theorem of Fatou stating that if F is the Poisson integral of $f \varepsilon L^1$, then f equals the non-tangential limit of F a.e., on the circle. These results generalize to other eigenfunctions of Δ: The representation theorems to the case of eigenvalues $c \geq -\frac{1}{4}$, and Fatou's theorem to all $c \varepsilon \mathbb{C}$ except $c < -\frac{1}{4}$. The only change is that both in the definition of the Hardy spaces and in Fatou's theorem F has to be divided by the appropriate spherical function ϕ_λ, in this case a Legendre function of $|z|$ (cf. [Mi 1], [Mi 2] and §4 here).

A Poisson representation for <u>all</u> eigenfunctions of Δ, for <u>any</u> $c \varepsilon \mathbb{C}$ was given by Helgason in 1970 [He 4] in the form

$$\int_T P(z,b)^{\frac{1}{2}+\alpha} \, d\Phi(b) \tag{1.5}$$

where $c = \alpha^2 - \frac{1}{4}$, $\mathrm{Re}\,\alpha \geq 0$ and Φ is an analytic functional (this is the same thing as a hyperfunction on T).

We might mention that $P(z,b)$ and its powers, which we use in the integral representations are indeed cocycles in the sense (1.3), with $B = T$. This is obvious once we notice that $P(g \cdot o,b) = \frac{dg \cdot b}{db}$ where db is the Haar measure of T.

Example C. Let $X = U \times U$ with the product of the Poincaré metrics (which is the Bergman metric of the bidisc), and let G be the product of the G's belonging to U as in Example B. The algebra $D(X,G)$ has now the two generators

$$\Delta_j = (1-|z_j|^2)^2 \frac{\partial^2}{\partial z_j \partial \bar{z}_j} \qquad (j = 1,2)$$

and the strongly harmonic functions are simply harmonic in each variable separately.

The Poisson representation theorems generalize without change, the kernel now is the product $P(z_1,b_1)P(z_2,b_2)$ of the kernels for U.

For the Fatou-type theorems, however, an interesting new phenomenon occurs. If F is the Poisson integral of $f \in L^p(T^2)(p>1)$, then $\ell im\, F(z_1,z_2) = f(b_1,b_2)$ a.e., provided that z_j tends to b_j non-tangentially $(j = 1,2)$. But as Marcinkiewicz and Zygmund showed [Z, Ch. XVII] in the case $p = 1$ this fails in general, and remains true only under the added condition

$$0 < c < \frac{1-|z_1|}{1-|z_2|} < C < \infty$$

("restricted non-tangential convergence").

The weakly harmonic functions are now the solutions of

$$\Delta F = (\Delta_1+\Delta_2)\ F = 0 \quad .$$

This is a larger class than the strong harmonics, and we have here an example showing that in order to deal with the eigenvalue zero we really need to know something about the case of other eigenvalues:

For $\lambda = (\lambda_1,\lambda_2) \in \mathbb{R}^2$ on the circle $\lambda_1^2 + \lambda_2^2 = \frac{1}{2}$ we set

$$P_b^\lambda(z) = P^\lambda(z,b) = P_{b_1}(z_1)^{\frac{1}{2}+\lambda_1} \cdot P_{b_2}(z_2)^{\frac{1}{2}+\lambda_2} \quad .$$

Now (1.4) shows that $\Delta P_b^\lambda = 0$ for any b. More generally, if λ is on the circle $\lambda_1^2 + \lambda_2^2 = \frac{1}{2} + c$ $(c \geq -\frac{1}{2})$, then $\Delta P_b^\lambda = cP_b^\lambda$. By a special case of a theorem of Karpelevič (cf. §4) every positive solution of $\Delta F = cF$ $(c \geq -\frac{1}{2})$ admits a unique integral representation

$$F(z) = \int P^\lambda(z,b)\, d\mu(b,\mu) \tag{1.6}$$

with a measure μ on $T^2 \times S_c^+$, where S_c^+ denotes the part in the positive quadrant of the circle $\lambda_1^2 + \lambda_2^2 = \frac{1}{2} + c$.

It is an open question whether a similar result is true for <u>any</u> eigenfunction F, if one replaces μ by some appropriate kind of generalized function and, maybe, S_c^+ by its complexification. Also, nothing is known about the case $c \not> -\frac{1}{2}$.

There is a Fatou type theorem known for (1.6) in the case where $\mu(b,\lambda) = f(b,\lambda)dbd\lambda$ with bounded f. Defining

$$\Phi_c(z)=\int_{T^2 \times S_c^+} P^\lambda(z,b)dbd\lambda$$

we have that the limit of F/Φ_c along the geodesic line $(b_1 \cdot \tanh \lambda_1 t, b_2 \cdot \tanh \lambda_2 t)$ or in any tube of constant diameter around it is $f(b,\lambda)$ a.e., as $t \to +\infty$. (Note that when $\lambda_1 = \lambda_2$ this is exactly restricted nontangential convergence to b.) This is a special case of a theorem of Linden [Ln], see also §4. What happens for more general f is not known, and approaching the question through the usual method of trying to construct maximal operators majorizing the Poisson integral seems quite difficult. Perhaps a study of the trajectories of the diffusion process associated with Δ together with a geometric argument of the type of [B-D] would lead here to a Fatou type result even for Poisson integrals of measures. The very recent results of John Taylor [T] about the complex ball seem to support this conjecture (cf. also [D]).

We can very well use the example of the bidisc to illustrate the following theorem of Furstenberg [Fu 1] (see also [Be] with an incomplete proof): <u>On a symmetric space every bounded weakly harmonic function is strongly harmonic</u>.

We prove the following more general and more illuminating statement, due to Michelson [Mi 2]: For $\lambda \in S_c^+$ $(c \geq -\frac{1}{2})$, let

$$\phi_\lambda(z) = \int_{T^2} P^\lambda(z,b)db \tag{1.7}$$

in accordance with our earlier general considerations. (Note that $\phi_{(1/2,1/2)}(z) \equiv 1$.)

Now if $\Delta F = cF$ and $|F| \leq M\phi_\lambda$ with some constant M, then we claim that

$$F(z) = \int_{T^2} P\ (z,b)f(b)db$$

with some f bounded by M. (So F is a <u>joint</u> eigenfunction of $D(X,G)$...)

In the proof we may assume that F is real. We observe that (1.7) means that in the unique representation of ϕ_λ in the form (1.6) the measure is $db \otimes \delta_\lambda$. Since $M\phi_\lambda - F$ and $F - M\phi_\lambda$ also have representations of type (1.6), it follows that F itself has such a representation with measure

$f(b)db \otimes \delta_\lambda$, which proves the theorem.

This proof (for c = 0) is due to Karpelevič [Ka] and extends unchanged to the general case of any Riemannian symmetric space.

§2. The Case of $\mathbb{R}^n$ and a Finite Reflection Group K

In this section we study in detail Example A of §1, in the case where K is finite. This case can be treated by elementary methods but still it shows in some form most of the important phenomena that occur in the more general cases. We will be able to obtain the most complete results in the case where K is a finite group generated by reflections (for brevity: a "reflection group"). These groups have been much studied by Chevalley, Steinberg and others; there is a good expository article on them by Flatto [Fl]. We recall that the Weyl group of a symmetric space is always a reflection group.

We have $X = \mathbb{R}^n$, $G = \mathbb{R}^n \times_s K$. The action of K extends by linearity to $\mathbb{C}^n$. We denote the orbit of a point $\lambda \in \mathbb{C}^n$ under K by $K\cdot\lambda$. We say that λ is regular if $K \to K\cdot\lambda$ is a bijection, i.e., if the stabilizer of λ is $\{e\}$. We denote the algebra of complex-valued polynomials on $\mathbb{R}^n$ by S, the subalgebra of K-invariant polynomials by J, the set of elements of J without constant term by J_o. If $p \in S$, we write

$$\partial(p) = p(\frac{\partial}{\partial x_1}, \dots, \frac{\partial}{\partial x_n}) \quad .$$

It is clear that $D(X,G) = \partial(J)$, $D_+(X,G) = \partial(J_o)$.

Every strongly harmonic function in this case is a polynomial [Sg], we denote the set of all of them by H. By using the Hilbertian inner product $(p|q) = \partial(p)\overline{q}(0)$ on S (which goes back to Ernst Fischer in 1911) one shows that $S = SJ_o + H$ (SJ_o is the ideal generated by J_o), and hence $S = JH$ [He 2].

If K is a reflection group one even has $S \simeq J \otimes H$ (this is essentially contained in [Ch] and [Sg]) and, more or less equivalently, every harmonic polynomial is uniquely determined by its restriction to any regular K-orbit.

Any character of J extends to one of S, since S is integral over J [Sg], [He 1, Ch. X]. Hence every character of J is of the form $p \to p(\lambda)$ with some fixed $\lambda \in \mathbb{C}^n$. Of course, if $k \in K$, $\lambda' = k\cdot\lambda$ gives the same character: The characters are given by K-orbits in $\mathbb{C}^n$. (Note that this is a bijective correspondence, since J separates the K-orbits in $\mathbb{C}^n$: An interpolation formula gives an element of S, 0 on $K\cdot\lambda_1$ and 1 on $K\cdot\lambda_2$; averaging on K gives such an element in J.)

So the system of equations for the joint eigenfunctions corresponding to the character given by λ (or $K\cdot\lambda$) is

$$\partial(p)F = p(\lambda)F \quad (p \in J) \quad . \tag{2.1}$$

Clearly, $x \to e^{(\lambda'|x)}$ is a solution of (2.1) for all $\lambda' \in K\cdot\lambda$. (It is also a cocycle in the sense of (1.3) with $B = K\cdot\lambda$ and $\mathbb{R}^n$ acting trivially on B.) Given a function f on $K\cdot\lambda$ it is a reasonable definition to call the function

$$F(x) = \frac{1}{|K\cdot\lambda|} \sum_{\lambda'\in K\cdot\lambda} f(\lambda')e^{(\lambda'|x)} \tag{2.2}$$

the λ-Poisson-integral of f, denoted $F = PI^{\lambda}(f)$. (We denoted by $|K\cdot\lambda|$ the number of elements in $K\cdot\lambda$.)

If K is a reflection group, it is known [Sg] that the dimension of the solution space of (2.1) equals $|K|$ for any λ. PI^{λ} is always an injective map (by linear independence of the exponentials), hence it follows that if λ is regular, and only then, every solution of (2.1) arises as a λ-Poisson integral. (One could also define the Poisson integral of certain distributions on $\mathbb{R}^n$ with support in $K\cdot\lambda$ and in this way get all solutions for any λ. We prefer the present definition which involves only the set $K\cdot\lambda$ and no ambient structure.)

In any case, the unique invariant eigenfunction ϕ_λ of the general theory is $PI^{\lambda}(1)$.

If λ is not regular, it is known [Sg] that all solutions of (2.1) arise in the form

$$F(x) = \sum_{\lambda'\in K\cdot\lambda} p_{\lambda'}(x)e^{(\lambda'|x)} \tag{2.3}$$

where $p_{\lambda'}$ is a strongly harmonic polynomial relative to $K_{\lambda'}$, the stabilizer of λ' in K.

One can see from this that all positive solutions of (2.1) arise for any λ in the form $PI^{\lambda}(f)$ with some $f > 0$. (Indeed, now λ must be real, so we may assume that each $p_{\lambda'}$ is real. By the mean value theorem for $K_{\lambda'}$, $p_{\lambda'}(y+t\lambda') = p_{\lambda'}(y)$ for all $t \in \mathbb{R}$. Now, if some $p_{\lambda'}$ in (2.3) is not constant, then again by the mean value theorem it necessarily assumes a negative value $p_{\lambda'}(y)$. Setting $x = y + t\lambda'$ in (2.3) with large t we get a contradiction to $F > 0$.)

The analogue of Fatou's theorem for any finite K appears now in a very simple form. Suppose $K_{Re\lambda} = K_\lambda$ (i.e., $K_{Re\lambda} = K_{Im\lambda}$, which is certainly true if $Re\lambda$ is regular), and let $F = PI^{\lambda}(f)$. It is immediate from (2.2) and the definition of ϕ_λ that

$$\lim_{t\to\infty} \frac{F}{\phi_\lambda}(t\, Re\lambda') = f(\lambda') \quad . \tag{2.4}$$

(Without the condition on λ, e.g., when λ is purely imaginary, this does not work, and getting f back from F becomes a problem of Fourier inversion. Of course, extending F by analyticity to $\mathbb{C}^n$ and taking a limit along $x = t\lambda'$ would work in any case.)

We can strengthen our Fatou theorem by allowing the variable to tend to infinity in a less restricted way than along the ray $t \cdot \mathrm{Re}\lambda'$ (just as in the unit disc we can consider nontangential instead of radial convergence). In fact, by the same proof as for (2.4) we have

$$\ell im \frac{F(x)}{\phi_\lambda(x)} = f(\lambda') \tag{2.5}$$

if x tends to ∞ (i.e., $\|x\| \to \infty$) in such a way that

$$(\mathrm{Re}\lambda'|x) - (\mathrm{Re}\lambda''|x) \to \infty$$

for each $\lambda'' \in K \cdot \lambda$, $\lambda'' \neq \lambda'$. In other words, if we denote by C, the intersection of the halfspaces $(\mathrm{Re}\lambda' - \mathrm{Re}\lambda''|x) > 0$ $(\lambda'' \neq \lambda')$, then $C_{\lambda'}$ is a convex cone, it induces a partial ordering of $\mathbb{R}^n$, and (2.5) is valid exactly if $x \to \infty$ in the sense of this partial ordering.

All this is especially nice if K is a reflection group. In this case the fixed hyperplanes of the reflections contained in K divide $\mathbb{R}^n$ up into simplicial cones, called fundamental chambers, which are now given independently of λ, and on the set of which K acts simply transitively. For λ' regular, $C_{\lambda'}$ is now just the chamber containing $\mathrm{Re}\lambda'$; in general $C_{\lambda'}$ is the union of those chambers that contain $\mathrm{Re}\lambda'$ in their closure.

It is interesting to see what happens if in (2.5) we let x tend to infinity parallel to a face C_o (of codimension possibly > 1) of a fundamental chamber C. We suppose, as we may, $\mathrm{Re}\lambda \in C$. Let X^o be the subspace of $X = \mathbb{R}^n$ orthogonal to C_o and K^o the subgroup of K fixing C_o pointwise. Then K^o is a reflection group acting on X^o. Let λ^o be the projection of λ onto C_o. Now it is easy to show that for all $x^o \in X^o$ and for h in the subspace generated by C_o and tending to infinity in the partial order induced by C_o,

$$\ell im_{h \to \infty} \frac{PI^\lambda(f)}{\phi_\lambda}(h + x^o) = \frac{PI^{\lambda^o}(f|K^o \cdot \lambda^o)}{\phi_{\lambda^o}}(x^o) \quad . \tag{2.6}$$

This is a precise analogue of the results on the behavior of Poisson integrals near the boundary components in a Furstenberg-Satake compactification of a symmetric space of non-compact type (cf. §4). One can in the present case too define a compactification of $\mathbb{R}^n$ by adjoining all the possible X^o-s (for each of the chambers) as "faces at infinity." One gets a kind of infinitely large polytope onto which every function $PI^\lambda(f)/\phi_\lambda$ extends continuously. The construction can be modified so as to give smaller compactifications which still serve the purpose for the appropriate non-regular λ.

There is also an analogue of Furstenberg's theorem described in Example C of §1:

If $\lambda \in \mathbb{R}^n$, $c = ||\lambda||^2$, $\Delta F = cF$ and $|F| \leq M\phi$, then $F = PI^\lambda(F)$ with $|f| \leq M$. (If $\lambda = 0$, this is just Liouville's theorem.) This is a special case of our analogous theorem in §3, a proof will be indicated there.

Now we raise the following question which will have a very interesting analogue in §4. We know that for non-regular λ, PI^λ gives only a subclass of the λ-eigenfunctions of $D(X,G)$; we may try to characterize this subclass by some additional (non G-invariant) differential equations. For the following we restrict ourselves to the case of a reflection group K.

For any λ, let J_λ be the kernel of the character $p \to p(\lambda)$ of J. Then $J = J_\lambda + \mathbb{C}1$, and it is clear that F is a solution of (2.1) if and only if $\partial(J_\lambda)$, or what amounts to the same, $\partial(SJ_\lambda)$ annihilates F.

It is easy to see (cf. [K-R]) that $S = SJ_\lambda \oplus H$. We may ask whether in addition to $\partial(SJ_\lambda)$ there is some part of $\partial(H)$ that annihilates every $PI^\lambda(f)$. Since for any $p \in S$ we have $\partial(p)e^{(\lambda'|x)} = p(\lambda')e^{(\lambda'|x)}$, it is clear that, denoting by H^λ the set of polynomials in H that vanish on $K\cdot\lambda$, we will have $\partial(H^\lambda)$ annihilate every $PI^\lambda(f)$.

If λ is regular, we remarked before that the elements of H are determined by their restriction to $K\cdot\lambda$, so $H^\lambda = \{0\}$. If $\lambda = 0$, clearly H^λ is the set of all harmonic polynomials with constant term zero. These are the trivial extreme cases. In the general case one sees by (2.3) and by a dimension count that the class of functions of form $PI^\lambda(f)$ is exactly characterized by the system of differential equations

$$\partial(p)F = 0 \quad (p \in SJ_\lambda \oplus H^\lambda) \quad . \tag{2.7}$$

§3. The Cartan Motion Groups

Some of what was said in §2 is valid also for $X = \mathbb{R}^n$, $G = \mathbb{R}^n \times_s K$, where K is any closed subgroup of $O(n)$. For any such K we still use the notations S, J, J_o, H, J_λ as before. So $D(X,G)$ is still just $\partial(J)$. For a character $D(X,G) \to \mathbb{C}$ given by $\partial(p) \to p(\lambda)$ with some $\lambda \in \mathbb{C}^n$ we are still interested in the solutions of (2.1). At least in the case of the Cartan motion groups all characters are still of the form $\partial(p) \to p(\lambda)$, by a theorem of Chevalley to be quoted later. However, the correspondence with K-orbits is not bijective anymore, since J does not separate K-orbits in $\mathbb{C}^n$. (of course, in $\mathbb{R}^n$ it does [He 1, p. 434], by the Weierstrass approximation theorem.)

We define the λ-Poisson integral of a function f given on $K\cdot\lambda$ by

$$F(x) = PI^{\lambda}(f)(x) = \int_{K\cdot\lambda} f(\lambda')e^{(\lambda'|x)}d\lambda' \qquad (3.1)$$

where $d\lambda'$ is the normalized K-invariant measure. Certainly every λ-Poisson integral is a solution of (2.1), and we again have $\phi_\lambda = PI^{\lambda}(1)$.

The simplest case is $K = SO(n)$ which has been thoroughly studied. Here weak and strong harmonicity coincide. We discuss this case first.

A special case of a theorem of Karpelevič [Ka] gives all positive solutions of $\Delta F = cF$; they exist only for $c \geq 0$, and, choosing $\lambda \in \mathbb{R}^n$ such that $c = \|\lambda\|^2$, they have a unique representation in the form $F = PI^{\lambda}(\mu)$ with a positive measure μ on $K\cdot\lambda$ (i.e., on the sphere of radius $c^{1/2}$).

The proof of Karpelevič uses Martin's method. The substitution $F(x) \to e^{(\lambda|x)}F(x)$ transforms the equation $(\Delta-c)F = 0$ into an equation without constant term, so Martin's method applies. The extremal ("minimal") solutions are therefore limits of $G(x,y_\nu)/G(o,y_\nu)$, where G is the Green's function and $\|y_\nu\| \to \infty$. Working with subsequences and using the fact that $G(x,y_\nu)$ is constant on spheres with center y_ν, one sees that the extremals are constant on parallel hyperplanes. The equation $\Delta F = cF$ then shows that they are of the form $e^{(\lambda'|x)}$, and the theorem follows by general principles [Me].

More recently it has been shown [Ha], [K-O] that the Poisson representation of the solutions of $\Delta F = cF$ extends to all $c \in \mathbb{C}$ and all F, if instead of a measure μ one uses a very general type of generalized function, even more general than hyperfunctions. For the case $n = 2$ Helgason [He 5] gave a particularly simple description of these generalized functions as functionals on certain spaces of holomorphic functions. We note that in this situation the choice of λ is important. E.g., $\lambda = 0$ and $(1,i,0,\ldots,0)$ give the same character of J, but only the latter choice gives a Poisson representation of all solutions of (2.1). All these results are proved using Fourier analysis on SO(n). The eigenfunction F is decomposed into a sum of eigenfunctions transforming as the various irreducible representations of SO(n); for each term an integral representation is proved, and finally everything is put back together again. All this involves non-negligible technical problems.

The analogue of Fatou's theorem for SO(n) has been proved by O. Linden [Ln]. It says that if $F = PI^{\lambda}(f)$ with f integrable on the sphere $K\cdot\lambda$, then

$$\lim \frac{F(x)}{\phi_\lambda(x)} = f(\lambda')$$

provided that x tends to ∞ in any tube of constant diameter around the ray $t\lambda'$ $(o<t<\infty)$; this is here the natural notion of admissible convergence.

The proof is of course not as elementary as for finite K, but it follows classical lines. The Poisson kernel divided by ϕ_λ (which is now a Bessel function) is shown to be an approximate identity, which gives the result for continuous f. Then a maximal operator $f \to M(f)$ of weak type (1,1) is found which majorizes the Poisson integral in an admissible domain. Since any $f \in L^1$ can be written as $f = f_1 + f_2$ with f_1 continuous and $\|f_2\|_1$ small, the result follows. In the present case the standard Hardy-Littlewood maximal operator can be used, but in other analogous cases (cf. §4) the question of finding an appropriate $M(f)$ can become very difficult and interesting.

For the case of arbitrary K, some results can be derived from those listed above. Thus, if F is a positive solution of (2.1), it equals $PI^\lambda(\mu)$ with a positive measure μ on $K\cdot\lambda$: The proof is essentially an argument from [Ka, Thm. 17.11.1]. Also the Furstenberg-type theorem holds in the same formulation as in §2: For the proof one applies the Karpelevič result about $SO(n)$ to F and to $M\phi_\lambda$, and by the argument we wrote out in Example C of §1 one concludes that the support of the representing measure of F must be $K\cdot\lambda$. The Fatou-type result also remains true in the formulation of Linden, but this is now not the best possible notion of admissible convergence, which may be hard to find in general.

To get more complete results on this last question and on some others one should restrict oneself to the Cartan motion groups. They are defined as follows.

Let $X = G/K$ be a symmetric space of non-compact type, i.e., G a semisimple real Lie group without center and without compact factors, and K a maximal compact subgroup. The Lie algebra of G has the Cartan decomposition $\underline{g} = \underline{k} + \underline{p}$; here $\underline{p} \simeq \mathbb{R}^n$ and K acts linearly on $\underline{p}$ by the adjoint representation. The Cartan motion group associated to G/K is $\underline{p} \times_s K$. (Note that if $G = SO(n,1)$ this yields the classical case $\mathbb{R}^n \times_s SO(n)$.

Choosing a maximal abelian subalgebra $\underline{a}$ in $\underline{p}$, one has $\underline{p} = K\cdot\underline{a}$; the subgroup M^* fixing $\underline{a}$ as a set acts on it as a finite reflection group W (the Weyl group).

It is a classical theorem of Chevalley that the algebra J of K-invariant polynomials on $\underline{p} \simeq \mathbb{R}^n$ is mapped by restriction bijectively onto the algebra of W-invariant polynomials on $\underline{a}$. It follows that every character of $D(\underline{p}, \underline{p} \times_s K)$ is given by $p \to p(\lambda)$ with some $\lambda \in \underline{a}^{\mathbb{C}}$ (the complexification of $\underline{a}$). Using the further fact that a fundamental chamber $\underline{a}^+$ is a fundamental domain for the action of W on $\underline{a}$ we see that all characters are given by the points in $\underline{a}^+ + i\underline{a}$ (some of them still more than once).

It might be interesting to make a precise study of Fatou-type theorems for this case. It seems likely that a fairly strong notion of admissible convergence would work: x tending to ∞ in $\underline{a}$ (and its K-translates) with respect to the ordering induced by $\underline{a}^+$. Also, as suggested by the situation in both §2 and §4, one might get interesting limit theorems by tending to ∞ in the faces of $\underline{a}^+$ and in their K-translates.

Another question that might be well manageable in this case is the Poisson representation for joint eigenfunctions of the invariant operators. For positive eigenfunctions, and for Hardy-type spaces with real λ, this is known even for arbitrary K, as indicated above. But the question is much more difficult if there is no positivity assumption. The results about $\mathbb{R}^n \times_s SO(n)$ and Theorem 8 of [K-R] make it reasonable to believe that at least in the case of any Cartan motion group the family of all joint eigenfunctions will have a representation $PI^\lambda(T)$ with some appropriate class of generalized functions T. A way to approach this problem (and the next one as well) would be to use the methods of Ehrenpreis [E].

The case of a non-regular λ leads to the same question as at the end of §2. For Cartan motion groups it is again known that $S \cong J \otimes H$, $S = SJ_\lambda \oplus H$, and the elements of H are uniquely determined by their restriction to any regular K-orbit. (These are non-trivial facts, due to Kostant and Rallis [K-R].) Exactly as in §2 one sees that the set of differential operators $\partial(H^\lambda)$ annihilates every λ-Poisson integral (H^λ denotes the set of harmonic polynomials vanishing on $K\cdot\lambda$). It seems fairly clear that applying Fourier analysis on K one can also show the converse: $\partial(SJ_\lambda + H^\lambda)F = 0$ implies $F = PI^\lambda(T)$ with the appropriate kind of generalized function T.

Finally we should mention that besides the Cartan motion groups there are other choices of K that arise naturally in the representation theory of semisimple Lie groups. The λ-Poisson integral for some such examples has been studied by K. Gross and R. Kunze [G-K 2] in terms of their theory of generalized Bessel functions [G-K 1].

§4. Symmetric Spaces of Non-compact Type

As at the end of the last section, we consider $X = G/K$, with a Cartan decomposition of the semisimple Lie algebra $\underline{g} = \underline{k} + \underline{p}$, and $\underline{a} \subset \underline{p}$ maximal abelian. We write M for the centralizer of $\underline{a}$ in K, so $W \simeq M^*/M$. We choose a fundamental chamber $\underline{a}^+$ and a corresponding ordering of roots, we write ρ for the half-sum of the positive roots, $\underline{n}$ for the sum of the positive root spaces, A, N for the analytic subgroups corresponding to $\underline{a}$ and $\underline{n}$. $G = KAN$ is the Iwasawa decomposition, and for $g \in G$, $H(g)$ denotes the element of $\underline{a}$ such that $g \in K\cdot\exp H(g)\cdot N$.

The "maximal boundary" of Furstenberg (which should rather be called maximal distinguished boundary) is $B = G/MAN \simeq K/M$, a homogeneous space of both G and K; one can consider cocycles on it in the sense of (1.3). The simplest one of these is

$$P(g\cdot o,b) = \frac{d(g^{-1}\cdot b)}{db} \tag{4.1}$$

where db is the K-invariant measure.

By an ingenious simple proof Furstenberg [Fu 1] proved that every bounded strongly harmonic function F on X has a unique representation

$$F(x) = \int_B P(x,b)f(b)db \tag{4.2}$$

with bounded f. This easily extends to Hardy classes with $p > 1$ and to positive F.

Going about the matter more systematically it is not hard to prove [Fu 2] that all cocycles on $X \times B$ are given by

$$P^{\lambda}(g\cdot o,k\cdot b_o) = e^{-(\rho+\lambda)H(g^{-1}k)} \tag{4.3}$$

where λ is a complex linear function on $\underline{a}$, and b_o denotes the base point in B. The kernel P in (4.1) is equal to P^{ρ}. These cocycles, which are the basic building blocks in the Fourier analysis of X, are analogous to the ordinary exponentials in $\mathbb{R}^n$ [He 3].

By non-trivial results of Harish-Chandra, in which Chevalley's theorem quoted in §3 plays an important role, D(X,G) is isomorphic with J^W, the algebra of W-invariant polynomials on $\underline{a}$. We write this isomorphism as

$$D \to p_D \tag{4.4}$$

and we have, for all D,

$$D\,P^{\lambda}(x,b) = p_D(\lambda)P^{\lambda}(x,b) \quad . \tag{4.5}$$

(This formula shows why it is convenient to put "ρ" into the definition (4.3).)

At this point it is natural to discuss weakly harmonic functions, or more generally, solutions of $\Delta F = cF$. (One can show that in the homomorphism (4.4), $p_{\Delta}(\lambda) = \|\lambda\|^2 - \|\rho\|^2$.)

Karpelevič proved that this equation has positive solutions if and only if $c \geq -\|\rho\|^2$; such a solution has a unique representation

$$F(x) = \int_{B\times \underline{a}_c^+} P^{\lambda}(x,b)\,d\mu(b,\lambda) \tag{4.6}$$

where $\underline{a}_c^+$ denotes the intersection of the real sphere of radius $(c+\|\rho\|^2)^{1/2}$ with $\underline{a}^+$, and μ is a positive measure. The proof, although with some complications, follows the same lines as the argument sketched in §3 for the case of $\mathbb{R}^n$ to show that the extremal solutions must be of the form $P^\lambda(x,b)$. Then at the end one shows that the cocycle $P^\lambda(x,b)$ is extremal if and only if $\lambda \in \underline{a}^+$.

This theorem was illustrated for a special case in Example C of 1. The general form of the theorem of Linden mentioned there is that for bounded $f(b,\lambda)$ one has

$$f(k\cdot b_o,\lambda) = \lim \frac{F(x)}{\Phi_c(x)}$$

as x tends to ∞ along the geodesic $k\cdot\exp t\lambda$ $(0<t<\infty)$, or in a tube of constant diameter around it.

Of course, the open problems mentioned in connection with Example C are open in the general case, and it would be very interesting to solve them.

The theorem of Karpelevič can be used to prove Furstenberg's theorem (weak harmonicity + boundedness $\Rightarrow$ strong harmonicity), as explained in Example C. Furthermore, one can also easily deduce from it [Ka 17.11.1] that if $\lambda \in \underline{a}^+$, then every positive solution of the general eigenvalue equation

$$DF = p_D(\lambda)F \quad (D\in D(x,G)) \tag{4.7}$$

is a "λ-Poisson integral,"

$$F(x) = \int_B P^\lambda(x,b)d\mu(b)$$

with a positive measure μ; the solutions of (4.7) satisfying an H^p-condition $(p>1)$ are similarly characterized as λ-Poisson integrals $(\lambda\varepsilon\underline{a}^+)$ of L^p-functions [Mi 2].

These representation results have been generalized to arbitrary solutions of (4.7), for any $\lambda \in \underline{a}^+ + i\underline{a}$. (This means for any character of $D(X,G)$.) Instead of measures one has to take now the λ-Poisson integrals of hyperfunctions. This had been conjectured and proved in special cases by Helgason [He 4], [He 5], and despite serious difficulties a number of partial results were obtained in this direction by several authors, using Fourier analysis on K. The complete solution has been obtained by a radically different method, by extending to systems and applying the theory of differential equations with regular singular point. The hyperfunction T giving $F = PI^\lambda(T)$ appears as a generalized boundary value of F in an appropriate imbedding of X [K-K]. This approach extends even to a large class of non-Riemannian symmetric spaces [O-S].

The map PI^λ is meaningful for any λ, not only for $\lambda \in \underline{a}^+ + i\underline{a}$. There is a general result of Helgason with important implications for representation theory about the question of injectivity of PI^λ. Here we just refer to [He 6] for this; we will return to the question in the more special context of harmonic functions, which is our primary concern.

Now we discuss Fatou-type theorems for strongly harmonic functions (and other solutions of (4.7)).

We distinguish two kinds of admissible convergence of $x \in X$ toward $b = k \cdot b_o \in B$. For unrestricted admissible convergence one sets $x = k \cdot \exp H \cdot c$ with c in some (arbitrary) fixed compact set in X, and lets $H \to \infty$ in $\underline{a}$ with respect to the ordering induced by $\underline{a}^+$. For restricted admissible convergence one has to choose once and for all an element H in the interior of $\underline{a}^+$, set $x = k \cdot \exp tH \cdot c$, and let $t \to \infty$. (These definitions specialize in Example C of §1 to unrestricted and restricted non-tangential convergence.)

If $F = PI^\lambda(f)$ with f bounded, and $\mathrm{Re}\lambda \in \mathrm{int}(\underline{a}^+)$ (and also in some slightly more general cases), Michelson [Mi 1] proved that F/ϕ_λ tends to $f(b)$ unrestrictedly for a.a. $b \in B$. He also proved the analogous statement with restricted convergence for all $f \in L^1$ in the special case of products of spaces of rank one. These results are probably not the best possible, but it seems hard to improve on them in this generality.

In the case of strongly harmonic functions there are more complete results. One knows [L1] that for every X there is a $p_o > 0$ such that the Fatou theorem holds unrestrictedly for all $f \in L^P$, $p > p_o$. But one does not know the best p_o in general; in all special cases where it is known it is equal to 1. For restricted convergence, however, there is a definitive result: Stein [St 2] showed by using an entirely new kind of maximal function that the convergence theorem holds for all $f \in L^1$, on every X.

One can go further and study the boundary behavior of Poisson integrals in the so-called maximal Furstenberg-Satake compactification $\overline{X}$ of X. The boundary of X in $\overline{X}$ contains B, but it also contains families of symmetric spaces of lower rank, called boundary components of X. (In the case of the bidisc, $\overline{X}$ is just the closure in $\mathbb{C}^2$, which, besides T^2, contains families of discs.) By tending to ∞ in the various faces of $\underline{a}^+$, one can again define natural notions of restricted and unrestricted convergence to the boundary components, and expect to prove that the limits exist and are harmonic functions on almost all of them analogously to the situation in (2.6).

This program is carried out in [Ko 2] for the case of bounded harmonic functions. The results remain true with unrestricted convergence for Poisson integrals of L^P-functions with sufficiently large p (unpublished). They can also be proved, by using and slightly extending Stein's methods [St 2], for all $p > 1$ with restricted convergence. The attempt to do this for $p = 1$ as well leads to the following open question about maximal functions.

The main result of [St 2] is that for a certain type of function P_δ on a nilpotent Lie group N with dilations, the map P^* defined on $L^1(N)$ by

$$(P^*f)(n) = \sup_{\delta>0} \left| f*P_\delta(n) \right|$$

is of weak type (1,1). For example, if $N = \mathbb{R}^2$,

$$P_\delta(x) = \frac{\delta^2}{(\delta^2+x_1^2)(\delta^2+x_1^2+x_2^2)} \tag{4.8}$$

is of the required type. Now let L be a compact measure-preserving group of automorphisms of N, commuting with the dilations. (E.g., in our example let $L = T$ acting by rotations.) For $f \in L^1(N\times L)$ write $f_\ell(n) = f(n,\ell)$. The question is whether the map M, from $L^1(N\times L)$ to functions on N, defined by

$$(Mf)(n) = \sup_{\delta>0} \int_L \left| f_\ell * P_\delta(\ell\cdot n) \right| d\ell$$

is of weak type (1,1). Even in the special example of (4.8) the question is open.

Our discussion of Fatou-type theorems and Poisson integrals for harmonic functions up to now has been in a way too special. We have been working only with B, while there exist other "distinguished boundaries" too. In fact, with any parabolic subgroup P of G one can form the space $B_P = G/P$, on which K again acts transitively. To each P there belongs also a Furstenberg-Satake compactification $\overline{X}_P$ of X, which contains B_P. Just as in (4.1), (4.2), one can define the Poisson integrals of functions on B_P; one gets in this way subclasses of the strongly harmonic functions.

As shown in [Ko 2], for each $\overline{X}_P$ and each type of boundary component there are natural notions of restricted and unrestricted admissible convergence, and all the Fatou-type results listed above for the case of $\overline{X}$ have their analogues on $\overline{X}_P$. In fact, the proofs in the general case go by reduction to the case of $\overline{X}$, exploiting the fact that B is in a natural way a fibre space over B_P.

The main interest of this detailed study lies probably in the fact that if X is a bounded symmetric domain (Cartan domain) in $\mathbb{C}^n$, then its closure relative to $\mathbb{C}^n$ is one of the $\overline{X}_P$, but in general not $\overline{X}$; its Bergman-Shilov boundary is the corresponding B_P. Our Fatou-type results therefore translate into geometric results about boundary behavior of holomorphic functions (in fact for larger classes of functions) on the Cartan domains (cf. [Ko 3]).

In connection with the Poisson integrals of functions or hyperfunctions defined on a B_P there is another interesting question: Try to characterize the class arising in this way (which, by the mean value theorem, is obviously a subclass of the strongly harmonic functions) by differential equations! One can show that this class is identical with the class of λ-Poisson integrals on B, with $\lambda = s\cdot\rho$ where s is an element of W depending on P. We are in fact in one of the situations described by Helgason [He 6] where PI^{λ} is not an injective map.

Our question now looks completely analogous to the question discussed in §§2,3, of characterizing the image of PI^{λ} for non-regular λ. It also has a similar answer obtained recently in [J-K], after some special cases had been settled in [K-M] and [Jo]:

For a function F defined on $X = G/K$ let $\tilde{F}$ denote its lift to G, i.e., $\tilde{F}(g) = F(g\cdot o)$. Let $\underline{U}$ be the algebra of left-invariant differential operators on G. Let $\underline{L}_P$ be the left ideal in $\underline{U}$ whose elements annihilate all functions $\tilde{F}$ such that F is the Poisson integral of a hyperfunction on B_P. ($\underline{L}_P$ can also be characterized in a more algebraic way.) Now, analogously to (2.7), we have the theorem that whenever F is a function on X such that $D\tilde{F} = 0$ for all $D \in \underline{L}_P$, then F is the Poisson integral of a hyperfunction on B_P.

This result can be put in a much more precise form in the case of bounded symmetric domains of tube type. This more precise form really means finding a nice simple set of generators for $\underline{L}_P$ in this case, but the result can be proved by quite direct methods. Let us recall that the bounded symmetric domains of Cartan with their Bergman metric are symmetric spaces of non-compact type. A distinguished subclass of them can be realized as tube domains over convex cones, these are said to be of the tube type (there are five classes of them, by the classification of simple Lie algebras). A long time ago some explicit calculations of L. K. Hua on one of the five classes implied that the Poisson integrals of functions on the Shilov boundary satisfy a certain set of second order differential equations, and E. M. Stein raised the question whether these equations characterize the Poisson integrals. This is the origin of the set of questions discussed here.

To state the result for symmetric domains of tube type: it is known that in this case the complexification of the $\underline{p}$ part of the Lie algebra $\underline{g}$ splits under the action of K into a direct sum of irreducible subspaces $\underline{p}^+$ and $\underline{p}^-$ which are dual to each other under the Killing form. Let $\{E_i\}$ be any basis of $\underline{p}^+$ and $\{E_i^*\}$ its dual basis in $\underline{p}^-$: For $A \in \underline{k}$ let $D^A = \Sigma\, E_i^*\cdot[A,E_i]$ in $\underline{U}$ Now a function F defined on the domain is the Poisson integral of a hyperfunction of the Shilov boundary if and only if

$$D^A\,\tilde{F} = 0$$

for all $A \in \underline{k}$. It is also easy to give an equivalent formulation of this result which involves a set of differential equations for F itself instead of the lift $\tilde{F}$.

Bibliography

[Be] F. A. Berezin, An analog of Liouville's theorem for symmetric spaces of negative curvature, Dokl. Akad. Nauk SSSR 125 (1959) 1187-1189. (In Russian)

[B-D] M. Brelot and J. L. Doob, Limites angulaires et limites fines, Ann. Inst. Fourier 13 (1963), 395-415.

[Ch] C. Chevalley, Invariants of finite groups generated by reflections, Amer. J. Math. 77 (1955), 778-782.

[D] A. Debiard, Espaces H^p au-dessus de l'espace hermitien hyperbolique, Thèse, Paris 1976.

[E] L. Ehrenpreis, Fourier analysis in several complex variables, Wiley-Interscience, New York 1970.

[Fl] L. Flatto, Invariants of finite reflection groups, to appear in Enseignement Math.

[Fu 1] H. Furstenberg, A Poisson formula for semisimple Lie groups, Ann. of Math. (2)77 (1963), 335-386. MR 26 #3820; errata MR 28, p. 1246.

[Fu 2] __________, Translation-invariant cones of functions on semisimple Lie groups, Bull. Amer. Math. Soc. 71 (1965), 271-326, MR 31 #1326.

[G-K 1] K. Gross and R. Kunze, Bessel functions and representation theory I, J. Funct. Analysis 22 (1976), 73-105.

[G-K 2] __________, A Poisson integral for Stiefel harmonics, Symposia Mathematica XXII, Academic Press (1977), 353-362.

[Go] R. Godement, Une généralisation du théorème de la moyenne pour les fonctions harmoniques, C. R. Acad. Sci. Paris 234 (1952), 2137-2139.

[G-S] R. Gundy and E. M. Stein, Manuscript in preparation.

[Ha] M. Hashizume, A. Kowata, K. Minemura & K. Okamoto, An integral representation of an eigenfunction of the Laplacian on the euclidean space, Hiroshima Math. J. 2 (1972), 535-545.

[He 1] S. Helgason, Differential Geometry and Symmetric Spaces. Academic Press, New York 1962.

[He 2] __________, Invariants and fundamental functions, Acta Math. 109 (1963), 241-258.

[He 3] __________, Duality and Radon transform for symmetric spaces, Amer. J. Math. 85 (1963), 667-692.

[He 4] __________, A duality for symmetric spaces with applications to group representations, Advan. Math. 5 (1970), 1-154.

[He 5] __________, Eigenspaces of the Laplacian; integral representations and irreducibility, J. Functional Analysis 17 (1974), 328-353.

[He 6] __________, A duality for symmetric spaces with applications to group representations, II. Differential equations and eigenspace representations, Advan. Math. 22 (1976), 187-219.

[Jo] K. D. Johnson, Differential equations and the Bergman-Shilov boundary on the Siegel upper halfplane, to appear in Ark. för Mat.

[J-K] K. D. Johnson and A. Korányi, The Hua operators on Hermitian symmetric domains of tube type, to appear.

[Ka] F. I. Karpelevič, The geometry of geodesics and the eigenfunctions of the Beltrami-Laplace operator on symmetric spaces, Trudy Moskow. Mat. Obsc. 14 (1965), 48-185 = Trans. Moscow Math. Soc. 1965, 51-199, Amer. Math. Soc., Providence, R. I., 1967. MR 37 #6876.

[Ko 1] A. Korányi, Harmonic functions on symmetric spaces, in "Geometry and Harmonic Analysis of Symmetric Spaces," Marcel Dekker Inc., New York 1972.

[Ko 2] __________, Poisson integrals and boundary components of symmetric spaces, Inventiones Math. 34 (1976).

[Ko 3] __________, Boundary behavior of holomorphic functions on bounded symmetric domains, Amer. Math. Soc. Proceedings of Symposia in Pure Math. 30 (1977), 291-295.

[K-M] A. Korányi and P. Malliavin, Poisson formula and compound diffusion associated to an overdetermined elliptic system on the Siegel halfplane of rank two, Acta Math. 134 (1975), 185-209.

[K-P] A. Korányi and R. B. Putz, Local Fatou theorem and area theorem for symmetric spaces of rank one, Trans. Amer. Math. Soc. 224 (1976), 157-168.

[K-K] M. Kashiwara, A. Kowata, K. Minemura, K. Okamoto, T. Oshima & M. Tanaka, Eigenfunctions of invariant differential operators on a symmetric space, Ann. of Math. 107 (1978), 1-39.

[K-R] B. Kostant and S. Rallis, Orbits and representations associated with symmetric spaces, Amer. J. Math. 93 (1971), 753-809.

[K-O] A. Kowata & K. Okamoto, Harmonic functions and the Borel-Weil theorem, Hiroshima Math. J. 4 (1974), 89-97.

[Ll] L. Lindahl, Fatou's theorem for symmetric spaces, Ark. för Mat. 10 (1972), 33-47.

[Ln] O. Linden, Fatou theorems for the eigenfunctions of the Laplace-Beltrami operator, Thesis, Yeshiva University 1977.

[M-M] Marie-Paule Malliavin and Paul Malliavin, Intégrales de Lusin-Calderon pour les fonctions biharmoniques, Bull. des Sci. Math. 101 (1977), 357-384.

[Me] P. A. Meyer, Processus de Markov: la frontière de Martin, Lecture Notes in Mathematics 77, Springer 1968.

[Mi 1] H. L. Michelson, Fatou theorems for eigenfunctions of the invariant differential operators on symmetric spaces, Trans. Amer. Math. Soc. 177 (1973), 257-274.

[Mi 2] __________, Generalized Poisson integrals and their boundary behavior, Amer. Math. Soc. Proceedings of Symposia in Pure Math. 26 (1973), 329-333.

[O-S] T. Oshima and J. Sekiguchi, Boundary value problem on symmetric homogeneous spaces, Proc. Japan Acad. Ser. A, 53 (1977), 81-84.

[R] A. Raugi, Fonctions harmoniques et théorèmes limites pour les marches aléatoires sur les groupes, Bull. Soc. Math. France 54 (1977), 5-118.

[St 1] E. M. Stein, Some problems in harmonic analysis suggested by symmetric spaces and semisimple Lie groups, Actes du Congrès International des Mathematiciens, Nice 1970.

[St 2] __________, Maximal functions: Poisson integrals on symmetric spaces, Proc. Natl. Acad. Sci. U.S.A. 73 (1976), 2547-2549.

[Sg] R. Steinberg, Differential equations invariant under finite reflection groups, Trans. Amer. Math. Soc. 112 (1964), 392-400.

[T] J. Taylor, A potential-theoretic approach to admissible convergence in the unit ball of $\mathbb{C}^n$, manuscript.

[Z] A. Zygmund, Trigonometric series, 2nd ed., Cambridge University Press, New York, 1959.

YESHIVA UNIVERSITY

NEW YORK

Proceedings of Symposia in Pure Mathematics
Volume XXXV, Part 1, 1979

POSITIVE HARMONIC FUNCTIONS VANISHING ON THE BOUNDARY OF CERTAIN DOMAINS IN $\mathbb{R}^{n+1}$

Michael Benedicks

ABSTRACT. The number of linearly independent generators of the cone of positive harmonic functions vanishing on a subset of a hyperplane in $\mathbb{R}^{n+1}$ is determined.

Let E be a closed genuine subset of the hyperplane $y = 0$ in $\mathbb{R}^{n+1}$. The points in $\mathbb{R}^{n+1}$ are denoted by (x, y), where $x \in \mathbb{R}^n$, $y \in \mathbb{R}$. Let $\Omega = \mathbb{R}^{n+1} \setminus E$. We assume that all points of E are regular for Dirichlet's problem in Ω. Let P be the cone of positive harmonic functions in Ω vanishing on E. Then the following holds:

THEOREM 1. The cone P is generated by either one or two linearly independent elements. ($\dim P = 1$ or 2.)

In terms of Martin's theory for positive harmonic functions this result may be stated:

The Martin boundary of Ω has either one or two infinite points.

A characterization of the two cases is given in

THEOREM 2. The following conditions are equivalent

(i) P is generated by one element;

(ii) all functions in P are symmetric with respect to the hyperplane $y = 0$;

(iii) for all functions $u \in P$, $u = o(|(x, y)|)$ as $|(x, y)| \to \infty$.

In an analogous manner we may also give equivalent characterizations of the two-dimensional case.

(I) P is generated by two elements;

AMS(MOS) subject classifications (1970). 31A05, 31B05, 31A15.

(II) there are non-symmetric functions in P;

(III) there exists a function $u \in P$ such that $u(x, y) \geq |y|$.

To determine for a given set E, whether dim P is one or two we introduce a "local harmonic measure" $\beta_E(x)$ as follows:

Let $0 < \alpha < 1$ and let $B_x(\alpha|x|)$ be the ball

$$B_x(\alpha|x|) = \{(\xi, \eta) | \xi \in \mathbb{R}^n, \quad \eta \in \mathbb{R}, \quad |\xi - x|^2 + \eta^2 \leq (\alpha|x|)^2\} .$$

Let $u^x(\xi, \eta)$ solve the Dirichlet problem

$$u^x(\xi, \eta) = \begin{cases} 1 & \text{on} \quad |\xi - x|^2 + \eta^2 = (\alpha|x|)^2 \\ 0 & \text{on} \quad E \cap B_x(\alpha|x|) . \end{cases}$$

$$\Delta u^x = 0 \quad \text{in} \quad B_x(\alpha|x|) \setminus E .$$

Then $\beta_E(x)$ is defined as $u^x(x, 0)$, and the distinction between the two cases is given by

THEOREM 3.

$$\int_{|x|\leq 1} \frac{\beta_E(x)}{|x|^n} dx < \infty \iff P \text{ has } \underline{\text{two}} \text{ generators } (\dim P = 2)$$

$$\int_{|x|\geq 1} \frac{\beta_E(x)}{|x|^n} dx = \infty \iff P \text{ has } \underline{\text{one}} \text{ generator } (\dim P = 1)$$

As a simple corollary we see that if there is a non-degenerate n-dimensional circular cone $K \subset \{y = 0\}$ such that $K \cap E = \phi$ then P has <u>one</u> generator.

The purpose of introducing the function $\beta_E(x)$ is that harmonic measures of this type may in many cases be fairly easily estimated. One such result is

LEMMA 1. Let E' be a closed subset of $\{x \in \mathbb{R}^n; |x| \leq R\}$, where all points of E' are regular for Dirichlet's problem in $\Omega' = \{(x, y) | x \in \mathbb{R}^n, y \in \mathbb{R}, |x|^2 + y^2 \leq R^2\} \setminus E$. Suppose that for some η, $h > 0$ and all points x_o, $|x_o| \leq R - h$.

$$|E' \cap S_{x_o}(h)| \geq \eta |S_{x_o}(h)| ,$$

where $|\cdot|$ denotes n-dimensional Lebesgue measure and $S_{x_o}(h) = \{x \in \mathbb{R}^n \mid |x - x_o| \leq h\}$. Let $\omega(x, y)$ solve the Dirichlet problem

$$\omega(x, y) = \begin{cases} 1 & \text{on} \quad |x|^2 + y^2 = R^2 \\ 0 & \text{on} \quad E' \end{cases}$$

$$\Delta\omega = 0 \quad \text{in} \quad \Omega' .$$

Then

$$\omega(0, 0) \leq C \frac{h}{\eta^3 R} ,$$

where C is a numerical constant.

Combining the results of Theorem 2 and Lemma 1 we obtain

THEOREM 4. If $|E \cap S_x(\alpha|x|^\gamma)| \geq C_1 > 0$ for some $\gamma < \frac{1}{4}$ and all $x \in \mathbb{R}^n$ then P has two generators.

There is no reason to believe that the constant $\frac{1}{4}$ is best possible.

We can also by these methods solve Problem 3.12 proposed by Professor B. Kjellberg in Professor W. Hayman's list of problems from the 1973 Canterbury symposium in Complex Analysis.

In our notation this problem may be stated as follows:

Assume that the set E has the property that there are numbers R and $\varepsilon > 0$, such that each ball in $\mathbb{R}^n$ of radius R contains a subset of E of n-dimensional Lebesgue measure ε. Has then every function $u \in P$ the representation

$$u(x, y) = cy + \varphi(x, y) , \quad y > 0 ,$$

where φ is bounded on $\mathbb{R}^{n+1}$ and $c > 0$? Our answer is affirmative.

Theorem 4 is not sharp. To give an idea where the limit between $\dim P = 1$ and $\dim P = 2$ goes we state the following theorem, which holds in the case $n = 1$ (the complex plane).

THEOREM 5. Let p be a real number, $p \geq 1$, and put

$$E = \bigcup_{m=-\infty}^{\infty} [\operatorname{sign}(m) \cdot |m|^p - d_m , \quad \operatorname{sign}(m) \cdot |m|^p + d_m] ,$$

where $\{d_m\}_{m=-\infty}^{\infty}$, $0 < d_m < \frac{1}{2}$, is a sequence of real numbers such that

$$\log d_k \sim \log d_m \quad \text{if} \quad k \sim m .$$

Then

(i) $\dim P = 1$ if and only if $\sum_{m \neq 0} - \frac{\log d_m}{m^2} = \infty$;

(ii) $\dim P = 2$ if and only if $\sum_{m \neq 0} - \frac{\log d_m}{m^2} < \infty$.

Further details and complete proofs may be found in the preliminary report [1] but are also intended to be published elsewhere.

REFERENCES

1. M. Benedicks, Positive harmonic functions vanishing on the boundary of certain domains in $\mathbb{R}^n$. Technical report no. 6 (1978), Institut Mittag-Leffler.

2. Y. Domar, On the existence of a largest subharmonic minorant of a given function. Ark. Mat. 3 (1957), 429-440.

3. S. Friedland and W.K. Hayman, Eigenvalue inequalities for the Dirichlet problem on spheres and the growth of subharmonic functions. Comm. Math. Helv. 51 (1976), 133-161.

4. W.K. Hayman, On the domains where a harmonic or subharmonic function is positive. In: Advances in complex function theory, Lecture Notes in Mathematics, 505, Springer Verlag, Berlin (1974).

5. H. Kesten, Positive harmonic functions with zero boundary value. Preprint.

6. B. Kjellberg, On certain integral and harmonic functions. Thesis, Uppsala, 1948.

7. ———, On the growth of minimal positive harmonic functions in a plane domain. Ark. Mat. 1 (1951), 347-351.

8. ———, Problem 3.12. Symposium of complex analysis, Canterbury, 1973, 163. Cambridge University Press (1974).

9. R.S. Martin, Minimal positive harmonic functions. Trans. Amer. Math. Soc. 49 (1941), 137-172.

INSTITUT MITTAG-LEFFLER
Auravägen 17
S-182 62 Djursholm
Sweden

Proceedings of Symposia in Pure Mathematics
Volume XXXV, Part 1, 1979

POSITIVE HARMONIC FUNCTIONS WITH ZERO BOUNDARY VALUES

Harry Kesten[1)]

Hunt and Wheeden [2] proved that for a bounded Lipschitz domain D in $\mathbb{R}^d$, $d \geq 2$, the Martin boundary is homeomorphic to the topological boundary, ∂D. The main conclusion of their result is that for each point $y \in \partial D$ there is a unique kernel function (with value one at a given point $x_0 \in D$). Here a kernel function at y is a positive harmonic function $K(\cdot)$ on D such that

$$(1) \qquad \lim_{\substack{x \to z \\ x \in D}} K(x) = 0 \quad \text{for all} \quad z \in \partial D \setminus \{y\}.$$

If $g(x,y)$ denotes the Green function of D, it also follows from [2] that the unique kernel function at y is given by the Martin function

$$(2) \qquad K(x,y) = \lim_{\substack{z \to y \\ z \in D}} \frac{g(x,z)}{g(x_0,z)},$$

(the existence of the limit in (2) is part of the conclusion).

We investigated to what extent these results remain valid if ∂D is not Lipschitz at y. The nature of our results is, that if D and ∂D satisfy additional restrictions at points different from y, then (2) still represents the unique kernel function at y.

The first class of domains D we considered are unbounded ones and we are interested in kernel functions at $y = \infty$, i.e., positive harmonic functions on D which vanish at each finite boundary point. In this case our sufficient conditions on D for $K(x,\infty)$ to be the unique kernel function at ∞ are rather messy. The principal condition is that at most points $z \in \partial D$ there exist balls $B_1 \subset D$ and $B_2 \subset \mathbb{R}^d \setminus D$ which are tangent to each other at z and such that their radii are not too small with respect to the "size of D near z." We refer the reader to [3] for the precise conditions. Here we content ourselves with describing two examples covered by our theorem.

<u>EXAMPLE 1</u>. D consists of the half strip

$$\{x = (x_1,x_2): x_1 > 0, \ |x_2| < 1\} \quad \text{in} \quad \mathbb{R}^2$$

AMS(MOS) subject classifications (1970). Primary 31B05, 60J50.

Key words and phrases. Postive harmonic functions, kernel function, Martin boundary, solid of revolution.

1) Research supported by the N.S.F.

from which the closed balls C_n with center $(2n,0)$ and fixed radius $r < 1$ have been removed.

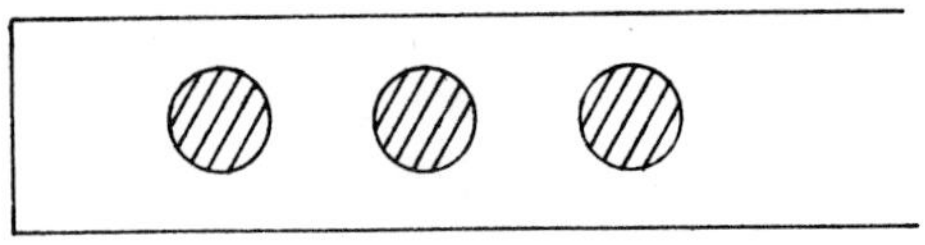

For this domain $K(0,\infty)$ is the unique kernel function at ∞. It can also be shown that if one takes the center of C_n at $(2rn+\delta_n,0)$ instead of $(2n,0)$, so that the width of the channel between C_n and C_{n+1} becomes $(\delta_{n+1}-\delta_n)$, then for $(\delta_{n+1}-\delta_n)$ tending to 0 sufficiently fast as $n\to\infty$, there exist at least two positive harmonic functions on D which vanish at each (finite) point of ∂D (see also Remark (iii)).

EXAMPLE 2. Let D be the domain in $\mathbb{R}^3$ between two cones defined by

$$D = \{x=(x_1,x_2,x_3): |x_1| < \tfrac{\pi}{2}+\delta\sqrt{x_2^2+x_3^2}\} .$$

For every $\delta > 0$ $K(\cdot,\infty)$ is the unique kernel function at ∞ for this D. On the other hand, for $\delta = 0$, the slab $\{x:|x_1| < \frac{\pi}{2}\}$ possesses the infinite class of kernel functions at ∞ of the form $K(x) = (\cos x_1)\exp(\alpha x_2+\beta x_3)$, $\alpha^2+\beta^2 = 1$. We do not know how many linearly independent kernel functions at ∞ exist for the intermediate domains

$$D = \{x: |x_1| < \tfrac{\pi}{2}+\delta(x_2^2+x_3^2)^{\gamma}\},$$

if $0 < \gamma < \frac{1}{2}$. ///

For "solids of revolution" one can give much better conditions. The theorems below answer a question raised by B. Dahlberg at this conference.

THEOREM A. Let $0 < a < \infty$ and $f:[0,a] \to [0,\infty)$ a lower semi continuous function satisfying

$$f(0) = \liminf_{t\downarrow 0} f(t) = 0 \quad \text{and}$$

$$\min\{f(t):\delta \le t \le a-\delta\} > 0 \quad \text{for all } 0 < \delta < \tfrac{a}{2} .$$

Let D be the domain

$$D = \{X = (x_1,\dots,x_d) \in \mathbb{R}^d: 0 < x_1 < a, x_2^2+\dots+x_d^2 < f^2(x_1)\},$$

and $g(x,y)$ its Green function. Then

$$K(x,0) = \lim_{\substack{y\to 0\\ y\in D}} \frac{g(x,y)}{g(x_0,y)} \tag{3}$$

exists. If all points of $\partial D\setminus\{0\}$ are regular for D^c, then $K(x,0)$ is the unique kernel function for D at 0.

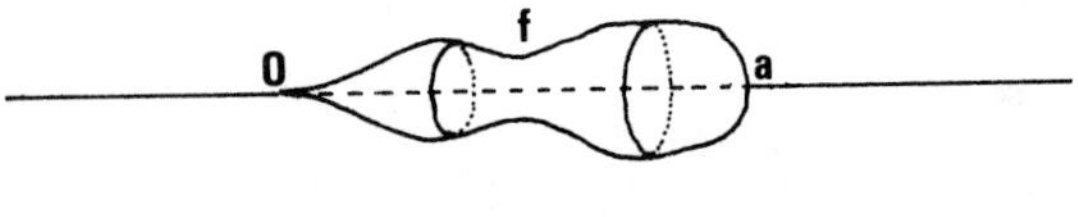

THEOREM B. Let $a_1, a_2 > 0$, $f:[0,a_1] \to [0,\infty)$ a continuous function and $k:[-a_2,a_1] \to [0,\infty)$ a lower semi continuous function. Assume the following conditions are satisfied

$$f(0) = 0, \quad f(t) > 0 \quad \text{for} \quad 0 < t \leq a_1 ,$$

$$f(t) \leq bt \quad \text{for some constant} \quad b > 0,$$

$$[\inf\{f(t):\tfrac{1}{2}p \leq t \leq 2p\}]^{-1}\sup\{f(t):\tfrac{1}{2}p \leq t \leq 2p\} \leq \gamma_0$$

$$\text{for some constant} \quad \gamma_0 > 0 \quad \text{and all} \quad 0 < p \leq \frac{a_1}{2} ,$$

$$\inf\{f(t):-a_2+\delta \leq t \leq 0\} > 0 \quad \text{and}$$

$$\inf\{k(t)-f(t):0 \leq t \leq a_1-\delta\} > 0 \quad \text{for} \quad \delta > 0.$$

Let D be the domain

$$D = \{x \in \mathbb{R}^d:[-a_2 \leq x_1 \leq 0 \quad \text{and} \quad x_2^2+\ldots+x_d^2 < k^2(x_1)]$$

$$\text{or } [0 < x_1 < a_1 \quad \text{and} \quad f^2(x_1) < x_2^2+\ldots+x_d^2 < k^2(x_1)]\},$$

and assume that all points of $\partial D\backslash\{0\}$ are regular for D^c. If $g(x,y)$ denotes the Green function on D, then the limit $K(x,0)$ in (3) exists and is the unique kernel function for D at 0.

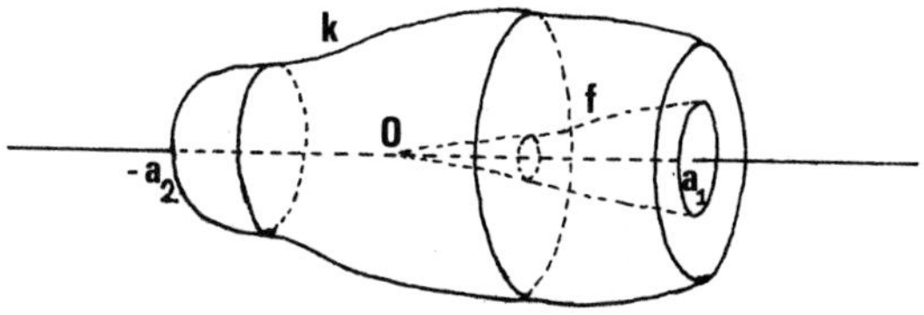

REMARKS: (i) By means of a Kelvin transformation the domains of Theorems A and B can be mapped into certain unbounded domains. These theorems then yield uniqueness of the kernel function for the transformed domains.

(ii) Our proofs use probabilistic techniques to estimate $g(x,y)$ and certain harmonic measures. For unbounded D we also apply our results to characterize the expectation of the first exit time from D of a Brownian motion $\{X_t\}_{t\geq 0}$, as well as to prove the existence of limits of the form

$$\lim_{\substack{x\to\infty \\ x\in D}} \frac{P^x\{X_\cdot \text{ hits } K \text{ in } A \text{ before leaving } D\}}{P^x\{X_\cdot \text{ hits } K \text{ in } B \text{ before leaving } D\}}$$

(K compact, A, B Borel subsets of K). Note that $P^x\{X_\cdot$ hits K in A before leaving $D\}$ is the harmonic measure of A relative to x and the set $D\backslash K$.

(iii) Michael Benedicks [1] has also determined the number of the kernel functions at ∞ for a class of unbounded domains. His class of domains is, however, quite distinct from ours. Dr. Benedicks has informed the author, though, that his methods (after a conformal map) do handle Example 1, and even give precise estimates on the rate at which $(\delta_{n+1}-\delta_n)$ must tend to 0 for the existence of two kernel functions at ∞.

REFERENCES

[1] M. Benedicks, Positive harmonic functions vanishing on the boundary of certain domains in $\mathbb{R}^n$, Report no.5, 1978, Institut Mittag-Leffler.

[2] R. A. Hunt and R. L. Wheeden, Positive harmonic functions on Lipschitz domains, Trans. Amer. Math. Soc. 147 (1970), 507-527, MR 43 #547.

[3] H. Kesten, Positive harmonic functions with zero boundary values, to appear.

CORNELL UNIVERSITY

Proceedings of Symposia in Pure Mathematics
Volume XXXV, Part 1, 1979

HARMONIC FUNCTIONS WITH BMO BOUNDARY VALUES

Umberto Neri

I shall review here(*) some joint work, done mainly with Professor E. B. Fabes, on the Dirichlet problem for Laplace's equation when the domain D is a half-space, and then when D is a Lipschitz domain in $\mathbb{R}^{n+1}$. Inspired by the classical work with L^p data, we sought to characterize the harmonic functions having boundary values with bounded mean oscillation. When our results are viewed from a historical perspective, they may be seen as a series of footnotes to some deeper ideas of Lennart Carleson, Charles Fefferman, and B. E. J. Dahlberg.

1. SOME BACKGROUND AND RESULTS FOR THE HALF-SPACE. To begin with, let us recall the description of the harmonic functions $u(x,t)$, in a half-space

$$D = \mathbb{R}^{n+1}_+ = \{(x,t): x \in \mathbb{R}^n, \quad 0 < t < \infty\} \text{ with } \partial D \cong \mathbb{R}^n,$$

with boundary data in $L^p(\mathbb{R}^n)$, $1 \le p \le \infty$. Given an $f \in L^p(\mathbb{R}^n)$, we form its Poisson integral

$$u(x,t) = [P_t * f](x) = \int_{\mathbb{R}^n} P_t(x-y)f(y)dy \tag{1}$$

where $P_t(x) = t^{-n}P(x/t)$ and $P(x) = c_n[1+|x|^2]^{-(n+1)/2}$. As is well-known, this absolutely convergent integral defines a harmonic function in D which reproduces $f(x)$ a.e., as $t \to 0$, since $\{P_t\}$ is an approximate identity. If $p = \infty$, then $|u(x,t)| \le \|f\|_\infty$ is bounded. Conversely, any *bounded* harmonic $u(x,t)$ such that $u(x,t) \to f(x)$ a.e., as $t \to 0$, must equal $P_t * f$, by the Maximum Principle. Hence, the solution of our problem for L^∞ data consists of all bounded harmonic functions, and all such functions are the Poisson integrals of their boundary values.

If $p < \infty$, taking $L^p(dx)$ norms and using Minkowski's inequality for integrals we find from (1) that $\|u(\cdot,t)\|_p \le \|P_t\|_1\|f\|_p = \|f\|_p$ for each $t > 0$. Hence, for such $u = P_t * f$, we have

$$\sup_{t>0} \|u(\cdot,t)\|_p \le \|f\|_p . \tag{2}$$

Following an idea of G. H. Hardy and turning (2) into a defining property,

AMS(MOS) subject classification (1970): Primary 31B25, 35J05.

(*) This paper is based on my seminar of July 14, 1978 at the A.M.S. Summer Institute in Williamstown, Mass. The seminar was dedicated to the memory of Guido Stampacchia who died in Paris last May.

we are lead to consider all harmonic functions $u(x,t)$ such that

$$(H_p) \qquad \sup_{t>0} \|u(\cdot,t)\|_p < \infty, \qquad 1 \le p \le \infty.$$

For $p = \infty$, the condition means that u is bounded and we are back to the previous case. For $p < \infty$, letting $u_k(x,t) = u(x,t+1/k)$ and $f_k(x) = u(x,1/k)$, we still have that u_k is bounded and, for each $k \in \mathbb{N}$, satisfies

$$(3) \qquad u_k(x,t) = [P_t * f_k](x)$$

with all $\|f_k\|_p \le C$ by (H_p). Hence, if $1 < p < \infty$, a weak compactness argument (see [16], p. 200) yields an $f \in L^p(\mathbb{R}^n)$ such that $u = P_t * f$. For $p = 1$ on the other hand, condition (H_1) is satisfied by any $u = P_t * d\mu$ if μ is a bounded measure. However, the modern theory of H^p spaces (amply featured in these Proceedings) takes care, in interesting ways, of all cases $0 < p \le 1$.

The simple formula for $P_t(x)$ shows that the Poisson integral in (1) is absolutely convergent whenever

$$(P) \qquad \int_{\mathbb{R}^n} |f(x)|(1+|x|^{n+1})^{-1}dx < \infty,$$

a condition clearly satisfied by $\log|x|$, the "worst" function in BMO, and (if $\alpha<1$) by $|x|^\alpha$, the typical homogeneous function in $\mathrm{Lip}(\alpha)$. Predictably, see [9] page 142, condition (P) is satisfied by all $f \in \mathrm{BMO}(\mathbb{R}^n)$. In fact, we found in §2 of [4] that (P) also holds for any $f \in E^{\alpha,p}(\mathbb{R}^n)$, a function space introduced by G. Stampacchia which coincides with BMO for $\alpha = 0$ and is equivalent to $\mathrm{Lip}(\alpha)$ if $0 < \alpha \le 1$. The stage was set for characterizing all such Poisson integrals.

Looking back at the conditions (H_p) when $p < \infty$, we see that they describe a *mixed norm* in the (x,t) variables. Following an idea of Carleson, let us replace such mixed norms by $L^p(d\mu)$ norms

$$\|u\|_{p,d\mu} = \left\{\int_0^\infty \int_{\mathbb{R}^n} |u(x,t)|^p \, d\mu(x,t)\right\}^{1/p}$$

and ask: for which positive measures μ on $\mathbb{R}^{n+1}_+$ can we prove the estimates

$$(4) \qquad \|P_t * f\|_{p,d\mu} \le A_p\|f\|_p$$

for all $f \in L^p(\mathbb{R}^n)$? Carleson [1] found that, for $p > 1$, such estimates hold if and only if μ satisfies the following *Carleson Property*: there is a constant $A > 0$ such that

$$(\text{C.P.}) \qquad \mu(Q \times [0,\delta]) \le A|Q|$$

for all cubes Q in $\mathbb{R}^n$ with side-length $\delta > 0$. (See also [16], page 236). These measures μ have been called *Carleson measures*.

More recently, looking at the *Littlewood-Paley measures* μ given by

(L.-P.) $$d\mu(x,t) = |\nabla u(x,t)|^2 \, t \, dxdt$$

where $u = P_t * f$, C. Fefferman proved (as part of a greater theorem in [9]) that, for $f \in L^2(\mathbb{R}^n)$, $d\mu$ is a Carleson measure if and only if f is in BMO. Close inspection of the result showed that the hypothesis $f \in L^2$ was not needed. Moreover, it suggested to me that one should consider *all* harmonic functions $u(x,t)$ whose corresponding (L.-P.) measures $d\mu$ satisfy (C.P.). Denoting this class by HMO $(\mathbb{R}_+^{n+1})$, in the sense of harmonic mean oscillation, Fefferman result said that if $f \in$ BMO then $(P_t * f) \in$ HMO. In February 1974, while visiting Professor Antoni Zygmund in Chicago, I conjectured the following converse.

THEOREM. Any $u \in \mathrm{HMO}(\mathbb{R}_+^{n+1})$ has a.e. limits $f \in \mathrm{BMO}(\mathbb{R}^n)$ and representation $u = P_t * f$. Hence (with equivalence of norms) we have

(5) $$\mathrm{HMO}(\mathbb{R}_+^{n+1}) = P_t * [\mathrm{BMO}(\mathbb{R}^n)] .$$

That is, HMO characterizes all Poisson integrals of BMO functions.

This result was proved soon after by E. Fabes and myself. Some of the proofs in it were later simplified, with the aid of R. Johnson, and extended to hold also for data in the $E^{\alpha,p}$ classes mentioned above. The proofs appeared, fractionally, in [4] and [5], and were partially included in [14], Chapter IV. The main steps are as follows: first, any $u \in$ HMO satisfies the pointwise estimates $|\nabla u(x,t)| \leq Ct^{-1}$. Then, with u_k and f_k as above, one shows that formula (3) still holds and that $\{u_k\}$ is a bounded sequence in HMO. Finally, with more labor, one deduces that $\{f_k\}$ is a bounded sequence in BMO. Examining ∇u_k and noting that the first partials of the Poisson kernel are in the pre-dual of BMO, the conclusion follows by familiar arguments.

Even more recently, Carleson-measure conditions have been used by P. Sjögren [15], N. Varopoulos [17], Garnett-Jones [10], and others. Earlier, studying the initial value problem for the heat equation with BMO data, Fabes and Neri [6] characterized the solutions by means of an appropriate Carleson property for certain parabolic (L.-P.) measures.

2. HARMONIC FUNCTIONS IN LIPSCHITZ DOMAINS. The boundary values of harmonic functions in a Lipschitz domain D were first studied by Hunt and Wheeden in [12], further explored in Dahlberg's papers [2] and [3], and recently reviewed in [11]. Two major conclusions emerged from Dahlberg's work:

(i) harmonic measure and (Lebesgue) surface measure are equivalent on ∂D;

(ii) the Dirichlet Problem for Laplace's equation is solvable with data $f \in L^p(\partial D)$ if, and only if, $p \geq 2$.

In view of these results, Fabes and I set out to solve the Dirichlet Problem for harmonic functions with BMO data on the boundary of a (possibly unbounded) Lipschitz domain. It is possible to reduce the problem to the case of an unbounded domain D, in $\mathbb{R}^{n+1}$, lying above the graph of a uniformly Lipschitz function $\varphi: \mathbb{R}^n \to \mathbb{R}$; that is, $D = \{(x,t): x \in \mathbb{R}^n$ and $t > \varphi(x)\}$. Its boundary is now a parametrized Lipschitz surface $S = \partial D$, with surface measure $d\sigma$ which is equivalent to n-dimensional measure dx. We have a Poisson kernel P_D which is the Radon-Nikodym derivative of harmonic measure: i.e., for a.e. $P_o \in D$ and $Q \in S$,

$$d\omega(P_o,Q) = P_D(P_o,Q)\, d\sigma(Q).$$

These harmonic measures, with pole at P_o, are probabilities on ∂D. That is, for each $P_o \in D$, $P_D(P_o,\cdot)\, d\sigma(\cdot)$ has integral one on S.

Around each $Q_o = (x_o,\varphi(x_o))$ in $S = \partial D$, we may consider the surface-balls $B_\delta = \{(x,\varphi(x)): |x - x_o| \leq \delta\}$ and define the space $BMO(S,d\sigma)$ useing quadratic averages over such B_δ, relative to the surface measure $d\sigma = d\sigma(x)$. With each harmonic function $u(x,t)$ in D, we associate now the Littlewood-Paley measure μ given by

$$d\mu(x,t) = |\nabla u(x,t)|^2\, [t-\varphi(x)]d\sigma(x)dt$$

and we say that $u \in HMO(D)$ whenever there is a constant $A > 0$ such that

$$\mu[T(B_\delta)] \leq A\sigma(B_\delta) \tag{6}$$

for all surface balls B_δ and corresponding *Carleson-tubes*

$$T(B_\delta) = \{(x,t): |x-x_o| \leq \delta \text{ and } 0 < t - \varphi(x) \leq 2\delta\}.$$

We have found that each $f \in BMO(S,d\sigma)$ has a harmonic extension (or, Poisson integral)

$$u(P) = [P_D(f)](P) = \int_S f(Q)P_D(P,Q)d\sigma(Q) \tag{7}$$

at all $P = (x,t) \in D$ and its associated measure μ satisfies the Carleson property (6) with constant A dominated only by the BMO norm of f. Conversely, each $u \in HMO(D)$ must have (nontangential) boundary values f on $S = \partial D$ belonging to $BMO(S,d\sigma)$ and characterizing u through formula (7). Thus, in analogy to (5) above, we have that

$$HMO(D) = P_D[BMO(S)].$$

These results were first proved for the case $n = 2$, using conformal mapping methods as in [13], and have been written up in [7] with some modifications suggested by B. Dahlberg. The proofs of the general $(n+1)$-dimensional case will appear soon in [8].

REFERENCES

[1] L. Carleson, Interpolation of bounded analytic functions and the corona problem, Ann. of Math. *76* (1962), 547-559.

[2] B. E. J. Dahlberg, Estimates of harmonic measure, Arch. Rat. Mech. Anal. *65* (1977), 275-288.

[3] ________________, On the Poisson integral for Lipschitz and C^1 domains, Studia Math. (to appear).

[4] E. Fabes, R. Johnson and U. Neri, Spaces of harmonic functions..., Indiana U. Math. J. *25* (1976), 159-170.

[5] ________, Green's formula and a characterization of the harmonic functions with BMO traces, Ann. Univ. Ferrara, Sez. VII, *21* (1975), 147-157.

[6] E. B. Fabes and U. Neri, Characterization of temperatures with initial data in BMO, Duke Math. J. *42* (1975), 725-734.

[7] _______________________, Harmonic functions with BMO traces on Lipschitz curves, Univ. of Maryland Tech. Rep 78-41 (July 1978).

[8] _______________________, Dirichlet Problem with BMO data in Lipschitz domains, (to appear).

[9] C. Fefferman and E. M. Stein, H^p spaces of several variables, Acta Math. *129* (1972), 137-193.

[10] J. B. Garnett and P. W. Jones, The distance in BMO to L^∞, Ann. of Math. *108* (1978), 373-393.

[11] R. Hunt, Harmonic measure and estimates of Green's function, Bull. A.M.S. *84* (1978), 194-200.

[12] R. Hunt and R. Wheeden, On the boundary values of harmonic functions, Trans. A.M.S. *132* (1968), 307-322.

[13] C. Kenig, Weighted H^p spaces on Lipschitz domains, Univ. of Chicago Ph. D. Thesis, March 1978.

[14] H. M. Reimann and T. Rychener, Funktionen beschränkter mittlerer Oszillation, Springer-Verlag Lect. Notes in Math., Vol. 487, 1975.

[15] P. Sjögren, Weak L_1 characterizations of Poisson integrals, Green potentials, and H^p spaces, Trans. A.M.S. *233* (1977), 179-196.

[16] E. M. Stein, Singular integrals and differentiability properties of functions, Princeton Univ. Press, Princeton, N.J., 1970.

[17] N. Th. Varopoulos, BMO functions and the $\bar{\partial}$-equation, Pacific J. of Math. *71* (1977), 221-273.

UNIVERSITY OF MARYLAND, COLLEGE PARK, MARYLAND 20742

Proceedings of Symposia in Pure Mathematics
Volume XXXV, Part 1, 1979

L^p-CAPACITARY INTEGRALS WITH SOME APPLICATIONS

David R. Adams[1]

1. CAPACITARY INTEGRALS. If C is a measure on R^n, then it is well-known that for $q > 0$,

$$\int |f(x)|^q dC(x) = \int_0^\infty C(\{x: |f(x)| > t\})dt^q$$

for any C-measurable function f. However, the integral on the right is well-defined for other set functions C, namely capacities. Our starting point then is to use that integral as our definition for the "L^q-capacitary integral of f with respect to C." Choquet [C] discusses some aspects of the case $q = 1$, C = Newtonian capacity. See also [Fu] and [An]. The capacities we have in mind are the Riesz capacities $R_{\alpha,p}$, $\alpha > 0$, $1 < p < n/\alpha$, defined on the σ-algebra of all subsets of R^n. They are given by: $R_{\alpha,p}(A) = \inf||f||_p^p$, where the infimum is over all non-negative $f \in L^p(R^n)$ for which $I_\alpha f(x) \equiv \int|x - y|^{\alpha-n}f(y)dy \geq 1$ on the set $A \subset R^n$. Here $||f||_p$ denotes the usual L^p-norm of f over R^n. The capacity $R_{1,2}$ is equivalent to the classical Newtonian capacity, $n > 2$.

Our main interest in these capacities lies with the integrals

$$J_{\alpha,p}(g) \equiv \int_0^\infty R_{\alpha,p}(\{x: |g(x)| > t\})dt^p$$

and certain of their applications. Later for simplicity we will write $J(\cdot) = J_{1,2}(\cdot)$ in sections 6 and 7.

Throughout we will write: "property P holds (α, p)--q.e. (quasi-everywhere)" meaning that property P holds except for a set of $R_{\alpha,p}$ capacity zero.

AMS(MOS) subject classification numbers (1970): Primary 31C15, 35J20, 38J35

[1]The author was partially supported by National Science Foundation grant MCS78-02698.

The interested reader is referred to [M] and [AM_1] where many of the aspects of the Riesz capacities, as well as the analogous Bessel capacities, are discussed in detail.

2. SUBLINEAR FUNCTIONALS. Although the functionals $J_{\alpha,p}(g)^{1/p}$ are not necessarily sublinear in g, they are nevertheless equivalent to functionals that are. Abusing the notation a bit, we define: $R_{\alpha,p}(g) = \inf||f||_p^p$ where the infimum is taken over all non-negative functions $f \in L^p(R^n)$ for which $I_\alpha f(x) \geq |g(x)|$, (α, p)--q.e., g a given function on R^n. Then from [A_3] it follows that

$$1/4 \cdot R_{\alpha,p}(g) \leq J_{\alpha,p}(g) \leq c \cdot R_{\alpha,p}(g)$$

where c is a constant independent of g. (The first inequality is a consequence of monotonicity and countable subadditivity of the capacity $R_{\alpha,p}$, while the second is a consequence of the capacitary strong type estimates found in [A_2] for integral α and most recently, for fractional α, in [D].) Thus, in particular, we have $R_{\alpha,p}(X_A) = R_{\alpha,p}(A)$, for any set $A \subset R^n$, X_A = indicator function of A. Also note that if $J_{\alpha,p}(g) < \infty$, then there is an $f \in L_+^p$ such that $|g(x)| \leq I_\alpha f(x)$, (α, p)--q.e., for if we truncate g, to say g_N, then $J_{\alpha,p}(g_N) \leq J_{\alpha,p}(g)$ and there is a sequence $f_N \in L_+^p$ such that $I_\alpha f_N \geq |g_N|$ with f_N tending to $f \in L_+^p$ weakly. This then gives $I_\alpha f \geq |g|$, (α, p)--q.e., by the Banach-Saks theorem.

In section 6 we write, for simplicity, $R = R_{1,2}$, both as a capacity and as a sublinear functional.

3. THE SPACES $L^p(R_{\alpha,p})$ AND DUALITY. A function g defined (α, p)--q.e. on R^n is said to be (α, p) quasi-continuous ((α, p)--q.c.) there, if for every $\varepsilon > 0$ there is an open set $G \subset R^n$ such that $R_{\alpha,p}(G) < \varepsilon$ and g restricted to the complement of G is continuous. Using this definition, we consider now some Banach function spaces and their dual spaces that arise using the functional $J_{\alpha,p}(\cdot)$.

We denote by $L^p(R_{\alpha,p})$ the space of all functions g (α, p)--q.c. on R^n and for which $J_{\alpha,p}(g) < \infty$. It is not hard now to check that $L^p(R_{\alpha,p})$ is a Banach space with $R_{\alpha,p}(\cdot)^{1/p}$ as its norm. Furthermore, the space $C_o(R^n)$, the continuous functions with compact support, is dense in $L^p(R_{\alpha,p})$.

Next consider the space $E_{\alpha,p}$ of all Radon measures μ on R^n such that $||I_\alpha|\mu|\,||_{p'} < \infty$, $p' = p/(p-1)$. Here $|\mu|$ is the total variation measure. The notation $E_{\alpha,p}$ for this space is to suggest "finite energy." Indeed, when $p = 2$, $||I_\alpha\mu||_2^2 = c_\alpha\int I_{2\alpha}\mu d\mu$, the classical energy integral.

The following result is due to K. Hansson [H] in the case $p = 2$.

THEOREM 3.1: *For* $p \geq 2$, $E_{\alpha,p}$ *is the dual of* $L^p(R_{\alpha,p})$.

Indeed, if we set $\Lambda(g) = \int g d\mu$, where $g \in L^p(R_{\alpha,p})$ and $\mu \in E_{\alpha,p}$, then

$$|\Lambda(g)| \leq \int I_\alpha f d|\mu| = \int f I_\alpha |\mu| dx$$
$$\leq ||f||_p ||I_\alpha|\mu| \,||_{p'},$$

for some $f \in L^p$, $f \geq 0$. Hence $||\Lambda|| \leq ||I_\alpha|\mu| \,||_{p'}$.

For the converse, we first note that if Λ is a continuous linear functional on $L^p(R_{\alpha,p})$, then $\Lambda(g) = \int g d\mu$ for all $g \in C_o$, μ a Radon measure. This can be extended by the density of C_o to give

$$\int f d|\mu| \leq ||\Lambda|| R_{\alpha,p}(f)^{1/p}$$

for all non-negative $f \in L^p(R_{\alpha,p})$. We next choose $f = U_{t,R} = I_\alpha(I_\alpha \mu_{t,R})^{1/(p-1)}$, where $\mu_{t,R} = |\mu|$ restricted to that part of the ball $B_R(0)$ (radius R, center zero) contained in the set

$$\{x: I_\alpha(I_\alpha|\mu|)^{1/(p-1)} \leq t\}.$$

$U_{t,R} \in L^p(R_{\alpha,p})$ and

$$||I_\alpha \mu_{t,R}||_{p'} \leq ||\Lambda|| \; ||I_\alpha \mu_{t,R}||_{p'}^{p'/p} ,$$

since $R_{\alpha,p}(U_{t,R}) \leq ||I_\alpha \mu_{t,R}||_{p'}^{p'}$. Hence $||I_\alpha \mu_{t,R}||_{p'} \leq ||\Lambda||$.

We conclude the argument by showing that $\mu_{t,R} \to |\mu|$ weakly. To this end, we note that if $E_t = \{x: U(x) > t\}$, $U = I_\alpha(I_\alpha|\mu|)^{1/(p-1)}$, then

$$\mu_R(E_t) \leq ||\Lambda|| \cdot R_{\alpha,p}(E_t)$$
$$\leq ||\Lambda|| \cdot t^{1-p}|\mu|(B_R(0))$$

by 4.4 of [AM_2] when $p \geq 2$.

Hence $||I_\alpha|\mu| \,||_{p'} \leq ||\Lambda||$ which gives the result. Theorem 3.1 remains open for $p < 2$.

4. TRACES OF SOBOLEV FUNCTIONS. If u is a Sobolev function in $W^{m,p}(\Omega)$, i.e. the distribution derivatives of u of orders $\leq m$ belong to $L^p(\Omega)$, $\Omega \subset R^n$, then it is well-known that when $mp < n$, u belongs to $L^q(M)$, M a smooth compact d dimensional manifold of Ω, provided $q \leq dp/(n - mp)$, and $mp > n - d$. Here $L^q(M)$ is understood as integration over M with respect to the d dimensional "surface" measure. Now u can be locally represented as a Riesz potential $I_m f$ in Ω, where $f \in L^p$. Hence we might try to characterize the sets M and the "surface" carried measures μ for which such a result is possible for the class of Riesz potentials.

Using the functional $J_{\alpha,p}$, we easily have for $q \geq p$

$$\left(\int |g|^q d\mu\right)^{1/q} \leq c \cdot J_{\alpha,p}(g)^{1/p}$$

for all μ measurable g (c a constant independent of g) if and only if $\mu(K) \leq c' \cdot R_{\alpha,p}(K)^{q/p}$ for all compact sets K (c' independent of K). Hence

THEOREM 4.1: *If* $q \geq p > 1$ *and* $\mu(K) \leq c \cdot R_{\alpha,p}(K)^{q/p}$ *for all compact sets* K*, then*

$$\left(\int |I_\alpha f|^q d\mu\right)^{1/q} \leq c' \cdot ||f||_p .$$

If $q > p$ in the above, then the condition on μ can be simplified. In particular μ can be a "d dimensional" measure.

THEOREM 4.2: *If* $q > p$ *and* $\mu(B_r(x)) \leq c \cdot r^d$, $d = (n - \alpha p)q/p$, *for all* $r > 0$ *(*$B_r(x)$ *a ball centered at* x *of radius* r*), then the conclusion of Theorem 4.1 remains valid.*

This last theorem gives the result stated earlier with M = support of μ. Also it should be noted that the restriction $q > p$ here is necessary since it is always possible to find a d dimensional measure supported on a set M of positive finite Hausdorff measure of dimension $d = n - \alpha p$. But such a set always has $R_{\alpha,p}$ capacity zero. Hence there is a non-negative function $f \in L^p$ such that $I_\alpha f(x) = +\infty$ on M.

The two previous theorems can be found in [A_2] and [A_1] respectively.

5. EXTENSIONS OF SOBOLEV FUNCTIONS. Suppose we are given a function u defined on R^n for which $u \in W^{m,p}(\Omega)$, Ω domain in R^n. We seek a function $U \in W^{m,p}(R^n)$ for which $U = u$ (m, p)--q.e. on Ω. Of course if $\partial\Omega$ is smooth, then it is well-known that such a U exists. However, if Ω has "cusps" at the boundary, then the existence of U puts certain "growth" restrictions on u at the boundary of Ω. Or if u is very large near $\partial\Omega$, then again the existence of U forces Ω to have "cusps" at the boundary to compensate. Hence we seek a general relationship between u and $\partial\Omega$ which implies the existence of such a U. This can be achieved, locally, using the functional $J_{m,p}(\cdot)$.

The following result can be found in [A_3]. Here $D^i u$ represents any derivative of u of order i.

THEOREM 5.1: *If* Ω *is a bounded measurable set in* R^n *with* $D^m u \in L^p(\Omega)$ *and*

$$|D^i u(x)|^\alpha \cdot |u(x)|^\beta \le c_i < \infty$$

for all $x \notin \Omega$, $\alpha = mp/i$, $\beta = p - \alpha - np/(n - mp)$, $1 \le i \le m$, $1 < p < n/m$, *then* $J_{m,p}(u \cdot X_\Omega) < \infty$ *implies that there is a function* $U \in W^{m,p}(R^n)$ *such that* $U = u$ (m, p)*--q.e. on* Ω. X_Ω *is the indicator function of* Ω.

Finally, we remark that the concept of "bounded point evaluations" is closely related to the above result. In particular, in [FP] the authors show that the existence of bounded point evaluations on Ω for solutions u of linear partial differential equations of order m in L^p is equivalent to the condition $J_{m,p}(uX_\Omega) < \infty$. Such a condition allows for approximation in L^p by solutions to partial differential equations in a neighborhood of Ω.

6. HARMONIC OBSTACLE PROBLEM. Let Ω be a bounded domain in R^n. For us in this section an "obstacle" for the Laplacian Δ in Ω will be a function $\psi(x)$ defined on Ω, $\max_\Omega \psi > 0$, and $\psi < 0$ near $\partial\Omega$. The obstacle problem for Δ is then to find a function $u \in K_1 = \{v \in H_o^1(\Omega): v \ge \psi$ $(1,2)$--q.e. on $\Omega\}$ such that u minimizes the Dirichlet integral $\int |\nabla v|^2 dx$ over all $v \in K_1$. In other words, we are to solve the variational inequality: find $u \in K_1$ such that

$$\int \nabla u \cdot \nabla(v - u)\,dx \ge 0$$

for all $v \in K_1$. See [LS] for a detailed discussion of this problem. H_o^1 is the closure of $C_o^\infty(\Omega)$ in the $W^{1,2}$ norm. If $K_1 \ne \emptyset$, then it is well-known that there is a unique solution u to this problem. The idea then is to deduce further regularity properties of u from the corresponding properties of ψ, e.g. integrability, boundedness, continuity and differentiability. There is a limit to the regularity that u will inherit from ψ: $\psi \in C^2(\Omega)$ implies only that the second derivatives of u are bounded in Ω while examples show that they can be discontinuous there; see [BK].

Our aim here is to point out a fundamental role played by the functional $J(\cdot)$ in this problem.

THEOREM 6.1: *If* ψ *is an obstacle for* Δ *in* Ω, *then* $K_1 \ne \emptyset$ *iff* $J(\psi^+) < \infty$.

This result then indirectly guarantees us a solution to the obstacle problem for a very general class of obstacles. Furthermore, the solution u satissatisfies

$$\int_\Omega |\nabla u|^2 dx \le c \cdot J(\psi^+)$$

for some constant c independent of ψ and u. Clearly, $\psi \in W^{1,2}(\Omega)$ with $\partial\Omega$ smooth implies $J(\psi^+) < \infty$. Theorem 6.1 follows from the relationship between the functionals $J(\cdot)$ and $R(\cdot)$ as outlined in section 2.

Finally, we show how integrability results for the solution u can be obtained from those of ψ using other functionals related to J. In particular, we introduce the functionals

$$P_{rs}(f) \equiv \int_0^\infty \{tR(|f| > t)^{1/r}\}^s \frac{dt}{t} .$$

Clearly $P_{22} = J$ and the general isoperimetric inequality between R and n-dimensional Lebesgue measure gives: $P_{rs}(f) < \infty \Rightarrow f \in L^s(R^n)$ when $r = s(1 - 2/n)$.

Now if we write T for the operator from "obstacles to solutions," i.e. $u = T\psi$, then T is sublinear, for according to [LS], $T\psi$ is the least super harmonic function that dominates ψ on Ω. Also by this fact, we have $||T\psi||_\infty \le ||\psi||_\infty$. And, on the other hand, we have $||T\psi||_{2^*} \le c \cdot J(\psi)$, $2^* = 2n/(n - 2)$. Hence interpolation in the spirit of Marcinkiewicz seems to be in order. This leads to

THEOREM 6.2: *If* ψ *is an obstacle for* Δ *in* Ω, *then* $T\psi \in L^p(\Omega) \cap H_o^1(\Omega)$ *whenever* $P_{rp}(\psi^+) < \infty$, $r = p(1 - 2/n)$, $2^* < p < \infty$.

Note, in particular, that if ψ is a radially decreasing obstacle on, say, the unit ball B of R^n, then $\psi^+ \in L^p(B)$ iff $P_{rp}(\psi^+) < \infty$. On the other hand, it is possible to find (radial) obstacles in $L^2(R) \cap L^q(B)$ for every $q < \infty$, $\psi \notin L^\infty(B)$, and for which $T\psi \notin L^p(B)$, for any $p > 2^*$. Also there are radial obstacles $\psi \in C^\infty(B - \{0\}) \cap H_o^1(B)$ for which $\psi \in L^q(B)$ for all $q < s = 2p(1 - 1/n - 1/p)$, but $T\psi \notin L^p(B)$. Note that $p < s$ when $2^* < p$.

7. BIHARMONIC OBSTACLE PROBLEM. Here again Ω is a bounded domain in R^n and ψ is the "obstacle" for Δ^2 in Ω, i.e. a function, now for simplicity, of class $W^{2,2}$ with compact support in Ω. The problem is to find a function $u \in K_2 = \{v \in H_o^2(\Omega): v \ge \psi \ (2, 2)\text{--q.e. on } \Omega\}$ which minimizes $\int(\Delta v)^2 dx$ over all $v \in K_2$ or equivalently a $u \in K_2$ such that

$$\int \Delta u \Delta(v - u)dx \ge 0$$

for all $v \in K_2$. If $\psi \in C_o^2(\Omega)$, then we know that the unique solution $u \in C^1(\Omega)$ with locally bounded second derivatives and, moreover, $u \in W^{3,2}_{loc}(\Omega)$. See [$F_1$] and [$F_2$]. The observation we make here is that by using the techniques of [CF], we can get an a priori bound on the $W^{3,2}$-norm of u, locally, by the quantity $J(\Delta\psi)$. So if we mollify the original $\psi \in W^{2,2}(\Omega)$ to $\psi_\varepsilon \in C_o^2(\Omega)$

and then pass to the limit, we get the following extension of Frehse's result $[F_1]$.

THEOREM 7.1: *If* ψ *is an obstacle for* Δ^2 *in* Ω, *then*

$$||u||^2_{W^{3,2}(\Omega_o)} \le C(\Omega_o) \cdot J(\Delta\psi)$$

where $\Omega_o \subset\subset \Omega$ *and* u *is the variational solution to the biharmonic obstacle problem described above.*

Note that if $\Delta\psi \in L^\infty(\Omega)$, then $J(\Delta\psi) < \infty$. Also if $\psi \in W^{3,2}(\Omega)$ and $\partial\Omega$ is smooth, then $J(\Delta\psi) < \infty$.

COROLLARY 7.2: *If* $J(\Delta\psi) < \infty$, *then* $R_{1,2}(C) > 0$, *where* $C \equiv \{x \in \Omega: u(x) = \psi(x)\}$, *the coincidence set.*

Of course C is defined only up to sets of $R_{1,2}$-capacity zero when $n \ge 6$.

To briefly outline the rest of the proof of Theorem 7.1, we first write (cf. [CF])

$$W(x) = -\gamma\int|x - y|^{2-n}\zeta(y)d\mu(y) + \alpha(x),$$

where $W(x)$ is the upper semi-continuous representation of $\Delta u(x)$, $\alpha(x) \in C^1(B_{R/4}(x_o))$, $x_o \in \Omega$, and $\zeta \in C_o^\infty(B_R(x_o))$ with $\zeta = 1$ on $B_{R/2}(x_o)$, $B_R(x_o) \subset \Omega$. γ is a constant and μ is a positive measure on Ω which in the sense of distributions satisfies $\Delta^2 u = \mu$ on Ω. Furthermore, we have the estimates

$$||\alpha||_{L^\infty(B_{R/4})} \le c_R||\Delta u||_{L^2(\Omega)}$$

and

$$||\nabla\alpha||_{L^2(B_{R/4})} \le c_R||\Delta u||_{L^2(\Omega)}$$

Finally, from [CF], we need

$$W(x) \ge \Delta\psi(x)$$

on support of μ inside Ω. Then since $J(\Delta\psi) < \infty$, there is a function $f \in L^2_+(R^n)$ such that $I_1f(x) \ge |\Delta\psi(x)|$ (1,2)--q.e., hence

$$\gamma\int_{B_{R/4}(x_o)} |x - y|^{2-n}\, d\mu(y) \le I_1f(x) + \alpha(x)$$

μ --a.e. on the support of μ contained in $B_{R/4}(x_o)$. So upon integrating this inequality with respect to μ restricted to $B_{R/4}(x_o)$, we get

$$\gamma' \cdot ||I_1\mu_{B_{R/4}}||_2^2 \le \int I_1\mu_{B_{R/4}} fdx + \int |\alpha| d\mu_{B_{R/4}}$$

from the Riesz composition formula, and then

$$||I_1\mu_{B_{R/4}}||_2^2 \le c \cdot J(\Delta\psi) + ||\alpha||_{L^\infty(B_{R/4})} \mu(B_{R/4})$$

by a proper choice of f. But, by our estimates of α and the fact that $\mu(B_{R/4}) \le c_R ||\Delta u||_{L^2(\Omega)}$, we get simply

$$||I_1\mu_{B_{R/4}}||_2^2 \le c \cdot J(\Delta\psi).$$

The result then follows since the third derivatives of u can all be bounded locally by a potential of the form $I_1\mu_B + \beta$, where β is locally bounded.

REFERENCES

[A_1] Adams, D. R., A trace inequality for generalized potentials, Studia Math 48(1973), 99-105.

[A_2] __________, On the existence of capacitary strong type estimates in R^n, Ark. Mat. 14(1976), 125-140.

[A_3] __________, Sets and functions of finite L^p-capacity, Ind. U. Math. J., to appear.

[AM_1] __________, Meyers, N. G., Bessel Potentials. Inclusion relations among classes of exceptional sets, Ind. U. Math. J. 22(1973), 873-905.

[AM_2] __________, __________, Thinness and Wiener criteria for non-linear potentials, Ind. U. Math. J. 22(1972), 169-197.

[An] Anger, B., Representation of capacities, Math. Amm. 229(1977), 245-258.

[BK] Brezis, H., Kinderlehrer, D., The smoothness of solutions to non-linear variational inequalities, Ind. U. Math. J. 23(1974), 831-844.

[C] Choquet, G., Theory of capacities, Ann. Inst. Fourier, 5(1953), 131-295.

[CF] Caffarelli, L., Friedman, A., The obstacle problem for the biharmonic operator, to appear.

[D] Dahlberg, B., Regularity properties of Riesz potentials, to appear.

[F_1] Frehse, J., Zum Differenzierbarkeitsproblem bei Variationsungleichungen höherer Ordnung, Hamburg Univ. Math. Sem. Abhand. 36(1971), 140-149.

[F_2] __________, On the regularity of the solution of the biharmonic variational inequality, Manuscripts Math. 9(1973), 91-103.

[FP] Fernström, C., Polking, J. C., Bounded point evaluations and approximation in L^p by solutions of elliptic partial differential equations, to appear.

[Fu] Fuglede, B., Capacity as a sublinear functional generalizing an integral, Mat. Fys. Medd. Dan. Vid. Selsk. 38, no.7(1971), 44pp.

[H] Hansson, K., Imbedding theorems of Sobolev type in potential theory, Linköping Univ., 1977.

[LS] Lewy, H., Stampacchia, G., On the regularity of the solution of a variational inequality, Comm. Pure App. Math. 22(1969), 153-188.

[M] Meyers, N. G., A theory of capacities for potentials of functions in Lebesgue classes, Math. Scand. 26(1970), 255-292.

DEPARTMENT OF MATHEMATICS
UNIVERSITY OF KENTUCKY
LEXINGTON, KENTUCKY 40506

Proceedings of Symposia in Pure Mathematics
Volume XXXV, Part 1, 1979

SOBOLOV SPACES, THE NAVIER-STOKES EQUATIONS AND CAPACITY

Victor L. Shapiro* and Grant V. Welland

1. INTRODUCTION. Let Ω be a bounded domain in Euclidean 3-space, $\mathbb{R}^3$, - and let $u = (u_1,u_2,u_3)$ be a vector with each u_i in $W^1_s(\Omega)$ and let p a function in $L_s(\Omega)$. Here $W^1_s(\Omega)$ is the space of functions in $L_s(\Omega)$ which also have first order distribution derivatives which are in $L_s(\Omega)$.

Our interest is in the removability of singularities for the nonlinear stationary Navier-Stokes equations

$$(1.1)\qquad \nu\Delta u_i - u_j \frac{\partial u_i}{\partial x_j} - \frac{\partial p}{\partial x_i} = 0 \quad i = 1, 2, 3$$

$$\text{and} \quad \frac{\partial u_i}{\partial x_j} = 0$$

where ν is a positive constant and the summation over repeated indices convention is observed. A distribution solution of (1.1) in Ω_1, an open subset of Ω, is a vector u and a function p for which the following holds:

$$(1.2)\qquad \int [\nu\, u_i\Delta\phi + u_i u_j \frac{\partial \phi}{\partial x_j} + p \frac{\partial \phi}{\partial x_i}]\ dx = 0 \quad i = 1, 2, 3,$$

and $\int [u_j \frac{\partial \phi}{\partial x_j}]\ dx = 0$ for ϕ in $C_0^\infty(\Omega_1)$.

We will say the pair (u,p) belongs to $WL_s(\Omega)$ if p is in $L_s(\Omega)$ and u_i is in $W^1_s(\Omega)$, i = 1, 2, 3. A compact subset of Ω, K, will be called a removable set for the class $WL_s(\Omega)$ provided the following holds with $\Omega_1 = \Omega - K$:

If (u,p) is in $WL_s(\Omega)$ and (1.2) holds for every $\phi \in C_0^\infty(\Omega_1)$ then (1.2) holds for every ϕ in $C_0^\infty(\Omega)$.

By capacity we will mean the capacity of [6]. We recall some facts which can be found in [1] and [6]. Let s' be the conjugate to s, i.e. $s' = s(s-1)^{-1}$ and let s' be in the range $1 < s' < 3$. From [1] we have the following definitions and facts. Let g_1 be the L_1^+ function in $\mathbb{R}^3$ which is the inverse Fourier transform of $\hat{g}_1(x) = (1+|x|^2)^{-\frac{1}{2}}$, $x \in \mathbb{R}^3$ and let $L_{1,s'} = g_1(L_{s'}) = \{g_1 * f \mid f \in L_{s'}\}$.

*This research was partially supported by NSF Grant MCS 76-02163.

Definition 1. $B_{1,s'}(A) = \inf||f||_{s'}^{s'}$ where the infimum is taken over functions in $L_{s'}^{+}$, for which $(f*g_1)(x) \geq 1$ on A and $A \subset \mathbb{R}^3$.

Definition 2. $C_{1,s'}(K) = \inf||\Psi||_{1,s'}^{s'}$ where the infimum is taken over functions Ψ in $C_0^\infty(\mathbb{R}^3)$ for which $\Psi(x) \equiv 1$ in a neighborhood of the compact set K and the norm is the usual $L_{1,s'} = L_1^{s'}$ norm see [10, Chapter V].

Theorem I. [1] For all compact sets $K \subset \mathbb{R}^3$, $B_{1,s'}(K) \sim C_{1,s'}(K)$

Theorem II. [6, p.276] Let Z be an analytic set. Suppose Z has a capacity $B_{1,s'}(Z)$ different from zero. Then there is a function f_1 in $L_{s'}^{+}$, and a positive Radon measure μ_1 satisfying the following properties.

(1.3) $$||f_1||_{s'} = 1$$

(1.4) μ_1 is concentrated on Z with total mass $||\mu_1|| = 1$

If $f_0 = (B_{1,s'}(Z))^{1/s'} f_1$ and $\mu_0 = (B_{1,s'}(Z))^{1/s'} \mu_1$ then

(1.5) $$f_0(y)^{s'-1} = (B_{1,s'}(Z))^{1/s} \int g_1(x-y)\,d\mu_0(x)$$

almost everywhere and

(1.6) μ_0 is concentrated on the set $Z \cap \{x \mid \int g_1(x-y) f_0(y)\,dy = 1\} = B$ and $B_{1,s'}(Z) = B_{1,s'}(B)$.

Finally if $\sigma_\rho = \sigma_\rho(x_0)$ is the ball of radius ρ centered at x_0 there is a positive constant c independent of ρ such that

(1.7) $c^{-1}\rho^{3-s'} \leq B_{1,s'}(\sigma_\rho) \leq c\rho^{3-s'}$. For (1.7) see [6,p283].

2. REMOVABLE SINGULARITIES. The main purpose of this paper is to extend results of [7] and [8]. In [9, Theorem 3] it is shown that for s > 3/2 a distribution solution agrees almost everywhere with a classical solution of (1.1). Since we will be in the case s > 3/2 we will use the term solution to mean distribution solution or classical solution according to the context with the understanding that the two concepts are in essential agreement.

The removability of a singularity will follow by an application of Theorem I and can be considered a corollary of the result of Adams and Polking. The proof of necessity in the theorem below is somewhat novel in that the proof uses Schauder's fixed point theorem and not the Leray-Schauder degree theorem as is common in the WL_2 case, see [5] chapter 5.

We shall establish the following theorem.

Theorem. Let $3/2 < s < \infty$. A necessary and sufficient condition that a compact set K be removable for $WL_s(\Omega)$ is that $B_{1,s'}(K) = 0$.

We first prove that a compact set of zero capacity is removable and assume Ω has the cone property, otherwise we use [2], lemma 4.22. For this purpose set $\Omega_1 = \Omega - K$ and assume (1.2) holds and let ϕ be in $C_0^\infty(\Omega)$. Let $r = 3s(3-s)^{-1}$ if $3/2 < s < 3$. By the Sobolev embedding theorem we have $u \in L_r$ and hence $u_i u_j \in L_\sigma$ where $\sigma^{-1} = 2r^{-1}$. In the case $3/2 < s < 3$ it follows that $\sigma > s$. In the case $s \geq 3$, $u_i \in L_p$ for all $p < \infty$ and hence we can assert $u_i u_j \in L_\sigma$ for some $\sigma < s$. Since (1.2) holds we only need to show

$$\int [-\nu \frac{\partial u_i}{\partial x_j}\frac{\partial(\Psi\phi)}{\partial x_j} + u_i u_j \frac{\partial(\Psi\phi)}{\partial x_j} + p\frac{\partial(\Psi\phi)}{\partial x_i}]\, dx \tag{2.1}$$

can be made arbitrarily small by varying Ψ, where Ψ is a function of the type in definition 2. For this purpose we note that

$$\left|\int \frac{\partial u_i}{\partial x_j}\frac{\partial(\Psi\phi)}{\partial x_j}\, dx\right| \leq \left|\left|\frac{\partial u_i}{\partial x_j}\right|\right|_s (||\Psi||_{s'}\left|\left|\frac{\partial\phi}{\partial x_j}\right|\right|_\infty + \left|\left|\frac{\partial\phi}{\partial x_j}\right|\right|_{s'}||\Psi||_\infty), \tag{2.2}$$

$$\left|\int u_i u_j \frac{\partial(\Psi\phi)}{\partial x_j}\, dx\right| \leq ||u_i u_j \chi_{K_1}||_s (||\Psi||_{s'}\left|\left|\frac{\partial\phi}{\partial x_j}\right|\right|_\infty + \left|\left|\frac{\partial\Psi}{\partial x_j}\right|\right|_{s'}||\phi||_\infty), \tag{2.3}$$

where $K_1 \supset K$ is a compact set containing the support of Ψ and we may assume the measure of K_1 is bounded by a fixed constant that does not vary with Ψ, and

$$\left|\int p\frac{\partial(\Psi\phi)}{\partial x_i}\, dx\right| \leq ||p||_s (||\Psi||_{s'}\left|\left|\frac{\partial\phi}{\partial x_j}\right|\right|_\infty + \left|\left|\frac{\partial\Psi}{\partial x_j}\right|\right|_s ||\phi||_\infty). \tag{2.4}$$

From [3], p221 we have

$$||\Psi||_{s'} \leq C||\nabla\Psi||_{s'} \tag{2.5}$$

After collecting inequalities (2.2) through (2.5) it follows that (2.1) is dominated by a constant times $||\nabla\Psi||_{s'}$. By an application of the fact that K has zero capacity and definition 2, (2.1) follows.

3. A COUNTER-EXAMPLE. Although the counter-example for which we now demonstrate existence can be presented in the context of $\mathbb{R}^3$, it is more convenient to work with Fourier series. We will now introduce further notation. Let $B(x,r)$ be the open ball centered at x with radius r and, for ease of notation, we will assume $\Omega \subset B(0,1)$. Next, let Ω be a set of positive capacity. We shall specify further requirements for Z later. Let $\Omega_1 = \Omega - Z$ be open.

We apply a number of familiar operators from the theory of singular integrals and harmonic analysis. If f is an L_1 function on $T_3 = \{x : -\pi \leq x_j < \pi, j = 1,2,3\}$ and m an integral lattice point we shall set

$$\hat{f}(m) = (2\pi)^{-3}\int_{T_3} f(x)e^{-i(m\cdot x)}\, dx. \tag{3.1}$$

If f is a distribution and P_t is the Poisson kernel $f*P_t$ is a function in L_1 and we define

$$\hat{f}(m) = (f*P_t)\hat{}(m)\cdot e^{t|m|} \tag{3.2}$$

Hence we define the Fourier coefficients for the following convolution operators

$$\hat{R}_j(m) = \begin{cases} im_j|m|^{-1}, & |m| \neq 0 \\ 0, & |m| = 0 \end{cases} \tag{3.3}$$

$$\hat{\Lambda}^{-1}(m) = \begin{cases} i|m|^{-1}, & |m| \neq 0 \\ 0, & |m| = 0 \end{cases} \tag{3.4}$$

$$\hat{\Delta}^{-1}(m) = \begin{cases} -|m|^{-2}, & |m| \neq 0 \\ 0, & |m| = 0 \end{cases} \tag{3.5}$$

With the above notation we define two more operators which operate on smooth vector functions on T_3, $\phi = (\phi_1,\phi_2,\phi_3)$, and can be extended to larger classes of functions:

$$S\phi = (R_jR_k + \delta_{jk}I)\,\Delta^{-1}\phi_k \tag{3.6}$$

$$A\phi = S((\phi\cdot\nabla)\phi) \tag{3.7}$$

We require a number of lemmas which will express the required properties of S and A for our problem. Let ${}_0W^1_s$ be the space of functions, f in W^1_s, for which $\hat{f}(0) = 0$ and let ${}_0\mathcal{W}^1_s$ be the space of vector functions $W = (W_1,W_2,W_3)$ with $W_j \in {}_0W^1_s$ and for which div(w) = 0. We have ${}_0\mathcal{W}^1_s$ is a Banach space.

Lemma 1. The operator S maps ${}_0\mathcal{W}^i_s$ continuously into ${}_0\mathcal{W}^{i+2}_s$ for $1 < s < \infty$. For a vector W in ${}_0\mathcal{W}^1_s$ the operator A becomes Sv with $v_i = \frac{\partial}{\partial x_k}(W_iW_k)$, $s>3/2$.

Lemma 2. The operator A can be extended by continuity to ${}_0\mathcal{W}^1_s$ for $s > 3/2$ such that $A\colon {}_0\mathcal{W}^1_s \to {}_0\mathcal{W}^1_s$ is compact.

Proof. Let $W \in {}_0\mathcal{W}^1_s$. Then $W \in L_r$ with $r = 3s(3-s)^{-1}$ if $3/2 < s < 3$ and $\nabla W \in L_s$ so that $(W\cdot\nabla)W \in L_p$ with $1/p = 1/s + 1/r = 2/s - 1/3$. Then, by lemma 1 and the definition of S in (3.6), $AW \in W^2_p$ and div(AW) = 0. Again, by the Sobolov embedding theorem $AW \in {}_0\mathcal{W}^1_\sigma$ where $1/\sigma = 1/p - 1/3$. If $\sigma > s$ then ${}_0\mathcal{W}^2_p$ is compactly imbedded in ${}_0\mathcal{W}^1_s$, see [2] for these results. This condition is satisfied since $s > 3/2$. This is true with $s \geq 3$ as well, as an easy check will show.

We now make a remark which is crucial in the following. We have from the proof of lemma 2 that

(3.8) $$||AW||_{W^1_s} \leq C||S(W\cdot\nabla W)||_{W^2_p} \leq C||W||_{L^r}\cdot||\nabla W||_{L^s} \leq C||W||^2_{W^1_s}.$$

Lemma 3. There exists an operator C such that if $\hat{J}_1(m) = (1+|m|^2)^{-\frac{1}{2}}$ then

(3.9) $$-i\Lambda^{-1} = J_1 + CJ_1 \text{ and}$$

(3.10) $$C \text{ is continuous on } L_s;\ 1 < s < \infty.$$

The operator J_1 is obtained as the periodization of the operator g_1 described in the introduction. This is accomplished by a standard application of the Poisson summation formula, a procedure which offers no difficulties since $g_1(x)$ is positive and decreases exponentially as $|x|$ tends to infinity.

From a measure μ_0 we construct a measure η_0 with the property that $\hat{\eta}_0(0) = 0$. Let $\phi_0 \geq 0$ be in C_0^∞, with support in $B(0,2) - B(0,3/2)$ and such that $\int \phi_0(x)dx = 1$. Let

(3.11) $$d\eta_0(x) = d\mu_0(x) - ||\mu_0||\phi_0(x)dx.$$

In what follows, μ_0 will be the capacitary distribution for a set Z of positive capacity and as in (1.4) we will have $||\mu_0|| = (B_{1,s'}(Z))^{1/s'}$.

We will require a set $Z\subset B(0,1)$ of positive capacity, for which we can estimate its capacity. The existence of such Z is assured by (1.7). With this remark we are now prepared to apply Schauder's fixed point theorem to obtain a distribution solution of (1.2) in $\Omega - Z$.

Schauder's Theorem [4]. Let X be a Banach space and $B\subset X$ be a bounded closed convex set. If $A:B\to B$ is a compact operator there is a point x_0 in B such that $Ax_0 = x_0$.

Let f_0 and μ_0 be the function and measure given after (1.4) for the set Z. Assume that $B^{1/s'}_{1,s'}(Z) = \varepsilon$, where $\varepsilon > 0$ is a value yet to be determined.

The operator $\nu^{-1}S$ is the solution operator for the linear problem, [8,p.6],

(3.12) $$\nu\,\Delta\,u_i - \frac{\partial p}{\partial x_i} - f_i = 0 \quad i = 1, 2, 3;\ \frac{\partial u_k}{\partial x_k} = 0.$$

Let $f_1 = (B_{1,s'}(Z))^{1/s}d\eta_0$ and $f_2 = f_3 = 0$; then a solution for (3.12) is $u = \nu^{-1}Sf = \nu^{-1}(R_jR_k+\delta_{jk}I)\,\Lambda^{-1}\,(\Lambda^{-1}f)$. By (3.9), (3.10) and (1.5) we have that $\Lambda^{-1}f$ belongs to L_s and S(f) belongs to ${}_0W^1_s$. The pressure p is obtained by

(3.13) $$p = -i(R_1,R_2,R_3)\cdot\Lambda^{-1}f \text{ and hence}$$

is clearly in L_s.

To solve the nonlinear problem

$$(3.14)\qquad \nu\,\Delta\, u_i - u_j \frac{\partial ui}{\partial x_j} - \frac{\partial p}{\partial x_i} - f_i = 0, \quad i = 1, 2, 3$$

we find a fixed point for the operator

$$(3.15)\qquad Aw = \nu^{-1}S[(w\cdot\nabla)w+f] = \nu^{-1}[Aw+Sf].$$

From (3.8) and the above remarks there is a constant $C > 0$ such that

$$(3.16)\qquad ||Aw||_{W^1_s} \le C(||w||^2_{W^1_s}+\varepsilon).$$

If we choose $\varepsilon < (4C^2)^{-1}$ it can be shown that A maps the ball $B_1 = \{w\in {}_0W^1_s : ||w||_{W^1_s} \le \left(1+\sqrt{1-4\varepsilon C^2}\right)(2C)^{-1}\}$ into itself. Since by lemma 2, A is compact, A is compact. By an application of Schauder's theorem the operator A has a fixed point in B_1. If we denote by W this fixed point and define p by

$$(3.17)\qquad p = -i(R_1,R_2,R_3)\,\Lambda^{-1}\,((W\cdot\nabla)W+f)$$

we obtain a solution, (W,p), for (3.14) with W and p in the appropriate spaces and f is zero in $\Omega_1 = \Omega - Z$.

We consequently conclude that for any function ξ in $C_0^\infty(\Omega)$

$$(3.18)\qquad \int [\nu W_1\Delta\xi+p\,\frac{\partial\xi}{\partial x_1} + W_kW_j\,\frac{\partial\xi}{\partial x_j}]\;dx = \int \xi d\eta_0$$

$$= \int \xi d\mu_0$$

$$(3.19)\qquad \int [\nu M_k\Delta\xi+p\,\frac{\partial\xi}{\partial x_k} + W_kW_j\,\frac{\partial\xi}{\partial x_j}]\;dx = 0,\ k = 2, 3.$$

Hence we have on $\Omega_1 = \Omega - Z$, with the above W and p, a distribution solution to equation (1.2) with the f_i in (1.2) identically zero.

We claim that Z is not a removable set for the class $WL_s(\Omega)$. If it were we would have for any ξ in $C_0^\infty(\Omega)$

$$(3.20)\qquad \int [\nu W_1\Delta\xi+p\,\frac{\partial\xi}{\partial x_1} + W_1W_j\,\frac{\partial\xi}{\partial x_j}]\;dx = 0$$

and hence from (3.18) that

$$(3.21)\qquad \int \xi d\mu_0 = 0 \text{ for all } \xi \text{ in } C_0^\infty(\Omega).$$

The equality (3.21) implies μ_0 is the trivial Radon measure thus contradicting $||\mu_0|| = B_{1,s'}(Z)^{1/s'} > 0$. Therefore Z is not removable.

REFERENCES

[1] David R. Adams and John C. Polking, The equivalence of two definitions of capacity, Proc. of the AMS 37(1973) 529-534.

[2] R. A. Adams, Sobolev spaces, Academic Press, New York (1975).

[3] L. Bers, F. John and M. Schechter, Partial differential equations, Lectures in Appl. Math., Vol. 3, Interscience, New York (1964).

[4] J. S. Cronin Fixed points and topological degree in nonlinear analysis, Providence, AMS, (1964).

[5] O. A. Ladyzenskaja, Mathematical problems in the dynamics of a viscous incompressible flow, Fizmatgiz, Moscow, 1961; English rev. ed, Gordon and Breach, New York (1969).

[6] Norman G. Meyers, Theory of capacity for potentials & functions in Lebesque classes, Math. Scand. 26 (1970) 255-292.

[7] Victor L. Shapiro, Capacity and the nonlinear Navies-Stokes equations, SIAM J. Math. Appl. 4 (1973) 329-343.

[8] ________________, Positive Newtonian capacity and stationary viscous incompressible flow, Indiana U. Math. J. 24 (1974) 1-16.

[9] ________________, Isolated singularities for solutions of the nonlinear stationary Navier-Stokes equations. Trans. of the AMS 187 (1974) 335-363

[10] E. M. Stein, Singular integrals and differentiality properties of functions, Princeton University Press (1970).

Proceedings of Symposia in Pure Mathematics
Volume XXXV, Part 1, 1979

APPROXIMATION IN L^p BY ANALYTIC AND HARMONIC FUNCTIONS

Lars Inge Hedberg

Let $K \subset C$ be compact, let $L^p(K)$ denote L^p with respect to plane Lebesgue measure restricted to K, and let $L^p_a(K)$ be the subspace of $L^p(K)$ consisting of functions analytic in K^0, the interior of K. Then it is well understood when the rational functions with poles off K are dense in $L^p_a(K)$. For example, for $p = 2$ one has the following result, due to V. P. Havin [2] and T. Bagby [1].

<u>Theorem 1</u>. The rational functions are dense in $L^2_a(K)$ if and only if one of the following equivalent conditions is satisfied:

(a) $\text{Cap}(G \setminus K) = \text{Cap}(G \setminus K^0)$ for all open G.

(b) The subset $E \subset \partial K$ where $\complement K$ is thin has $\text{Cap}\, E = 0$.

Here Cap means logarithmic capacity, or the equivalent Bessel capacity, which is defined for compact F by

$$\text{Cap}\, F = \inf\{\int |\omega|^2 dxdy + \int |\nabla\omega|^2 dxdy;\ \omega \in C_0^\infty,\ \omega \geq 1 \text{ on } F\},$$

and for general E by $\text{Cap}\, E = \sup\{\text{Cap}\, F;\ F \subset E,\ F \text{ compact}\}$.

A set is thin (effilé) at a point z_0 if

$$\int_0 \text{Cap}(B(z_0,\delta) \cap E)\delta^{-1} d\delta < \infty, \quad \text{where}$$

$B(z_0,\delta)$ denotes the disk with center at z_0 and radius δ.

There are similar results for $1 \leq p < \infty$, in particular the rational functions are always dense in $L^p_a(K)$ for $1 \leq p < 2$. See the papers quoted above and also [3].

It is surprising at first sight that the corresponding problem for approximation in L^p by functions that are harmonic in a neighborhood of

---

AMS(MOS) subject classifications (1970). 30A82, 31A05, 31B05, 31A15, 31B15, 31B30, 31C15, 46E35, 46E15, 46E20.

K is much more difficult. Even for $p = 2$ it is not completely solved, and for the purposes of this talk I will limit myself to this case.

The difficulty is connected with the fact that harmonic functions are solutions of a second order partial differential equation, whereas the Cauchy-Riemann equation is of order one. To explain this more precisely we need some more notation. K will now be a compact set in R^d, $d \geq 2$, and $L_h^2(K)$ will denote the subspace of $L^2(K)$ that consists of functions harmonic in K^0. As usual in PDE theory we denote the Sobolev spaces $W_1^2(R^d)$ and $W_2^2(R^d)$ by H^1 and H^2. I.e., a function $f \in H^i$ if and only if f and its weak derivatives of order $\leq i$ are in L^2, and

$$||f||_{H^i}^2 = \sum_{0 \leq |\alpha| \leq i} \int_{R^d} |D^\alpha f|^2 dx .$$

The closure in H^i of the test functions with support in an open set G will be denoted $H_0^i(G)$.

Then it is well known that $L^2(K) = L_h^2(K) \oplus \Delta(H_0^2(K^0))$, i.e., a function g in L^2 is orthogonal to $L_h^2(K)$ if and only if $g = \Delta\phi$, where $\phi \in H_0^2(K^0)$. In the same way g is perpendicular to $L_h^2(\Omega)$ for all open $\Omega \supset K$ if and only if $g = \Delta\phi$, where $\phi \in H_0^2(\Omega)$ for all $\Omega \supset K$. But it is easy to see that this is the same thing as saying that ϕ vanishes on $\complement K$. We have proved the following theorem [7].

<u>Theorem 2</u>. The harmonic functions on K are dense in $L_h^2(K)$ if and only if $\bigcap H_0^2(\Omega) = H_0^2(K^0)$, where the intersection is taken over all open Ω containing K.

If $K \subset C$ one obtains in a similar way the following theorem, which is used in proving Theorem 1.

<u>Theorem 3</u>. The rational functions are dense in $L_a^2(K)$ if and only if $\bigcap_{\Omega \supset K} H_0^1(\Omega) = H_0^1(K^0)$.

We recall that $H^2(R^d) \subset C(R^d)$ if $d = 2$ or 3, but that there is no such inclusion for $d \geq 4$. It follows immediately that if $K^0 = \emptyset$ the harmonic functions are always dense in $L^2(K)$ in dimensions 2 and 3, and for $d \geq 4$ a simple necessary and sufficient condition can be given in terms of capacities. (See Polking [5].) If K has an interior, matters turn out to be more complicated, however.

We first observe that if $\phi \in H^2$, then grad $\phi \in H^1$. But it is well known that functions in H^1 can be defined quasi-everywhere (q.e.), i.e., except on sets of capacity zero. If $\phi \in H_0^2(K^0)$, and if we for simplicity assume that $d = 2$ or 3, then it is easily seen that $\phi(x) = 0$ on ∂K, and that grad $\phi(x) = 0$ q.e. on ∂K. The converse problem, to decide whether, for a given K, every ϕ in H^2 that vanishes off K is in $H_0^2(K^0)$, can therefore be looked upon as having two parts.

I. Suppose $\phi \in H^2$, $\phi = 0$ on $\complement K$. Is $\phi = 0$ on ∂K and is grad $\phi = 0$ q.e. on ∂E ?

II. Suppose $\phi \in H^2$, $\phi = 0$ on $\complement K^0$ and grad $\phi = 0$ q.e. on $\complement K^0$. Is ϕ in $H_0^2(K^0)$?

If K has this second property we say that the closed set $\complement K^0$ has the H^2 spectral synthesis property.

<u>Theorem 4</u>. In R^2 and R^3 all closed sets have the H^2 spectral synthesis property.

The rather complicated proof appears in [4]. Whether the result extends to higher dimensions is still unknown, but there are weak sufficient conditions which guarantee spectral synthesis and also extend to $p \neq 2$. See [4].

The corresponding property in H^1 (and W_1^p, $1 \leq p < \infty$) is true for all closed sets [1]. This is not difficult to prove, and depends on the fact that these spaces are closed under truncation. When higher derivatives are involved one can no longer use this method, and this is what makes the situation for harmonic functions more complicated.

Theorem 4 has the following immediate consequence.

<u>Corollary</u>. For any open bounded G in the complex plane $L_h^2(G)$ is spanned by functions of the form

$$\int \log \frac{1}{|\zeta - z|}\, d\mu(\zeta), \quad \int \frac{\zeta - x}{|\zeta - z|^2}\, d\mu(\zeta), \text{ and } \int \frac{\eta - y}{|\zeta - z|^2}\, d\mu(\zeta)$$

where $\text{supp}\,\mu \subset \complement G$. Similarly if $G \subset R^3$.

We return to I. If $\phi \in H^2$, then grad $\phi \in H^1$. Moreover, Theorems 1 and 3 show that either of the conditions (a) or (b) in Theorem 1 is necessary and sufficient for $H_0^1(K^0)$ to equal $\bigcap_{\Omega \supset K} H_0^1(\Omega)$. Theorem 4 therefore gives the following sufficient condition for approximation.

Theorem 5. If $K \subset R^2(R^3)$ and if (a) or (b) is satisfied, then the harmonic functions on K are dense in $L_h^2(K)$.

Note that in contrast to the analytic case this condition is no longer necessary. In order to see this it is sufficient to consider the case $K^0 = \emptyset$. In fact, if $\phi \in H^2$ and $\phi = 0$ off K , then $\phi \equiv 0$, and thus also grad $\phi \equiv 0$, without any further assumptions on K .

In view of this it is reasonable to ask: Is there any $K \subset R^2(R^3)$ such that the harmonic functions are not dense in $L_h^2(K)$? That is, does the presence of an interior really make a difference? The answer is yes [4].

Theorem 6. There are compact sets $K \subset R^2(R^3)$ such that the harmonic functions are not dense in $L_h^2(K)$.

Since the proof is easy we give it here in the case of R^2 (R^3 is easier).

It is enough to construct a set K and a function $\phi \in H^2$ such that $\text{supp}\ \phi \subset K$ and $\nabla\phi(x) \neq 0$ on a subset of ∂K with positive capacity.

Denote the closed unit disk in R^2 by B_0 and the interval $\{x \in R^2; |x_1| \leq 1/2,\ x_2 = 0\}$ by I . We shall choose suitable disjoint disks B_k , $k = 1, 2, \ldots$, $B_k = \{x;\ |x - x_k| < r_k\}$, $x_k \in I$, and set $K = B_0 \setminus (\bigcup_{k=1}^{\infty} B_k)$.

Let $R_k > r_k$, and let $\chi_k \in C^\infty(0,\infty)$ be such that $\chi_k(r) = 1$ for $0 \leq r \leq r_k$, $\chi_k(r) = 0$ for $r \geq R_k$, $0 \leq \chi_k \leq 1$, and $|D^j\chi_k(r)| \leq Ar^{-|j|}(\log R_k/r_k)^{-1}$, $|j| = 1, 2$. Such functions can easily be constructed. Let $\psi_k(x) = \chi_k(|x - x_k|)$, and choose a function $\phi_0 \in C_0^\infty(B_0)$ such that $\phi_0(x) = x_2$ in a neighborhood of I .

It is easily verified that $\int |D^j(\phi_0\psi_k)|^2 dx \leq A(\log R_k/r_k)^{-1}$, $|j| = 2$, if R_k is small enough. Now choose $\{R_k\}$ so that

$\sum_{k=1}^{\infty} R_k < 1/2$, and $\{x_k\}$ so that the balls $\{x; |x - x_k| \le R_k\}$ are disjoint and dense in I . Finally, choose r_k so that $\sum_{k=1}^{\infty} (\log R_k/r_k)^{-1} < \infty$, and set $\phi = \phi_0(1 - \sum_1^{\infty} \psi_k)$. Clearly $\phi \in H^2$, and supp $\phi \subset K$. But every $x \in I$ that is not in one of the balls $\{x; |x - x_k| \le R_k\}$ is a boundary point of K . On the line perpendicular to I through such a point we have $\phi = \phi_0$, and thus $\frac{\partial \phi(x)}{\partial x_2} = 1$ at these boundary points. Since the set of such points has positive one-dimensional Lebesgue measure, ϕ has the desired properties, which proves Theorem 6.

Thus, in order to find a necessary and sufficient condition one needs to find something that takes into account the fact that the function grad ϕ in H^1 is not an arbitrary H^1 function, but the gradient of an H^2 function. One candidate is the following capacity, which is defined for compact F in a fixed ball B by

$$N(F) = \inf\{\int |\Delta\omega|^2 dx;\ \omega \in C_0^\infty(B),\ \omega \equiv 1 \text{ in a neighborhood of } F\} .$$

The condition $\omega \equiv 1$ near K (instead of $\omega \ge 1$ which leads to a classical capacity with respect to a kernel which is essentially $|x|^{4-d}$), makes this capacity difficult to work with, but this is also what makes it a candidate.

In fact, a necessary and sufficient condition for approximation, similar to (a) but using $N(\cdot)$, has been given by È. M. Saak [6], but the proof applies only for $d \ge 5$ and does not extend to $p \ne 2$.

Finally, we note that the approximation problem for harmonic functions in L^2 can be reformulated as a stability problem for the Dirichlet problem for the biharmonic equation $\Delta(\Delta u) = 0$, which perhaps also gives an explanation for the difficulty [7].

REFERENCES

1. T. Bagby, Quasi Topologies and Rational Approximation, J. Functional Analysis 10 (1972), 259-268.

2. V. P. Havin, Approximation in the Mean by Analytic Functions, Dokl. Akad. Nauk SSSR 178 (1968), 1025-1028.

3. L. I. Hedberg, Non-linear Potentials and Approximation in the Mean by Analytic Functions, Math. Z. 129 (1972), 299-319.

4. L. I. Hedberg, Two Approximation Problems in Function Spaces, Ark. Mat. 16 (1978), 51-81.

5. J. C. Polking, Approximation in L^p by Solutions of Elliptic Partial Differential Equations, American J. Math. 94 (1972), 1231-1244.

6. È. M. Saak, A Capacity Condition for a Domain with a Stable Dirichlet Problem for Higher Order Elliptic Equations, Mat. Sb. 100 (142) (1976) 201-209.

7. I. Babuška, Stability of the domain with respect to the fundamental problems in the theory of partial differential equations, mainly in connection with the theory of elasticity I, II (Russian). Czechoslovak Math. J. 11(86)(1961), 76-105, and 165-203.

UNIVERSITY OF STOCKHOLM

SWEDEN

Proceedings of Symposia in Pure Mathematics
Volume XXXV, Part 1, 1979

ON THE HELSON-SZEGÖ THEOREM AND A RELATED CLASS OF MODIFIED TOEPLITZ KERNELS

Mischa Cotlar and Cora Sadosky[1]

ABSTRACT. The Helson-Szegö theorem is sharpened, extended to the case of different measures and to several dimensions. It is also considered as a special case of a general Bochner theorem. The connections with Toeplitz forms are studied.

I. INTRODUCTION AND SUMMARY. We here emphasize some connections between the Helson-Szegö theorem, the Bochner theorem and Toeplitz forms. It follows in particular that the Helson-Szegö theorem can be viewed as a special case of a generalized Herglotz-Bochner theorem.

In Section II we present the main idea and state the basic results. These include the characterization of pairs of positive measures μ, ν for which the Hilbert transform is continuous from $L^2_\nu(\mathbf{T})$ to $L^2_\mu(\mathbf{T})$. In Section III we give the proofs of the theorems and indicate their relations with Toeplitz forms. In Section IV we sketch some generalizations, particularly to the several-dimensional case. In Section V we state some problems for the modified Toeplitz forms introduced in Section III.

II. THE HELSON-SZEGÖ CHARACTERIZATION AS A HERGLOTZ-BOCHNER THEOREM. Given f, a function, and μ, a measure, defined on $\mathbf{T}$, $\hat{f}(k)$ and $\hat{\mu}(k)$ denote their Fourier coefficients, and Hf denotes the Hilbert transform of f, defined by $\widehat{Hf}(k) = -i\hat{f}(k)$ for $k \geq 0$, $\widehat{Hf}(k) = i\hat{f}(k)$ for $k < 0$.

For a fixed $M \geq 1$, let $\mathcal{R}_M = \mathcal{R}_M(\mathbf{T})$ be the class of all positive measures μ on $\mathbf{T}$, which satisfy the Riesz inequality

$$\int |Hf|^2 d\mu \leq M \int |f|^2 d\mu \tag{1}$$

for such M and all trigonometric polynomials $f(t) = \sum_{k=-N}^{N} c_k e^{ikt}$.

AMS(MOS)subject classification (1970). Primary 42A40, 47B35; Secondary 44A25.

[1]Research partially supported by the National Science Foundation under grant.

Setting $f_+ = \Sigma_0^N c_k e^{ikt}$, $f_- = f - f_+$, we get $Hf = -if_+ + if_-$, and (1) can be rewritten as

(2) $$\int f_+\bar{f}_+ d\mu_{11} + \int f_+\bar{f}_- d\mu_{12} + \int f_-\bar{f}_+ d\mu_{21} + \int f_-\bar{f}_- d\mu_{22} \geq 0 \quad ,$$

for all such f_+, f_-, and

(3) $$\mu_{11} = \mu_{22} = (M - 1)\mu, \qquad \mu_{12} = \mu_{21} = (M + 1)\mu \quad .$$

A pair of positive measures μ, ν is called a Riesz-M pair if

(1a) $$\int |Hf|^2 d\mu \leq M \int |f|^2 d\nu$$

for all trigonometric polynomials f. The inequality (1a) can also be rewritten as (2), provided that

(3a) $$\mu_{11} = \mu_{22} = M\nu - \mu, \qquad \mu_{12} = \mu_{21} = M\nu + \mu \quad .$$

If a 2×2 matrix of measures $\underline{\underline{\mu}} = (\mu_{\alpha\beta})$, $\alpha, \beta = 1, 2$, satisfies

(2a) $$\Sigma_{\alpha,\beta=1}^2 \int f_\alpha \bar{f}_\beta d\mu_{\alpha\beta} \geq 0$$

for all pairs f_1, f_2 of analytic and antianalytic polynomials, we write $\underline{\underline{\mu}} \gg 0$. Thus, μ, ν is a Riesz-M pair iff $\underline{\underline{\mu}} \gg 0$ for $\underline{\underline{\mu}} = (\mu_{\alpha\beta})$ given by (3a).

Observe that if $\underline{\underline{\nu}} = (\nu_{\alpha\beta})$, $\alpha, \beta = 1, 2$, is given by

(4) $$\nu_{11} = \mu_{11}, \qquad \nu_{22} = \mu_{22}, \qquad \nu_{12} = \bar{\nu}_{21} = \mu_{12} + h(t)dt \quad ,$$

$h \in H^1$ (i.e., $\hat{h}(k) = 0$ for $k < 0$), then $\underline{\underline{\mu}} \gg 0$ is equivalent to $\underline{\underline{\nu}} \gg 0$. In such case we write $\underline{\underline{\mu}} \sim \underline{\underline{\nu}}$.

If $\underline{\underline{\mu}}$ satisfies (2a) for all pairs f_1, f_2 of continuous functions, we write $\underline{\underline{\mu}} > 0$. This is equivalent (see Section II) to the numerical 2×2 matrix $(\mu_{\alpha\beta}(E))$ being positive definite for every set $E \subset \mathbb{T}$. Clearly, $\underline{\underline{\mu}} > 0$ implies $\underline{\underline{\mu}} \gg 0$.

Let $\mathcal{R} = \cup_M \mathcal{R}_M$. Helson and Szegö [9] proved that $\mu \in \mathcal{R}$ iff $d\mu = \omega(t)dt$ with

(5) $$\omega = \exp(u + \tilde{v}), \qquad u \in L^\infty, \qquad \|v\|_\infty < \pi/2 \quad ,$$

where $\tilde{v}$ is the conjugate function of v.[2]

They also showed that if (c_k) is the sequence of Fourier coefficients of $\exp(-i\,\widetilde{\log \omega})$, then $\omega(t)dt \in \mathcal{R}$ iff for some $\rho < 1$ there is a solution

[2] The conjugate function $\tilde{v}$ and the Hilbert transform Hv, defined at the beginning of this section, differ only in the fact that $(\tilde{v})^\wedge(0) = 0$ for any integrable v.

to the problem: find f such that

(6) $$\hat{f}(k) = c_k \quad \text{for} \quad k \geq 0 \quad \text{and} \quad \|f\|_\infty \leq \rho \; .$$

Thus the Helson-Szegö theorem is related to the moment problem (6), whose study was started by Nehari [14] and developed by Adamjan, Arov and Krein [1].

Condition (5) is of little use to decide whether a given measure μ belongs to $\mathcal{R}$. Furthermore, the proof given in [9] relies on special properties of the analytic functions in the circle and does not extend to higher dimensions. Hunt, Muckenhoupt and Wheeden [10] proved that $\mu \in \mathcal{R}$ iff $d\mu = \omega(t)dt$ and $\omega \in A_2$, i.e.,

(7) $$\left(\frac{1}{|I|}\int_I \omega(t)dt\right)\left(\frac{1}{|I|}\int \omega^{-1}(t)dt\right) \leq A$$

for all intervals I, where $|I|$ is the (Lebesgue) measure of I. Condition (7) is good for deciding whether a given μ belongs to $\mathcal{R}$ and extends to $\mathbf{R}^n$ [4] as well as to L^p, $p \neq 2$ [10], leading to important applications.

However, the Helson-Szegö theorem is interesting for its relation to moment and function theories, and especially, because condition (5) gives a constructive representation of all $\mu \in \mathcal{R}$. Instead, condition (7) does not provide such a representation, which was lacking in the case $\mathbf{R}^n$, $n > 1$.[3]

Let us remark that even condition (7) may become of little use if μ is given, not by its density, but by its Fourier transform $\hat{\mu}$, as happens in Prediction Theory. Furthermore, conditions (5) and (7) do not characterize $\mathcal{R}_M$ for a fixed M.

These considerations rise the two following problems:

(a) give a representation of all measures $\mu \in \mathcal{R}_M$ by a method independent of the special properites of $\mathbb{T}$, that extends to higher dimensions;

(a_1) the same for the Riesz-M pairs;

(b) characterize the measures $\mu \in \mathcal{R}_M$ in terms of their Fourier transforms $\hat{\mu} = \mathcal{F}(\mu)$. More precisely,

(b_1) give a necessary and sufficient condition for a given bounded sequence (m_k), $m_{-k} = \overline{m}_k$, to belong to $\mathcal{F}(\mathcal{R}_M)$, i.e.

$$m_k = \hat{\mu}(k), \quad k = 0, \pm 1, \pm 2, \ldots, \quad \text{for some} \quad \mu \in \mathcal{R}_M \quad ,$$

and

---

[3] During the AMS 1978 Summer Institute in Williamstown, we learned about the extension of the Helson-Szegö theorem to $\mathbf{R}^n$ proved by Garnett and Jones [7]. This important result gives different necessary and sufficient conditions for a measure to belong to $\mathcal{R}(\mathbf{R}^n)$, but does not provide a complete description of it.

(b_2) give a constructive representation of all such sequences $(m_k) \in \mathcal{F}(\mathcal{R}_M)$.

Consider problem (a) first. Observe that condition (5) implies that there exist two positive constants c_1, c_2 such that $\cos v(t) \geq c_1$ and $e^{u(t)} + e^{-u(t)} \leq c_2$. Hence, if $h = -c \exp(\tilde{v} - iv)$, c a positive constant, $h \in H^1$ and $|h| \leq cc_2\omega$, $|\mathrm{Re}\, h| = ce^{-u}\omega \cos v \geq (cc_1/c_2)\omega$. Therefore we can choose c and M large enough as to have

$$(8) \qquad -4M\omega^2(t) - 2\mathrm{Re}\, h(t)(M + 1)\omega(t) - |h(t)|^2 \geq 0, \quad \forall t \in \mathbf{T} .$$

Thus condition (5) implies that there is a function $h \in H^1$ and a constant M such that (8) holds. On the other hand, (8) says that the determinant of the matrix $(\nu_{\alpha\beta}(t))$, given by $\nu_{11}(t) = \nu_{22}(t) = (M - 1)\omega(t)$, $\nu_{12}(t) = \overline{\nu_{21}(t)} = (M + 1)\omega(t) + h(t)$, is positive for all t. Thus the matrix $(\nu_{\alpha\beta}(t))$ is positive definite for all t, and therefore $\underline{\underline{\nu}} > 0$. If $\underline{\underline{\mu}} = (\mu_{\alpha\beta})$, $\alpha, \beta = 1, 2$, is given by (3) then $\underline{\underline{\mu}} \sim \underline{\underline{\nu}}$ and also $\underline{\underline{\mu}} \gg 0$. Thus condition (8) implies that $d\mu = \omega(t)dt \in \mathcal{R}_M$. We shall prove that the converse is also true.

THEOREM A. $\mu \in \mathcal{R}_M$ iff $d\mu = \omega(t)dt$ and ω satisfies (8) for some $h \in H^1$.

Moreover, we shall prove

THEOREM A_1. μ, ν is a Riesz-M pair iff

(8a) (i) $\mu(E) \leq M\nu(E)$, $\forall E \subset \mathbf{T}$

and

(ii) $\exists h \in H^1$ such that

$$(8b) \qquad -4M\mu(E)\nu(E) - 2\mathrm{Re}\, h(\chi_E)(\mu(E) + M\nu(E)) - |h(\chi_E)|^2 \geq 0 ,$$

$\forall E \subset \mathbf{T}$ (where $h(\chi_E) = \int_E h(t)dt$).

Observe that (i) and (ii) imply that at least μ must be an absolutely continuous measure, and that if $\nu = \mu$ then $d\mu = \omega(t)dt$ with $\omega \neq 0$ a.e.. Therefore for $\nu = \mu$, Theorem A_1 reduces to Theorem A.

We have already remarked that if ω satisfies condition (5), then it also satisfies condition (8) for some M. The converse also holds, as observed in [2]. Assume that ω is such that there is $h \in H^1$ for which (8) is satisfied. It follows easily from (8) that $\omega/|h| \in L^\infty$ and $|\pi - \arg h| \leq \arctan(M - 1)/2\sqrt{M} < \pi/2$. Therefore, choosing $u = \log(\omega|h(0)|/|h|)$ and $v = \pi - \arg h$, we get that ω satisfies (5), since then $\tilde{v} = \log(|h|/|h(0)|)$. The conclusion of this equivalence is that Theorem A includes and sharpens the Helson-Szegö theorem. The elementary proof of Theorem A and A_1 that is given in Section II provides, therefore, a new and simple proof of it.

Condition (8) is not only a sharper version of the Helson-Szegö condition (5) but it also extends naturally, as (8b), to the case $\nu \neq \mu$, that could not be treated by (5) or (7) (c.f. [13]). Thus Theorem A_1 gives an answer, for the case $p = 2$, for a question posed at the AMS 1978 Summer Institute in Williamstown (problem 7 of [12]).

Since for μ, ν to be a Riesz-M pair is equivalent to $\underline{\underline{\mu}} \gg 0$ for $\underline{\underline{\mu}} = (\mu_{\alpha\beta})$ given by (3a), Theorem A_1, and then also Theorem A, is a special case of the more general

THEOREM B. If $\underline{\underline{\mu}} = (\mu_{\alpha\beta})$ is a 2×2 matrix of measures, then $\underline{\underline{\mu}} \gg 0$ iff there exists $\underline{\underline{\nu}}$ such that $\underline{\underline{\nu}} \sim \underline{\underline{\mu}}$ and $\underline{\underline{\nu}} > 0$.

In Section III we give a simple proof of Theorem B, using only the Hahn-Banach theorem and general properties of analytic functions.

We have also shown that ω satisfies (5) iff it satisfies (8) for some M, thus obtaining a simplified proof of the Helson-Szegö theorem.

Let us remark that the simplest measure 2×2 matrices $\underline{\underline{\nu}} = (\nu_{\alpha\beta}) > 0$ are those of the form

$$\underline{\underline{\nu}}_{tc} = (c_{\alpha\beta}\delta_t) \quad , \tag{9}$$

where $(c_{\alpha\beta})$ is a numerical positive definite matrix and δ_t is the Dirac measure at t. Such $\underline{\underline{\nu}}_{tc}$ are called <u>elementary p.d. matrices</u>.

We have $\underline{\underline{\mu}} \sim \underline{\underline{\nu}}$ iff $\mu_{\alpha\alpha}(f) = \nu_{\alpha\alpha}(f)$, $\alpha = 1, 2$, for all $f \in C(\mathbf{T})$ and $\mu_{12}(f) = \nu_{12}(f)$ only for $f \in H_0^\infty = \{g \in H^\infty : \hat{g}(0) = 0\}$, since $f_1 \cdot f_2 \in H_0^\infty$ whenever $f_1 \in H^\infty$ and $f_2 \in \overline{H_0^\infty}$. Thus, modulo $\sim$, the matrices $\underline{\underline{\mu}}$ are to be considered as functionals on the space $C \times H_0^\infty \times \overline{H_0^\infty} \times C \subset C \times C \times C \times C$. We have then the following variant of Theorem B.

THEOREM B_1. $\underline{\underline{\mu}} \gg \underline{0}$ iff $\underline{\underline{\mu}}$ is the limit, in the weak-* topology of the dual of $C \times H_0^\infty \times \overline{H_0^\infty} \times C$, of convex combinations of elementary p.d. matrices $\underline{\underline{\nu}}_{tc}$.

Theorems B and B_1 give a solution to problem (b), as follows. Problem (b) is a moment problem for the class $\mathcal{R}_M$. In the case of the ordinary moment problem, the Herglotz-Bochner theorem (or the Bochner-Schwartz theorem in $\mathbf{R}^n$) tells us

(i) a given bounded sequence (m_k), $m_{-k} = \overline{m}_k$, satisfies $m_k = \hat{\mu}(k)$ for some positive measure μ iff the associated kernel

$$B_m(j, k) = m(j - k) = m_{j-k}$$

is positive definite (p.d.), and

(ii) a kernel of the form $B_m(j, k) = m(j - k)$ is p.d. iff it has an integral representation

$$B_m(j, k) = \int_{\mathbf{T}} B_t(j, k)d\sigma(t), \quad \sigma \geq 0 \tag{10}$$

where $B_t(j, k) = \exp(it(k - j))$, so that B_m is the limit (in a certain weak topology) of convex combinations of the elementary kernels B_t.

This suggests the following problems:

(c) assign to each sequence $(m_k) = m$ a kernel $K_m(j, k)$, $j, k = 0, \pm 1, \pm 2, \ldots,$ so that correspondence $m \longleftrightarrow K_m$ is 1-1, and

(c_1) $m = \hat{\mu}$ for $\mu \in \mathcal{R}_M$ iff K_m is p.d.,

(c_2) K_m is p.d. iff it is the limit (in a certain topology) of convex combinations of "elementary" p.d. kernels K_t, $t \in \mathbf{T}$.

Thus, any solution of (c_1), (c_2) will give also a "reasonable" solution of (b_1), (b_2).

Now, <u>a solution of</u> (c_1) <u>is easily obtained from</u> (2). In fact, since the $f_{\pm}$ in (2) satisfy

$$\hat{f}_{\pm}(k) = \hat{f}(k)\chi_{\pm}(k)$$

where $\chi_+(k) = 1$ for $k \geq 0$ and zero otherwise and $\chi_+ + \chi_- = 1$, (2) can be rewritten, transforming Fourier $f = \Sigma\, c_k e^{ikt}$, as

$$\Sigma^{\infty}_{j,k=-\infty} K(j, k)\hat{f}(j)\overline{\hat{f}(k)} = \Sigma_{j,k} K(j, k)c_j\bar{c}_k \geq 0 \tag{2b}$$

where

$$K(j, k) = \Sigma^2_{\alpha,\beta=1} \hat{\mu}_{\alpha\beta}(j - k)\chi_\alpha(j)\chi_\beta(k) \tag{11}$$

with $\chi_1 = \chi_+$, $\chi_2 = \chi_-$ and $(\mu_{\alpha\beta})$ given by (3).

If $m = \hat{\mu}$, we denote the kernel K in (11) by K_m.

Since (2b) holds for any finite sequence (c_k),

$$\mu \in \mathcal{R}_M \text{ iff } K_m \text{ in (11) is a p.d. kernel .} \tag{12}$$

This suggests to assign to every arbitrary bounded sequence $m = (m_k)$, $m_{-k} = \bar{m}_k$, the kernel

$$K_m(j, k) = \Sigma^2_{\alpha,\beta=1} m_{\alpha\beta}(j - k)\chi_\alpha(j)\chi_\beta(k) \tag{13}$$

with

$$m_{11} = m_{22} = (M - 1)m, \qquad m_{12} = m_{21} = (M + 1)m \tag{13a}$$

for a given $M > 0$.

It is easy to show (see Section II) that the correspondence $m \longleftrightarrow K_m$

is 1-1, and that if K_m is p.d then $m = \hat{\mu}$ for a positive measure μ, so that by (9) this μ belongs to $\mathcal{R}_M$. We thus obtain the following solution of (c_1) and (b_1):

PROPOSITION 1. [5] A bounded sequence $m = (m_k)$, $m_{-k} = \overline{m}_k$, satisfies $m = \hat{\mu}$ for some $\mu \in \mathcal{R}_M$ if and only if the associated kernel K_m given in (13), (13a) is p.d., or equivalently, if and only if all the principal minors of the matrix $(K_m(j, k))$ are nonnegative.

The simplest kernels of the form (13) are obtained taking $m_{\alpha\beta}(k) = c_{\alpha\beta} \exp(ikt)$, where $c_{\alpha\beta}$ are constants, $\alpha, \beta = 1, 2$. In this case the corresponding kernel will be p.d. iff $(c_{\alpha\beta})$ is a 2×2 p.d. numerical matrix. Such kernels are called elementary p.d. kernels. Thus an elementary p.d. kernel is of the form

$$K_t(j, k) = \Sigma^2_{\alpha,\beta=1} c_{\alpha\beta}\chi_\alpha(j)\chi_\beta(k)e^{i(j-k)t} \tag{14}$$

with $(c_{\alpha\beta})$ a p.d. numerical matrix.

If $Q_{\alpha\beta} = \{(j, k) : \chi_\alpha(j)\chi_\beta(k) \neq 0\}$ then

$$K_t(j, k) = c_{\alpha\beta}e^{i(j-k)t} \quad \text{for} \quad (j, k) \in Q_{\alpha\beta} \quad . \tag{14a}$$

One of our aims is to prove that (c_2) holds if K_m is of the form (13) with $m_{\alpha\beta}$ given by (13a), and the K_t's are given by (14).

THEOREM C. (i) A kernel $K = K_m$, of the form (13), (13a) is p.d. iff it is the limit of convex combinations of elementary p.d. kernels K_t of form (14).

(ii) More precisely, such kernels have an integral representation of the Herglotz-Bochner type, i.e.,

$$K(j, k) = \int_{\mathbf{T}} K_t(j, k)d\sigma(t), \quad \sigma \geq 0 \tag{15}$$

where $K_t = K_{t,c}$, $c = (c_{\alpha\beta})$, is of the form (14) and $c_{\alpha\beta} = c_{\alpha\beta}(t)$.

Observe that the elementary p.d. kernels K_t are of the form (13) but they do not satisfy (13a). Instead, in Theorem C, K is supposed to be of the form (13) and satisfy (13a). Actually, the proof of Theorem C applies even when K does not satisfy (13a), provided the $m_{\alpha\beta}$'s are Fourier transforms of measures. More precisely, we say that $K(j, k)$ is a modified Toeplitz kernel if there are four complex measures $\mu_{\alpha\beta}$, $\alpha, \beta = 1, 2$, in $\mathbf{T}$, such that $\mu_{21} = \overline{\mu_{12}}$ and $K(j, k)$ is given by (8a) for $m_{\alpha\beta} = \hat{\mu}_{\alpha\beta}$. In this case we write $K = K_{\underline{\underline{\mu}}}$, where $\underline{\underline{\mu}} = (\mu_{\alpha\beta})$.

In particular, if the $\mu_{\alpha\beta}$'s are given by (3), then $K_{\underline{\underline{\mu}}} = K_m$, $m = \hat{\mu}$ is the kernel in (11).

Theorem C holds for any modified Toeplitz kernel $K = K_{\underline{\underline{\mu}}}$, even when the $\mu_{\alpha\beta}$'s are not associated with a measure $\mu \in \mathcal{R}_M$:

THEOREM C_1. The sequences $m = \hat{\mu}$, $\mu \in \mathcal{R}_M$, are those bounded sequences whose associated kernels K_m are p.d. modified Toeplitz kernels, and a kernel is a p.d. modified Toeplitz kernel iff it has the integral representation (15).

For each 2×2 matrix $\underline{\underline{\mu}} = (\mu_{\alpha\beta})$ there is a kernel $K_{\underline{\underline{\mu}}}$ given by (15) or, equivalently, by

(15a)
$$K_{\hat{\underline{\underline{\mu}}}}(j, k) = m_{\alpha\beta}(j - k) \quad \text{for} \quad (j, k) \in Q_{\alpha\beta}$$
$$m_{\alpha\beta} = \hat{\mu}_{\alpha\beta}, \qquad \alpha, \beta = 1, 2 \ .$$

But now the correspondence $\underline{\underline{\mu}} \longrightarrow K_{\underline{\underline{\mu}}}$ is not 1-1 anymore. It is easy to see that $K_{\underline{\underline{\mu}}} = K_{\underline{\underline{\nu}}}$ iff $\underline{\underline{\mu}} \sim \underline{\underline{\nu}}$, so that the correspondence $\underline{\underline{\mu}} \longmapsto K_{\underline{\underline{\mu}}}$ is 1-1 (mod $\sim$). Since (2b) is equivalent to (2),

(16)
$$\underline{\underline{\mu}} \gg 0 \quad \text{iff} \quad K_{\underline{\underline{\mu}}} \quad \text{is p.d.} \ .$$

Similarly, $K_{\underline{\underline{\mu}}}(j, k)$ converges pointwise to zero iff $\underline{\underline{\mu}} \longrightarrow 0$ in the weak-* topology of the dual of $C \times H_0^\infty \times \overline{H_0^\infty} \times C$, and $K = K_t$ is of the form (14) iff $K = K_{\underline{\underline{\nu}}}$ with $\underline{\underline{\nu}} = \underline{\underline{\nu}}_{tc}$. From these remarks follows that the first part of Theorem C_1 can be restated as Theorem B_1.

The second part of Theorem C_1 says that the p.d. kernel $K = K_{\underline{\underline{\mu}}}$ has the representation (15). Since

$$K(j, k) = m_{\alpha\beta}(j - k) \quad \text{for} \quad (j, k) \in Q_{\alpha\beta}$$

and

$$K_t(j, k) = c_{\alpha\beta}(t)e^{i(j-k)t} \quad \text{for} \quad (j, k) \in Q_{\alpha\beta} \ ,$$

$$m_{\alpha\beta}(j - k) = \int_{\mathbf{T}} c_{\alpha\beta}(t)e^{i(j-k)t}d\sigma(t)$$
$$= \hat{\sigma}_{\alpha\beta}(j - k) \qquad \text{for} \quad (j, k) \in Q_{\alpha\beta} \ ,$$

where $d\sigma_{\alpha\beta} = c_{\alpha\beta}(t)d\sigma(t)$. Thus the matrix $\underline{\underline{\sigma}} = (\sigma_{\alpha\beta})$ satisfies $K_{\underline{\underline{\sigma}}} = K = K_{\underline{\underline{\mu}}}$ and $\underline{\underline{\sigma}} \sim \underline{\underline{\mu}}$. And since $(c_{\alpha\beta}(t))$ is a numerical p.d. matrix for each t, and $\sigma \geq 0$, it is clear that $\underline{\underline{\sigma}}$ satisfies (15), i.e., $\underline{\underline{\sigma}} > 0$. Therefore the second part of Theorem C_1 can be restated as Theorem B.

Thus Theorems B, B_1 and C_1 are different ways to state the Herglotz-Bochner theorem for modified Toeplitz kernels. In Theorem B_1 it is stated in terms of the matrices $\underline{\underline{\mu}}$ (mod $\sim$), and in Theorem C_1, in terms of the associated Toeplitz kernels $K_{\underline{\underline{\mu}}}$. In Section II, Toeplitz kernels are interpreted in terms of Toeplitz forms or Wiener-Hopf equations. In Section III we indicate briefly how these two approaches can be used to extend the above to $\mathbf{R}$ and $\mathbf{R}^n$. In the case of $\mathbf{R}^n$, the associated Toeplitz kernels $K_{\underline{\underline{\mu}}}$ satisfy an invariance property with respect to the differentiation operators which allows us to establish the generalized Bochner theorem C_1 by means of the general theory of invariant kernels developed in [3]. But this method is not elementary and does not apply to the periodic case. Instead, the matrices $\underline{\underline{\mu}}$ (mod $\sim$) satisfy a "Schur property" which applies to general situations and provides an elementary treatment adapting better to the periodic case.

REMARK. If $\mathcal{M}$ is the set of all matrices of measures $\underline{\underline{\mu}} \gg 0$, then $\mathcal{R}_M$ can be identified with the subset $\mathcal{M}_M = \{\underline{\underline{\mu}} \in \mathcal{M} : \underline{\underline{\mu}} = (\mu_{\alpha\beta})$ given by (3a)$\}$. Theorem C_1 describes the "simple" elements out of which the cone $\mathcal{M}$ is built and the Helson-Szegö theorem comes as a corollary of this description. The analytic description of the "simple" elements of $\mathcal{M}_M$ is a more difficult problem, which was solved by R. Arocena [2].

III. PROOFS, SCHUR'S PROPERTY, AND MODIFIED TOEPLITZ FORMS. We consider 2×2 matrices $\underline{\underline{\mu}} = (\mu_{\alpha\beta})$, $\alpha, \beta = 1, 2$, whose elements are (complex) measures in $\mathbf{T}$, and vector functions $\underline{\underline{f}} = (f_1, f_2) \in C(\mathbf{T}) \times C(\mathbf{T})$. Set

(17) $$\underline{\underline{\mu}}(\underline{\underline{f}}, \underline{\underline{f}}) = \langle \underline{\underline{\mu}}, (\underline{\underline{f}}, \underline{\underline{f}})\rangle = \Sigma^2_{\alpha,\beta=1} \int f_\alpha \bar{f}_\beta d\mu_{\alpha\beta} = \Sigma^2_{\alpha,\beta=1} \mu_{\alpha\beta}(f_\alpha \bar{f}_\beta) .$$

Again, we write $\underline{\underline{\mu}} > 0$ if $\underline{\underline{\mu}}(\underline{\underline{f}}, \underline{\underline{f}}) \geq 0$ for all $\underline{\underline{f}} \in C \times C$. Setting $f_1 = \lambda_1 f$, $f_2 = \lambda_2 f$, λ_1, λ_2 constants, $f \in C$, and letting $f \downarrow \chi_E$, we obtain that

(18) $$\underline{\underline{\mu}} > 0 \text{ iff } (\mu_{\alpha\beta}(E)) \text{ is a numerical p.d. matrix, } \forall E \subset \mathbf{T} .$$

If $d\mu_{\alpha\beta} = \omega_{\alpha\beta} dt$, then

(18a) $$\underline{\underline{\mu}} > 0 \text{ iff } (\omega_{\alpha\beta}(t)) \text{ is a numerical p.d. matrix, } \forall t \in \mathbf{T} .$$

Condition (18) is equivalent to

(18b) $$\mu_{11}(E) \geq 0, \quad \mu_{22}(E) \geq 0 \quad \text{and} \quad |\mu_{12}(E)|^2 \leq \mu_{11}(E)\mu_{22}(E) ,$$

$\forall E \subset \mathbf{T}$. In particular, $\underline{\underline{\mu}} > 0$ implies that μ_{11} and μ_{22} are positive measures.

We say that $\underline{\underline{\mu}} > 0$ <u>with respect to</u> $\mathcal{H}$ and write $\underline{\underline{\mu}} > 0$ (w.r. $\mathcal{H}$) or $\underline{\underline{\mu}} \gg 0$, if

(19) $$\underline{\underline{\mu}}(\underline{\underline{f}}, \underline{\underline{f}}) \geq 0 \text{ for all } \underline{\underline{f}} \in H^\infty \times \overline{H^\infty_0} .$$

Observe that if $\varphi \in C(\mathbb{T})$ then

(20) $$\varphi \geq 0 \text{ iff } \exists(\varphi_n) \subset H^\infty \text{ (or } (\psi_n) \subset \overline{H_0^\infty}) \text{ such that } \varphi_n\overline{\varphi}_n \text{ (respectively, } \psi_n\overline{\psi}_n) \text{ converges to } \varphi \text{ in } C(\mathbb{T}) .$$

In fact, φ is the limit of $f_n\overline{f}_n$, where the $f_n = \Sigma_{-N}^{N} c_k e^{ikt}$ are trigonometric polynomials, and it is enough to set $\varphi_n = \exp(iNt)f_n$.

Let us show that $\underline{\underline{\mu}} \gg 0$ iff

(21) $$\mu_{11} \geq 0, \qquad \mu_{22} \geq 0, \qquad \text{and}$$

(21a) $$|\mu_{12}(f_1\overline{f}_2)|^2 \leq \mu_{11}(|f_1|^2)\mu_{22}(|f_2|^2) \text{ for all } (f_1, f_2) \in H^\infty \times \overline{H_0^\infty} .$$

Replacing $\underline{\underline{f}} = (f_1, f_2)$ by $(\lambda_1 f_1, \lambda_2 f_2)$ in (19), we obtain that $(\mu_{\alpha\beta}(f_\alpha\overline{f}_\beta))$ is a p.d. matrix, for all $\underline{\underline{f}} \in H^\infty \times \overline{H_0^\infty}$. This implies that (21a) holds, and that $\mu_{\alpha\alpha}(|f_\alpha|^2) \geq 0$, for such f_α, $\alpha = 1, 2$. But (20), then, implies (21). Conversely, (21) and (21a) imply (19).

If $\underline{\underline{\mu}}(\underline{\underline{f}}, \underline{\underline{f}}) = \underline{\underline{\nu}}(\underline{\underline{f}}, \underline{\underline{f}})$ for all $\underline{\underline{f}} \in C \times C$, then $\underline{\underline{\mu}} = \underline{\underline{\nu}}$. On the other hand, $\underline{\underline{\mu}}(\underline{\underline{f}}, \underline{\underline{f}}) = \underline{\underline{\nu}}(\underline{\underline{f}}, \underline{\underline{f}})$ only for $\underline{\underline{f}} \in H^\infty \times \overline{H_0^\infty}$ iff

(22) $$\mu_{\alpha\alpha} = \nu_{\alpha\alpha}, \quad \alpha = 1, 2, \quad \text{and} \quad \mu_{12} = \nu_{12} + h, \quad h \in H^1 .$$

In fact, letting $\underline{\underline{f}} = (f_1, 0)$, we have $\mu_{11}(|f_1|^2) = \nu_{11}(|f_1|^2)$ for all $f_1 \in H^\infty$, so that, by (20), $\mu_{11}(\varphi) = \nu_{11}(\varphi)$ for all $0 \leq \varphi \in C$. Hence $\mu_{11} = \nu_{11}$ and, similarly $\mu_{22} = \nu_{22}$. Therefore

$$\mu_{12}(f_1\overline{f}_2) = \nu_{12}(f_1\overline{f}_2) \text{ for all } (f_1, f_2) \in H^\infty \times \overline{H_0^\infty} ,$$

and thus $\mu_{12}(\psi) = \nu_{12}(\psi)$ for all $\psi \in H_0^\infty$. Hence $\mu_{12} - \nu_{12} = h \in H^1$. If $\underline{\underline{\mu}}$ and $\underline{\underline{\nu}}$ are related by (22) we write $\underline{\underline{\mu}} \sim \underline{\underline{\nu}}$. Thus the correspondence $\underline{\underline{\mu}} \mapsto \underline{\underline{\mu}}(\underline{\underline{f}}, \underline{\underline{f}})$, which assigns to each matrix a form on $H^\infty \times \overline{H_0^\infty}$, is 1-1 (mod $\sim$). Each $\underline{\underline{\mu}}$ is a functional on $C \times C \times C \times C$, but as a form $\underline{\underline{\mu}}(\underline{\underline{f}}, \underline{\underline{f}})$, $\underline{\underline{\mu}}$ must be considered mod $\sim$, i.e. as a functional on $C \times H_0^\infty \times \overline{H_0^\infty} \times C$.

Letting $f_1 = 1$ in (21a), we get that if $\underline{\underline{\mu}} \gg 0$ then

(23) $$\left(\|\mu_{12}\|_{H_0^\infty}\right)^2 \leq \|\mu_{11}\| \; \|\mu_{22}\|_{H_0^\infty} \leq \|\mu_{11}\| \; \|\mu_{22}\| ,$$

so that if $\underline{\underline{\mu}} \gg 0$ then $\underline{\underline{\mu}} \sim \underline{\underline{\nu}}$ with $\|\nu_{12}\|^2 \leq \|\mu_{11}\| \; \|\mu_{22}\|$.

A classical result of Schur says that if $\underline{\underline{a}}$ and $\underline{\underline{b}}$ are two p.d. numerical matrices then the matrix $\underline{\underline{c}}$, $c_{\alpha\beta} = a_{\alpha\beta}b_{\alpha\beta}$, is also p.d., or equivalently,

(24) $$\underline{\underline{a}} \text{ is p.d. iff } (\underline{\underline{a}}, \underline{\underline{b}}) = \Sigma_{\alpha,\beta}\, a_{\alpha\beta}b_{\alpha\beta} \geq 0, \quad \forall \underline{\underline{b}} \text{ p.d.} .$$

From (24) and (18) it follows that, for every $\underline{\underline{\mu}}$ and every $\underline{\underline{\varphi}} = (\varphi_{\alpha\beta})$, $\varphi_{\alpha\beta} \in C(\mathbf{T})$, $\alpha, \beta = 1, 2$,

(25) $$\underline{\underline{\mu}} > 0 \text{ and } \underline{\underline{\varphi}} > 0 \text{ imply } \langle \underline{\underline{\mu}}, \underline{\underline{\varphi}} \rangle = \Sigma_{\alpha,\beta} \int \varphi_{\alpha\beta} d\mu_{\alpha\beta} \geq 0$$

or, equivalently,

(25a) $$\underline{\underline{\mu}} > 0 \text{ iff } \langle \underline{\underline{\mu}}, \underline{\underline{\varphi}} \rangle \geq 0 \text{ whenever } \underline{\underline{\varphi}} > 0 .$$

(Here $\underline{\underline{\varphi}} > 0$ means that the 2×2 numerical matrix $(\varphi_{\alpha\beta}(t))$ is p.d., for all t.)

Let A_1 and A_2 be two closed subspaces of $C(\mathbf{T})$, $\mathcal{A} = (A_1, A_2)$ and let $\mathcal{E}_{\mathcal{A}} = C \times A_1\overline{A}_2 \times \overline{A}_1 A_2 \times C$ be the subspace of $C \times C \times C \times C$ of all quadruples of the form $(f, \varphi_1\overline{\varphi}_2, \varphi_2\overline{\varphi}_1, g)$, for $f, g \in C$, $\varphi_1 \in A_1$, $\varphi_2 \in A_2$.

We write $\underline{\underline{\mu}} > 0$ <u>with respect to</u> $\mathcal{A}$ if $\underline{\underline{\mu}}(\underline{\underline{f}}, \underline{\underline{f}}) \geq 0$ whenever $\underline{\underline{f}} \in A_1 \times A_2$. $\mathcal{A}$ <u>satisfies the Schur property</u> if

(26) $$\underline{\underline{\mu}} > 0 \text{ with respect to } \mathcal{A} \text{ and } \underline{\underline{\varphi}} = (\varphi_{\alpha\beta}) \in \mathcal{E}_{\mathcal{A}}, \quad \underline{\underline{\varphi}} > 0 \text{ imply } \langle \underline{\underline{\mu}}, \underline{\underline{\varphi}} \rangle \geq 0 .$$

For example, if $A_1 = A_2 = C$, then $\underline{\underline{\mu}} > 0$ with respect to $\mathcal{A}$ iff $\underline{\underline{\mu}} > 0$, and (25a) expresses that this $\mathcal{A}$ has the Schur property.

PROPOSITION 2. Let $\mathcal{A} = (A_1, A_2)$ be a pair of subspaces of $C(\mathbf{T})$ such that the function 1 can be written as $1 = |\varphi_1|^2 = |\varphi_2|^2$, $\varphi_j \in A_j$, $j = 1, 2$. Then the following conditions are equivalent:

(i) $\mathcal{A}$ satisfies the Schur property,

(ii) $\underline{\underline{\mu}} > 0$ with respect to $\mathcal{A}$ iff $\langle \underline{\underline{\mu}}, \underline{\underline{\varphi}} \rangle \geq 0$ whenever $\underline{\underline{\varphi}} \in \mathcal{E}_{\mathcal{A}}$, $\underline{\underline{\varphi}} > 0$,

(iii) $\underline{\underline{\mu}} > 0$ with respect to $\mathcal{A}$ iff $\underline{\underline{\mu}}$ is the limit, in the weak-* topology of the dual of $C \times C \times C \times C$, of convex combinations of matrices of the form $\underline{\underline{\mu}}_t + h \sim \underline{\underline{\mu}}_t$, where $\underline{\underline{\mu}}_t$ are the elementary p.d. matrices of form $(c_{\alpha\beta}\delta_t)$ and $\underline{\underline{h}} = (0, h_{12}, \overline{h}_{12}, 0)$, h_{12} orthogonal to $A_1\overline{A}_2$,

(iv) $\underline{\underline{\mu}} > 0$ with respect to $\mathcal{A}$ if $\underline{\underline{\mu}}$ is the limit, in the weak-* topology of the dual of $\mathcal{E}_{\mathcal{A}}$, of convex combinations of elementary p.d. matrices $\underline{\underline{\mu}}_t$,

(v) $\underline{\underline{\mu}} > 0$ with respect to $\mathcal{A}$ iff there exists a matrix $\underline{\underline{\nu}} > 0$ such that $\underline{\underline{\mu}} \sim \underline{\underline{\nu}}$ (with respect to $\mathcal{E}_{\mathcal{A}}$),

(vi) $\underline{\underline{\varphi}} \in \mathcal{E}_{\mathcal{A}}$ and $\underline{\underline{\varphi}} > 0$ implies that $\underline{\underline{\varphi}}$ is the limit, in the weak-* topology of $C \times C \times C \times C$, of convex combinations of matrices of the form $(\varphi_{\alpha\beta} = \varphi_\alpha\overline{\varphi}_\beta)$, $(\varphi_1, \varphi_2) \in A_1 \times A_2$.

PROOF. (i) implies (ii): Clear from the definitions.

(ii) implies (iii): Assume (ii) and $\underline{\underline{\mu}} > 0$ with respect to $\mathcal{A}$. Let us prove that $\underline{\underline{\mu}}$ is in the closure of the cone generated by $\underline{\underline{\mu}}_t + \underline{\underline{h}}$. If this were not the case, there would exist an open convex set containing the (translated) linear manifold $\underline{\underline{\mu}} + L$, $L = \{\underline{\underline{h}} = (h_{\alpha\beta}),\ \alpha = 1, 2 : h_{11} = h_{22} = 0,$ $h_{12} = \overline{h}_{21}$ vanishes on $A_1\overline{A}_2\}$, and not containing any element of the convex set generated by $\underline{\underline{\mu}}_t$. By the separation theorem for convex sets, there would then exist a matrix function $\underline{\underline{\varphi}} \in C \times C \times C \times C$ such that $\langle \underline{\underline{\mu}} + \underline{\underline{h}}, \underline{\underline{\varphi}} \rangle \leq c_1$ for all $\underline{\underline{h}} \in L$, and $c \leq \langle \underline{\underline{\nu}}, \underline{\underline{\varphi}} \rangle$ for any $\underline{\underline{\nu}}$ convex combination of $\underline{\underline{\mu}}_t$'s, $c_1 < c$. Since L is a linear subspace, $\underline{\underline{\varphi}}$ vanishes on L, so that $h_{12}(\varphi_{12}) = 0$ for all $h_{12} \perp A_1\overline{A}_2$. Hence $\varphi_{12} \in A_1\overline{A}_2$. Thus $\underline{\underline{\varphi}} \in \mathcal{E}_{\mathcal{A}}$ and, since $(1, 0, 0, 0) \in \mathcal{E}_{\mathcal{A}}$, we may choose $\underline{\underline{\varphi}} \in \mathcal{E}_{\mathcal{A}}$ in such a way as to satisfy $\langle \underline{\underline{\mu}}, \underline{\underline{\varphi}} \rangle < 0$, while $\langle \underline{\underline{\nu}}, \underline{\underline{\varphi}} \rangle \geq 0$ for every $\underline{\underline{\nu}} = \underline{\underline{\mu}}_t$. Taking $\underline{\underline{\mu}}_t = (\lambda_1\overline{\lambda}_1\delta_t, \lambda_1\overline{\lambda}_2\delta_t, \lambda_2\overline{\lambda}_1\delta_t, \lambda_2\overline{\lambda}_2\delta_t)$, $\langle \underline{\underline{\nu}}, \underline{\underline{f}} \rangle = \Sigma_{\alpha,\beta}\, f_{\alpha\beta}(t)\lambda_\alpha\overline{\lambda}_\beta \geq 0$, $\forall t$, $\forall \lambda_1, \lambda_2$. Hence $\underline{\underline{f}} > 0$ and $\underline{\underline{f}} \in \mathcal{E}_{\mathcal{A}}$ and, by the Schur property, $\langle \underline{\underline{\mu}}, \underline{\underline{f}} \rangle \geq 0$ which is a contradiction.

(iii) implies (iv): Clear since every $\underline{\underline{h}}$ in (iii) satisfies $\underline{\underline{h}} \perp \mathcal{E}_{\mathcal{A}}$.

(iv) implies (v): Let $\underline{\underline{\mu}} > 0$ with respect to $\mathcal{A}$. By (iv), $\underline{\underline{\mu}}$ is the limit (in the dual of $\mathcal{E}_{\mathcal{A}}$) of $\{\underline{\underline{\nu}}^\gamma\}$, $\underline{\underline{\nu}}^\gamma > 0$. By (18b) ν_{11}^γ and ν_{22}^γ are positive measures and, since $(1, 0, 0, 1) \in \mathcal{E}_{\mathcal{A}}$, we may assume that $\|\nu_{\alpha\alpha}^\gamma\| \leq c$, $\alpha = 1, 2$. By (18b), then, we have that all $\nu_{\alpha\beta}^\gamma$ are uniformly bounded. Passing to a subnet, $\{\nu_{\alpha\beta}^\gamma\}$ converges in the weak-* topology of the dual of $C \times C \times C \times C$ to a $\underline{\underline{\nu}}$, and $\underline{\underline{\nu}} > 0$. Since $\{\underline{\underline{\nu}}^\gamma\}$ converges in the weaker topology of the dual of $\mathcal{E}_{\mathcal{A}}$ to $\underline{\underline{\mu}}$, $\underline{\underline{\mu}} - \underline{\underline{\nu}} = 0$ on $\mathcal{E}_{\mathcal{A}}$. Conversely, if $\underline{\underline{\nu}} > 0$ and $\underline{\underline{\mu}} - \underline{\underline{\nu}} = 0$ on $\mathcal{E}_{\mathcal{A}}$, then $\Sigma_{\alpha\beta}\, \mu_{\alpha\beta}(\varphi_\alpha\overline{\varphi}_\beta) = \underline{\underline{\mu}}(\varphi_\alpha\overline{\varphi}_\beta) = \underline{\underline{\nu}}(\varphi_\alpha\overline{\varphi}_\beta)$ $= \Sigma_{\alpha\beta}\, \nu_{\alpha\beta}(\varphi_\alpha\overline{\varphi}_\beta) \geq 0$ for every $(\varphi_1, \varphi_2) \in A_1 \times A_2$, since $(\varphi_\alpha\overline{\varphi}_\beta) \in \mathcal{E}_{\mathcal{A}}$ and hence $\underline{\underline{\mu}} > 0$ with respect to $\mathcal{A}$.

(v) implies (vi): Assume (v), $\underline{\underline{\varphi}} \in \mathcal{E}_{\mathcal{A}}$, $\underline{\underline{\varphi}} > 0$ but $\underline{\underline{\varphi}}$ is not in the closure of the cone generated by $(\varphi_\alpha\overline{\varphi}_\beta)$, $(\varphi_1, \varphi_2) \in A_1 \times A_2$. Then there exists a matrix $\underline{\underline{\mu}} > 0$ on that cone, and $\langle \underline{\underline{\mu}}, \underline{\underline{\varphi}} \rangle < 0$. Therefore $\underline{\underline{\mu}} > 0$ with respect to $\mathcal{A}$ and, by (v), $\underline{\underline{\mu}} - \underline{\underline{\nu}} = 0$ on $\mathcal{E}_{\mathcal{A}}$, for $\underline{\underline{\nu}} > 0$. But $\langle \underline{\underline{\mu}}, \underline{\underline{\varphi}} \rangle = \langle \underline{\underline{\nu}}, \underline{\underline{\varphi}} \rangle \geq 0$ since $\underline{\underline{\nu}} > 0$ and $\underline{\underline{\varphi}} > 0$ (see (25)).

(vi) implies (i): Assume (vi) and $\underline{\underline{\mu}} > 0$ with respect to $\mathcal{A}$ and let us prove that $\langle \underline{\underline{\mu}}, \underline{\underline{\varphi}} \rangle \geq 0$ whenever $\underline{\underline{\varphi}} \in \mathcal{E}_{\mathcal{A}}$ and $\underline{\underline{\varphi}} > 0$. By (vi), $\underline{\underline{\varphi}}$ is the limit of convex combinations of elements $(\varphi_\alpha\overline{\varphi}_\beta)$, $(\varphi_1, \varphi_2) \in A_1 \times A_2$, so that $\Sigma\, \mu_{\alpha\beta}(\varphi_\alpha\overline{\varphi}_\beta) \geq 0$, and hence $\langle \underline{\underline{\mu}}, \underline{\underline{\varphi}} \rangle \geq 0$. This completes the proof.

PROPOSITION 3. Assume that $\mathcal{A} = (A_1, A_2)$ has the following properties: (i) $1 = |\varphi_1|^2 = |\varphi_2|^2$, $\varphi_\alpha \in A_\alpha$, $\alpha = 1, 2$. (ii) $\varphi_2, \psi_2 \in A_2$ imply $\varphi_2\psi_2 \in A_2$ and $\overline{A}_2 \subset A_1$. (iii) If $\varphi_2 \in A_2$ and does not vanish, then $1/\varphi_2$ is the weak limit, in $C(\mathbf{T})$, of elements of A_2. (iv) Any function $f \geq 0$, $f \in C$, is the (strong) limit in $C(\mathbf{T})$ of elements of the form $\varphi_1\overline{\varphi}_1$, $\varphi_1 \in A_1$. Then $\mathcal{A}$ satisfies the Schur property.

PROOF. By (vi) of Proposition 2, it is enough to prove that if $\underline{\underline{\varphi}} \in \mathcal{E}_{\mathcal{A}}$, $\underline{\underline{\varphi}} > 0$, then φ is the limit of sums of matrices $\underline{\underline{f}} = (f_{\alpha\beta})$, $f_{\alpha\beta} = f_\alpha \overline{f}_\beta$, $f_\alpha \in A_\alpha$, $\alpha = 1, 2$. Since $(\varphi_{\alpha\beta}(t))$ is a p.d. matrix for all t, we may express the quadratic form $\Sigma_{\alpha,\beta}\, \varphi_{\alpha\beta}(t)\lambda_\alpha\overline{\lambda}_\beta$ as a sum of "squares" by the Lagrange method, as follows. Observe that, by (18a), $\varphi_{12} = \overline{\varphi}_{21}$, and $\varphi_{11} \geq 0$, $\varphi_{22} \geq 0$. We may assume φ_{11}, φ_{22} strictly positive, so that

$$\Sigma_{\alpha,\beta}\, \varphi_{\alpha\beta}\lambda_\alpha\overline{\lambda}_\beta = \frac{1}{\overline{\varphi}_{11}}(\varphi_{11}\lambda_1 + \varphi_{21}\lambda_2)\overline{(\varphi_{11}\lambda_1 + \varphi_{21}\lambda_2)} + \left(\varphi_{22} - \frac{\varphi_{21}\overline{\varphi}_{21}}{\overline{\varphi}_{11}}\right)\lambda_2\overline{\lambda}_2 ,$$

or, equivalently,

$$(27) \qquad (\varphi_{\alpha\beta}) = \frac{1}{\overline{\varphi}_{11}}(\varphi_{11}\overline{\varphi}_{11}, \varphi_{11}\overline{\varphi}_{21}, \overline{\varphi}_{11}\varphi_{21}, \varphi_{21}\overline{\varphi}_{21}) + \left(0, 0, 0, \frac{\overline{\varphi}_{11}\varphi_{22} - |\varphi_{12}|^2}{\overline{\varphi}_{11}}\right).$$

By the properties of $\mathcal{A}$, $1/\overline{\varphi}_{11}$ is the limit of elements of the form $\varphi_1\overline{\varphi}_1$, $\varphi_1 \in A_1$ and, by definition of $\underline{\underline{\varphi}} \in \mathcal{E}_{\mathcal{A}}$, $\varphi_{21} = \psi_1\overline{\psi}_2$, $\psi_\alpha \in A_\alpha$, $\alpha = 1, 2$. Thus, the first term of the sum in (27) is the limit of matrices of the form $\underline{\underline{f}} = (f_{\alpha\beta})$, $f_{\alpha\beta} = f_\alpha\overline{f}_\beta$, where $f_1 \in A_1$, $f_2 = \psi_1\overline{\psi}_2/\overline{\varphi}_1$, so by (ii) and (iii), $f_2 \in A_2$. Therefore, $(f_1, f_2) \in A_1 \times A_2$. Since $\underline{\underline{\varphi}} > 0$, $\varphi_{11}\varphi_{22} - |\varphi_{12}|^2 \geq 0$, and the non-zero component of the second term of the sum in (27) is also the limit of elements of the form $\psi\overline{\psi}$, $\psi \in A_2$. This completes the proof.

From Properties 2 and 3 we obtain immediately the

PROOFS OF THEOREMS B AND C. It is clear that $\mathcal{H} = (H^\infty, \overline{H_0^\infty})$ satisfies properties (i)-(iv) of Proposition 3 (see (20)), therefore $\mathcal{H}$ has the Schur property. Furthermore, $\varphi_1 \in H^\infty$ and $\varphi_2 \in \overline{H_0^\infty}$ imply $\varphi_1\overline{\varphi}_2 \in H_0^\infty$, so that $\underline{\underline{f}} \in \mathcal{E}_{\mathcal{H}}$ iff $f_{12} \in H_0^\infty$, and $\underline{\underline{h}}$ is orthogonal to $\mathcal{E}_{\mathcal{H}}$ iff $h_{11} = h_{22} = 0$ and $h_{12} \in H^1$. Thus, in the case $\mathcal{A} = \mathcal{H}$, Proposition 2 yields Theorems B and C.

PROOF OF THEOREM A_1. Let μ and ν be a pair of positive measures and let $\underline{\underline{\mu}} = (\mu_{\alpha\beta})$ be given by (3a). We want to show that $\underline{\underline{\mu}} \gg 0$ iff conditions (8a) and (8b) are satisfied. As $\underline{\underline{\mu}} \gg 0$ means $\underline{\underline{\mu}} > 0$ with respect to $\mathcal{H} = (H^\infty, \overline{H_0^\infty})$, by Theorem B, this happens iff there exists an $h \in H^1$ such that $\underline{\underline{\nu}} = (\nu_{\alpha\beta})$, $\nu_{11} = \mu_{11} = M\nu - \mu = \mu_{22} = \nu_{22}$, $\nu_{12} = \mu_{12} + h = M\nu + \mu + h$, $\nu_{21} = \overline{\nu}_{12} = M\nu + \mu + \overline{h}$, $\underline{\underline{\nu}} > 0$. But $\underline{\underline{\nu}} > 0$ is equivalent, by (18b), to $M\nu(E) - \mu(E) \geq 0$ and $|M\nu(E) + \mu(E) + h(\chi_E)|^2 \leq |M\nu(E) - \mu(E)|^2$, for all $E \subset \mathbf{T}$, which amounts to (8a) and (8b).

REMARK. As was already observed after the statement of Theorem A_1, if E has Lebesgue measure zero, h being integrable, condition (8b) entails $\mu(E)\nu(E) = 0$ and, by (8a), μ is absolutely continuous. Thus, in the case $\mu = \nu$, we obtain Theorem A, that includes the Helson-Szegö theorem.

PROOF OF PROPOSITION 1. By what was said in (12), it only remains to show that if we assign to each sequence $m = (m_k)$, $m_{-k} = \overline{m}_k$, the kernel K_m given by (13) and (13a), then $m \mapsto K_m$ is 1-1, and that if K_m is p.d. then $m = \hat{\mu}$ for some positive measure μ. That $m \mapsto K_m$ is 1-1 follows immediately from the relation

(28) $$K_m(2k, k) = m_k(M - 1)(\chi_+(k) + \chi_-(k)) \quad ,$$

so that the kernels K of the type $K = K_m$ are characterized by the relation

(29) $$K(j, k) = \frac{1}{M-1}(\chi_+(j-k) + \chi_-(j-k))^{-1}K(2j-2k, j-k)$$
$$\cdot [(M-1)\chi_+(j)\chi_+(k) + (M+1)\chi_+(j)\chi_-(k) + (M+1)\chi_-(j)\chi_+(k)$$
$$+ (M-1)\chi_-(j)\chi_-(k)] \quad .$$

Now assume that K_m is p.d., so for every sequence (λ_k) of finite support,

$$\Sigma_{j,k} K_m(j, k)\lambda_j\overline{\lambda}_k = \Sigma_{j,k} \Sigma_{\alpha\beta} m_{\alpha\beta}(j - k)\chi_\alpha(j)\chi_\beta(k)\lambda_j\overline{\lambda}_k \geq 0 \quad .$$

Letting $\lambda_k = 0$ for $k < 0$, we get $\Sigma_{j,k} m_{11}(j - k)\lambda_j\overline{\lambda}_k \geq 0$ for every finite sequence (λ_k), with support in $k \geq 0$. Since $m_{11}(j - k)$ remains invariant under the change of variables $j \mapsto j + \ell$, the preceding inequality holds for every finite sequence (λ_k). Thus, $m_{11}(j) = (M - 1)m_j$ is a p.d. sequence and, therefore, there is a positive measure μ for which $m_k = \hat{\mu}(k)$.

<u>We have thus proved the theorems stated in Section I</u> and now we are going to interpret them in terms of the modified Toeplitz forms.

Let $\mathbb{Z}_+ = \{n \in \mathbb{Z},\ n \geq 0\}$, $\mathbb{Z}_- = \{n \in \mathbb{Z},\ n < 0\}$, $L_0 = L_0(\mathbb{Z}) = \{m = (m_k)$, sequence of finite support$\}$, $L_\pm(\mathbb{Z}) = \{m \in L_0,\ \text{supp } m \subset \mathbb{Z}_\pm\}$, $L^2_\pm(\mathbb{Z}) = \{m \in L^2(\mathbb{Z})$, $\text{supp } m \subset \mathbb{Z}_\pm\}$.

Let us recall that the classical Wiener-Hopf equation is

(30) $$f(x) + \int_{-\infty}^{\infty} a(x - y)f(y)dy = g(x)$$

with $\hat{a} \in L^\infty$, $f, g \in L^2_+(\mathbb{R})$, and can be rewritten as an equation in $L^2(\mathbb{R})$:

(30a) $$\chi_+(x)f(x) + \int a(x - y)\chi_+(y)f(y)dy = \chi_+(x)g(x) \quad ,$$

for $\chi_+(x) = \chi_{\mathbb{R}_+}$. The discrete analogue of (30a) is the Toeplitz equation in

$L^2(\mathbb{Z})$:

(30b) $$\Sigma_{j=-\infty}^{\infty} a(j-k)\chi_+(j)\chi_+(k)f_j = \chi_+(k)g_k \quad .$$

Accordingly, we assign to each sequence $a = (a_k)$ the kernels

(31) $$L_a(j, k) = a(j-k), \quad T_a(j, k) = a(j-k)\chi_+(j)\chi_+(k) \quad ,$$

the quadratic forms

(31a) $$L_a(f, f) = \Sigma\,\Sigma_{j,k=-\infty}^{\infty} L_a(j, k)f_j\bar{f}_k$$
$$T_a(f, f) = \Sigma\,\Sigma_{j,k=-\infty}^{\infty} T_a(j, k)f_j\bar{f}_k$$

and the operators

(31b) $$(L_af)(k) = \Sigma_{j=-\infty}^{\infty} L_a(j, k)f_j$$
$$(T_af)(k) = \Sigma_{j=-\infty}^{\infty} T_a(j, k)f_j \quad .$$

Now (30b) can be rewritten as

$$T_af = g \quad .$$

If $a = (a_k)$ is a general sequence, the operators L_a, T_a do not act on $L^2(\mathbb{Z})$, but the quadratic forms (31a) are well defined for any $f \in L_0$. We consider also 2×2 matrix Toeplitz operators which correspond to 2×2 systems of Toeplitz equations. To each 2×2 matrix sequence $\underline{\underline{a}} = (a_{\alpha\beta})$, $\alpha, \beta = 1, 2$, $a_{\alpha\beta} = a_{\alpha\beta}(k)$, $k \in \mathbb{Z}$, we assign the matrix kernels

(32) $$L_{\underline{\underline{a}}}(j, k) = (a_{\alpha\beta}(j-k)), \quad T_{\underline{\underline{a}}}(j, k) = (a_{\alpha\beta}(j-k)\chi_+(j)\chi_+(k))$$

and the corresponding quadratic forms

(32a) $$L_{\underline{\underline{a}}}(\underline{\underline{f}}, \underline{\underline{f}}) = \Sigma_{\alpha,\beta}\,\Sigma_{j,k}\, a_{\alpha\beta}(j-k)f_\alpha(j)\overline{f_\beta(k)}$$

(32b) $$T_{\underline{\underline{a}}}(\underline{\underline{f}}, \underline{\underline{f}}) = \Sigma_{\alpha,\beta}\,\Sigma_{j,k}\, a_{\alpha\beta}(j-k)\chi_+(j)\chi_+(k)f_\alpha(j)\overline{f_\beta(k)}$$

defined for $\underline{\underline{f}} = (f_1, f_2) \in L_0 \times L_0$.

If $a_{\alpha\beta} = \hat{\mu}_{\alpha\beta}$, for $\mu_{\alpha\beta}$ measures on $\mathbb{T}$, then these forms define operators $L_{\underline{\underline{a}}}$, $T_{\underline{\underline{a}}}$ on $L^2(\mathbb{Z}) \times L^2(\mathbb{Z})$, and the Toeplitz operator $T_{\underline{\underline{a}}}$ acts on $L^2 \times L^2$ as well as on $L^2_+ \times L^2_+$.

We consider only hermitian sequences a, $a^*(k) = \overline{a(-k)} = a(k)$, and hermitian matrices $\underline{\underline{a}} = (a_{\alpha\beta})$, $a_{21} = a^*_{12}$.

The operators L_a, $a \in L^1(\mathbb{Z})$, form a commutative C^*-algebra, and $c = a * b$ implies $L_c = L_aL_b$, but this is not true for the operators T_a.

However, the Toeplitz operators T_a and, more generally, the Toeplitz forms still retain the following properties:

I. $T_a(f, f) = T_b(f, f), \ \forall f \in L_0$ iff $L_a(f, f) = L_b(f, f), \ \forall f \in L_0$
iff $a = b$.

II. $T_a(f, f) \geq 0, \ \forall f \in L_0$ iff $L_a(f, f) \geq 0, \ \forall f \in L_0$.

III. If $a_\pm \in L_\pm$ then $L_c = L_{a_+} L_b L_{a_-}$ implies $T_c = T_{a_+} T_b T_{a_-}$.

Properties I, II and III underly the theory of Toeplitz or Wiener-Hopf equations (cf. [10]), and since by II, $T_a(f, f) \geq 0$ is equivalent to $a = (a_k)$ be a positive definite sequence, Bochner's theorem applies to the forms T_a.

Properties I, II and III adapt for the matrix case of $T_{\underline{\underline{a}}}$. If $\underline{\underline{a}} = (a_{\alpha\beta})$, $a_{\alpha\beta} = \hat{\mu}_{\alpha\beta}$, $\alpha, \beta = 1, 2$, property II can be stated as

$$\underline{\underline{\mu}} > 0 \text{ iff } T_{\underline{\underline{a}}}(\underline{\underline{f}}, \underline{\underline{f}}) \geq 0, \ \forall \underline{\underline{f}} \in L_0 \times L_0 \tag{33}$$

$$\text{iff } L_{\underline{\underline{a}}}(\underline{\underline{f}}, \underline{\underline{f}}) \geq 0, \ \forall \underline{\underline{f}} \in L_0 \times L_0 .$$

To prove (33) observe that, for $(f_1, f_2) \in L_+ \times L_+$, the expressions (32b) reduce to

$$\Sigma_{\alpha,\beta} \Sigma_{j,k} a_{\alpha\beta}(j - k) f_\alpha(j) \overline{f_\beta(k)} = \Sigma_{\alpha,\beta} \mu_{\alpha\beta}(g_\alpha \bar{g}_\beta)$$

where $f_\alpha = \hat{g}_\alpha$, so that, if $T_{\underline{\underline{a}}}(\underline{\underline{f}}, \underline{\underline{f}}) \geq 0$, letting $f_\alpha = \lambda_\alpha f_+$, $f_+ \in L_+$, $g_\alpha = \lambda_\alpha g_+$, $g_+ \in H^1$, we obtain that

$$\Sigma_{\alpha,\beta} \mu_{\alpha\beta}(|g_+|^2) \lambda_\alpha \bar{\lambda}_\beta \geq 0, \quad \forall \lambda_1, \lambda_2 .$$

Thus if $|g_+|^2 \downarrow \chi_E$, we get $\Sigma_{\alpha,\beta} \mu_{\alpha\beta}(E) \lambda_\alpha \bar{\lambda}_\beta \geq 0$, for all E, hence, $\underline{\underline{\mu}} > 0$.

The Toeplitz forms or operators $T_{\underline{\underline{a}}}$ are obtained by "projecting" the convolution operator $L_{\underline{\underline{a}}}$ on $L^2_+ \times L^2_+$. Now the decomposition $f = f_+ + f_-$, $f_\pm = \chi_\pm(k) f(k)$, establishes a natural isomorphism between $L^2(\mathbb{Z})$ and $L^2_+(\mathbb{Z}) \times L^2_-(\mathbb{Z})$, and assigns to each operator on L^2 an operator on $L^2_+ \times L^2_-$, given by a 2×2 matrix of operators. Therefore it is natural to consider, instead of projections on $L^2_+ \times L^2_+$, projections on $L^2_+ \times L^2_-$, and to introduce the following modified (or skew) Toeplitz operators $S_{\underline{\underline{a}}}$.

For each 2×2 matrix $\underline{\underline{a}} = (a_{\alpha\beta}(k))$, $\alpha, \beta = 1, 2$, we define the modified Toeplitz matrix kernel

(34) $$S_{\underline{\underline{a}}}(j, k) = \begin{pmatrix} a_{11}(j-k)\chi_+(j)\chi_+(k) & a_{12}(j-k)\chi_+(j)\chi_-(k) \\ a_{21}(j-k)\chi_+(k)\chi_-(j) & a_{22}(j-k)\chi_-(j)\chi_-(k) \end{pmatrix}$$

and the corresponding form $S_{\underline{\underline{a}}}(\underline{\underline{f}}, \underline{\underline{f}})$, $\underline{\underline{f}} \in L_0 \times L_0$.

If $a_{\alpha\beta} = \hat{\mu}_{\alpha\beta}$, in particular, if $a_{\alpha\beta} = \hat{F}_{\alpha\beta}$, $F_{\alpha\beta} \in L^\infty(\mathbf{T})$, then the corresponding operator $S_{\underline{\underline{a}}}\underline{\underline{f}}$ acts on $L^2 \times L^2$ as well as on $L^2_+ \times L^2_-$, and $S_{\underline{\underline{a}}}$ is the "projection" of $L_{\underline{\underline{a}}}$ on $L^2_+ \times L^2_-$.

If $a_{\alpha\beta} = \hat{\mu}_{\alpha\beta}$, then from (34) it is clear that

(35) $$S_{\underline{\underline{a}}}(\underline{\underline{f}}, \underline{\underline{f}}) = K_{\underline{\underline{\mu}}}(\underline{\underline{f}}, \underline{\underline{f}}) = \Sigma_{j,k}\, K_{\underline{\underline{\mu}}}(j, k) f_j \bar{f}_k \quad ,$$

where $K_{\underline{\underline{\mu}}}$ is the scalar modified Toeplitz kernel of Section I. Thus, to each modified Toeplitz matrix kernel $S_{\underline{\underline{a}}}$ corresponds a modified Toeplitz scalar kernel $K_{\underline{\underline{\mu}}}$, such that both kernels define the same quadratic form. However, the correspondence $S_{\underline{\underline{a}}} \longrightarrow K_{\underline{\underline{\mu}}}$ is not 1-1, and from what we have seen about the kernels $K_{\underline{\underline{\mu}}}$, and Theorems B and C, we obtain for $S_{\underline{\underline{a}}}$ the following analogue of properties I and II:

Ia. If $\underline{\underline{a}} = (a_{\alpha\beta}) = (\hat{\mu}_{\alpha\beta})$, $\underline{\underline{b}} = (b_{\alpha\beta}) = (\hat{\nu}_{\alpha\beta})$, then $S_{\underline{\underline{a}}}(\underline{\underline{f}}, \underline{\underline{f}}) = S_{\underline{\underline{b}}}(\underline{\underline{f}}, \underline{\underline{f}})$, $\forall\underline{\underline{f}}$ iff

(36) $$a_{11} = b_{11},\ a_{22} = b_{22},\ a_{12} - b_{12} \in L_+,\ a_{21} - b_{21} \in L_- \ .$$

IIa. $S_{\underline{\underline{a}}}(\underline{\underline{f}}, \underline{\underline{f}}) \geq 0$, $\forall\underline{\underline{f}}$ iff $\exists\underline{\underline{b}}$ such that

(37) $$S_{\underline{\underline{a}}}(\underline{\underline{f}}, \underline{\underline{f}}) = S_{\underline{\underline{b}}}(\underline{\underline{f}}, \underline{\underline{f}}),\ \forall\underline{\underline{f}} \text{ and } L_{\underline{\underline{b}}}(\underline{\underline{f}}, \underline{\underline{f}}) \geq 0,\ \forall\underline{\underline{f}} \ .$$

We write $\underline{\underline{a}} \sim \underline{\underline{b}}$, $S_{\underline{\underline{a}}} \sim S_{\underline{\underline{b}}}$ if $S_{\underline{\underline{a}}}(\underline{\underline{f}}, \underline{\underline{f}}) = S_{\underline{\underline{b}}}(\underline{\underline{f}}, \underline{\underline{f}})$, $\forall\underline{\underline{f}}$. We write $\underline{\underline{a}} > 0$ if $L_{\underline{\underline{a}}}(\underline{\underline{f}}, \underline{\underline{f}}) \geq 0$, $\forall\underline{\underline{f}}$ and $\underline{\underline{a}} \gg 0$ if $S_{\underline{\underline{a}}}(\underline{\underline{f}}, \underline{\underline{f}}) \geq 0$, $\forall\underline{\underline{f}}$. Thus, Ia says that the quadratic forms $S_{\underline{\underline{a}}}$ and $S_{\underline{\underline{b}}}$ are equal iff $\underline{\underline{a}} \sim \underline{\underline{b}}$, and IIa says that $\underline{\underline{a}} \gg 0$ iff $\underline{\underline{a}} \sim \underline{\underline{b}}$ with $\underline{\underline{b}} > 0$.

In summary, by projecting on $L^2_+ \times L^2_-$ instead of on $L^2_+ \times L^2_+$, we get the modified Toeplitz operators $S_{\underline{\underline{a}}}$ for the ordinary ones, $T_{\underline{\underline{a}}}$. But this modification changes the character of the operators, and the properties I and II of the Toeplitz operators are no longer valid for the modified ones. However, the Theorems B and C tell us that the modified forms still satisfy substitute properties Ia and IIa. Thus Theorems B and C provide substitutes, for the operators $S_{\underline{\underline{a}}}$, of basic principles underlying the Teoplitz-Wiener-Hopf equations.

Let us still observe that, for a fixed $\underline{\underline{a}} \gg 0$, the corresponding $\underline{\underline{b}}$ in (37) is in general not uniquely determined. Letting $a_{jk} = \hat{F}_{jk}$, $b_{12} = \hat{G}$, we see that there are as many such $\underline{\underline{b}}$ as solutions G of the moment problem

(38) $$\hat{G}(k) = a_{12}(k) \quad \text{for} \quad k \geq 0, \qquad |G| \leq (F_{11}F_{22})^{1/2} \quad .$$

Problem (38) is similar to problem (6), but here $(F_{11}F_{22})^{1/2}$ is not in general a constant. This leads also to a problem of "regions of stability": given $\underline{\underline{b}} > 0$ find all $\underline{\underline{c}} > 0$ with $\underline{\underline{c}} \sim \underline{\underline{b}}$.

IV. SOME GENERALIZATIONS. Let us first extend Theorems A-C to $\mathbf{R}$ and $\mathcal{R}_M = \mathcal{R}_M(\mathbf{R})$. Though in such case, the measures $\mu \in \mathcal{R}_M$ are not finite, they still are of slow growth, since Hf must be in $L^2(\mu)$ for f equal to the characteristic function of an interval. Thus, we may restrict ourselves to measures $\mu \geq 0$ satisfying

(39) $$\int_{\mathbf{R}} (1 + |x|)^{-2} d\mu < \infty \quad ,$$

i.e. to measures of the form $d\mu = (1 + |x|)^2 d\nu$, $\nu \geq 0$ and finite. Let C_2 be the space of all functions of the form $\varphi(x) = (1 + |x|)^{-2}\psi(x)$, with $\psi \in C_\infty$, continuous and vanishing at infinity, and define in C_2 the norm $\|\varphi\| = \|\psi\|_\infty$. Then the measures of (39) are in the dual of the normed space C_2, and the norm of μ is the integral in (39). Working with C_2 instead of C_∞, only minor changes are necessary to adapt the proofs of Propositions 2, 3 and Theorems A and B to the case of $\mathbf{R}$.

As was already remarked at the end of Section I, to treat (in $\mathbf{R}$), Proposition 1 and Theorem C and C_1, it is more convenient to use the associated kernel $K_{\underline{\underline{\mu}}}$ instead of the matrix $\underline{\underline{\mu}}$. Now, in the place of $(\hat{\mu}(k))$ we have to write $\hat{\underline{\underline{\mu}}}(x) = m(x)$, where m is in general a tempered distribution. By transforming Fourier (2), we get, as in (2a) that for all $f \in C_2$,

(40) $$\iint K(x, y)\hat{f}(x)\overline{\hat{f}(y)}dxdy \geq 0 \quad ,$$

where $K = K_{\hat{\mu}}$ is the associated kernel given by

(41) $$K_{\hat{\mu}}(x, y) = \Sigma_{\alpha,\beta=1}^2 \hat{\mu}_{\alpha\beta}(x - y)\chi_\alpha(x)\chi_\beta(y) \quad ,$$

$(\mu_{\alpha\beta})$ given as in (3). (40) says that $\mu \in \mathcal{R}_M$ iff $K_{\hat{\mu}}$ is p.d., and, as in Section I, we obtain Proposition 1, namely, $m = \hat{\mu}$, $\mu \in \mathcal{R}_M$ iff K_m (defined in a form similar to (13) and (13a)) is p.d.

However, both definition (41) and this proof must be made precise, since $m = \hat{\mu}$ and $m_{\alpha\beta} = \hat{\mu}_{\alpha\beta}$ are distributions now, so that $\hat{\mu}_{\alpha\beta}(x - y)\chi_+(x)\chi_+(y)$ does not make sense in general. To make the definition of $K = K_m$ precise, observe that the integral (40) is well defined and its value coincides with (2), in the case when $\hat{f}$ has support away from zero and $f \in \mathcal{S}$ or $f \in C_2$. Setting $B(\hat{f}, \bar{\hat{g}}) = M\int f\bar{g}d\mu - \int (Hf)\overline{(Hg)}d\mu$ we get a continuous bilinear form which determines, by Schwartz' theorem, a kernel $K(x, y) \in \mathcal{D}'(\mathbf{R} \times \mathbf{R})$. By the above remark, this kernel K coincides with K_m, $m = \hat{\mu}$ (defined formally by (41)) on the functions $\hat{f}(x)\overline{\hat{g}(y)}$, with $\hat{f}$, $\hat{g}$ having supports away from zero. Thus (41) and (40) have precise sense. What matters is that K_m is well defined in each open quadrant $Q_{\alpha\beta}$, that is in the open domain $G = \cup Q_{\alpha\beta}$ obtained by deleting the coordinate axes in $\mathbf{R} \times \mathbf{R}$, and in each of these quadrants $K_m(x, y)$ coincides with $m_{\alpha\beta}(x - y) = \hat{\mu}_{\alpha\beta}(x - y)$. Therefore, if $D = i(d/dx)$, then K_m satisfies, in the open domain G, the relation

(42)
$$D_x K_m(x, y) = \overline{D}_y K_m(x, y) \quad .$$

If $K(x, y)$ is a kernel defined in a domain $G \subset \mathbf{R} \times \mathbf{R}$ and A is a (differential) operator, we say that K is A-symmetric if (42) holds (with A instead of D). $K_t(x, y)$ is an elementary A-kernel if $A_x K_t(x, y) = \overline{A}_y K_t(x, y) = tK_t(x, y)$ and K_t is p.d.. For instance, if $G = \mathbf{R} \times \mathbf{R}$ and $A = D$, then K is A-symmetric iff it is of the form $K = B_m(x, y) = m(x - y)$, and K_t is A-elementary iff $K_t = B_t = \exp(it(x - y))$. In such case, the Bochner-Schwartz theorem says that formula (10) holds. More generally, (10) is true for other A-symmetric kernels under quite general hypothesis which were given in detail by Berezanskii in his book [3] and were extended by his collaborators [15]. In our case $G = (\mathbf{R}^n - \{0\}) \times (\mathbf{R}^n - \{0\})$, $A = D$ and $K = \hat{\mu}_{\alpha\beta}(x - y)$ in the quadrant $Q_{\alpha\beta}$, $\hat{\mu}_{\alpha\beta}$ being a second order derivative of a continuous function. It is easy to see that now the elementary kernels are just the K_t's given by (44a) (with variables (x, y) in place of (j, k)) (cf. [3], pp. 651-656). Hence, applying the general Bochner-Schwartz theorem we get Theorems C and C_1 for the case of $\mathcal{R}_M(\mathbf{R})$ and modified Toeplitz kernels in $\mathbf{R}$. Though the representation (10) is now valid only in the open quadrants, by what we saw in Section I, this is enough to deduce Theorem B and its consequences.

REMARK. This proof of Theorem C, based on the D-symmetry of K, does not apply to $\mathbf{T}$ and uses the general non-elementary theory of A-symmetric kernels. On the contrary, the proof given in Section II based on the Schur property of the matrix measure $\underline{\underline{\mu}}$ is elementary and applies both to $\mathbf{T}$ and $\mathbf{R}$.

The notion of (scalar) modified Toeplitz kernels and the above considerations extend to higher dimensions as follows. For simplicity, we consider only the case of $\mathbf{R}^2$.

Fix N disjoint cones Γ_j, such that $\mathbf{R}^2 = \Gamma_1 \cup \ldots \cup \Gamma_N$, and let χ_j be the characteristic function of Γ_j. We assign to each $N \times N$ matrix $\underline{\underline{\mu}} = (\mu_{\alpha\beta})$, $\alpha, \beta = 1, \ldots, N$ ($\mu_{\alpha\beta}$ complex measures in $\mathbf{R}^2$ satisfying (39)), the modified Toeplitz kernel

(43) $$K_{\underline{\underline{\mu}}}(x, y) = \Sigma_{\alpha,\beta=1}^{N} \hat{\mu}_{\alpha\beta}(x - y)\chi_{\alpha}(x)\chi_{\beta}(y) \quad ,$$

so that $K_{\underline{\underline{\mu}}}$ coincides with $\hat{\mu}_{\alpha\beta}(x - y)$ in each

$$Q_{\alpha\beta} = \{(x, y) \in \mathbf{R}^2 \times \mathbf{R}^2 : x \in \Gamma_{\alpha}, \quad y \in \Gamma_{\beta}\} \quad .$$

Therefore, K is symmetric with respect to the differential operators D_{x_1}, D_{x_2} in the domain $G = \cup Q_{\alpha\beta}$. It is easy to see that the elementary kernels $K_t(x, y)$ (i.e. those that verify $D_{x_j}K_t = \overline{D}_{y_j}K_t = t_j K_t$, $j = 1, 2$, $t = (t_1, t_2)$) are now of the form

(44) $$K_t(x, y) = K_{tc}(x, y) = \Sigma_{\alpha,\beta=1}^{N} c_t(\alpha, \beta)\exp(it(x - y))\chi_{\alpha}(x)\chi_{\beta}(y) \quad .$$

Applying Berezanskii's theory we get the following generalization of Theorem C_1:

PROPOSITION 4. The modified Toeplitz kernel $K_{\underline{\underline{\mu}}}$, $\underline{\underline{\mu}} = (\mu_{\alpha\beta})$, $\alpha, \beta = 1, \ldots, N$, is p.d. iff there is a positive measure σ in $\mathbf{R}^2$ and, for each $t \in \mathbf{R}^2$, an elementary kernel K_t of the form (44) such that

(45) $$K_{\underline{\underline{\mu}}}(x, y) = \int_{\mathbf{R}^2} K_t(x, y)d\sigma(t) \quad .$$

As before, we write $\underline{\underline{\nu}} = (\nu_{\alpha\beta}) > 0$ if $\Sigma_{\alpha,\beta=1}^{N} \nu_{\alpha\beta}(f_{\alpha}\overline{f}_{\beta}) \geq 0$ for every system $f = (f_1, \ldots, f_N) \in C_2 \times \ldots \times C_2$. As in Section II, we deduce from Proposition 4 that, if $K_{\underline{\underline{\mu}}}$ is p.d. then there exists a matrix $\underline{\underline{\nu}}$ such that

(46) $$\underline{\underline{\mu}} \sim \underline{\underline{\nu}} \quad \text{and} \quad \underline{\underline{\nu}} > 0 \quad ,$$

where $\underline{\underline{\mu}} \sim \underline{\underline{\nu}}$ means that $K_{\underline{\underline{\mu}}} = K_{\underline{\underline{\nu}}}$ in $G = \cup Q_{\alpha\beta}$.

Let us outline the idea of how to deduce from (45) and (46) an analogue of the Helson-Szegö theorem for $\mathbf{R}^2$.

If $d\mu = \omega(t)dt$, $\omega \geq 0$, is a measure in $\mathbf{R}^2$, we write $\mu \in \mathcal{K}_M$ if

(47) $$\int_{\mathbf{R}^2} |Hf|^2 d\mu \leq M \int_{\mathbf{R}^2} |f|^2 d\mu \quad ,$$

where H is the Riesz operator defined by $(Hf)^{\wedge}(x) = e^{-ix'}\hat{f}(x)$, $x' = x/|x|$. For $x \in \mathbf{R}^2$, x' will denote the point $x/|x|$ of $\mathbf{R}^2$, as well as the corresponding point on the sphere S^1 in $\mathbf{R}^2$. Replacing $|Hf(t)|^2$ by

$$\iint \hat{f}(x)\overline{\hat{f}(y)}e^{-ix'}e^{iy'}e^{ixt}e^{-iyt}dxdy$$

and $|f(t)|^2$ by the similar expression, (47) becomes

(48) $$\iint K_m(x, y)\hat{f}(x)\overline{\hat{f}(y)}dxdy \geq 0, \quad m = \hat{\mu}$$

where

(49) $$K_m(x, y) = (M - e^{i(y'-x')})m(x - y), \quad m = \hat{\omega} = \hat{\mu} .$$

Setting

(50) $$\mu_{x'y'}(t) = (M - e^{i(y'-x')})\mu(t)$$

we have an "infinite dimensional" matrix $\underline{\underline{\mu}} = (\mu_{x'y'})$, $(x', y') \in S^1 \times S^1$, called the associated matrix of μ, the analogue of $(\mu_{\alpha\beta})$ of (3).

Also $K_m(x, y) = \hat{\mu}_{x'y'}(x - y)$ in each angle

(51) $$A_{x'y'} = \{\xi \in \mathbf{R}^2 : \xi = \lambda_1 x' - \lambda_2 y', \quad \lambda_1, \lambda_2 > 0\} .$$

Thus, to each measure μ, we associate the infinite matrices $(\mu_{x'y'})$ and $(m_{x'y'} = \hat{\mu}_{x'y'})$, and the "infinite dimensional" modified Toeplitz kernel K_m of (49), (51). $\mu \in \mathcal{R}_M$ iff K_m satisfies (48), i.e., iff K_m is p.d..

The idea is now to approximate the "infinite dimensional" K_m given by the infinite matrix $(\mu_{x'y'})$ by finite dimensional ones, of the type $K_{\underline{\underline{\mu}}}$, $\underline{\underline{\mu}} = (\mu_{\alpha\beta})$, $\alpha, \beta = 1, \ldots, N$, as in (43).

Let

(52) $$K_m^{\varepsilon}(x, y) = (M + \varepsilon - e^{i(y'-x')})\hat{\mu}(x - y) .$$

Assuming $\mu \in \mathcal{R}_M$, K_m^{ε} is strictly p.d.. Divide $\mathbf{R}^2$ into N equal cones $\Gamma_1, \ldots, \Gamma_N$, $\Gamma_1 \cup \ldots \cup \Gamma_N = \mathbf{R}^2$, so that the oscillation of $e^{ix'}$ is very small in each Γ_j. Let $K_m^{\varepsilon N}$ be the kernel obtained from K_m^{ε} by replacing $e^{i(y'-x')}$ in (52) by the function $e_N(x, y) = \Sigma_{\alpha,\beta=1}^{N} e^{ic_\alpha} e^{-ic_\beta} \chi_\alpha(x)\chi_\beta(y)$, where $c_1, \ldots, c_N$ are constants. It is not hard to verify (using the slow growth of μ and the strict positive definiteness of K_m^{ε}) that, for large N, the new kernels $K_m^{\varepsilon N}$ are still p.d., and that $K_m^{\varepsilon N} \longrightarrow K_m^{\varepsilon}$ for $N \longrightarrow \infty$. But $e_N(x, y)$ is constant in each $\Gamma_\alpha \times \Gamma_\beta$, $\alpha, \beta = 1, \ldots, N$, so that $K_m^{\varepsilon N}$ is a finite dimensional Toeplitz kernel of type (43), $K_m^{\varepsilon N} = K_{\underline{\underline{\mu}}_N}$.

By (46) we have $\underline{\underline{\mu}}_N \sim \underline{\underline{\nu}}_N > 0$, and for each N, the kernel has a representation of the form (45) with $c_t(\alpha, \beta) = c_t^N(\alpha, \beta)$. Letting $N \longrightarrow \infty$, $\varepsilon \longrightarrow 0$, the p.d. matrices $c_t^N(\alpha, \beta)$, $\alpha, \beta = 1, \ldots, N$, will "converge" to a p.d. kernel $c_t(x', y')$ on $(x', y') \in S^1 \times S^1$. Thus we obtain for $K_m = \lim K_m^{\varepsilon N}$ the representation

(53) $$K_m(x, y) = \int_{\mathbf{R}^2} c_t(x', y')e^{it(y-x)}d\sigma(t), \quad \sigma \geq 0 ,$$

where, for each $t \in \mathbf{R}^2$, $c_t(x', y')$ is a p.d. kernel in $S^1 \times S^1$.

The $\underline{\underline{\nu}}^N = (\nu_{\alpha\beta}^N) > 0$, $\alpha, \beta = 1, \ldots, N$, will converge to an infinite dimensional matrix $\underline{\underline{\nu}}$ (of measures in t, depending on parameters x', y'), $\underline{\underline{\nu}} = (\nu_{x',y'})$. This $\underline{\underline{\nu}}$ is p.d. in (x', y') and $\underline{\underline{\mu}} \sim \underline{\underline{\nu}}$, i.e.

$$\mu_{x'y'} = \nu_{x'y'} + a(x', y', t) \quad ,$$

where $\underline{\underline{a}} = (a_{x'y'})$, $a_{x'y'}(t) = a(x', y', t)$ has null associated kernel, $K_{\underline{\underline{a}}} = 0$. This motivates the following definition.

Observe that if h is a measure in $\mathbf{R}^1$ such that $h \in H^1(\mathbf{R}^1)$, and if $a_{x'y'}$ $(x' \in S^0,\ y' \in S_0 = \{-1, +1\})$ is defined by

$$a_{11} = a_{-1(-1)} = 0, \qquad a_{1(-1)} = \bar{h}, \qquad a_{(-1)1} = h$$

then $\hat{a}_{x'y'}(\lambda) = 0$ in the ray $\lambda = \lambda_1 x' - \lambda_2 y'$, $\lambda_1, \lambda_2 \geq 0$. Now if $(a_{x'y'}(t) = \overline{a_{y'x'}(t)})$ is an "infinite dimensional" matrix in $\mathbf{R}^2$, $(x', y') \in S^1 \times S^1$, we say that

$$\underline{\underline{a}} \in H(\mathbf{R}^2) \text{ if } \hat{a}_{x'y'}(\xi) = 0 \text{ for } \xi \in A_{x'y'} \text{ (see (51)) if } x' \neq y'$$
$$\text{and } 0 \text{ does not belong to the support of } \hat{a}_{x'y'}, \text{ if } x' = y'.$$

Then we shall have that $\underline{\underline{\nu}} = \underline{\underline{\mu}} + \underline{\underline{a}}$, for $\underline{\underline{a}} \in H(\mathbf{R}^2)$, and we thus arrive at the following analogue of the Helson-Szegö theorem:

THEOREM D. Let $d\mu = \omega(t)dt$; $\mu \in \mathcal{R}_M(\mathbf{R}^2)$ iff there is a matrix function $\underline{\underline{a}} = (a_{x'y'}(t))$, $(x', y') \in S^1 \times S^1$, $t \in \mathbf{R}^2$, such that $\underline{\underline{a}} \in H(\mathbf{R}^2)$, and $\underline{\underline{\nu}} = \underline{\underline{\mu}} + \underline{\underline{a}} = (\nu_{x'y'}(t))$ is p.d. in $(x', y') \in S^1 \times S^1$, for each $t \in \mathbf{R}^2$. Here $\underline{\underline{\mu}} = (\mu_{x'y'})$ is given by (50).

The unit sphere S^0 of $\mathbf{R}$ has two points, so that $x', y' = -1, 1$, and $\underline{\underline{\nu}}, \underline{\underline{\mu}}, \underline{\underline{a}}$ are 2×2 matrices in the one-dimensional case, thus Theorem D reduces to Theorem A, the sharp version of the Helson-Szegö theorem.

The proofs of Proposition 4 and Theorem D sketched above are based on the symmetry properties of the kernel with respect to the differential operators D_j and do not apply to $\mathbf{T}^2$. Let us indicate the idea of how Proposition 4 can be derived in such case by using the method of Section II. For $\mathbf{T}^2$, Theorem D also follows from Proposition 4.

Let $\Gamma = \{\Gamma_1, \ldots, \Gamma_N\}$, χ_j, $j = 1, \ldots, N$, be as before. Let $\mathcal{H} = \{\mathcal{H}_1, \ldots, \mathcal{H}_N\}$ be N subspaces of C_2. To each N vector $\underline{\underline{\varphi}} = (\varphi_1, \ldots, \varphi_N) \in \mathcal{H}$ we associate the $N \times N$ matrix

(54) $$\underline{\underline{\varphi}} \cdot \underline{\underline{\varphi}} = \underline{\underline{\nu}} = (\nu_{\alpha\beta}) \text{ given by } \nu_{\alpha\beta} = \varphi_\alpha \overline{\varphi_\beta} \quad .$$

Let $\mathcal{E}$ be the set of all linear combinations of the matrices (54). We consider $\mathcal{E}$ as a subspace of $(C_2)^{N\times N}$. We say that $\underline{\underline{\mu}} > 0$ with respect to $\mathcal{H}$, and write $\underline{\underline{\mu}} \gg 0(\mathcal{H})$ if

(55) $$\Sigma_{\alpha,\beta=1}^N \mu_{\alpha\beta}(\varphi_\alpha \overline{\varphi_\beta}) = \langle \underline{\underline{\mu}}, \underline{\underline{\varphi}} \cdot \underline{\underline{\varphi}} \rangle \geq 0, \quad \forall \underline{\underline{\varphi}} \in \mathcal{H} \ .$$

Let $\underline{\underline{\nu}}_t = \underline{\underline{\nu}}_{tc}$ be the elementary matrix of the form $\underline{\underline{\nu}}_t = (c_{\alpha\beta}\delta_t)$, where $(c_{\alpha\beta})$ is a p.d. $N \times N$-numerical matrix. Assume that $\mathcal{H}$ satisfies the properties:

(i) for each $\alpha = 1, \ldots, N$, $1 = |e_\alpha|^2$, $e_\alpha \in \mathcal{H}_\alpha$,

(ii) given $\varepsilon > 0$, $z \in \mathbb{R}^2$ and an arbitrarily small neighborhood V of z, there is $\eta \in \mathcal{H}_\alpha$ such that $\eta(z) = 1$, $|\eta(t)| \leq 2$ for all t and $|\eta(t)| < \varepsilon$ for $t \notin V$.

PROPOSITION 5. Let $\mathcal{H}$ satisfy (i) and (ii) and let $\underline{\underline{\mu}} > 0$ with respect to $\mathcal{H}$. Then $\underline{\underline{\mu}}$ is the limit, in the weak-* topology of the dual of $\mathcal{E}$, of convex combinations of elementary $\underline{\underline{\nu}}_t$'s.

The following is a sketch of the proof, whose idea consists in approximating $\mathcal{H}$ by an $\mathcal{H}_1$ that has the Schur property: if $\underline{\underline{\mu}} \gg 0\ (\mathcal{H})$ and $\underline{\underline{f}} \in \mathcal{E}(\mathcal{H}_1)$, $\underline{\underline{f}} > 0$ then $\underline{\underline{\mu}}(\underline{\underline{f}}) \geq 0$.

Let $\underline{\underline{e}}^1 = (e_1, 0, \ldots, 0) \in \mathcal{H}, \ldots, \underline{\underline{e}}^N = (0, 0, \ldots, e_N) \in \mathcal{H}$, let $L \subset \mathcal{H}$ be a finite set containing $\underline{\underline{e}}^1, \ldots, \underline{\underline{e}}^N$ and let $\mathcal{E}_L$ be the set of all linear combinations of matrices of form (54) with $\underline{\underline{\varphi}} \in L$. Let $L = \{\underline{\underline{\varphi}}_1, \ldots, \underline{\underline{\varphi}}_p\}$. For each $\underline{\underline{\varphi}}_k \in L$ we fix a point z_k, so that the z_k's are distinct points and are all in the same small neighborhood. Then, using property (ii), we modify the values of $\underline{\underline{\varphi}}$ in z_k in such a way to obtain $\underline{\underline{\psi}}$ such that among the vectors $\underline{\underline{\psi}}_j(z_k)$ there are only N independent vectors and $\underline{\underline{\psi}}_k(z_k)$ is independent of the others, for $k = 1, \ldots, p$. Let $\mathcal{E}'_L$ be the set of linear combinations of the matrices of form (54) with $\underline{\underline{\varphi}}$ equal to one of the $\underline{\underline{\psi}}_j$. If $\Sigma_{j=1}^p c_j \underline{\underline{\psi}}_j \underline{\underline{\psi}}_j \geq 0$ then all $c_j \geq 0$, since we can choose $\underline{\underline{\lambda}} = (\lambda_1, \ldots, \lambda_N)$ so that $(\underline{\underline{\psi}}_j(z_k), \underline{\underline{\lambda}}) = 0$ for $j \neq k$, and $= 1$ for $j = k$, hence

$$\Sigma\, c_j (\underline{\underline{\psi}}_j(z_k)\underline{\underline{\psi}}_j(z_k), \underline{\underline{\lambda}}) = \Sigma\, c_j |(\underline{\underline{\psi}}_j(z_k), \underline{\underline{\lambda}})|^2 = c_k \geq 0 \quad .$$

Therefore, $\underline{\underline{f}} = \Sigma_{j=1}^p c_j \underline{\underline{\psi}}_j \underline{\underline{\psi}}_j \geq 0$ and $\underline{\underline{\mu}} \gg 0\ (\mathcal{H})$ imply $\underline{\underline{\mu}}(\underline{\underline{f}}) \geq 0$. Thus $\mathcal{E}'_L$ has the Schur property and, as in the proof of Proposition 2, we can find a convex combination $\underline{\underline{\nu}} = \underline{\underline{\nu}}(L')$ of elementary matrices such that $\underline{\underline{\nu}} = \underline{\underline{\mu}}$ on the elements of $\mathcal{E}'_L$. Letting $L' \longrightarrow L$ and then $p \longrightarrow \infty$, we obtain $\underline{\underline{\mu}}$ as the limit of such $\underline{\underline{\nu}}$, in the weak-* topology of the dual of $\mathcal{E}$ = linear span of $\cup_L \mathcal{E}_L$.

The corollary (46) of Proposition 4 can be interpreted as the property IIa (see (37)) for Γ-<u>modified matrix Toeplitz forms</u> $S_{\underline{\underline{a}}}(x, y)$, defined by

$$S_{\underline{\underline{a}}}(x, y) = (a_{\alpha\beta}(x - y)\chi_\alpha(x)\chi_\beta(y)), \quad \alpha, \beta = 1, \ldots, N$$

where $\underline{\underline{a}} = (a_{\alpha\beta}(x))$ is a $N \times N$ matrix given for the system of cones $\Gamma_1, \ldots, \Gamma_N$ (for $N = 2$ compare with (34)).

V. FURTHER REMARKS. 1. The moment problem (6) is a generalization of the classical Fejér-Schur problem, which corresponds to the case when only a finite number of the c_k in (6) are $\neq 0$. The Fejér-Schur problem consists in finding a function $F \in B_\rho$ (i.e., F analytic in $|z| < 1$, $\|F\|_\infty < \rho$) with N prescribed Taylor coefficients. It is closely related to the Carathéodory problem of finding $F \in \mathcal{C}$ (i.e., F analytic in $|z| < 1$, $\operatorname{Re} F > 0$) with N prescribed coefficients. The solutions of these problems are given in terms of the Toeplitz forms associated with (c_k) and their eigenvalues. In circuit theory the functions $F \in \mathcal{C}$ (defined in the disc or in the half-plane), with real coefficients, are called impedances, and the operators L_a, such that a is a real sequence with support in $\mathbf{Z}_+$ (or in $\mathbf{R}_+$) and such that $L_{a+a^*} > 0$, are called passive operators. The impedances are the Taylor or Laplace transforms of the passive operators. The passive matrices $L_{\underline{\underline{a}}}$ are defined similarly. Replacing $\mathbf{R}_+$ by a cone Γ in $\mathbf{R}^n$ and $\mathbf{R}_+^2$ by the tube T_Γ, we obtain a general notion of Γ-passive operators $L_{\underline{\underline{a}}}$. Vladimirov [16] (cf. [11]) extended the above relation between impedances and passive systems to the general Γ-case. If in the definitions of passive operators and Γ-impedances, we replace the condition $L_{\underline{\underline{a}}+\underline{\underline{a}}^*} > 0$ by the condition $S_{\underline{\underline{a}}+\underline{\underline{a}}^*} > 0$, we obtain a corresponding notion of $(\Gamma_1, \ldots, \Gamma_N)$-passive systems which has several variants. For instance, we have the subclass of the especial $\mathbf{Z}_+$-impedances, which are associated with measures $\mu \in \mathcal{R}_M$, and otherwise we have the generalized $S_{\underline{\underline{a}}}$ matrix impedances. The techniques of circuit theory can be applied to the study of the moment problem (38) and related questions. It would be of special interest to indicate differential operators which are not passive in the ordinary sense but are passive in the modified $S_{\underline{\underline{a}}}$ sense.

2. In the case of continuous functions $\hat{a}_{\alpha\beta} \in C(\mathbf{T})$, the operator $S_{\underline{\underline{a}}}$, $\underline{\underline{a}} = (a_{\alpha\beta})$, can be identified with a Calderón-Zygmund operator in L^2, so that $S_{\underline{\underline{a}}}$ has a Calderón-Zygmund symbol. But it also has the more refined symbol $\underline{\underline{a}}$, determined modulo $\sim$. It may be interesting to consider the relations between $S_{\underline{\underline{a}}}$ and its symbol (mod $\sim$), as connected to a Wiener-Hopf treatment of singular equations.

3. The classical results on Toeplitz forms (asymptotic distribution of eigenvalues, extension of positive forms and related moment problems, invertibility criteria for Toeplitz operators) can be considered for the modified Toeplitz and $(\Gamma_1, \ldots, \Gamma_N)$-Toeplitz forms.

4. Finally, it would be interesting to extend the above Bochner type theorems to the Wightman moment problem in Field Theory.

REFERENCES

1. V. M. Adamjan, D. Z. Arov and M. G. Krein, Infinite Hankel matrices and generalized Carothéodory-Fejér problems, Functional An. and Appl. 2 (1968), 1-19.

2. R. Arocena, to appear.

3. Iu. Berezanskii, Eigenfunctions expansions of selfadjoint operators, Translations AMS (1968). (Kiev, 1965. In Russian).

4. R. R. Coifman and C. Fefferman, Weighted norm inequalities for maximal functions and singular integrals, Studia Math. 51 (1974), 241-250.

5. M. Cotlar and C. Sadosky, Transformée de Hilbert, théorème de Bochner et le problème des moments, I, C. R. Acad. Sc. Paris 285 (1977), 433-436.

6. M. Cotlar and C. Sadosky, Transformée de Hilbert, théorème de Bochner et le problème des moments, II, C. R. Acad. Sc. Paris 285 (1977), 661-664.

7. J. B. Garnett and P. W. Jones, The distance in BMO to L^∞, Ann. of Math. 108 (1978), 373-393.

8. I. C. Gohberg and I. A. Fel'dman, Convolution equations and projection methods for their solutions, Transl. Amer. Math. Soc. (1974).

9. H. Helson and G. Szegö, A problem in prediction theory, Am. Math. Pura Appl. 51 (1960), 107-138.

10. R. A. Hunt, B. Muckenhoupt and R. L. Wheeden, Weighted norm inequalities for the conjugate function and Hilbert transform, Trans. Amer. Math. Soc. 176 (1973), 227-252.

11. A. Korányi and L. Pukánsky, Holomorphic functions with positive real part on polycylinders, Trans. Amer. Math. Soc. 108 (1963), 449-459.

12. B. Muckenhoupt, Weighted norm inequalities for classical operators, these Proceedings.

13. B. Muckenhoupt and R. L. Wheeden, Two weight function norm inequalities for the Hardy-Littlewood maximal function and the Hilbert transform, Studia Math. 55 (1976), 279-294.

14. Z. Nehari, On bounded bilinear forms, Ann. of Math. 65 (1957), 153-162.

15. Nguyen Txi Tam Bak, Ukrainski Math. J. 22 (1970), 247-252.

16. V. Vladimirov, Math. Sb. 93 (1974), 9-17; 94 (1974), 499-515; 98 (1975).

Universidad Central de Venezuela
Caracas, Venezuela

Institute for Advanced Study
Princeton, New Jersey

Proceedings of Symposia in Pure Mathematics
Volume XXXV, Part 1, 1979

SOME SHARP INEQUALITIES FOR CONJUGATE FUNCTIONS

Albert Baernstein II[1]

ABSTRACT. We present non-probabilistic proofs of Burgess Davis's theorems about the best constants in Kolmogorov's conjugate function inequalities. For the L^p inequality, $0 < p < 1$, our proof is based on considerations involving a certain maximal type subharmonic function, which has previously been used to solve extremal problems elsewhere in function theory, for example, in the theory of univalent functions. We also discuss some related results about the behavior of the conjugate function when the original function is rearranged.

1. INTRODUCTION. We consider a real valued integrable function f on the unit circle T, and denote its conjugate function by $\tilde{f}$. The two following results are due to Kolmogorov [7].

WEAK 1-1 INEQUALITY: For every $t > 0$,

$$|\{z \in T: |\tilde{f}(z)| \geq t\}| \leq C\|f\|_1 \, t^{-1} .$$

L^p INEQUALITY: $\tilde{f} \in L^p(T)$ _for every_ $p \in (0,1)$, _and_ $\|\tilde{f}\|_p \leq C_p\|f\|_1$, $0 < p < 1$.

Here $|E|$ denotes the 1-dimensional Lebesgue measure of a set $E \subset T$, $|T| = 2\pi$, while

$$\|f\|_p = \left(\frac{1}{2\pi}\int_0^{2\pi} |f(e^{i\theta})|^p \, d\theta\right)^{1/p} .$$

The values of C and C_p found by Kolmogorov were not the smallest possible ones. The problem of finding the _best possible_ constants was solved only very recently, by Burgess Davis [5], [6]. Davis's proofs are probabilistic.

AMS(MOS) subject classifications (1970). Primary 30A40, 30A42, 30A78, 42A40; Secondary 30A70.

[1]Research supported by the National Science Foundation under grant MCS77-01156.

They make extremely imaginative use of the theory of stopping times for Brownian motion, and of other concepts from probability theory.

Burkholder [4] raised the question of finding non-probabilistic proofs of Davis's theorems. Here I will show how this can be done. The paper [3] contains complete proofs, along with some related results.

For the weak 1-1 inequality my proof amounts to a translation of Davis's proof from probabilistic ideas into the more "classical" notions of potential theory and analytic function theory. This proof is short, and is given here essentially in its entirety. The proof of the L^p inequality, on the other hand, is much more complicated. Davis's method, which I found truly amazing, involves reducing the conjugate functions problem to a certain optimal stopping problem for sums of independent random variables. Probabilistic things seem to be going on there that have no counterpart in classical potential or function theories. My method appears to be completely different. It is based on considerations involving a certain maximal type function, one of whose chief virtues is that it is subharmonic. This method originated in the study of functions meromorphic in the plane [1], and has later been applied in various other places, for example, in the theory of univalent functions [2].

Along the way to Davis's theorem I found a theorem about the behavior of the conjugate function when the original function is rearranged. This result, which seems to be of independent interest, is stated in full in §3. Here we state just a special case.

COROLLARY. *Suppose* $f \in L^1(T)$ *and let* g *denote its symmetric decreasing rearrangement. Then* $\|\tilde{f}\|_p \le \|\tilde{g}\|_p$ *for* $1 \le p \le 2$ *and* $\|\tilde{g}\|_p < \|\tilde{f}\|_p$ *for* $2 \le p \le \infty$.

Bearing in mind that the area under the graphs is the same for $|\tilde{f}|^2$ and $|\tilde{g}|^2$, the corollary may be regarded as asserting that the graph of $|\tilde{g}|^2$ is always longer and flatter than that of $|\tilde{f}|^2$. This phenomenon is quite visible in simple situations, for example when $g(e^{i\theta}) = \cos\theta$ is rearranged so as to be smooth except for a jump discontinuity at one point, in which case we have $\|\tilde{g}\|_\infty = 1$, $\|\tilde{f}\|_\infty = \infty$.

2. THE BEST POSSIBLE WEAK 1-1 INEQUALITY. Let S denote the horizontal strip $\{w \in \mathbb{C} : |\operatorname{Im} w| < 1\}$ in the complex plane $\mathbb{C}$, and denote by G the conformal mapping from the unit disk D onto S with $G(0) = 0$. Thus $G(z) = \frac{2}{\pi} \log \frac{1+z}{1-z}$. Define the constant Θ by

$$\Theta^{-1} = \|\operatorname{Re} G(e^{i\theta})\|_1 = \frac{1}{2\pi} \int_0^{2\pi} \left|\frac{2}{\pi} \log\left|\frac{1+e^{i\theta}}{1-e^{i\theta}}\right|\right| d\theta .$$

THEOREM 1 (B. Davis [5]). For $f \in L^1(T)$,

$$|\{z \in T : |\tilde{f}(z)| \geq t\}| \leq 2\pi\,\Theta\,\|f\|_1\, t^{-1} .$$

Equality holds here for $t = 1$ and $f = \operatorname{Re} G$. Given any t, we can achieve equality by taking $f = t \operatorname{Re} G$. Thus $2\pi\,\Theta$ is the smallest constant for which the weak 1-1 inequality is true.

For the proof of the theorem, we introduce the function $\varphi(w)$, defined in C as follows: $\varphi(w) = |\operatorname{Re} w|$ for $w \notin S$, while inside S, φ is the harmonic function with boundary values $|\operatorname{Re} w|$. We also require $\varphi(w) = 0(w)$ as $w \to \infty$, and then φ is uniquely determined. It is continuous and subharmonic in the whole plane.

The inequalities

(1) $$\varphi(iv) \leq \varphi(0),\ \varphi(u) \geq \varphi(0) \quad (u, v \text{ real})$$

are easily established, by direct computation or otherwise. Thus the inequality $\varphi(w) \leq \operatorname{Re} w + \varphi(0)$ holds on the boundary of the right half strip $S^+ = S \cap (\operatorname{Re} w > 0)$ and, since both functions are harmonic in S^+, the inequality is still true in S^+. Moreover, $\varphi(w) = \varphi(-w)$, and hence we have shown that

(2) $$\varphi(w) \leq |\operatorname{Re} w| + \varphi(0),\ w \in S .$$

Let $E = \{z \in T : |\tilde{f}(z)| \geq 1\}$, $E' = T - E$, and consider the function $F(z) = f(z) + i\tilde{f}(z)$, which is analytic in D, with $F(0) = f(0)$ real. Note that $z \in E$ if and only if $F(z) \notin S$. Thus $\varphi(F(z)) = |\operatorname{Re} F(z)|$ for $z \in E$, while, by (2), $\varphi(F(z)) \leq |\operatorname{Re} F(z)| + \varphi(0)$ for $z \in E'$. Hence

$$\varphi(F(z)) \leq |\operatorname{Re} F(z)| + \varphi(0)\chi_{E'}(z),\ z \in T.$$

Now $\varphi \circ F$ is subharmonic in D, so, taking the mean value over T, and also using (1), we obtain

$$\varphi(0) \leq \varphi(F(0)) \leq \|f\|_1 + \varphi(0)\frac{1}{2\pi}|E'|, \quad \text{and hence} \quad |E| \leq 2\pi\,\varphi(0)^{-1}\|f\|_1 .$$

Finally, $\varphi \circ G$ is harmonic in D, while $\varphi(G(e^{i\theta})) = |\operatorname{Re} G(e^{i\theta})|$. Hence

$$\varphi(0) = \varphi(G(0)) = \frac{1}{2\pi}\int_{-\pi}^{\pi} |\operatorname{Re} G(e^{i\theta})|\,d\theta = \Theta^{-1} ,$$

which gives the desired weak 1-1 inequality for $t = 1$. To obtain it for arbitrary t, simply consider $t^{-1}f$.

3. CONJUGATE FUNCTIONS AND REARRANGEMENTS. We again consider a real valued function $f \in L^1(T)$, along with its Poisson integral $f(z)$, $z \in D$, its conjugate function $\tilde{f}$, and the analytic function $F = f + i\tilde{f}$ with $F(0)$ real. By g we shall denote the symmetric decreasing rearrangement of f on T.

Thus, $g(e^{i\theta}) = g(e^{-i\theta})$, $-\pi$ 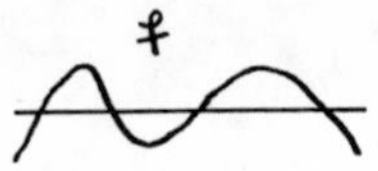 π $-\pi$ 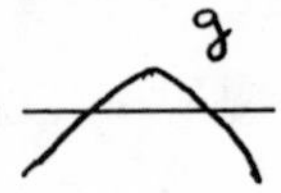 π $g(e^{i\theta})\downarrow$ as $\theta\uparrow$

on $[0,\pi]$, and, for every real t, $|\{z \in T : g(z) > t\}| = |\{z \in T : f(z) > t\}|$.
Let $G = g + i\tilde{g}$.

THEOREM 2. For $0 < r \leq 1$,

(a) $$\int_0^{2\pi} |F(re^{i\theta})|^p \, d\theta \leq \int_0^{2\pi} |G(re^{i\theta})|^p \, d\theta \ , \ 0 < p \leq 2 \ ,$$

(b) $$\int_0^{2\pi} |\tilde{f}(re^{i\theta})|^p \, d\theta \leq \int_0^{2\pi} |\tilde{g}(re^{i\theta})|^p \, d\theta \ , \ 1 \leq p \leq 2 \ ,$$

(c) $$\int_0^{2\pi} |f(re^{i\theta})|^p \, d\theta \leq \int_0^{2\pi} |g(re^{i\theta})|^p \, d\theta \ , \ 1 \leq p < \infty \ .$$

Inequality (c) is easy to prove by classical techniques. For $1 < p \leq 2$, (c) can be combined with M. Riesz's inequality to obtain bounds of the form $\|\tilde{f}\|_p \leq C_p \|\tilde{g}\|_p$, $\|F\|_p \leq C'_p \|G\|_p$ (but this will not give the sharp bounds $C_p = C'_p = 1$). For $0 < p \leq 1$ the only result in the literature along the lines of (a), or (b) I have come across is the theorem of Stein and Weiss [8], which asserts that for <u>characteristic functions</u> f of sets $E \subset T$, $\tilde{f}$ and $\tilde{g}$ have the same distribution function, so that in this case equality holds in (b) for $r = 1$ and $0 < p < \infty$.

As R. O'Neill pointed out to me, inequality (b) for $p = 1$ and $r = 1$ is an easy consequence of the Stein-Weiss theorem together with the fact used below that $\tilde{g}(e^{i\theta})$ is an odd function of θ which is non-negative for $0 \leq \theta \leq \pi$. However, for $0 < p < 1$ it is not clear how to prove by previously known techniques any inequality at all of the form $\|F\|_p \leq C_p \|G\|_p$.

M. Essén and D. Shea have studied uniqueness questions associated with Theorem 2. During the course of their study they observed that my proof shows that when $2 \leq p < \infty$ and $r = 1$ inequalities (a) and (b) are alwyas <u>reversed</u>. (For $0 < r < 1$ this is no longer necessarily the case.) For $0 < p < 1$ the status of (b) and (c) remain in doubt (except for equality in (c) when $r = 1$).

Theorem 2 is proved in detail in [3]. Here we give only an outline of the proof. To keep things simple, we will consider only inequality (a), for $r = 1$, and will also assume that F and G are <u>bounded</u> in D.

The first step is to transfer the problem from the z-plane to the w-plane. This is accomplished by means of the formula

(3) $$\int_{-\pi}^{\pi} |F(e^{i\theta})|^p \, d\theta = p^2 \int_{-\infty}^{\infty} \int_{-\infty}^{\infty} N(w)|w|^{p-2} du\,dv \ .$$

Here N is Nevanlinna's weighted counting function,

$$N(w) = \Sigma\{\log \frac{1}{|z_j|} : F(z_j) = w , \quad |z_j| < 1\} .$$

Since F is bounded, N has compact support. One way to prove the formula is to apply Green's theorem to the differentiated mean

$$r \frac{d}{dr} \int_0^{2\pi} |F(re^{i\theta})|^p \, d\theta ,$$

make the change of variables $w = F(z)$, and then integrate with respect to r.

Denote the counting function for G by $\overline{N}$. Since, when $0 < p \leq 2$, $|w|^{p-2} = |u+iv|^{p-2}$ is a symmetric decreasing function of v for fixed u, integration by parts in (3) shows that the p-norm inequality we want will follow from the inequalities

$$\int_{-v}^{v} N(u+is)ds \leq \int_{-v}^{v} \overline{N}(u+is)ds, \quad v > 0, \quad u \in (-\infty,\infty) . \tag{4}$$

At this point the geometry of the situation comes to our aid. It is not hard to show that the symmetrized function G is <u>univalent</u> in D. Moreover, $G(D)$ is <u>Steiner symmetric</u> with respect to the real axis, i.e. each vertical line intersects $G(D)$ either not at all or in a single interval symmetric about the real axis. It follows that $\overline{N}(w)$ is simply the <u>Green's function</u> of $G(D)$, with pole at $G(0) = F(0)$.

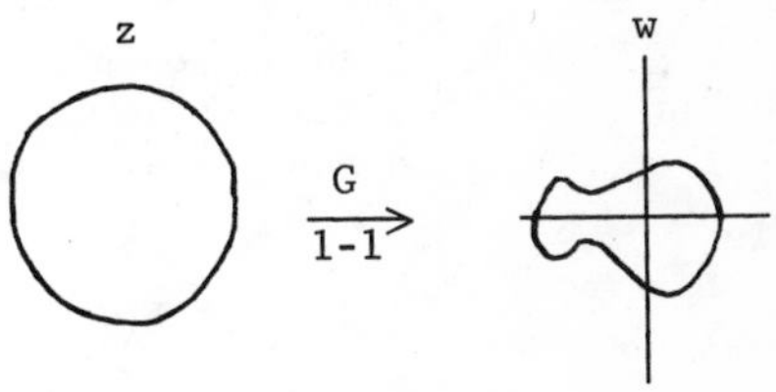

Form now, for w in the upper half plane H,

$$N^*(u+iv) = \sup_E \int_E N(u+is)ds$$

where the sup is taken over all sets $E \subset (-\infty,\infty)$ with $|E| = 2v$. Using the symmetry of the domain $G(D)$, one shows that $\overline{N}(u+is)$ is a symmetric decreasing function of s for every u. It follows that

$$\overline{N}^*(u+iv) = \int_{-v}^{v} \overline{N}(u+is)ds .$$

Hence, inequality (4) is a consequence of the stronger inequality

$$N^*(w) \leq \overline{N}^*(w), \quad w \in H . \tag{5}$$

To prove (5), we first use a theorem of Lehto, which asserts that N is subharmonic in $\mathbb{C}$, except for a logarithmic pole at $w = F(0)$.

Then, using an adapted version of a theorem of mine about *-functions, it follows that

$$N^*(w) + 2\pi(\operatorname{Re} w - F(0))^+ \quad \underline{\text{is subharmonic in}} \quad H.$$

On the other hand, it turns out that the corresponding function with N^* replaced by $\bar{N}^*$ is harmonic in $G(D)^+ = G(D) \cap H$. Thus, the difference $d = N^* - \bar{N}^*$ is subharmonic in $G(D)^+$ and is continuous with compact support in the closure of H.

To finish up, we need one more formula,

$$\int_{-\infty}^{\infty} N(u+is)ds = \int_0^{2\pi} [f(e^{i\theta})-u]^+ \, d\theta - 2\pi(f(0)-u)^+ .$$

Since f and g are equidistributed, the right hand side is the same for g as for f, and from this one concludes that $d(w) \le 0$ when w sits "above" $G(D)^+$. Using this, together with the maximum principle, one shows that in fact $d(w) \le 0$ throughout H, and this is inequality (5).

4. THE BEST POSSIBLE L^p INEQUALITY.

As with the weak 1-1 inequality, the extremal case comes from a certain conformal map. This time, however, the extremal "function" is no longer a function, but degenerates to a singular measure. Let ν_0 be the two-point measure on T supported on $\{-1,1\}$ with mass $\frac{1}{2}$ at $z = 1$ and mass $-\frac{1}{2}$ at $z = -1$. The analytic function associated with ν_0 is

$$G_0(z) = \tfrac{1}{2}\left(\frac{1+z}{1-z} - \frac{1-z}{1+z}\right) = \frac{2z}{1-z^2}, \quad z \in D .$$

This function maps D univalently onto the complex plane with two slits along the imaginary axis removed, one going from i to $+i\infty$, the other from $-i$ to $-i\infty$. Define Θ_p by

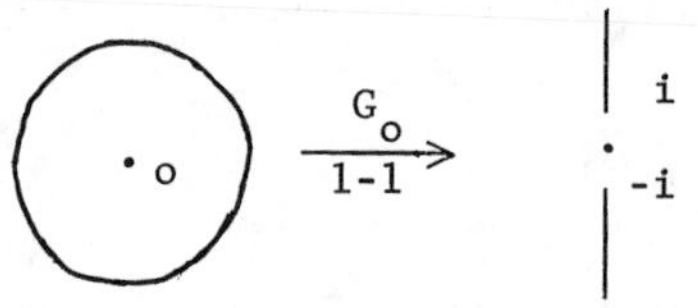

$$\Theta_p^p = \frac{1}{2\pi}\int_0^{2\pi} |G_o(e^{i\theta})|^p \, d\theta = \frac{1}{2\pi}\int_0^{2\pi} |\operatorname{Im} G_o(e^{i\theta})|^p \, d\theta = \frac{1}{2\pi}\int_0^{2\pi} |\sin\theta|^{-p} \, d\theta .$$

THEOREM 3 (B. Davis [6]). For $f \in L^1(T)$ and $0 < p < 1$, $\|\tilde{f}\|_p \le \Theta_p \|f\|_1$.

Consideration of $f(z) = \operatorname{Re} G_o(Rz)$ for R slightly less than one shows that Θ_p is the smallest constant for which the L^p inequality holds.

We deduce Davis's theorem from a more general result. For $0 \le b \le 1$ let ν_b be the two point measure supported on $\{-1,1\}$ with mass $\frac{1}{2}(1+b)$ at $z = 1$

and mass $-\frac{1}{2}(1-b)$ at $z = -1$. Denote by G_b the associated analytic function. Then G_b also univalently maps D onto the plane with two symmetric slits along the imaginary axis removed, and $G_b(0) = b$.

THEOREM 4. Suppose that $f \in L^1(T)$, that $\|f\|_1 = 1$, and that $f(0) = b$. Then the conclusion of Theorem 2 holds, with the G there replaced by G_b.

When $r = 1$, the right hand side in (c) must be replaced by 1 for $p = 1$, ∞ for $p > 1$.

In this theorem it seems possible that (a) and (b) might still hold for $2 < p < \infty$, and that (b) might hold for $0 < p < 1$ (although (c) does not, in general). We point out that this theorem, and also Theorems 1 and 3, is still true if the functions f on T are replaced by real valued measures on T.

To obtain Theorem 3 from Theorem 4, use part (a) with $r = 1$, and then show (it's easy) that, for p fixed, $\int_0^{2\pi} |G_b(e^{i\theta})|^p \, d\theta$ is largest when $b = 0$.

Davis's technique also gives a result which takes $f(0) = b$ into account. However, his method does not seem to give any information about means on circles $|z| = r$ with $r < 1$.

Theorem 4 can be deduced quite easily from Theorem 2, although this is not the way we do it in [3]. By Theorem 2, we may assume that f is symmetric decreasing on T. As before, the integral means inequalities (a), (b), (c) follow from

$$N^*(w) = \int_{-v}^{v} N(u+is)ds \le \int_{-v}^{v} \overline{N}(u+is)ds = \overline{N}^*(w)$$

where $\overline{N}$ denotes now the counting function of G_b. This last inequality is proved by exploiting the subharmonicity of $N^* - \overline{N}^*$ in the singly slit half plane $G_b(D) \cap \{\mathrm{Im}\, w > 0\}$, and by use of the formula

$$\int_{-\infty}^{\infty} N(is)ds = \tfrac{1}{2} \int_0^{2\pi} |f(e^{i\theta})| d\theta - \pi |f(0)| .$$

REFERENCES

1. A. Baernstein, Proof of Edrei's spread conjecture, Proc. London Math. Soc. 26 (1973), 418-434. MR 51 #10629.

2. A. Baernstein, Integral means, univalent functions, and circular symmetrization, Acta Math. 133 (1974), 139-169. MR 54 #5456.

3. A. Baernstein, Some sharp inequalities for conjugate functions, Indiana Univ. Math. J., to appear.

4. D. Burkholder, Harmonic analysis and probability, in Studies in Harmonic Analysis, edited by J. Marshall Ash, 1976.

5. B. Davis, On the weak type (1,1) inequality for conjugate functions, Proc. Amer. Math. Soc. 44 (1974), 307-311. MR 50 #879.

6. B. Davis, On Kolmogorov's inequalities $\|\tilde{f}\|_p \leq C_p\|f\|_1$, $0 < p < 1$, Trans. Amer. Math. Soc. 222 (1976), 179-192. MR 54 #10967.

7. A. N. Kolmogorov, Sur les fonctions harmoniques conjuguées et la series de Fourier, Fund. Math. 7 (1925), 23-28.

8. E. Stein & G. Weiss, An extension of a theorem of Marcinkiewicz and some of its applications, J. Math. Mech. 8 (1959), 263-284. MR 21 #5888.

DEPARTMENT OF MATHEMATICS, WASHINGTON UNIVERSITY, ST. LOUIS, MISSOURI 63130

Proceedings of Symposia in Pure Mathematics
Volume XXXV, Part 1, 1979

CONSTRUCTIONS FOR BMO($\mathbb{R}$) AND $A_p(\mathbb{R}^n)$

by

Peter W. Jones

ABSTRACT. A constructive method is presented for decomposing a function in BMO($\mathbb{R}$) into a bounded function plus the Hilbert Transform of a bounded function. We also answer in the affirmative a question of B. Muckenhoupt on the factorization of A_p weights.

This paper treats two problems in the study of BMO function. A real valued function φ is in BMO($\mathbb{R}^n$) if

$$\|\varphi\|_* = \sup_I \frac{1}{|I|} \int_I |\varphi - \varphi_I| dx < \infty,$$

where φ_I denotes the mean value of φ over I, and the supremum is taken over all cubes $I \subset \mathbb{R}^n$. C Fefferman has proved that BMO($\mathbb{R}^n$) is the dual space of $H^1(\mathbb{R}^n)$, the first Hardy class. A corollary of this duality theorem is that every $\varphi \in$ BMO($\mathbb{R}$) can be represented as $\varphi = u + Hv$, where $\|u\|_\infty, \|v\|_\infty \leq c_0\|\varphi\|_*$. (Here H denotes the Hilbert transform.) See [3] for details. The splitting $\varphi = u + Hv$, for $\varphi \in$ BMO($\mathbb{R}$) can also be deduced by merging the results of Helson and Szegö [6] and Hunt, Muckenhoupt, and Wheeden [8]. Both of these methods of decomposing BMO functions are nonconstructive, relying as they do upon existence theorems from functional analysis. At the Williamstown conference a constructive method was presented for decomposing BMO($\mathbb{R}$) functions. We present below an outline of the proof. Details will appear elsewhere.

Let $\varphi \in$ BMO($\mathbb{R}$) have compact support. N. Th. Varopoulos [10], [11] has found a (constructive) method of building an extension $\Phi(x,y)$ to $\mathbb{R}^2_+$ such that

(i) $\lim_{y \to 0} \Phi(x,y) = \varphi(x)$, a.e. (dx), and

(ii) $\| |\nabla\Phi| dx\, dy\|_c \leq c_0\|\varphi\|_*$.

AMS(MOS) subject classifications (1977). Primary 30A78, 42A40; Secondary 30A76, 30A80.

Here $\|\cdot\|_C$ denotes the Carleson norm of a measure μ on $\mathbb{R}^2_+$,

$$\|\mu\|_C = \sup_I \frac{1}{|I|}|\mu|(I\times(0,|I|),$$

where the above supremum is taken over all intervals I. (If $\|\mu\|_C < \infty$ then μ is said to be a Carleson measure.) Let $\bar\partial = \frac{1}{2}(\frac{\partial}{\partial x} + i\frac{\partial}{\partial y})$. If we can constructively solve the equation $\bar\partial F = \bar\partial\Phi$ and keep $F(x,0) \in L^\infty$, then we will be done. For then $\varphi(x) - F(x,0)$ are the boundary values of a holomorphic function and $H(\varphi-F(\cdot,0)) = -i(\varphi-F(\cdot,0))$. Taking imaginary parts in the last equation we obtain $\varphi = \mathrm{Re}F(\cdot,0) + iH(\mathrm{Im}(F(\cdot,0)))$.

We now indicate how to solve constructively the equation $\bar\partial F = \mu$ with $\|F(x,0)\|_\infty \le c_0\|\mu\|_C$.

Call a sequence $\{z_j\} = \{x_j + iy_j\}$ of points in $\mathbb{R}^2_+$ an I,δ sequence if

$$\inf_j \prod_{k\neq j} \left|\frac{z_j-z_k}{z_j-\bar z_k}\right| \ge \delta > 0.$$

Let $\{z_j\}$ be an I,δ sequence and let B_0 be the Blaschke product with simple zeros on the sequence $\{z_j\}$. Then by Cauchy's theorem it is easy to see that $\bar\partial(\frac{1}{B_0}) = \Sigma\varepsilon_j y_j\delta_{z_j}$, where $c_1 \le |\varepsilon_j| \le \frac{c_2}{\delta}$. (Here δ_{z_j} denotes the Dirac measure at z_j.) Suppose $\{\alpha_j\} \in \ell^\infty$, $\|\{\alpha_j\}\|_{\ell^\infty} \le 1$ and suppose $\mu = \Sigma\alpha_j y_j\delta_{z_j}$. Such a measure is called an I,δ measure. Using a constructive method due to Earl [2] we can find a Blaschke product B_1 such that $\bar\partial(C(\delta)B_1/B_0) = \mu$, where $C(\delta)$ is a constant depending only on δ. Then $\|C(\delta)B_1(x)/B_2(x)\|_\infty = C(\delta)$. So we have solved $\bar\partial F = \mu$, $F(x,0) \in L^\infty$, in the case where μ is an I,δ measure. Now fix δ and suppose that μ is a Carleson measure with support inside some cube Q, and suppose $\|\mu\|_C = 1$. Then by using a method developed by Hoffman [7] we can prove that $\varepsilon(\delta)\mu$ is in the weak star closed convex hull of I,δ meausres, where $\varepsilon(\delta)$ is a constant depending only on δ. Thus we can solve our equation $\bar\partial F = \mu$, $\|F(x,0)\|_\infty \le c_0\|\mu\|_C$ by taking convex combinations of functions of the form $C(\delta)\ B_1/B_2$ where B_1 and B_2 are Blaschke products. This completes the discussion of our first topic.

At this meeting Benjamin Muckenhoupt (see [9]) asked the following question: If $w(x)$ is a positive weight in $A_p(\mathbb{R}^n)$, $p > 1$, is it true that $w = w_1(w_2)^{1-p}$ where $w_1,w_2 \in A_1(\mathbb{R}^n)$? See [1] or [8] for the definition of A_p and A_1. We announce here the following affirmative answer to Muckenhoupt's question. Details will appear elsewhere.

THEOREM. $w \in A_p(\mathbb{R}^n)$, $p > 1$, *if and only if* $w = w_1(w_2)^{1-p}$ *for some* $w_1,w_2 \in A_1(\mathbb{R}^n)$.

On $\mathbb{R}$ some corollaries can be obtained by combining the theorem with the results of [6] and [8]. The theorem can also be used to rather easily deduce the major results of [5]. (See also John Garnett's article, these proceedings.) This is not surprising since the technical obstacles which must be overcome to prove the theorem are much the same as those problems encountered in [5]. The proof of the theorem is rather long and uses some of the techniques of [5]. We point out that the corresponding dyadic version of the theorem is true and not terribly hard to prove. One first notices that if $w \in A_p$ (dyadic) then $w \in A_{p-\varepsilon}$ (dyadic) for some $\varepsilon > 0$. By using a Calderón-Zygmund stopping time argument, we may write the logarithm of w, φ, as a weighted sum of characteristic functions of dyadic intervals plus an L^∞ error, term, $\varphi = \Sigma\alpha_j\chi_{I_j} + u$. Let

$$\varphi_1 = \sum_{\alpha_j>0} \alpha_j\chi_{I_j}, \quad \varphi_2 = \sum_{\alpha_j<0} \alpha_j\chi_{I_j}.$$

Then if the stopping time argument is done adroitly, e^{φ_1} and $e^{\varphi_2/1-p}$ are in A_1 (dyadic). We invite the reader to supply the details for the argument in the dyadic case.

REFERENCES

1. R. Coifman and C. Fefferman, "Weighted norm inequalities for maximal functions and singular integrals," Studia Math., T.LI. (1974), 241-250.
2. J. P. Earl, "On the interpolation of bounded sequences by bounded functions," J. London Math. Soc. (2) 2 (1970), 544-548.
3. C. Fefferman and E. M. Stein, "H^p spaces of several real variables," Acta Math. 129 (1972), 137-193.
4. J. Garnett, these proceedings.
5. ———— and P. W. Jones, "The distance in BMO to L^∞," to appear in Annals of Math.
6. H. Helson and G. Szegö, "A problem in prediction theory," Ann. Math. Pura Appl. 51 (1960), 107-138.
7. K. Hoffman, "Bounded analytic functions and Gleason parts," Annals of Math. 86 (1967), 74-111.
8. R. A. Hunt, B. Muckenhoupt, and R. L. Wheeden, "Weighted norm inequalities for the conjugate function and Hilbert transform," T.A.M.S. 176 (1973), 227-251.
9. B. Muckenhoupt, these proceedings.
10. N. Th. Varopoulos, "BMO functions and the $\bar{\partial}$ equation," Pacific J. Math. 71 (1977), 221-273.
11. ————, "A remark on BMO and bounded harmonic functions," Pacific J. Math. 73 (1977), 257-259.

Proceedings of Symposia in Pure Mathematics
Volume XXXV, Part 1, 1979

STRUCTURE OF SOME SUBALGEBRA OF L^∞ OF THE TORUS

Sun-Yung A. Chang[1]

ABSTRACT. It is proved that the algebra generated by H^∞ (the Hardy space) and C (the continuous functions) on the torus can be written as the linear span of four closed subspaces of L^∞. Some positive indications are given to generalize this type of structure for other subalgebras of L^∞ containing H^∞ on the torus.

Let T and T^2 denote the unit circle and the torus $= T \times T$ in the complex plane $\mathbb{C}$ and $\mathbb{C}^2$ respectively. Let $L^\infty(T)$ and $L^\infty(T^2)$ denote the corresponding Lebesgue spaces of essentially bounded functions, and $H^\infty(T)$ and $H^\infty(T^2)$ the corresponding Hardy subspaces. The subalgebras of $L^\infty(T)$ which contain $H^\infty(T)$ properly has been studied most recently (cf., [12], [13]). They enjoy many interesting properties. In particular, it is known that each such algebra is the linear span of $H^\infty(T)$ and some C^*-algebra ([2]). The phenomenon was first pointed out by D. Sarason [10], who proved that the closed algebra of $L^\infty(T)$ generated by $H^\infty(T)$ and continuous functions $C(T)$ is in fact the linear span $H^\infty(T) + C(T)$. This structural property of subalgebra between $L^\infty(T)$ and $H^\infty(T)$ helps us to understand their maximal ideal spaces (cf. [14],[3],[9]). This fact is also in close connection with the structure of some closed subspaces of functions of bounded mean oscillation (BMO) ([7], [14], [3]). In this note, we will generalize Sarason's work from the unit circle to the torus by proving a similar structure theorem for the closed subalgebra $\langle H^\infty(T^2), C(T^2)\rangle$ of $L^\infty(T^2)$ generated by $H^\infty(T^2)$ and $C(T^2)$. We will also state some property of functions in $L^\infty(T^2)$ which suggest similar results for other subalgebras of $L^\infty(T^2)$ containing $H^\infty(T^2)$.

Let (z_1, z_2) be the two independent variables in the torus T^2. Define C_1 to be the closed algebra of $L^\infty(T^2)$ with generators of the form $u(z_1)h(z_2)$ where $u(z_1)$ is continuous in z_1 and $h(z_2)$ is in H^∞ of z_2. Let C_2 be similarly defined.

AMS (MOS) subject classifications. Primary 30A76, 43A40; Secondary 46J35.

[1]Research supported in part by the National Science Foundation under Grant MCS 77-16281.

THEOREM 1. $\langle H^\infty(T^2), C(T^2)\rangle = H^\infty(T^2) + C_1 + C_2 + C(T^2)$.

REMARK: Rudin [10] has proved that the linear span $H^\infty(T^2) + C(T^2)$ is a closed subspace of $L^\infty(T^2)$, but is not an algebra.

In the following, we will sketch the proof of Theorem 1.

Step 1: For $i = 1,2$; $C_i + C(T^2)$ is a closed subalgebra in $L^\infty(T^2)$. This could be proved by a method similar to the one variable proof that $H^\infty(T) + C(T)$ is a closed algebra of $L^\infty(T)$ (cf. [12]).

Step 2: For $i = 1,2$; $H^\infty(T^2) + C_i$ is a closed subspace in $L^\infty(T^2)$.

Proof: Define the operator $\Lambda_r: L^\infty(T^2) \to L^\infty(T^2)$ by sending each $f(z_1,z_2)$ to $(\Lambda_r f)(z_1,z_2) = \frac{1}{2\pi}\int_0^{2\pi} \frac{1-r^2}{|e^{it}-rz_1|^2} f(e^{it},z_2)\,dt$. Then

(1) $\|\Lambda_r f\|_\infty \le \|f\|_\infty$ for each $f \in L^\infty(T^2)$.

(2) If $h \in H^\infty(T^2)$; then $\Lambda_r h \in H^\infty(T^2) \cap C_1$. To see this, observe that $h \in H^\infty(T^2)$ implies that the function $\Lambda_r h(z_1,z_2)$ is equicontinuous in z_2 with respect to z_1, hence the cesaro means σ_n (which are in C_1) of $\Lambda_r h(z_1,z_2) = \sum_0^\infty h_m(z_2)z_1^m$ with respect to z_1 converges to $\Lambda_r h$ uniformly in $L^\infty(T^2)$ as $n \to \infty$.

(3) If $f \in C_1$ then $\|\Lambda_r f - f\|_\infty \to 0$ as $r \to 1$.

Using (1), (2), (3), we can again apply the argument in [12].

Step 3: To finish the proof we need only to show that $\langle H^\infty(T^2), C(T^2)\rangle \subset H^\infty(T^2) + C_1 + C_2 + C(T^2)$. To see this latter fact it suffices to show that

(*) Given $h \in H^\infty(T^2)$; $\bar{z}_1^n \bar{z}_2^m h \in H^\infty(T^2) + C_1 + C_2 + C(T^2)$ boundedly independent of n, m, i.e., there exists some constant K independent of n, m and functions g, u_1, u_2, u belonging to $H^\infty(T^2)$, C_1, C_2, $C(T^2)$ respectively with the supremum norm of each function bounded by $K\|h\|_\infty$ and $\bar{z}_1^n \bar{z}_2^m h = g + u_1 + u_2 + u$.

Since $\bar{z}_1^n h \in H^\infty(T^2) + C_1$ by inspection, we can apply the result in step 3 to conclude $\bar{z}_1^n h \in H^\infty(T^2) + C_1$ boundedly independent of n. Repeat the same argument using results in step 2 and 1, we then get (*).

The above theorem suggests that to study subalgebras of $L^\infty(T^2)$, we should consider the behavior of the algebra in the four directions z_1, $\bar{z}_1$, z_2, $\bar{z}_2$ independently. As the results ([14], [2]) in the unit circle suggest, any attempt to study the structure of subalgebras of L^∞ containing H^∞ should correspond to a parallel structural problem of some closed subspaces of BMO (cf., C. Fefferman and E. Stein, [7] for BMO). For example,

The following proposition is easy to verify.

PROPOSITION 1. Suppose A is a C^*-algebra containing constants in $L^\infty(T)$, Then $H^\infty(T) + A$ is a closed subspace of $L^\infty(T)$ if and only if $A + P(A)$ is a closed subspace of BMO(T), where P is the projection $L^2(T) \to H^2(T)$.

As mentioned by C. Fefferman [6], R. Coifman and G. Weiss [5], it is an interesting problem to generalize the fundamental theorems about BMO in [7] to the torus. The basic difficulty was indicated by an ingenious counterexample of L. Carleson [1] concerning Carleson measures. (We will state the result below.) In the following we will report some effort in this direction.

For each point $z = re^{i\theta_0}$ in the unit disk D, let I_z denote the arc $\{e^{i\theta} \mid |\theta-\theta_0| < 1-r\}$ and for each arc $I \subset T$, let $S(I)$ denote the region $\{z \in D \mid I_z \subset I\}$. For each function $f \in L^p(T^2)$, let $F(r_1e^{i\theta_1}, r_2e^{i\theta_2}) = P_{r_2} * (P_{r_1} * f)(\theta_1, \theta_2)$ denote its bi-harmonic extension to the bi-disc D^2; where P_{r_i} is the Poisson kernel. Applying the reasoning in [7] in each variable θ_1, θ_2 independently, we can prove

PROPOSITION 2. For a function $f \in L^2(T^2)$, the following are equivalent:

(a) There is a constant K such that for each arc $I, J \subset T$, there exists functions $f_1(\theta_1)$, $f_2(\theta_2)$ with

$$\frac{1}{|I|\times|J|} \iint_{I\times J} |f(\theta_1,\theta_2) - f_1(\theta_1) - f_2(\theta_2)|^2 \, d\theta_1 \, d\theta_2 \le \text{constant } K < \infty.$$

(b) Define

$$|\nabla F(z_1,z_2)|^2 = \left(\left|\frac{\partial^2 F}{\partial z_1 \partial z_2}\right|^2 + \left|\frac{\partial^2 F}{\partial z_1 \partial \bar{z}_2}\right|^2 + \left|\frac{\partial^2 F}{\partial \bar{z}_1 \partial z_2}\right|^2 + \left|\frac{\partial^2 F}{\partial \bar{z}_1 \partial \bar{z}_2}\right|^2 \right)(z_1,z_2).$$

Then

$$\sup_{I,J} \frac{1}{|I|\times|J|} \iint_{S(I)\times S(J)} |\nabla F(z_1,z_2)|^2 \log\frac{1}{|z_1|} \log\frac{1}{|z_2|} \, dz_1\, d\bar{z}_1\, dz_2\, d\bar{z}_2 < \infty.$$

Let $P_{i,j}$ $(i,j = \pm 1)$ denote the projection from $L^2(T^2)$ to the closed subspace of $L^2(T^2)$ spanned by $\{z_1^{in}, z_2^{jm}$, n, m non-negative integers$\}$. Since the gradient ∇ is invariant under projections $P_{i,j}$, it follows from the proposition that the functions in $\sum_{i,j=1} P_{i,j}(L^\infty)$ satisfy conditions in (a), (b). One doubts conditions (a), (b) could characterize the space $\sum_{i,j=1} P_{i,j}(L^\infty)$, because of the work of Carleson [1]. Carleson constructed an example to show that measures μ in D^2 satisfy the condition in (b), i.e.

(b') $\mu(S(I) \times S(J)) \leq C(|I| \times |J|)$ for all arcs $I, J \subset T^2$

may not have the property

(c) $$\iint_{D^2} |F(z_1,z_2)|^p \, d\mu \leq C_p \iint_{T^2} |f(\theta_1,\theta_2)|^p \, d\theta_1 \, d\theta_2 \quad \forall f \in L^p(T^2)$$

$$1 < p < \infty.$$

Most recently, R. Fefferman [8] proved that functions satisfying condition (a) may not be in $L^4(T^2)$. However, if one considers arbitrary open, connected set $U \subset T^2$, instead of rectangles $I \times J$ and defines the corresponding region $S(U)$ in D^2 to be $\{(z_1,z_2) \in D^2 \mid I_{z_1} \times I_{z_2} \subset U\}$, one has

PROPOSITION 3. Let μ be a positive measure on D^2, then μ is bounded in $L^p(T^2)$, i.e., condition (c) holds if and only if

(d) $\mu(S(U)) \leq C|U|$ for all connected, open sets $U \subset T^2$.

THEOREM 2. For a function $f \in \sum_{i,j=1} P_{i,j}(L^\infty)$; the measure $d\mu_f(z_1,z_2) = |\nabla F(z_1,z_2)|^2 \log \frac{1}{|z_1|} \log \frac{1}{|z_2|} \, dz_1 \, d\bar{z}_1 \, dz_2 \, d\bar{z}_2$ satisfies condition (d).

There is strong evidence the condition in Theorem 2 should characterize the space $\sum_{i,j=\pm 1} P_{i,j}(L^\infty)$. Details of the above results will appear in [4].

REFERENCES

1. L. Carleson, A counter example for measures bounded on H^p for the bi-disc, Institut Mittag-Leffler, report No. 7, 1974.

2. S.Y. Chang, Structure of subalgebras between L^∞ and H^∞, Trans. Amer. Math. Soc. 227(1977), 319-332.

3. S.Y. Chang, A characterization of Douglas algebras, Acta Math. 137 80-89.

4. S.Y. Chang, Carleson measure on bi-disc, to appear in Annals of Math.

5. R. Coifman and G. Weiss, Extensions of Hardy spaces and their use in harmonic analysis, Bull. Amer. Math. Soc. No. 4, 83(1977), 569-645.

6. C. Fefferman, Harmonic Analysis and H^p spaces, MAA Studies, vol. 13, 38-75.

7. C. Fefferman and E. Stein, H^p spaces of several variables, Acta Math. 129(1972), 137-193.

8. R. Fefferman, A note on BMO functions of two variables, (preprint).

9. D. Marshall, Subalgebras of L^∞ containing H^∞, Acta Math. 137 91-98.

10. W. Rudin, Spaces of type $H^\infty + C$, Annales de l'Institute Fourier 25, 1(1975), 99-125.

11. D. Sarason, An attendem to "past and future," Math. Scand. 301(1972), 62-64.

12. D. Sarason, Algebras of functions on the unit circle, Bull. Amer. Math. Soc. 79(1973), 286-299.

13. D. Sarason, Algebras between L^∞ and H^∞, Spaces of Analytic Functions, Kristansand Norway, Springer Lecture Notes 512, 1975.

14. D. Sarason, Functions of vanishing mean oscillation, Trans. Amer. Math. Soc. 207(1975), 391-405.

UNIVERSITY OF MARYLAND
MATHEMATICS DEPARTMENT
COLLEGE PARK, MARYLAND
20742

Proceedings of Symposia in Pure Mathematics
Volume XXXV, Part 1, 1979

A GEOMETRIC CONDITION WHICH IMPLIES BMOA

David A. Stegenga[1]

The space BMOA is the collection of holomorphic functions on the unit disc D which are in the Hardy space H^1 and whose boundary values belong to the space of functions with bounded mean oscillation, BMO, of John and Nirenberg [7].

We give a geometric condition for the range of a holomorphic function which insures membership in BMOA. This result was independently discovered by Hayman and Pommerenke [6]. We give several applications of this result and sketch a proof of the sufficiency using a distributional characterization of bounded mean oscillation.

Let $D(w,r)$ denote the open disc centered at w of radius r. Denote by $\text{Cap}\,(E)$ the logarithmic capacity of a set E.

THEOREM. Let Ω be a connected open subset of the plane and assume that

$$(*) \qquad \inf_{w\in\Omega} \frac{\text{Cap}\,(D(w,r)\setminus\Omega)}{r} = \delta > 0$$

holds for some positive numbers r,δ. There is a constant $c = c(\delta)$ depending on δ but not on r or Ω such that a function f is in BMOA and the BMOA norm $||f||_* \leq c(\delta)r$ whenever f is holomorphic on D and $f(D)$ is contained in Ω.

Conversely, if Ω does not satisfy (*) then there is a holomorphic function with $f(D)$ contained in Ω but f is not in BMOA.

We now give some geometric conditions of a more elementary nature which imply BMOA. Let $m_w(t)$ denote the Lebesgue measure of the set of numbers r, $0 \leq r \leq t$, for which the circle $|z - w| = t$ is contained in Ω.

COROLLARY 1. The condition

$$\sup_{w\in\Omega} \frac{m_w(r)}{r}\, d < 1$$

AMS(MOS) subject classification (1970). Primary 30A78, 30A70; Secondary 30A44.

[1]The author is partially supported by a grant from the National Science Foundation.

implies (*) with $\delta \geq \frac{1}{4}(1 - d)$.

PROOF. The circular projection mapping z into $|z|$ decreases distances and hence decreases capacity, see Pommerenke's book [9, chapter 10]. Taking w to be the origin we see that the circular projection of $D(w,r)\backslash\Omega$ is a set whose complement in the interval $[0,r]$ has measure $m_w(r)$. Since the capacity of a linear set is at least one quarter of its length the result follows.

COROLLARY 2. If the vertical cross-sectional measures of $f(D)$ are bounded by d then f is in BMOA and $||f||_* \leq cd$ for some constant c independent of f .

PROOF. Take $\Omega = f(D)$ and $r = d$ then $D(w,r)\backslash\Omega$ contains a linear set of measure at least equal to d .

COROLLARY 3. For w in Ω , let $\Omega_w(r)$ be the component of $D(w,r) \cap \Omega$ containing w . If

$$\sup_{w\in\Omega} \frac{\text{area } (\Omega_w(r))}{\pi r^2} = d^2 < 1$$

for some $r > 0$ then (*) holds with $\delta \geq \frac{1}{4}(1 - d)$.

PROOF. A calculation shows that $\pi m_w(r)^2 \leq \text{area } (\Omega_w(r))$ and hence the result follows from Corollary 1.

We remark that Corollary 1 can be applied, in particular, when Ω does not contain circles of arbitrary large radii centered in Ω . If Ω is simply connected this yields a result of Pommerenke [8] that a univalent function f is in BMOA if and only if $f(D)$ contains no discs of arbitrary large radii.

Corollary 2 contains a result of Baernstein [3]. If f is a non-vanishing univalent function on D then the vertical cross-sectional measures of $\log f(D)$ are bounded by 2π and hence $\log f$ is in BMOA.

Finally, a special case of Corollary 3 is the case that area $(f(D))$ is finite. In this case we also obtain $||f||_* \leq c \ (\text{area } f(D))^{1/2}$. Since BMOA is contained in H^p for all $p < \infty$, see [7], this generalizes the results in [2, Theorem 1] and [4].

We now indicate a proof of the theorem. Let E be a compact subset of $D(\alpha,r)$ and B be the unbounded component of the complement of E . For $t > r$, let V_t be the harmonic function on $D(\alpha,t) \cap B$ which is one on $|z - \alpha| = t$ and zero on the boundary of B . We will assume $\alpha \in B$.

LEMMA. If $\text{Cap } E \geq \delta r$ where $\delta > 0$ then

$$V_t(\alpha) \leq \frac{\log \delta^{-1}}{\log t/r}$$

PROOF. We will assume that $\alpha = 0$ and that the boundary of B consists of a finite number of Jordan curves. Let $\text{Cap } E = e^{-\gamma}$ so that $\gamma \le \log 1/\delta r$. Let g be the Green's function of B with pole at ∞. It is known that $g(z) - \log|z|$ is harmonic at ∞ and $\lim_{|z|\to\infty} (g(z) - \log|z|) = \gamma$, see [1].

We easily obtain $g(z) \ge \log|z|/r$ for $|z| \ge r$ and hence $V_t(z) \le g(z)(\log t/r)^{-1}$ on $B \cap D(0,t)$. Extending g to be zero on the complement of B we have a subharmonic function on $|z| \le r$ and hence

$$V_t(0) \le \frac{g(0)}{\log t/r} \le (\log t/r)^{-1} \left[\frac{1}{2\pi}\int_0^{2\pi} g(re^{i\theta})d\theta\right]$$

$$= (\log t/r)^{-1} [\gamma + \log r] \le \frac{\log 1/\delta}{\log t/r}$$

which proves the lemma.

PROOF OF THE THEOREM. Since any subdomain of Ω will also satisfy (*) it suffices to assume that f is holomorphic in a neighborhood of the closure of D and $\Omega = f(D)$. This involves a standard argument using the functions $f_s(z) = f(sz)$ for $0 < s < 1$. We may also assume that $r = 1$ since a standard dilation argument we complete the general case.

Assuming these simplifications put $w = f \circ \phi(0)$ and $g = f \circ \phi$ where ϕ is a Möbius transformation of the disc. Observe that $g(D) = \Omega$. Let $t > 1$ and $\delta' < \delta$. Let E be a compact subset of $D(w,r)\backslash\Omega$ with $\text{Cap } E \ge \delta'$. By combining the previous lemma with Theorem 4 [5] we obtain, after letting δ' tend to δ, that

$$m\{e^{i\theta} : |f \circ \phi(e^{i\theta}) - f \circ \phi(0)| > t\} \le \frac{\log \delta^{-1}}{\log t} \tag{1}$$

where m is normalized Lebesgue measure on the circle.

Let $I_\alpha = \{e^{i\theta} : |\theta| \le \alpha\}$ and take $\phi(z) - (z + r)/(1 + rz)$ where $1 - r = m(I_\alpha)$ in (1). By a change of variable argument we obtain for $I = I_\alpha$

$$\frac{m\{e^{i\theta} \in I : |f(e^{i\theta}) - C_I| > t\}}{m(I)} \le c\,\frac{\log \delta^{-1}}{\log t} \tag{2}$$

where c is a constant independent of I and f and C_I is a function depending on I and f. After a rotation we obtain (2) for all arcs I.

Taking t large enough we obtain

$$\sup_I \frac{m\{e^{i\theta} \in I : |f(e^{i\theta}) - C_I| > t\}}{m(I)} < \frac{1}{2}$$

and applying a result in [10] we obtain that f is in BMO and $||f||_* \le c(\delta)$.

This completes the sufficiency argument, the necessity of (*) can be found in [6].

CONCLUDING REMARKS. It can be shown that (*) can be replaced with

$$\inf_{w \in \Omega} \frac{\mathrm{Cap}(D(w,r)\setminus\Omega_w(r))}{r} > 0$$

which is easier to apply in some situations. In addition, if Ω satisfies (*) then r can be chosen so that the corresponding δ is 1/2. This results in the best norm estimate, namely, the $\sup ||f||_*$ where f maps D into Ω is comparable to r.

Finally, we observe that there are in general no restriction on the range of a function in BMOA. In particular, there is a function f in BMOA with $f(D)$ being the entire complex plane. To see this we use the Riemann mapping theorem to produce a function which is at worst two-to-one, takes on all values, and whose schlicht discs have radii bounded by one. Simply take an overlapping spiral. By a theorem of Pommerenke [8, Theorem 1] any such function will be in BMOA.

REFERENCES

1. L.V. Ahlfors, Complex Analysis, 2nd ed., McGraw-Hill Book Co., New York, 1966.

2. H. Alexander, B.A. Taylor, and J.L. Ullman, "Areas of projections of analytic sets", Inv. Math., 16 (1972), 335-341.

3. A. Baernstein, "Univalent functions and the class BMO", Mich. Math. J., 23 (1976), 217-223.

4. L. Hansen, "The Hardy class of a function with slowly-growing area", Proc. Amer. Math. Soc., 45 (1974), 409-410.

5. W.K. Hayman, and A. Weitsman, "On the coefficients and means of functions omitting values", Math. Proc. Camb. Phil. Soc., 77 (1975).

6. W.K. Hayman, and Ch. Pommerenke, "On analytic functions of bounded mean oscillation", Bull. Lon. Math. Soc., 10 (1978), 219-224.

7. F. John, and L. Nirenberg, "On functions of bounded mean oscillation", Comm. Pure Appl. Math., 14 (1961), 415-426.

8. Ch. Pommerenke, "Schlichte funktionen und analytische funktionen von beschränkter mittlerer oszillation", Comm. Math. Helv., 152 (1977), 591-602.

9. ________________, Univalent Functions, Vandenhoeck and Ruprecht, Göttingen, 1975.

10. J. Stromberg, "Bounded mean oscillation with Orlicz norms and duality of Hardy spaces", Bull. Amer. Math. Soc., 82 (1976).

INDIANA UNIVERSITY
BLOOMINGTON, INDIANA 47401

Proceedings of Symposia in Pure Mathematics
Volume XXXV, Part 1, 1979

PROOF OF THE BEURLING-MALLIAVIN THEOREM BY DUALITY AND HARMONIC ESTIMATION

Paul Koosis

1. Let us recall that an entire function $F(z)$ is said to be of exponential type if

$$|F(z)| \leq \mathfrak{C} \exp C|z|$$

for all z and some finite C. The infimum of the constants C for which such an estimate holds is called the type of $F(z)$.

An important theorem, due to Beurling and Malliavin [1], runs as follows:

If $F(z)$ is entire, of exponential type, and if

$$(*) \qquad \int_{-\infty}^{\infty} \frac{\log^+|F(x)|}{1+x^2}\, dx < \infty,$$

then, for any $\eta > 0$ there exists a non-zero entire function $g(z)$ of exponential type $\leq \eta$ (a "multiplier" for F) with $(1+|F(x)|)|g(x)|$ bounded on $\mathbb{R}$.

This result is definitive because its converse is also true - is, in fact, well-known and classical.

In this lecture, I will try to explain the ideas on which a new proof of the Beurling-Malliavin theorem, outlined in [2], is based.

Easy reductions show that it is enough, assuming (*), to obtain, for each $\eta > 0$, an entire $h \not\equiv 0$ of exponential type $\leq \eta$ with

$$(\dagger) \qquad \int_{-\infty}^{\infty} \frac{\sqrt{|F(x)|}\;|h(x)|}{1+x^2}\, dx < \infty$$

(no logarithm here!). Indeed, from such an h one easily constructs an entire function $g \not\equiv 0$, of slightly larger exponential type, with $|F(x)||g(x)|^2$ bounded on $\mathbb{R}$, and this turns out to be sufficient.

2. Here is a simple duality argument. One may always take $F(z)$ to be even and satisfy $F(x) \geq 1$, $x \in \mathbb{R}$. Under these conditions, consider in $L_2(-\infty, \infty)$, the cone

$$\mathcal{K} = \mathcal{P} + \frac{e^{2i\eta x}}{F(x)} H_2,$$

where $\mathcal{P}$ is the set of non-negative L_2-functions and H_2 refers to the upper half plane.

THEOREM. If $\mathcal{K}$ is <u>not dense</u> in L_2, there is a non-zero entire h of exponential type $\le \eta$ satisfying $(\dagger)$.

SKETCH OF PROOF. $\mathcal{K}$ not dense implies the existence of a non-zero $Q \in L_2$ with $\Re \int_{-\infty}^{\infty} Q(x)k(x)dx \ge 0$, $k \in \mathcal{K}$. $\mathcal{P} \subseteq \mathcal{K}$ implies

$$(1) \qquad \Re Q(x) \ge 0, \quad x \in \mathbb{R},$$

and $\frac{e^{2i\eta x}}{F(x)} H_2 \subseteq \mathcal{K}$ implies

$$(2) \qquad \frac{e^{2i\eta x}}{F(x)} Q(x) \in H_2.$$

Put $Q(x)/F(x) = q(x)$; from (2), $q \in e^{-2i\eta x}H_2$, and from (1) $\Re q(x) \ge 0$ on $\mathbb{R}$. Let ψ, $-\pi/2 \le \psi(x) \le \pi/2$, be such that $q(x)e^{-i\psi(x)} \ge 0$ on $\mathbb{R}$, take a harmonic conjugate, $\tilde{\psi}$, of ψ, and write $p(x) = q(x)\exp[\tilde{\psi}(x) - i\psi(x)]$. Then $p(x) \ge 0$ on $\mathbb{R}$. As is well-known, $(x+i)^{-3}\exp[\tilde{\psi}(x) - i\psi(x)] \in H_{2/3}$, so since $q \in e^{-2i\eta x}H_2$,

$$(3) \qquad (x + i)^{-4}\, p(x) \in e^{-2i\eta x}\, H_{1/2},$$

by Hölder's inequality (with exponents 4 and 4/3).

The fact that $p \ge 0$ on $\mathbb{R}$ together with (3) implies by a known theorem (due both to Helson and Sarason and to Neuwirth and Newman) that $p(z)$ <u>extends across</u> $\mathbb{R}$ <u>by Schwarz reflection</u>. Thus, $p(z)$ is <u>entire</u>, and, by (3), easily seen to have exponential type $\le 2\eta$. Positivity of $p(x)$ on $\mathbb{R}$ and the relation $\int_{-\infty}^{\infty} [\sqrt{p(x)}\,/(1+x^2)]dx < \infty$ gives us the factorization $p = hh^*$ ($h^*(z) = \overline{h(\bar{z})}$) with an entire h of exponential type $\le \eta$.

But $Q = Fq$ is <u>also</u> in L_2, so $F(x)p(x)/(x^2+1)^2 \in L_{1/2}$ by the same Hölder argument used in proving (3). This last statement is the same as $(\dagger)$.

3. In order to prove the theorem of Beurling and Malliavin it suffices, by the theorem of the preceding section, to show that the cone $\mathcal{K}$ defined there is <u>not dense</u> in L_2. We can gain some understanding of why this should be so by first solving the easier problem of proving that a certain cone $\mathcal{K}_1$, <u>containing</u> $\mathcal{K}$ <u>and</u> - with possible exception of a countable set of values of the parameter η - <u>contained in</u> $\mathcal{K}$'s <u>closure, is not equal to</u> L_2. We take

$$\mathcal{K}_1 = \mathcal{P} + \frac{e^{2i\eta x}}{F(x)}\mathcal{H},$$

where $\mathcal{H}$ is the set of functions $f(z)$ analytic in $\Im z > 0$, having the following properties:

f(z) is locally H_2 up to $\mathbb{R}$ and its boundary values $f(x)$ satisfy $f(x)/F(x) \in L_2(-\infty,\infty)$;

for each $\varepsilon > 0$ there is a C_ε such that

$$\int_0^1 |f(rz)|dr \leq C_\varepsilon \exp \varepsilon|z|, \ \Im z > 0.$$

No non-zero $q \in L_2$ with $q(x) \leq 0$ on $\mathbb{R}$ can belong to $\mathcal{K}_1$. Otherwise we would have functions $p(x) \geq 0$ and $f \in \mathcal{H}$ with

$$(4) \qquad e^{2i\eta x} f(x) = F(x)[q(x) - p(x)] \leq 0, \quad x \in \mathbb{R}.$$

The right side, hence the left, is $\not\equiv 0$, and, since $f(x)$ is locally H_2 up to $\mathbb{R}$, $e^{2i\eta z} f(z)$ can be analytically continued across $\mathbb{R}$ by Schwarz reflection. We obtain thereby an entire function $\Phi(z)$ which is easily seen to be of <u>exponential type</u>. We have, however, for $y > 0$,

$$|\Phi(\pm iy)| \leq C'_\varepsilon \exp(\varepsilon y - 2\eta y),$$

where $\varepsilon > 0$ is arbitrary and C'_ε depends on ε. Taking $\varepsilon < 2\eta$, we see that $\Phi(z) \equiv 0$ - contradicting $f \not\equiv 0$ - <u>because an entire function of exponential type which decreases exponentially in both directions along a straight line must vanish identically</u>.

If we merely suppose $\Re q(x) \leq 0$ on $\mathbb{R}$ and $q \not\equiv 0$, we <u>still</u> cannot have $q \in \mathcal{K}_1$. For then the equality in (4) would hold, with the right side having <u>negative real part</u>, at least. Multiplying $e^{2i\eta z} f(z)$ by a suitable factor $\exp[\tilde{\psi}(z) - i\psi(z)]$ we would obtain a function $\Phi(z)$ analytic in $\Im z > 0$ with $\Phi(x) \leq 0$ on $\mathbb{R}$, and we could continue $\Phi(z)$ across $\mathbb{R}$ by Schwarz reflection (method of Section 2!). The above argument would still enable us to conclude that $\Phi \equiv 0$.

4. We want, however, to argue that $\mathcal{K}$ <u>cannot be dense in</u> L_2. <u>Assuming such density in order to reason by contradiction</u>, let us say that we can approximate each function $-\chi_N(x)F(x)$ by elements of $\mathcal{K}$ in L_2-norm; χ_N denoting the characteristic function of the interval $[-N,N]$. Using the sign $\approx$ to indicate approximate equality in L_2, we then have $p_N \in \mathcal{P}$ and $h_N \in H_2$ for each N such that

$$(5) \qquad -\frac{\chi_N F^2}{F} \approx \frac{p_N F + e^{2i\eta x} h_N}{F}.$$

With L a given fixed large number, take the means $\frac{1}{2L}\int_{-L}^{L} \text{thing}(x+t)dt$ of the <u>numerators</u> on both sides of (5). For the difference of these means we can deduce from (5), by routine manipulations with Schwarz' inequality,

$$(6) \qquad |P_N(x) - e^{2i\eta x} H_N(x)| \leq \varepsilon_N W(x), \quad x \in \mathbb{R}.$$

Here, $\varepsilon_N \xrightarrow[N]{} 0$, $P_N(x) \geq 0$, $H_N \in \mathcal{C} = H_\infty \cap C_0$, and $W(x) = \{\frac{1}{2L}\int_{-L}^{L} [F(x+t)]^2 dt\}^{1/2}$. It is also true that $P_N(x) \geq [W(x)]^2$ for $|x| \leq N - L$.

The idea now is to show that (6) is <u>impossible for sufficiently large</u> N.

Let $\mathcal{O}_N = \{x; |H_N(x)| > (1-\varepsilon_N)W(x)\}$ and $E_N = \mathbb{R} \sim \mathcal{O}_N$. The set $\mathcal{O}_N$ is open and bounded, and includes the large interval $(-N+L, N-L)$. By (6), $P_N(x) > (1-2\varepsilon_N)W(x)$ for $x \in \mathcal{O}_N$, so

$$\Re e^{2i\eta x} H_N(x) > 0, \quad x \in \mathcal{O}_N: \tag{7}$$

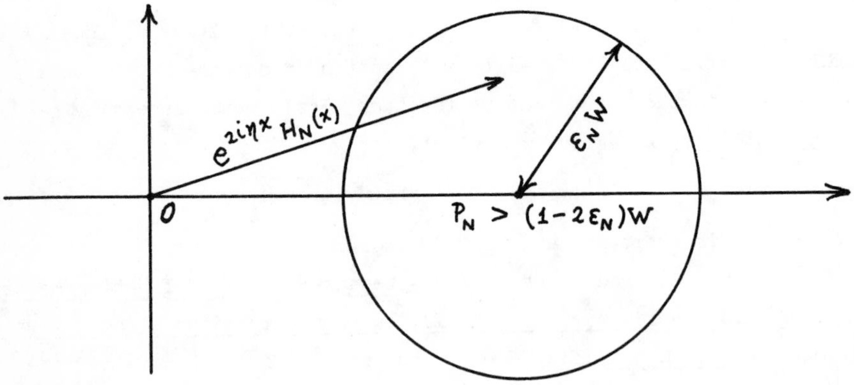

On E_N, $|H_N(x)| \le (1-\varepsilon_N)W(x)$.

To understand what is going on, let us suppose that instead of (7) we have the stronger condition

$$e^{2i\eta x} H_N(x) > 0, \quad x \in \mathcal{O}_N. \tag{7 bis}$$

Write simply $\mathcal{O}$, E and H for $\mathcal{O}_N$, E_N and H_N respectively, and take the domain $\mathcal{D} = \mathbb{C} \sim E$:

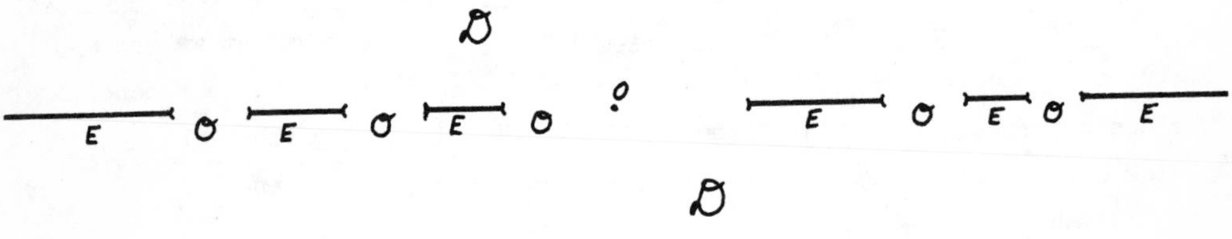

By (7 bis), $\Phi(z) = e^{2i\eta z} H(z)$ can be continued by Schwarz reflection across $\mathcal{O}$ into all of $\mathcal{D}$, and $\Phi(z)$ becomes $O(e^{-2\eta|y|})$ there. If $\mathcal{O}$ were all of $\mathbb{R}$, we would be back in the situation of Section 3, and could conclude that $\Phi(z) \equiv 0$ in contradiction to the relation $\Phi(0) \ge (1-\varepsilon_N)W(0)$.

But $\mathcal{O}$ is not all of $\mathbb{R}$. What is needed here is some quantitative estimate which goes over into the general principle invoked in Section 3 when $\mathcal{O}$ fills out $\mathbb{R}$.

Denote by $Y(z)$ the function which is harmonic and positive in $\mathcal{D}$, zero on $E = \partial\mathcal{D}$, and equal to $|y|$ plus a bounded quantity in $\mathcal{D}$. Let $d\omega(t,z)$ be harmonic measure (of dt) on E relative to $\mathcal{D}$, as seen from $z \in \mathcal{D}$. Since $|\Phi(x)| \le W(x)$ on E and $|\Phi(z)| \le O(e^{-2\eta|y|})$ in $\mathcal{D}$, we have

(8) $$\log|\Phi(0)| \leq -2\eta Y(0) + \int_E \log W(t)d\omega(t,0).$$

The idea is to show that (8) contradicts $\Phi(0) \geq (1-\varepsilon_N)W(0)$ if N is large.

From the inclusion $(-N+L, N-L) \subseteq \mathfrak{G} \subseteq \mathfrak{D}$ it follows easily (by the principle of extension of domain) that $Y(0) \geq N-L$, which of course tends to ∞ with N. The trouble is with the integral in (8).

5. In order to get a grip on the term $\int_E \log W(t)d\omega(t,0)$ which arises because $\Phi(z)$ is only analytic in $\mathfrak{D}$ instead of in all of $\mathbb{C}$ as was the case in Section 3, let us first observe that

$$[W(x)]^2 = \frac{1}{2L}\int_{-L}^{L} [F(x+t)]^2 dt$$

is also an entire function of exponential type, and therefore smooth on $\mathbb{R}$. Hence, since E has two infinite legs at both ends, it is obvious that the integral in question behaves, to a first approximation, like the ordinary Poisson integral of $\log W(t)$ for the upper half plane, i.e.,

(9) $$\int_E \log W(t)d\omega(t,0) \sim \int_{-\infty}^{\infty} \frac{\log W(t)dt}{1+t^2}.$$

It is at this point that property (*) of F comes into play.

The question is, what is the constant of proportionality in (9)? A first step towards the answer to this question is the

THEOREM. For $x_0 > 0$,

(10) $$\omega(\{|x| \geq x_0\} \cap E, 0) \leq \frac{Y(0)}{x_0}.$$

Time does not allow my giving any idea of the proof in this lecture.

If (10) could be differentiated with respect to x_0, we would obtain

$$\int_E \log W(t)d\omega(t,0) \leq Y(0)\int_E \frac{\log W(t)}{t^2} dt.$$

Then, since our construction ensures that $\int_E t^{-2}\log W(t)dt < \eta$ for large N if (*) holds, (8) would yield, for $\Phi_N(z) = \Phi(z)$, $\log|\Phi_N(0)| \to -\infty$, $N \to \infty$, although $\Phi_N(0) \geq (1-\varepsilon_N)\, W(0)$.

Direct differentiation of (10), is, however, certainly not possible, because $\frac{d\omega(t,0)}{dt}$ becomes infinite at any endpoint a of any component of E (in fact, like $|t-a|^{-1/2}$ as $t \to a$), and these components may be very numerous.

We have to take more account of the smoothness of $W(t)$. Now it is possible to arrange matters so that $W(x)$, besides being even (like F) and satisfying $W(x) \geq 1$ on $\mathbb{R}$, also has $W(0) = 1$. This is accomplished by using $1 + (\frac{x}{N-L})^2(W(x))^2$ in place of $(W(x))^2$; such replacement has the further advantage of rendering $\int_{-\infty}^{\infty} t^{-2} \log W(t)dt$ convergent, and small, if N is large. The new entire function, which we continue to denote by $(W(x))^2$, has now, on $\mathbb{R}$, the representation

$$(11) \qquad \frac{\log W(x)}{x} = - \int_0^\infty \log\left|\frac{x+t}{x-t}\right| \, d\left(\frac{\nu(t)}{t}\right),$$

with $\nu(t)$ a certain increasing function related to the distribution of the zeros of $[W(z)]^2$; $\nu(t)$ is $\mathcal{O}(t)$ for $t \geq 0$. (This formula is just a peculiar way of writing the Hadamard factorization for $[W(z)]^2$.)

DEFINITION. For functions $U(x) = \frac{\log W(x)}{x}$ having the representation (11),

$$\mathcal{E}\langle U(x)\rangle = \int_0^\infty \int_0^\infty \log\left|\frac{x+t}{x-t}\right| \, d\left(\frac{\nu(t)}{t}\right) d\left(\frac{\nu(x)}{x}\right)$$

is called the energy of U.

$\mathcal{E}\langle\ \rangle$ is a positive quadratic functional.

It turns out that $\mathcal{E}\langle\frac{\log W(x)}{x}\rangle$ is a correct measure for the smoothness of $W(x)$; a correct quantitative replacement for (9) is

$$(12) \qquad \int_E \log W(t)d\omega(t,0) \leq Y(0)\left[\int_0^\infty \frac{\log W(t)}{t^2}\, dt + \sqrt{\pi\mathcal{E}\langle\log W(t)/t\rangle}\right],$$

valid for even functions $W(x) \geq 1$ of the form (11).

The functional $\mathcal{E}\langle\ \rangle$ first appears in the work of Beurling and Malliavin [1]. I think (12) helps us to see why their proof of the theorem in Section 1 could be based on its use.

In [1], Beurling and Malliavin observed that for entire functions $[W(z)]^2$ of exponential type, and of the form considered here, the facts that $\nu(t)$ increases and remains $\mathcal{O}(t)$, and that $W(t) \geq 1$, already imply that $\mathcal{E}\langle\log W(t)/t\rangle < \infty$ provided that $J = \int_0^\infty t^{-2} \log W(t)dt$ is finite. Indeed, simple computations based on Jensen's formula show that if $[W(z)]^2$ is of type $2A$, say, $\mathcal{E}\langle\log W(t)/t\rangle \leq 2eJ(J+A)$. With (12) and (8), this yields

$$(13) \qquad \log|\Phi(0)| \leq Y(0)[-2\eta + J + \sqrt{2\pi eJ(J+A)}]\,.$$

Since $J \to 0$ as $N \to \infty$ (for $\log W(t)$ read $\frac{1}{2}\log[1 + (\frac{t}{N-L})^2(W(t))^2]$ in the above formula for J!) and $Y(0) \to \infty$, $N \to \infty$, (13) furnishes a contradiction to $\Phi(0) \geq (1-\varepsilon_N)W(0)$ for large N. Therefore (7bis) is incompatible with the relation $|H_N(x)| \leq W(x)$ on E_N if N is large.

6. In reality, however, the construction at the beginning of Section 4 did not furnish (7 bis), but only (7). For this reason, we must first take the function ψ, $-\pi/2 \leq \psi(x) \leq \pi/2$, with $\psi(x) \equiv 0$ on E, satisfying

$$e^{2i\eta x}\, H_N(x)e^{-i\psi(x)} \geq 0, \quad x \in \mathfrak{O},$$

and then put

$$\Phi_N(x) = e^{2i\eta x}\, H_N(x)e^{\tilde{\psi}(x)-i\psi(x)},$$

using a suitably defined harmonic conjugate $\tilde{\psi}$ of ψ. Inequality (8) of Section 4 is then replaced by

$$(14) \qquad \log|\Phi_N(b)| \leq -2\eta Y(b) + \int_E \log|H_N(t)|\,d\omega(t,b) + \int_E \tilde{\psi}(t)\,d\omega(t,b),$$

with b (instead of 0) chosen in $[-1,1]$ so as to have $|\Phi_N(b)| \geq e^{-3}W(b) \geq e^{-3}$.

Suppose we <u>know</u> that each component of $\mathfrak{O} = \mathfrak{O}_N$ has length $\geq \ell$, say. Then an estimate (of best possible magnitude) for the <u>third</u> term on the right in (14) is

$$(15) \qquad \int_E \tilde{\psi}(t)\,d\omega(t,b) \leq 1 + \frac{12\pi Y(b)}{\ell},$$

provided that N is large. The duality argument of Section 2 was carried out in L_2 <u>precisely in order</u> that the averaging operation $\frac{1}{2L}\int_{-L}^{L} \text{thing}(x+t)dt$ could be brought in at the beginning of Section 4. <u>It is this averaging which ensures that</u> $\ell \geq L$. For then, the $P_N(x)$ in (6), being such an average of a <u>positive</u> function, has a certain regularity in its behaviour. We can also take $F(x)$ to enjoy a mild regularity of the form $\log F(x') \leq e^{|x'-x|} \log F(x)$, by starting with $G(z+i)$ instead of $G(z)$, the entire function having all its zeros in $\Im z < 0$ and satisfying $G(z)G^*(z) = F(z)$. These regularity properties are enough to give us a bounded $\mathfrak{O}_N \supseteq (-N+L, N-L)$ <u>having each of its components of length</u> $\geq L$, with (7) holding on it. A side effect of the averaging is that we only have $|H_N(x)| \leq [W(x)]^{k_L}$ instead of $|H_N(x)| \leq W(x)$ on $E_N = \mathbb{R} \sim \mathfrak{O}_N$, k_L being a large number depending on L. Starting from (7) instead of (7 bis), we thus get, from (14) and (15),

$$(16) \qquad \log|\Phi_N(b)| \leq 1 + Y(b)\left[-2\eta + k_L J + k_L\sqrt{2\pi e J(J+A)} + \frac{12\pi}{L}\right],$$

with $J = \int_0^\infty t^{-2} \log[1 + (t/N-L)^2(W(t))^2]dt$ tending to 0 as $N \to \infty$. At the same time, $Y(b) \sim Y(0) \to \infty$. Taking L large enough to begin with, and then fixing it, we see that $\log|\Phi_N(b)| \to -\infty$ as $N \to \infty$, <u>contradicting</u> $|\Phi_N(b)| \geq e^{-3}$.

This contradiction shows that $\mathcal{K}$ is <u>not dense</u> in L_2 and hence proves the theorem of Beurling and Malliavin by the argument in Section 2.

References

1. Beurling, A. and Malliavin, P. On Fourier transforms of measures with compact support. Acta Math., 107 (1962), pp. 291-309.

2. Koosis, P. a) Harmonic Estimation in Certain Slit Regions and a Theorem of Beurling and Malliavin. Institut Mittag-Leffler, Report No. 4, 1978. 49 page preprint. b) Corrections to a), especially to Part I, Section 5. Laurel, Quebec, 1978. 7 page xeroxed brochure.

INSTITUT MITTAG-LEFFLER
DJURSHOLM, SWEDEN

MATHEMATICS DEPARTMENT
UNIVERSITY OF CALIFORNIA
LOS ANGELES, U.S.A.

Proceedings of Symposia in Pure Mathematics
Volume XXXV, Part 1, 1979

ZERO SETS OF ABSOLUTELY CONVERGENT TAYLOR SERIES

R. Kaufman

ABSTRACT. A^+ is the Banach algebra of Taylor series $f(z) = \sum_0^\infty a_n z^n$, normed by $\|f\| = \Sigma|a_n|$, and a closed set F on the unit circle is a <u>ZA-set</u> if $f(F) = 0$ for some $f \neq 0$ in A^+. In the theorem stated below, ϕ is any Möbius transformation of $|z| < 1$ onto itself (excepting rotations).

THEOREM A There exist ZA-sets F such that $\phi(F)$ is not a ZA-set.

The proof presented in [5] is unduly complicated; the complications are dictated by reliance on the following theorem of Carleson [1,4].

THEOREM B Let F be an M_o-set on the circle $|z| = 1$ (construed as the group $R/2\pi$). Then there is a sequence $S = \{1, z_2, \ldots, z_n, \ldots\}$ with $\lim z_n = 1$, so that $F \cdot S$ is not a ZA-set.

Our purpose is to draw a more definitive line between ZA-sets and the others.

THEOREM There exists a singular probability measure on $|z| = 1$ so that for every $f \in A^+$ with $f(0) = 1$

$$\int \{\log^+ |f|^{-1}\}^{1/2} \, d\mu \leq C(\|f\|),$$

a constant depending only on the norm of f. Moreover $\phi(s)$ is a ZA-set, S being the closed support of μ.

AMS subject classification (1970). Primary 42A28, 30A10; Secondary 42A72, 46J15.

COROLLARY $\mu(N) = 0$ for every ZA-set N, so that S is not covered by a countable union $\cup N_i$ of ZA-sets.

The corollary merely emphasizes that $S = \phi^{-1} \circ \phi(S)$ is very far from the class of ZA-sets, a phenomenon left in doubt by Theorems A and B.

(And real analysis) To avoid calligraphic problems we introduce the symbol $\Lambda(f) = \log^+ |f|^{-1}$, (sic!) and write Jensen's inequality in the form

$$\int \Lambda(f^{-1}) - \Lambda(f)|dz| \geq 2\pi \log |f(0)|,$$

whence $\int \Lambda(f) \leq 4\pi \log \|f\|$, when $f(0) = 1$. Let now $1 = e_1 > e_2 > \cdots e_n > 1/2$, and suppose that we have constructed a function $g_n(z)$ so that

$$\mathrm{I}_n \quad g_n \in C^\infty,\ 0 \leq g_n \leq n!, \quad \pi < \int g_n |dz| < 2\pi.$$

$$\mathrm{II}_n \quad \int \Lambda^{e_n}(f) g_n(z)|dz| < 4\pi \log \|f\| + 2 - n^{-1}$$

for all f in A^+ with $f(0) = 1$.

We construct g_{n+1} so that $0 \leq g_{n+1} \leq (n+1)g_n$ and $m\{g_{n+1}>0\} < m\{g_n>0\} \cdot 1/2$. (If we were not concerned with the mapping ϕ, we could be more temperate in the bounds on g_n.) Now II_{n+1} is obtained serendipitously for the case $(\log \|f\|)^{1-e_2} > n!$, or $\log \|f\| \geq A_{n+1}$. Henceforth, in dealing with step $n + 1$, we suppose $\log \|f\| \leq A_{n+1}$. Using Hölder's inequality again, we see that there is some large constant B_{n+1} so that

$$\int \Lambda^{e_{n+1}} - \int (\Lambda^{e_{n+1}} \wedge B_{n+1}) < (n+1)^{-4}$$

for all f in question, with the weight $g_{n+1} \leq (n+1)!$ So, writing (temporarily) $L = \Lambda^{e_{n+1}} \wedge B_{n+1}$, we are faced with the estimation of integrals

$$\int L(|f|)(g_n - g_{n+1})|dz|,\ \|f\| \leq A_{n+1},\ f(0) = 1.$$

For the first time we use the extra information on f: A^+ carries a norm stronger than the uniform, and A^+ is an algebra. L can be treated as a function on the w-plane, $w = u + iv$, and approximated to any degree of accuracy by a polynomial $p(u,v)$. The degree of accuracy is of course determined by the inequality $\int g_{n+1}|dz| < 2\pi$. Now $p(f) = p(\text{Re}f, \text{Im}f)$ can be estimated in the norm of A, the full algebra of absolutely convergent series in z and $\bar{z}$, with a bound depending only on p and A_{n+1}. But then $\int p(f(z))(g_{n+1}-g_n)|dz|$ can be controlled by the quantity $\max|\hat{g}_{n+1}(k)-\hat{g}_n(k)|$, often called the PF, or pseudofunction, norm of $g_{n+1} - g_n$.

(New Harmony) Let ψ be a function of class C^∞ on $|z| = 1$, $\psi \geq 0$, $\int \psi = 2\pi$; let $H(z)$ be a real function of class C^∞ such that $d^2H/d\theta^2$ has only isolated zeroes. The PF norm of $\psi(e^{imH}) - 1$ tends to 0 as $m \to +\infty$. Indeed this function is equal to $\Sigma' a_p e^{impH}$, a sum in which $\Sigma |a_p| < +\infty$ and $a_0 = 0$. In the integral $\int e^{impH} e^{-ik\theta} d\theta$ we exclude a small interval around each zero of $H_{\theta\theta}$ and find a bound $|mp|^{-1/2}$ for the remaining integral. Because g_n belongs to A we have also $\psi(e^{im\theta})g_n \to g_n$ in PF-norm.

Forgetting for a moment the Moreover of the theorem, we choose H as before and ψ so that $\psi(e^{i\theta}) = 0$ unless $|\theta| < 7n^{-1}$ (modulo 2π) and $\psi < n$. Then $\psi(e^{imH})g_n = g_{n+1}$ fulfills all the requirements of I_{n+1}, II_{n+1} when m is large.

The method just illustrated in a purely qualitative way has a long and mostly honorable history [6,8]. Recently it has been anatomized [7].

To prove the last part of the theorem we introduce a sequence $\{h_n\}$ of C^∞ functions, uniformly dense in the space of continuous, real-valued functions on $|z| = 1$. We choose H on $[0,2\pi]$ so that $H(2\pi) = H(0) + 2\pi$ and $e^{iH(\theta)} = \phi(e^{i\theta})$. Thus we can form $g_{n+1} = \psi(\exp imH-h_n)$. Fourier expansion of ψ yields the convergence $g_{n+1} \to g_n$ in PF-norm as $m \to +\infty$. The support of g_{n+1} is contained in the set

$$|\phi^m(z) - e^{ih_n}| < 7n^{-1}.$$

The weak*-limit of the sequence of measures $g_{n+1}|dz|$ is the measure μ; and its support S is transformed by ϕ onto a Kronecker set [3]. Then $\phi(S)$ is a "peak set" for the algebra A^+ [2], and peak sets are zero-sets.

(Marginalia) A small variation in the problem leads to an entirely different construction: we ask that S have h-measure 0 for a certain Hausdorff-measure function (z.B. $h(t) = t^{1/2}$), and the measure μ carried by S have the property named in the Corollary. The construction breaks down with the approximation of $\Lambda^{e_{n+1}}$ by the bounded function L, because the size of g_{n+1} cannot be predicted in advance. The circumvention sketched in the next paragraph relies explicitly on the Banach algebra A.

Let $G(u,v) = u^{-2}\sin^2 u + v^{-2}\sin^2 v$, so that $G(0,0) = 1$, $G \geq 0$, and G is the restriction to R^2 of an entire function on C^2. (Illuminati will observe that the more obvious choice $G(u,v) = (1+u^2+v^2)^{-1}$ leads to another detour.) (See [3, VI].) We shall find the measure μ, along with constants $B_n > 0$ so that

$$\int G(B_n f)d\mu \leq n^{-2} \quad \text{if} \quad f(0) = 1, \quad \|f\| \leq n.$$

In virtue of the holomorphy of G, the norm of $G(f)$ in A is bounded by a function of $\|f\|$ alone. Hence we can control $\int G(B_j f)(g_{n+1}-g_n)$, $1 \leq j \leq n - 1$, by means of the PF norm of $g_{n+1} - g_n$, with no reference to the size of g_{n+1}. To pass from g_n to g_{n+1}, we first replace g_n by $g_n^* \leq g_n$, an approximation in PF-norm whose support is sparse enough to obtain the metric property of S in the limit; then we pass to g_{n+1} with a factor $\psi(\exp imH-h_n)$ just as before.

(Problems) (i) Carleson's method can be used to find sets (of Lebesgue measure zero) that aren't null sets for many classes of analytic functions. For example: functions continuous on the disk $|z| \leq 1$, with Taylor series $\sum_0^\infty a_n z^n$ such that $\sum |a_n|^{3/2} < +\infty$. As this class has no (obvious) algebraic properties, our method fails.

(ii) Let $z_n = r_n e^{i\theta_n}$, with $0 < r_n < 1$, $\lim r_n = 1$, $\lim \theta_n = 0$. Can a power series in A^+ vanish at all z_n? Functions in A^+ have certain metrical properties in the domain $|z| < 1$, but their influence on the zero-sets is obscure. For example $\int_0^1 M(r,f')dr < +\infty$; $M(r,f')$ is the maximum modulus on $|z| = r$. If $|\theta_n| \le (1-r_n)^\alpha$ for some $\alpha > 0$, then the sequence $\{z_n\}$ and its limit is the zero-set of a function class C^∞.

REFERENCES

[1] L. Carleson, Sets of uniqueness for functions regular in the unit circle. Acta Math. 87 (1952), 325-345.

[2] S. W. Drury, Sur les ensembles de Dirichlet. C. R. Acad. Sci. Paris 272 (1971), A 1507-1509.

[3] J.-P. Kahane, Séries de Fourier Absolument Convergentes. Ergebnisse der Math. 50, 1970, Berlin.

[4] J.-P. Kahane et Y. Katznelson, Sur les algèbres de restrictions des séries de Taylor absolument convergents à un fermé du cercle. J. Analyse Math. 23 (1970), 185-197.

[5] R. Kaufman, Transformations of exceptional sets. (To appear)

[6] R. Kaufman, Topics on Kronecker sets. Ann. Inst. Fourier (Grenoble) 23 (1973), 65-74.

[7] T. W. Körner, On the theorem of Ivasev-Musatov I. Ann. Inst. Fourier (Grenoble) 27 (1977), 97-115.

[8] N. Wiener and A. Wintner, Fourier-Stieltjes transforms and singular infinite convolutions. Amer. J. Math. 60 (1938), 513-522.

University of Illinois at Urbana-Champaign

Urbana, Illinois 61801

Proceedings of Symposia in Pure Mathematics
Volume XXXV, Part 1, 1979

CAPACITY AND UNIFORM ALGEBRAS

J. Wermer

ABSTRACT. We study relations between finiteness of fibers in the maximal ideal space of a uniform algebra and analytic structure, describing work done jointly with B. Aupetit in [2] generalizing a result of E. Bishop.

Let M be a compact space and A an algebra of complex-valued continuous functions on M which contains the constants. We assume that A is closed under uniform convergence on M, separates points on M, and that every homomorphism of A into $\mathbb{C}$ is evaluation at some point of M. Let X be a closed subset of M such that $|f(y)| \leq \max_{x \in X} |f(x)|$ if $y \in M$ and $f \in A$. Then A is called a _uniform algebra on_ M and X is called a _boundary_ of A. Examples are: (i) M is the closed unit disk, A is the disk-algebra, X is the unit circumference. (ii) M is the solid cylinder $|z| \leq 1, 0 \leq t \leq 1$, $X = \{(z,t) \text{ in } M \mid |z| = 1\}$, $A = \{f \in C(M) \mid f(z,t) \text{ analytic in } |z| < 1$ for each $t\}$. (iii) M is the bidisk $|z| \leq 1$, $|w| \leq 1$, X is the torus $|z| = 1$, $|w| = 1$, $A = \{f \in C(M) \mid f \text{ analytic in } |z| < 1, |w| < 1\}$.

Let A be a uniform algebra on M and X a boundary of A and fix f in A. For each $\lambda \in \mathbb{C}$ the _fiber_ over λ, $f^{-1}(\lambda)$, is defined to be $\{y \in M \mid f(y) = \lambda\}$.

LEMMA 1: _Let_ W _be a component of_ $f(M)\setminus f(X)$. _For_ h _in_ A, _put_ $Z_h(\lambda) = \max_{f^{-1}(\lambda)} |h|$. _Then_ $\lambda \to \log Z_h(\lambda)$ _is subharmonic in_ W.

This is proved in [6] by a method of Oka. For f in A, λ in $\mathbb{C}\setminus f(X)$, we have in example (i) that each fiber is finite, in example (ii) with $f(z,t) = z$, each fiber is a line segment, and in example (iii), with $f = z$, each fiber is a closed disk. The finiteness of fibers characterizes analytic functions of one complex variable. This was discovered by Errett Bishop in [3]. We write # for "cardinality of".

THEOREM 1 ([3]): _Let_ A _be a uniform algebra on a space_ M, X _a boundary of_ A _and_ f _in_ A. _Choose a component_ W _of_ $\mathbb{C}\setminus f(X)$. _Assume that there exists a subset_ G _of_ W _having positive area such that_ $\# f^{-1}(\lambda) < \infty$ _for each_ λ _in_ G. _Then there exists an integer_ n _such that_ $\# f^{-1}(\lambda) \leq n$ _for each_ λ _in_ W, _and_ $f^{-1}(W)$ _can be made into a one-dimensional analytic space such that each_ h _in_ A _is analytic on_ $f^{-1}(W)$.

AMS(MOS) Subject Classifications (1970) Primary 31A15, 46J20.

Now let B be a Banach algebra, not necessarily commutative. For x in B, sp x denotes the spectrum of x. By the capacity of a plane set E, denoted $c^+(E)$, we shall mean the exterior logarithmic capacity of E. In [1] Bernard Aupetit proved the following:

THEOREM 1*: Let $\lambda \to \varphi(\lambda)$ be an analytic map of a domain D in $\mathbb{C}$ into B. Assume $\exists$ a subset E of D with $c^+(E) > 0$ such that for each λ in E, $\# \operatorname{sp} \varphi(\lambda) < \infty$. Then $\exists$ integer n such that $\# \operatorname{sp} \varphi(\lambda) \leq n$ for each λ in D.

The formal resemblance of Theorems 1 and 1* suggests the possibility of weakening the hypothesis in Theorem 1, and indeed Aupetit and I showed:

THEOREM 2 ([2]): The conclusions of Theorem 1 remain true if in the hypothesis on G "positive area" is replaced by "positive exterior capacity".

Furthermore, "positive capacity" is the correct condition. Starting with an arbitrary compact set E of capacity 0, Miss H.-P. Lee has shown (see [2]) that $\exists$ A, M, X, f, W with $E \subset W$, $\# f^{-1}(\lambda) = 1$ for λ in E and $\# f^{-1}(\lambda) = \infty$ for λ in $W \setminus E$. Also Herbert Alexander has given a different example (in a letter) to the same purpose, in which the algebra A is doubly generated. Certain set-valued mappings are involved in the proofs of Theorems 1* and 2. Let D be a plane domain and let K be a set-valued map which assigns to each λ in D a compact plane set $K(\lambda)$ such that

(1) If $\lambda_0 \in D$ and if U is a neighborhood of $K(\lambda_0)$, then $K(\lambda) \subset U$ if $|\lambda - \lambda_0|$ is small enough, and

(2) For each α, $|\alpha| = 1$, put $u_\alpha(\lambda) = \max\limits_{K(\lambda)} [\mathrm{Re}(\alpha z)]$. Then $\lambda \to u_\alpha(\lambda)$ is a subharmonic function in D.

We shall then say that K is a subharmonic set-valued function on D. If $b \in \mathbb{C}$ and λ is in D, then $bK(\lambda)$ denotes $\{bz \mid z \in K(\lambda)\}$.

THEOREM 3: Let K be a subharmonic set-valued map defined on a plane domain D. Assume that for each c in $\mathbb{C}$ the map: $\lambda \to e^{\lambda c} K(\lambda)$ is subharmonic. Assume $\exists$ a subset E of D with $c^+(E) > 0$ such that $\# K(\lambda) = 1$ for each λ in E. Then $\# K(\lambda) \leq 1$ for each λ in D.

PROOF (following [1]): We write diam S for the diameter of the set S; put $\delta(\lambda) = \operatorname{diam} K(\lambda) = \max\limits_{|\alpha|=1} \left[\max\limits_{z \in K(\lambda)} [\mathrm{Re}(\alpha z)] + \max\limits_{z \in K(\lambda)} [\mathrm{Re}(-\alpha z)] \right]$. Because of (1), δ is upper-semicontinuous. By (2), $\lambda \to \max\limits_{z \in K(\lambda)} [\mathrm{Re}(\alpha z)]$ is subharmonic, and similarly for $-\alpha$. Hence δ is a subharmonic fnnction on D. Similarly $\operatorname{diam}(e^{\lambda c} K(\lambda))$ is subharmonic in λ, so $\lambda \to |e^{\lambda c}| \delta(\lambda)$ is subharmonic. Since this holds for all c in $\mathbb{C}$, by a theorem of Rado $\log \delta$ is subharmonic. For λ in E, $\delta(\lambda) = 0$, so $\log \delta(\lambda) = -\infty$. Since $c^+(E) > 0$, this implies $\log \delta(\lambda) = -\infty$ in D and so $\delta(\lambda) = 0$ for all λ in D. Hence $\# K(\lambda) \leq 1$ for all λ in D, as asserted.

The following corollary of Theorem 3 can be used to prove Theorem 2 (see [2], pp. 389 ff.), and is also a special case of Theorem 2.

COROLLARY 1: Let A be a uniform algebra on M, X a boundary of A, f in A and W a component of $\mathbb{C}\setminus f(X)$. Suppose $\exists\, E \subset W$ with $c^+(E) > 0$ and with $\# f^{-1}(\lambda) = 1$ for each λ in E. Then $\# f^{-1}(\lambda) = 1$ for each λ in W. Also, for each h in A, $\lambda \to h(f^{-1}(\lambda))$ is an analytic function on W.

PROOF: Fix g in A. Put $K(\lambda) = g(f^{-1}(\lambda))$ for λ in W. $\lambda \to K(\lambda)$ is then a set valued map satisfying (1). Fix $c \in \mathbb{C}$, α with $|\alpha| = 1$. Put $h = e^{cf}g$. Then $h \in A$, and we have

$$\max_{z \in e^{\lambda c}K(\lambda)} (\operatorname{Re}(\alpha z)) = \max_{z \in e^{\lambda c} g(f^{-1}(\lambda))} (\log |e^{\alpha z}|)$$

$$= \max_{z \in (e^{cf}g)(f^{-1}(\lambda))} (\log |e^{\alpha z}|) = \max_{f^{-1}(\lambda)} (\log |e^{\alpha h}|).$$

By Lemma 1, then, $\lambda \to e^{\lambda c}K(\lambda)$ satisfies (2), as well as (1), so is a subharmonic set-valued function. Also $\# K(\lambda) = 1$ for each λ in E and $c^+(E) > 0$. So, by Theorem 3, $\# K(\lambda) = \# g(f^{-1}(\lambda)) = 1$ for each λ in W. This holds for every g in A and so $\# f^{-1}(\lambda) = 1$ for every λ in W. Next, let $h \in A$. Fix a in $\mathbb{C}$. By Lemma 1, $\lambda \to \log |h(f^{-1}(\lambda)) - a|$ is subharmonic in W. Let now φ be a function defined on a region Λ in $\mathbb{C}$. Assume that for each constant a the map: $\lambda \to \log |\varphi(\lambda) - a|$ is subharmonic. Then, by an elementary argument, either φ or $\bar{\varphi}$ is analytic in Λ. Thus $h(f^{-1}(\lambda))$ is analytic or antianalytic in W. Hence also $\lambda h(f^{-1}(\lambda)) = (fh)(f^{-1}(\lambda))$ is analytic or antianalytic in W. It follows that $\lambda \to h(f^{-1}(\lambda))$ is analytic in W, so we are done.

We also obtain a Corollary of Theorem 3 for an analytic map into a Banach algebra B. The role of Lemma 1 in this case is played by the following, due to Vesentini [5]: Denoting by $\rho(x)$ the spectral radius of x in B, if $\lambda \to \varphi(\lambda)$ is an analytic map of a domain D into B, then $\lambda \to \log \rho(\varphi(\lambda))$ is subharmonic in D.

COROLLARY 2: Let φ be an analytic map of D into B. Suppose $\exists\, E \subset D$ with $c^+(E) > 0$ and $\operatorname{sp} \varphi(\lambda) = 1$ for λ in E. Then $\# \operatorname{sp} \varphi(\lambda) \leqq 1$ for every λ in D.

PROOF: For λ in D, put $K(\lambda) = \operatorname{sp} \varphi(\lambda)$. Fix α with $|\alpha| = 1$. We have

$$\max_{z \in K(\lambda)} (\operatorname{Re}(\alpha z)) = \max_{z \in \operatorname{sp}\varphi(\lambda)} (\log |e^{\alpha z}|) = \log\left(\max_{z \in \operatorname{sp}\varphi(\lambda)} |e^{\alpha z}|\right).$$

By the spectral mapping theorem, the image of $\operatorname{sp}(\varphi(\lambda))$ under the map $z \to e^{\alpha z}$ equals $\operatorname{sp}(e^{\alpha\varphi(\lambda)})$. So $\max_{z \in \operatorname{sp}\varphi(\lambda)} |e^{\alpha z}| = \max_{\operatorname{sp}(e^{\alpha\varphi(\lambda)})} |w| = \rho(e^{\alpha\varphi(\lambda)})$, and so

$\max_{z \in K(\lambda)} \operatorname{Re}(\alpha z) = \log \rho(e^{\alpha\varphi(\lambda)})$. Since $\lambda \to e^{\alpha\varphi(\lambda)}$ is analytic, $\lambda \to \log \rho(e^{\alpha\varphi(\lambda)})$ is subharmonic by Vesentini's above mentioned result. Thus $\lambda \to \max_{z \in K(\lambda)} (\mathrm{Re}(\alpha z))$ is subharmonic. Hence the map $K: \lambda \to K(\lambda)$ satisfies (2). By well-known results about the spectrum, K also satisfies (1). Thus K is a subharmonic set-valued map. For c in $\mathbb{C}$, $e^{\lambda c}K(\lambda) = sp(e^{\lambda c}\varphi(\lambda))$ and $\lambda \to e^{\lambda c}\varphi(\lambda)$ is analytic. Hence also $\lambda \to e^{\lambda c}K(\lambda)$ is a subharmonic set-valued map. By Theorem 3, then, $\# \, sp(\varphi(\lambda)) \leqq 1$ for each λ in D, as asserted.

Finally, we raise a question regarding inner functions on the unit disk Δ. By an inner function $z \to f(z)$ on Δ we mean a bounded analytic function on Δ such that $\lim_{r \to 1} |f(re^{i\theta})| = 1$ a.e. The following result is due to Seidel and Frostman (see [4], Chapter 3, p. 37):

THEOREM 4: *If f is an inner function, then either f is a finite Blaschke product or every value in $|w| < 1$, except perhaps for a set of values of capacity 0, is assumed by f infinitely often in Δ.*

In other words, if $\# f^{-1}(\lambda) < \infty$ for every λ in a set $E \subset \{|w| < 1\}$ with E having positive capacity, then f is a finite Blaschke product, and so $\# f^{-1}(\lambda) \leqq n$ for all λ in the unit disk, and some fixed integer n. This result has the form of Theorem 2 above, the algebra in question being the algebra H^{∞} of all bounded analytic functions on Δ, with this difference: in theorem 2 the fiber $f^{-1}(\lambda)$ is taken in M, while in Theorem 4 $f^{-1}(\lambda)$ is taken in Δ, which is a proper subset of M. Are Theorems 2 and 4 both special cases of a single result on algebras of functions?

Interesting results related to the above are given in Kumagai's thesis [7].

A potential theoretic technique very much like the above was found by Dloussky, independently, and used for a problem of analytic continuation, in [8].

REFERENCES

1. B. Aupetit, Caractérisation spectrale des algèbres de Banach de dimension finie, J. Functional Analysis 26 (1977), 232-250.

2. B. Aupetit and J. Wermer, Capacity and uniform algebras, J. Functional Analysis 28 (1978), 386-400.

3. E. Bishop, Holomorphic completions, analytic continuations and the interpolation of semi-norms, Ann. of Math. 78 (1963), 468-500.

4. K. Noshiro, Cluster Sets, Springer Verlag, 1960.

5. E. Vesentini, On the subharmonicity of the spectral radius, Boll. Un. Mat. Ital. 4 (1968), 427-429.

6. J. Wermer, Subharmonicity and hulls, Pacific J. Math. 58 (1975), 283-290.

7. D. Kumagai, Uniform algebras and subharmonic function, Dissertation, Lehigh University, 1978.

8. G. Dloussky, Enveloppes d'holomorphie et prolongements d'hypersurfaces, Lecture Notes in Mathematics, vol. 578. Springer Verlag, Berlin, 1977, pp. 217-235.

DEPARTMENT OF MATHEMATICS
BROWN UNIVERSITY
PROVIDENCE, RI 02912

Proceedings of Symposia in Pure Mathematics
Volume XXXV, Part 1, 1979

FOLLOWING FUNCTIONS OF CLASS H^2

Douglas N. Clark[1]

ABSTRACT. Given a rational function f(z) with poles in $|z| < 1$, when is some function of f in H^2? More generally, given a positive integer m, when is there an H^2 function g such that, on the sets where $f(z) = \lambda$, we have $g(z) = p_\lambda(z)$, p_λ a polynomial of degree m-1?

1. Given sets A,B and functions f,g from A to B, we say g <u>follows</u> f if, for any two points $x,y \in A$, $f(x) = f(y)$ implies $g(x) = g(y)$; briefly: the fibres of g contain the fibres of f. An equivalent condition is that there exist some function $\mathcal{J}$ from B to B such that

$$g = \mathcal{J}(f) \quad , \tag{1}$$

but it is more interesting to add the restriction that one or more of the functions f,g,$\mathcal{J}$ belong to some special class. In this spirit, Stephenson, who introduced the term "following function" in [4], proved there that if $A = B = \{|z|< 1\}$ and if f is an inner function, then the functions g that follow f and belong to the class H^p are exactly the functions satisfying (1) with $\mathcal{J} \in H^p$. Similar theorems were proved with H^p replaced by the Smirnoff class N_* and by the class of meromorphic functions of bounded characteristic, [4, Theorem 3.2].

In a recent paper on Toeplitz operators [1], I was led to consider the case where f is a rational function mapping $|z| = 1$ to itself, B is the unit disk $\{|z| < 1\}$, A is the set of all z such that z and f(z) belong to B, and where a nonconstant $g \in H^2$

AMS(MOS) subject classifications (1970). Primary 30A78, 47B35.

[1]Research partially supported by an NSF grant.

following f is sought. Subsequently [2], there arose the more general problem of finding a nonconstant g of class H^2 that follows f (to order m). This means that on the sets where $f = \lambda$, instead of having $g = c$ (some constant, depending on λ), we have $g = p$ (some polynomial of degree m-1, depending on λ). There was one more generalization in [2], and we shall follow it here: f was only assumed to map $|z| = 1$ to a simple closed curve Γ, B was the interior of Γ and A was the set of all z such that $|z| < 1$ and $f(z) \in B$.

The analogue of (1) for g following f to order m is the existence of m functions $\mathcal{J}_0, \ldots, \mathcal{J}_{m-1}$ such that

$$g(z) = \mathcal{J}_0(f(z)) + \mathcal{J}_1(f(z))z + \ldots + \mathcal{J}_{m-1}(f(z))z^{m-1} .$$

To give a simple example of the concept of following to order m, suppose we have a rational function F(z), given by

$$F(z) = P(z)/Q(z) = P(z)/\left[\prod_{i=1}^{N}(z-\Delta_i)\prod_{i=1}^{M}(z-\Gamma_i)\right] \tag{2}$$

where $|\Gamma_i| < 1 < |\Delta_i|$. We shall show that if $m = M+1$, there always exist H^2 (in fact H^∞) functions that follow F, to order m (with B = interior of Γ, $A = F^{-1}(B) \cap \{|z| < 1\}$) . Indeed, the function

$$G(z) = P(z)/\prod_{i=1}^{N}(z-\Delta_i)$$

certainly lies in H^∞ (of the unit disk) and, on the set where $F(z) = \lambda$, we have

$$G(z) = \prod_{i=1}^{M}(z-\Gamma_i)F(z) = \lambda\prod_{i=1}^{M}(z-\Gamma_i).$$

2. Here is why I am interested in the problem of following to order m, and how it arose in [1] and [2]. Given a rational function (2) mapping $|z| = 1$ to a simple closed curve Γ, the Toeplitz operator T_F is defined (on $x \in H^2$) by

$$T_F x = \underline{P}Fx$$

where $\underline{P}$ is the projection of L^2 on H^2. It is well known from the general theory of Toeplitz operators that Γ is the essential spectrum of T_F and that $-\nu$ is the index of $T_F - \lambda I$ (off Γ) where ν

is the winding number of $t \to F(e^{it})$ about $\lambda \in \text{int}\,\Gamma$.

LEMMA. *An* H^2 *function* $k(z)$ *is orthogonal to the span of the eigenvectors of* T_F^* *if and only if, for all* λ *interior to* Γ, *there is a polynomial* p_λ *of degree at most* M-1 *such that*

$$[\Pi(z-\Gamma_i)k - p_\lambda(z)]/[P-\lambda Q] \in H^2 \tag{3}$$

where P,Q,M *and the* Γ_i *come from* (2).

Since $P-\lambda Q$ is a polynomial, it is clear that (3) holds for $\lambda \in \text{int}\,\Gamma$ if and only if the numerator of the left member of (3) vanishes whenever the denominator does, i.e., whenever $P = \lambda Q$, or $F = \lambda$. This is equivalent to saying that $\Pi(z-\Gamma_i)k$ follows F to order M. Furthermore, whenever some $x \in H^2$ follows F (to order M) one may, by subtracting off a polynomial of order M-1, obtain another H^2 function that follows F (to order M) and has the form $\Pi(z-\Gamma_i)k$.

Suppose now that the curve $\Gamma: t \to F(e^{it})$ is oriented positively. Thus as e^{it} moves around the circle clockwise, $F(e^{it})$ moves around Γ clockwise. If on some arc γ of the circle, $F(e^{it})$ retraces itself back over $F(\gamma)$, we shall say F backs up on γ. It was proved in [2] that, at least if Γ is an analytic curve, the eigenvectors of T_F^* span H^2 if and only if F never backs up. In the following two sections, we show how to construct H^2 functions that follow F to order M in case F backs up.

For more on the span of the eigenvectors of a rational Toeplitz operator, see Gambler [3].

3. First we describe a very straightforward method from [1] for constructing H^2 following functions in case M = 1 in (2). In this case, F backing up on an arc γ is equivalent to F omitting an arc α $(= F(\gamma))$ in $|z| < 1$. Pick two points $z_1, z_2 \in \gamma$ and let $\lambda_i = F(z_i) \in \alpha$. The function $G(z) = (F(z)-\lambda_1)/(F(z)-\lambda_2)$ is analytic in $|z| < 1$ but G does not lie in H^2 as it has poles (one at z_2, at least) on $|z| = 1$. But it does have another desirable property: in $|z| < 1$, $G(z)$ omits a ray from 0 to ∞. Therefore, $G(z)$ has a single valued εth power $G_\varepsilon(z) = G(z)^\varepsilon$ which follows F and, for suitable ε, lies in H^2.

By the lemma, we have just proved that if M = 1 in (2) and if F backs up, then the eigenvectors of T_F^* fail to span H^2.

For general M, the above method of constructing H^2 following functions fails. For suppose F backs up at z_0. $F(z) - F(z_0)$ does have at most M-1 zeros ($z_1,\dots,z_{M-1}$, say) in the disk, so there is an analytic function that follows F to order M, viz.

$$(4) \qquad \prod_{i=1}^{M-1} (z-z_i)[F(z)-F(z_0)]^{-1}.$$

However the εth power of the function (4) no longer follows F to order M (even if the function (4) omits a ray from 0 to ∞), as it takes the values of an εth root of a polynomial where F is constant.

The "reason" for the failure of the above method in case $M > 1$ seems to be that "taking εth powers" is too nonlinear. In the next section, we give a linear method for constructing following functions to order M, at least when Γ is an analytic curve.

4. For a rational function given by (2), write

$$P(z) - \lambda Q(z) = a\Pi(1-D_i(\lambda)z)\Pi(1-E_i(\lambda)z)\Pi(1-G_i(\lambda)z)$$

where $|D_i| < 1 = |G_i| < |E_i|$. The D_i, E_i and G_i may be thought of as functions of λ by fixing any piecewise continuous determinations. We consider the integral

$$(5) \qquad \int_\sigma\int \frac{x(\varphi,t)\psi(t)\Pi(1-\Delta_i^{-1}z)\,d\mu_t(\varphi)\,dt}{\Pi(1-D_i(F(e^{it}))z)\prod_{i=1}^{j+\nu}(1-G_i(F(e^{it}))z)(1-ze^{-i\varphi})}$$

where the G_i are arranged so that F backs up on the ranges of $G_1(\lambda),\dots,G_j(\lambda)$ (j may vary with λ) and F goes forward on the ranges of $G_{j+1}(\lambda),\dots,G_{2j+\nu}(\lambda)$. Furthermore, in (5), ψ is a suitable weight function (given explicitly in [2]), σ is a subset of the set Σ of all e^{it} where F backs up, such that $F(\sigma) = F(\Sigma)$, and $d\mu_t$ is the measure defined by

$$\int x(\varphi)\,d\mu_t(\varphi) = \sum_{k=j+\nu+1}^{2j+\nu} \rho_k(t)x(G_k(F(e^{it}))) ,$$

for suitable weight functions ρ_k (again defined in [2]). One of the main results of [2] is that the operator L (denoted L^*_{1F} in [2]) on

$$\mathcal{H} = \int L^2(d\mu_t(\varphi))\,dt$$

that sends $x(\varphi,t)$ to the integral (5) is bounded and invertible from $\mathcal{K}$ to $K^{\perp}$, the orthogonal complement of the eigenvectors of T_F^*. Indeed, L implements a similarity between $T_{F|_{K^{\perp}}}$ and a certain normal operator on $\mathcal{K}$. Thus for every $x \in \mathcal{K}$, $\Pi(z-\Gamma_i)Lx$ is in H^2 and is a following function, to order M. This last statement may be seen directly as follows. Choose z such that F(z) is interior to Γ. We have

$$\Pi(z-\Gamma_i)Lx$$

$$= \int_\sigma \Sigma \frac{\Pi(1-\Delta_i^{-1}z)\Pi(z-\Gamma_i)\psi(t)\rho_k(t)x(G_k(F(e^{it})),e^{it})dt}{\Pi(1-D_i(F(e^{it}))z)\prod_1^{j+\nu}(1-G_i(F(z))z)(1-G_{j+\nu+k}(F(e^{it}))z)}$$

$$= \int_\sigma \Sigma \frac{\psi(t)\rho_k(t)x(G_k(F(e^{it})),e^{it})\Pi(1-E_i(F(e^{it}))z) \prod_{k\neq i=1}^{j}(1-G_{j+\nu+i}(F(e^{it}))z)dt}{F(e^{it}) - F(z)}$$

It turns out that

$$\Pi(1-E_i(f(e_m^{it}))z) \prod_{k\neq i=1}^{j} (1-G_{j+\nu+i}(F(e^{it}))z) = \Sigma a_r z^r$$
$$n$$

is a polynomial (in z) of degree M-1, so that

$$(6) \qquad \Pi(z-\Gamma_i)Lx = \sum_{r=0}^{M-1} z^r \int_\sigma \Sigma \frac{\psi(t)\rho_k(t)x(G_k(F(e^{it})),e^{it})a_r(t)}{F(e^{it}) - F(z)} dt$$

and since the integrand in (6) agrees at each of the points where $F(z) = \lambda$, we see that $\Pi(z-\Gamma_i)Lx$ follows F to order M.

Thus there is a nonconstant H^2 following function (to order M) if and only if F backs up. Another consequence of this representation of following functions is that if for some point λ_0, there are j points $e^{i\theta}$ such that

$$(7) \qquad \text{F backs up at } e^{i\theta} \text{ and } F(e^{i\theta}) = \lambda_0$$

then there is a nonconstant H^2 function following F to order 1+M-j. Summing up, we have

THEOREM. *If* Γ *is an analytic curve and* d *is a positive integer, then there exists a nonconstant* H^2 *function following* F *to order* d

a) *in case* $d > M$, *always;*

b) *in case* $d = M$, *if and only if* F *backs up;*

c) *in case* $d < M$, *if for some* λ *there are* $1+M-d$ *points* $e^{i\theta}$ *satisfying* (7).

In case b), *all* H^2 *following functions (to order* d*) are given by* (6).

It would be most interesting to know the spectral significance (if any) of the H^2 functions that follow F to order $d < M$. This might yield a proof of the converse implication in c) of the Theorem.

REFERENCES

1. D. N. Clark, Sz.-Nagy-Foiaş theory and similarity for a class of Toeplitz operators, Proceedings of the 1977 Spectral Theory Semester, the Stefan Banach Institute, Warsaw (to appear).

2. ___________, On a similarity theory for rational Toeplitz operators (to appear).

3. Leonard C. Gambler, A study of rational Toeplitz operators, Ph.D. Dissertation, SUNY at Stony Brook, August, 1977.

4. Kenneth Stephenson, Functions which follow inner functions, Illinois J. Math. (to appear).

THE UNIVERSITY OF GEORGIA

Proceedings of Symposia in Pure Mathematics
Volume XXXV, Part 1, 1979

A HANKEL TYPE OPERATOR ARISING IN DEFORMATION THEORY

Richard Rochberg[1]

The approach to deformation theory of uniform algebras presented in [2] suggests that the following question should be studied. For which linear maps T of the disk algebra A into the continuous functions C is the range $T(A)$ an algebra?

To gain some insight into the problem we suppose that we have a one parameter family of such maps -- $T(\epsilon) = I + \epsilon S + O(\epsilon^2)$ -- and try to study S. Comparing powers of ϵ it is easy to check that $\hat{S}$, the map S followed by projection of C onto C/A, satisfies

$$\hat{S}(fg) = f\hat{S}(g) + g\hat{S}(f)$$

and hence there is some G $(G = \hat{S}(Z))$ such that for all f in A

$$\hat{S}(f) = f'G .$$

The problem is now to use the boundedness of $\hat{S}$ to gain some information about G. More precisely, let $\overline{P}$ be the projection onto antiholomorphic functions.

QUESTION: Suppose

$$f \longmapsto \overline{P}(f'G) \qquad (*)$$

is a bounded map from A into VMO, does this imply $\overline{P}(G')$ is in BMO?

This is a natural conjecture in light of the following result.

AMS(MOS) subject classifications (1970). Primary 46J15, 46J35, 42A40.

[1]Research supported by the National Science Foundation under grant MCS76-05789.

THEOREM: For $1 < p < \infty$, (*) is a bounded map of H^p to $\overline{H}^p$ if and only if $\overline{P}(G')$ is in BMO .

The theorem is proved by using standard results about Hankel operators and the Melin analysis techniques of Coifman and Meyer [1].

REFERENCES

1. R. Coifman and Y. Meyer, On commutators of singular integrals and bilinear singular integrals, Trans. Amer. Math. Soc. 212 (1975), 315-331.

2. R. Rochberg, Deformation of uniform algebras, Proc. Lond. Math. Soc., (to appear).

DEPARTMENT OF MATHEMATICS, WASHINGTON UNIVERSITY, ST. LOUIS, MISSOURI 63130

Proceedings of Symposia in Pure Mathematics
Volume XXXV, Part 1, 1979

REPRESENTATION THEOREMS FOR HOLOMORPHIC AND HARMONIC FUNCTIONS

R. R. Coifman[1] and R. Rochberg[1]

Let D be the unbounded realization of a Siegel domain of type 2 and $B(z,\zeta)$ its Bergman kernel function. The Bergman space $A^{p,r}(D)$ is defined as the space of holomorphic functions f on D for which

$$\|f\|_{p,r}^{p} = \int_D |f(z)|^p \, B(z,z)^{-r} \, dV(z) < \infty$$

where dV is the Euclidean volume element, $0 < p < \infty$, and r is greater than some negative constant $C(D)$ which depends on the domain D.

THEOREM: If $0 < p < \infty$, $r > C(D)$, $\theta > (p-2)(1+r)$ then there is a set of points $\{\zeta_i\} \subset D$ such that

a) If $f \in A^{p,r}(D)$ then

$$(*) \qquad f(z) = \sum \lambda_i \left(\frac{B^2(z,\zeta_i)}{B(\zeta_i,\zeta_i)}\right)^{\frac{1+r}{p}} \left(\frac{B(z,\zeta_i)}{B(\zeta_i,\zeta_i)}\right)^{\theta/p}$$

with

$$\sum |\lambda_i|^p \le C_{p,r} \|f\|_{p,r}^{p}$$

AMS(MOS) subject classifications (1970). Primary 30A31, 30A78, 32A07, 32A10.

[1]Research supported by the National Science Foundation under grants MCS75-02411 A03 and MCS76-05789.

b) Conversely if $\sum |\lambda_i|^p < \infty$ then $f(z)$ defined by (*) is in $A^p_{p,r}$ and

$$\|f\|^p_{p,r} \leq c_{p,r} \sum |\lambda_i|^p \quad .$$

The same result is true for the analogously defined spaces on the complex ball and polydisk and in fact is a new result even when D is the disk. An analogous result is true for harmonic functions defined on the ball in $\mathbb{R}^n$. There are also $p = \infty$ results involving BMO type spaces.

The usefulness of the result is based on the fact that many of the properties of functions in these spaces can be deduced from properties of the simple summands.

The basic idea of the proof is to start with certain integral representations formulas and then approximate the integrals with Riemann sums.

Full details and applications of the result will appear in a forthcoming paper.

DEPARTMENT OF MATHEMATICS, WASHINGTON UNIVERSITY, ST. LOUIS, MISSOURI 63130